Springer Series in

OPTICAL SCIENCES 88

founded by H.K.V. Lotsch

Springer
Berlin
Heidelberg
New York
Hong Kong
London
Milan
Paris
Tokyo

Physics and Astronomy ONLINE LIBRARY

http://www.springer.de/phys/

Springer Series in
OPTICAL SCIENCES

The Springer Series in Optical Sciences, under the leadership of Editor-in-Chief *William T. Rhodes*, Georgia Institute of Technology, USA, and Georgia Tech Lorraine, France, provides an expanding selection of research monographs in all major areas of optics: lasers and quantum optics, ultrafast phenomena, optical spectroscopy techniques, optoelectronics, quantum information, information optics, applied laser technology, industrial applications, and other topics of contemporary interest.
With this broad coverage of topics, the series is of use to all research scientists and engineers who need up-to-date reference books.

The editors encourage prospective authors to correspond with them in advance of submitting a manuscript. Submission of manuscripts should be made to the Editor-in-Chief or one of the Editors. See also http://www.springer.de/phys/books/optical_science/

Norbert Kaiser Hans K. Pulker (Eds.)

Optical Interference Coatings

With 197 Figures, 7 in Colour, and 50 Tables

Springer

Dr. Norbert Kaiser
Fraunhofer-Institut
Angewandte Optik und Feinmechanik
Winzerlaer Strasse 10
07745 Jena
Germany
E-mail: norbert.kaiser@iof.fraunhofer.de

Dr. Hans K. Pulker
Universität Innsbruck
Institut für Ionenphysik
Technikerstrasse 25
6020 Innsbruck
Austria

ISSN 0342-4111

ISBN 3-540-00364-9 Springer-Verlag Berlin Heidelberg New York

Library of Congress Cataloging-in-Publication Data applied for
Bibliographic information published by Die Deutsche Bibliothek
Die Deutsche Bibliothek lists this publication in the Deutsche Nationalbibliografie; detailed bibliographic data is available in the Internet at http://dnb.ddb.de

Springer-Verlag Berlin Heidelberg New York
a member of BertelsmannSpringer Science+Business Media GmbH

http://www.springer.de

Printed in Germany

Camera-ready by the authors
Cover concept by eStudio Calamar Steinen using a background picture from The Optics Project. Courtesy of John T. Foley, Professor, Department of Physics and Astronomy, Mississippi State University, USA.
Cover production: *design & production* GmbH, Heidelberg

Printed on acid-free paper 57/3141/di 5 4 3 2 1 0

Foreword

Thin film optics has a very long history. The origins are completely hidden from us but it is likely that prehistoric man would have recognized the colors produced by thin films of oil on water long before he constructed a lens. Gold was known from around 6000BC and could be beaten into very thin films and after mercury was discovered around 1600BC thin-films of amalgams became possible. Colors produced by materials in thin film form have fascinated workers through the ages and many descriptions of them have been written, but the most vivid account is still that of Newton in his Optiks published in 1704. Newton, however, went far beyond simple qualitative description. He related the colors so accurately to the thicknesses of the films that Thomas Young, some one hundred years later, was able to use Newton's results accurately to calculate the associated wavelengths of light. Fraunhofer shortly after recognized the antireflecting properties of a thin film of low refractive index. Fresnel and Poisson developed the multiple-beam theory of thin film interference and great progress in interference optics, including the understanding of interference in thin films, was made during the remainder of Nineteenth Century.

In spite of the antiquity of the subject and the advances of the Nineteenth Century applications were still largely lacking. Mirrors represented the major product. It was the second half of the Twentieth Century that saw thin film coating assume its critical importance in optics. Without optical coatings virtually none of the advances in optical instrumentation of the last half century would have been possible. Until the cold mirror, dangerous fires were common in film theatres. Lasers rely on optical coatings and could not have been developed without them. Astronomical observations depend on them. They are critical components of the latest gravitational-wave detectors. They form the pigment for color-variable inks that resist counterfeiting. They insulate buildings. They shorten the pulses in femtosecond lasers. They form sensitive biochemical and medical probes. They enable the detection of dangerous concentrations of gas. They manipulate the pulses in optical telecommunication. The list is as long as optics itself and applications continue to multiply.

Of course, we recognize that Nature was first with applications by a long way. Animals, insects and plants all make use of interference effects in thin films. The uses range from decorative and protective colors to reflecting and imaging systems in the eye. Newton was well aware of this and was of the opinion that all colors in Nature were due to structures of thin films. Studies of thin film structures in Nature laid much of the ground for later technological developments.

The principal, and traditional, role of a thin film interference coating is to modify the properties of an optical surface. That is, the coating modifies, in almost any imaginable way, the character of the light that is transmitted or reflected or absorbed. Most often the modification is limited to the specular properties without changing the direction of the light or introducing significant diffuse scattering. The behavior of the coating depends on the nature of the materials together with interference effects, which vary with wavelength, angle of incidence and polarization. The films are thin so that the interference fringes that result are sufficiently broad to be useful. In most cases, but not all, they are essentially isotropic. Calculation of the properties of a given thin-film structure is a multiple-beam interference problem that is well organized and understood. The inverse problem of finding a structure with given properties, coating design, is rather more complicated with no unique solution. There are powerful theoretical tools that assist, but experience and skill are still important. Once a design has been established there is the problem of actually producing a coating. The first requirement is a knowledge of the properties of materials, and not just on their own but in combination with other materials, and all in thin film form. The microstructure of thin films depends critically on the way in which they are produced and has an immense influence on their properties. At times thin films have been described as a fourth state of matter so far does their behavior regularly differ from similar materials in bulk. All kinds of characterization methods are needed in order to understand and to control the properties. Anyone who has looked at a bimetallic strip understands the complication of placing two thin layers of material together. But optical coatings have many thin layers in intimate contact and the problems of the bimetallic strip are multiplied many times over. A further requirement is a knowledge of deposition processes and particularly their relationship with the ultimate role of the coating. The interference properties depend on the thicknesses of the films, as well as other parameters, and these must be accurately controlled during their deposition. In spite of the fact that the primary reason for the coating is optical, nevertheless the coatings are by their nature usually on the outside of the component and they are invariably expected to contribute to an improvement of the environmental resistance also. Clearly with such a mixture of requirements, knowledge of the details of the actual application is indispensable.

One would imagine that with the unequaled range of applications of thin films in optics and the complications in their realization, they would represent the most important topic in any optics course. Those of us who are deeply involved in thin films find it difficult to understand why this is so far from the reality. It may be a mistaken view that the complexities of the subject defy any logical framework. Nothing could be further from the truth. The subject is, admittedly, complicated, but it obeys the same rules of chemistry and physics as any other science. Who better to explain the different facets of the subject than the team assembled here? Who better to learn from than these experts? The book ranges over all the important topics, from fundamentals to applications, each section written by well-known specialists of well-deserved international reputation. This is the time-slice across the subject that should be at hand on all thin-film practitioners' bookshelves.

May 2002

Angus Macleod
Thin Film Center Inc, Tucson, Arizona

Preface

This book covers the current area of interference coatings which are essential to all modern enabling optical technologies. It is addressed to newcomers and experts at engineering and university levels and to senior students of engineering and science, offering a comprehensive view of the field of optical coatings while reflecting its dynamic and rapidly changing nature. The 18 chapters, written and reviewed by experts, provide authoritative information on a broad range of topics including, but not limited to natural coatings, photonic structures, design, materials, film growth, deposition on small and large areas, characterization, stress, microcomponents, DUV/VUV/EUV coatings, ultrafast optics, displays, ultrasensitive fluorescence applications, laser-resistant coatings, coatings for free electron lasers, and for plastic optics. The basic and most current results of the entire field of optical coatings are presented in logical order and supplemented by extensive bibliographic material.

Naturally, our book starts with a chapter on natural coatings. In his introduction to this subject, Peter Vukusic writes, "For some time it has been known that nature makes extensive use of the fundamental way that light interacts with periodic structures". Thomas Mann´s last great novel "Doctor Faustus", first published in 1947, faces this reality in the following, genuinely artistic way: There were many times of an evening, when Adrian´s father would open his books with colorful illustrations of exotic moths and sea creatures, and we all – his sons and I, and on occasion Frau Leverkühn as well – would gather to gaze from over the back of his leather wing chair; and he would point with his forefinger at the splendors and oddities pictured there: the papilios and morphos of the tropics, flitting along, somber and radiant, in all the colors of the palette, configured and patterned with the most exquisite taste an artisan could ever invent – insects that in their fantastically exaggerated beauty eke out an ephemeral life and some of which native peoples consider to be evil spirits that bear malaria. The most glorious hue that they flaunted, an azure of dreamlike beauty, was, so Jonathan explained, not a real or genuine color, but was produced by delicate grooves and other variations on the scaly surface of their wings, a device in miniature that could exclude most of the light rays and bend others so that only the most radiant blue light reaches our eyes. "Look at that," I can still hear Frau Leverkühn say, "so it's a sham?" "Do you call the blue of the sky a sham?" her husband replied, leaning back to look up at her. "You can't tell me what pigment produces it, either." (Thomas Mann, "Doctor Faustus", translated from the German by John E. Woods, Vintage International, Vintage Books, A Division of Random House, Inc., New York (1999), p. 17).

On July 15 – 20, 2001, during the Optical Interference Coating Topical Meeting in Banff, Alberta, Canada, the idea to bring together authors from the exciting field of coatings was conceived. Later on, we were encouraged by Springer editor Claus Ascheron, Heidelberg, to wrap up everything together as a contribution to Springer's Series in Optical Sciences. Our thanks are due to our contributors for finding the time and energy to write the chapters in a very limited time scale to make the book as topical as possible. Angus Macleod and Ric Shimshock honoured our work by writing the foreword and an outlook. Claus Ascheron spent a lot of time helping us from the planning to the final proof. We are also grateful, on behalf of all authors, to our secretaries Ms. Petra Thomas and Ms. Ellen Kämpfer from the Fraunhofer Institute of Jena and Ms. Sigrid Cuneus from Springer, Berlin, for her administrative assistance in editing the manuscript, including Word formatting.

Jena and Innsbruck
March 2003

Norbert Kaiser
Hans Pulker

Table of Contents

Authors Index

Professor Günter Bräuer
Fraunhofer Institute for Thin Films and Surface Engineering
Bienroder Weg 54 E, 38108 Braunschweig, Germany
Phone: 0049 5 31 / 21 55 – 500
Fax: 0049 5 31 / 21 55 - 900
E-Mail: braeuer@ist.fhg.de

Professor Ludovic Escoubas
Ecole Nationale Superieure de Physique de Marseille
Institut Fresnel Dom. Univ. de St Jerome, 13397 Marseille cedex 20, France
Phone: 0033 4 91 28 80 68
Fax: 0033 4 91 28 80 67
E-Mail: ludovic.escoubas@enspm.u-3mrs.fr

Dr. Francois Flory
Ecole Nationale Superieure de Physique de Marseille
Institut Fresnel Dom. Univ. de St Jerome, 13397 Marseille cedex 20, France
Phone: 0033 4 91 288052
Fax: 0033 4 91 28 80 67
E-Mail: francois.flory@fresnel.fr

Dr. Martin Friz
Merck KgaA, Pigments-Optics
Mainzer Str. 41, 64579 Gernsheim, Germany
Phone: 0049 62 58 / 12 65 40
Fax: 0049 62 58 / 12 65 67
E-Mail: Martin.Friz@merck.de

Dr. Alexandre Gatto
Fraunhofer Institute for Applied Optics and Precision Engineering
Winzerlaer Str. 10, 07745 Jena, Germany
Phone: 0049 36 41 / 807-343
Fax: 0049 36 41 / 807-601
E-Mail: gatto@iof.fhg.de

Dr. François Y. Génin
Lawrence Livermore National Laboratory
National Ignition Facility
7000 East Ave., Livermore, CA 94550-9234, USA
Phone: 925-422-3562
Fax: 925-423-0792
E-Mail: genin1@llnl.gov

Dr. Norbert Kaiser
Fraunhofer Institute for Applied Optics and Precision Engineering
Winzerlaer Str. 1, 07745 Jena, Germany
Phone: 0049 36 41 / 807-321
Fax: 0049 36 41 / 807-601
E-Mail: kaiser@iof.fhg.de

Dr. Ferenc Krausz
Photonics Institute, Vienna University of Technology
Gusshausstr. 27/387, 1040 Wien, Austria
Phone: 0043 1 58801 38725
Fax: 0043 1 58801 38799
E-Mail: krausz@tuwien.ac.at

Dr. Jolanta-Ewa Klemberg-Sapieha
Department of Engineering Physics
Ecole Polytechnique
Box 6079, Station "Downtown", Montreal, Quebec H3C 3A7, Canada
Phone: 001 514 340-4099
Fax: 001 514 340-3218
E-Mail: jsapieha@mail.polymtl.ca

Dr. Friedrich König
Merck KgaA, Pigments-Optics
Mainzer Str. 41, 64579 Gernsheim, Germany
Phone: 0049 62 58 / 12 65 40
Fax: 0049 62 58 / 12 65 67
E-Mail: Friedrich.Koenig@merck.de

Dr. Li Li
Institute for Microstructural Sciences, National Research Council
1500 Montreal Rd., Ottawa, Ontario, K1A 0R6, Canada
Phone: 001 613 990-1714
Fax: 001 613 952-5711
E-Mail: Li.Li@nrc.ca

Dr. Ludvik Martinu
Department of Engineering Physics
Ecole Polytechnique
Box 6079, Station "Downtown", Montreal, Quebec H3C 3A7, Canada
Phone: 001 514 340-4099
Fax: 001 514 340-3218
E-Mail: lmartinu@mail.polymtl.ca

Professor Hans K. Pulker
University of Innsbruck - Institute of Ion Physics
Thin Film Technology
Technikerstrasse 25, 6020 Innsbruck, Austria
Phone: 00423-388-4785
Fax: 00423-388-5419
E-Mail: Hans.Pulker@uibk.ac.at

Professor Carl-G. Ribbing; Alt.
Dept. of Materials Science
Box 534; SE-751 21 Uppsala, Sweden
Phone: +46-18 471 3133
Fax: +46-18 500131
E-Mail: cgribb@foi.se

Dr. Hervé Rigneault
Institut Fresnel, ENSPM - Domaine Universitaire de St Jerome
13397 Marseille Cedex 20, France
Phone: +33 4 91 28 80 49
Fax: +33 4 91 28 80 67
E-Mail: herve.rigneault@fresnel.fr

Dr. Detlev Ristau
Laser Zentrum Hannover (LZH)
Hollerithallee 8, 30419 Hannover, Germany
Phone: 0049 511 / 27 88-240
Fax: 0049 511 / 27 88–100
E-mail: DR@LZH.de

Dr. Christopher J. Stolz
Lawrence Livermore National Laboratory
National Ignition Facility
7000 East Ave., Livermore, CA 94550-9234, USA
Phone: 925-422-3562
Fax: 925-423-0792
E-Mail: stolz1@llnl.gov

Mag. Georg N. Strauss
University of Innsbruck - Institute of Ion Physics, Thin Film Technology
Technikerstrasse 25, 6020 Innsbruck, Austria
Phone: 0043 512 507 6248
E-Mail: georg.strauss@uibk.ac.at

Dr. Gabriel Tempea
Photonics Institute, Vienna University of Technology
Gusshausstr. 27/387, 1040 Wien, Austria
Phone: 0043 1 58801 38725
Fax: 0043 1 58801 38799
E-Mail: tempea@tuwien.ac.at

Dr. Roland Thielsch
Southwall Europe GmbH
Southwallstr. 1, 01900 Großröhrsdorf, Germany
Phone: 0049 35952 44 343
Fax: 0049 35952 44 321
E-Mail: rthielsch@southwalleurope.de

Professor Alexander Vladimirovitch Tikhonravov
Research Computer Center MSU
Lenin's Hills, Moscow, 119899, Russia
Phone: (095) 939-5430
Fax: (095) 938-2136
E-Mail: tikh@tikh.srcc.msu.su

Dr. Peter Vukusic
BBSRC David Phillips RF, School of Physics, Exeter University
Exeter EX4 4QL, UK
Phone: (0)1392-264156
Fax: (0)1392-264111
E-Mail: P.Vukusic@ex.ac.uk

Dr. Friedrich Waibel
Unaxis Materials
Schlossweg 11, FL-9496 Balzers
Phone: 00423-3885004
Fax: 00423-3887102
E-Mail: friedrich.waibel@unaxis.com

Dr. Vladislav Yakovlev
Photonics Institute, Vienna University of Technology
Gusshausstr. 27/387, 1040 Wien, Austria
Phone: 0043 1 58801 38725
Fax: 0043 1 58801 38799
E-Mail: yakovlev@tuwien.ac.at

Dr. Sergey Yulin
Fraunhofer Institute for Applied Optics and Precision Engineering
Winzerlaer Str. 1, 07745 Jena, Germany
Phone: 0049 3641 807-241
Fax: 0049 3641 807-601
E-Mail: yulin@iof.fhg.de

Natural Coatings

Peter Vukusic

School of Physics, University of Exeter, Exeter, UK

1 Introduction

Nature is profligate in the way it uses light and colour. As the principle component of inter and intra-species communication, light supports stimuli associated with courtship, crypsis, predation, feeding and a wide range of other biological functions. Light is believed to be the stimulus behind the *Cambrian explosion*; the sudden and enormous diversification of life that accompanied the start of the Cambrian period over 500 million years ago. Evidence from this era suggests that the co-development of predator and prey colouration, concurrently with their visual systems, led to an explosion in evolution of lifeforms (Parker 1998, 1999). Light is believed to drive significant selection pressures that have been responsible for the subsequent development of all life; creating the astonishing diversity of photonic systems that are present in the world today.

In natural systems colour, or the absence of it, is facilitated by one of two distinct levels of animal morphology. By far the most common process by which colour is produced, involves natural chemicals or pigments. These lead to selective absorption or scattering of particular wavelengths due to electronic transitions that are specific to certain molecular groups. The history and background to the study of this regime of natural colour are not appropriate for discussion here: the interested reader is referred to the wealth of information existing in literature (Fox and Vevers 1960, Fox 1976). The alternate process, however, by which nature colours both the world around us, and objects which may possess no pigmentary constituents, is through its ingenious use of microstructure. It is the realm of natural photonics associated with such microstructure that is of primary interest to workers in optics and which will be discussed in this chapter.

For some time it has been known that nature makes extensive use of the fundamental way that light interacts with periodic structures. However, what has only recently become much clearer, is not just the existence and broad range of such structures, but specifically the ingenuity with which some of them are designed. However, where we might find both ingenuity and elegance in the microstructure of, say, a butterfly species, it pays to remember that it was produced to serve a purpose. Whether to effect ultra-high intensity reflection over limited range of angle; to optimise the light transmitted through a medium; or to effect complex polarisation signatures from surfaces; such systems were designed and are continually being adjusted by ever-present evolutionary selection pressures. It is not unreasonable to suggest that some natural optical systems present in the world today have been in development for hundreds of millions of years. Ingenuity in nature's optics is therefore not surprising.

Presentation of detailed descriptions of the photonics associated with all animal and plant phyla, is an enormous task; one that would require considerable space. This chapter therefore presents detailed and referenced summaries of such descriptions. In certain cases, however, extensive descriptions are presented. To assist with the organisation of these summaries, descriptions of optical systems will be loosely based on taxonomic and environmental divisions. Outlines of multilayered systems of water-based animal groups will be followed by those for multilayered systems in land-based animal systems, followed by those for plant groups. A description of a final category of non-layered microstructure completes the overall presentation.

Examples of microstructure-generated photonic effects abound in nature. Among the most startling and colourful examples are the intense iridescent hues of several species of exotic butterfly. Several *Morpho* species in particular, exhibit spectacular blue reflection from their dorsal wings. The eye-catching metallic gold or silver surfaces of some species of beetle are equally astonishing to view in bright daylight, especially considering their colouration results from all-dielectric periodic systems. Several species of fish display similar brilliance with their scales, exhibiting broad-angle silver reflection. More every-day iridescence in nature may be observed in particular bird feathers, noticeable, as the definition of iridescence implies, by a distinct change in colouration with viewing angle of the feather.

While these and many other species exhibit such structural colouration that is easily visible to the naked eye, many other examples exist wherein photonic effects may only be detected through the use of optical microscopy, specialised UV or IR-sensitive apparatus, or polarising filters. As a result of the limited wavelength-sensitivity, resolution capability and polarisation-sensitivity of the human visual system, many instances of natural photonics may go unnoticed.

2 Nature's Optical Materials

With few exceptions, materials which form the basis of natural photonic systems may be divided into three broad groups: guanine-based systems associated with marine species; cuticle-based systems associated with terrestrial species and cellulose-based systems associated with plant species. Since the physical and optical properties of a material itself contribute to the optical properties of the system it comprises, a discussion of the nature of the three system materials is appropriate.

The first of the three systems mentioned, guanine-based systems, are normally associated with marine species. A variety of tests (Denton 1970) have shown that the majority of reflecting elements found in fish are largely guanine-based. Guanine itself is found in the form of small elongated crystals that comprise nitrogenous compounds (named from the *guano* deposits from which the compound was first isolated). However, a similar compound, hypoxanthine, is also present in high quantities in the crystals found in some species (Denton 1970, 1971); the ratio of guanine to hypoxanthine appearing to control the shape of crystals formed.

A good-size body of work details investigations of the optical properties associated with these crystals. In a species of copepod, *Sappharina*, the crystals were found to be strongly birefringent, with a refractive index of 1.79 in the direction perpendicular to and 1.55 in the direction parallel to the broad surface of the crystal. Other work, on a species of fish, measures a refractive index of 1.85 and 1.81 for light perpendicularly incident on the long and short axes of the broad face of the crystal respectively and 1.46 for light incident parallel to the broad face. Such birefringence appears to indicate molecular orientation lying essentially parallel to the broad face of the crystals (Greenstein 1966). The exact mechanism by which guanine structures are crystallised is left for discussion elsewhere, but a brief mention is made of the following. A growing scale of a typical fish specimen exhibits three distinct layers: the first is an external bony layer laid down only at the edge of the scale; the second comprises a layer of collagenous fibre; and the third is a layer of cells which contain the reflecting guanine crystalline material. The guanine crystals are found intracellularly within this layer, existing in platelets formed from four to five crystals separated by cytoplasm of index approximately 1.33 (Fig. 1). Generally, the reflecting cells are inclined at specific angles with respect to the surface of its scale, often depending on its position on the body of the fish, with each reflecting cell held in place with respect to others by interdigitating ridges and pockets (Denton 1970).

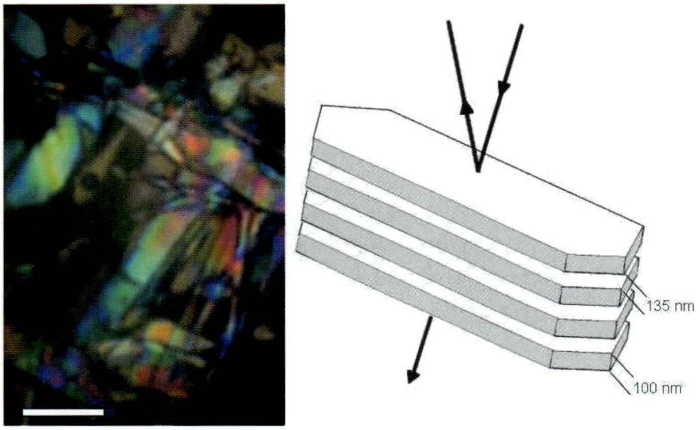

Fig. 1. (Left) Optical microscope image showing a collection of overlapping guanine crystals and incomplete platelets removed from a silver coloured scale of the common sardine *Harengula jaguana*, observed in reflection (scale bar 25 μm). (Right) Schematic diagram of a reflecting platelet showing a stack of guanine crystals and typical crystal dimension and spacing for iridescent red/orange colouration (after Denton 1970).

The second of the systems under consideration, associated mainly with terrestrial arthropods, comprises the material referred to as cuticle. This material is secreted by epidermal cells that, together with the cuticle, make up integument. In some taxa, this serves as an exoskeleton that dominates lifestyle by influencing physiology and behaviour. The manner in which cuticle structures are extruded by insect epidermal layers makes possible an enormous range of external structures and appendages and creates a diversity of function that, certainly, is a hallmark of arthropods. Such cuticular features are relied upon heavily by insect taxonomist for species identification. Apart from chemically derived pigmentary colouration, projections of cuticle that form insect scales give rise to structural colouration by interference, diffraction or scattering, each associated with highly periodic microstructure: this is centrally relevant to this chapter and will be addressed in detail later.

Generally, cuticle is secreted by a single layer of epithelial cells to form a thin outer epicuticle (comprising mainly lipids) and an inner procuticle. The procuticle, of primary interest to us here, comprises a system of chitin rods set in a matrix of proteins. The material chitin is the main fibrous component; it is a linear polysaccharide of N-acetyl-d-glucosamine (Phelps 1978) which forms rods of approximately 3 nm diameter. Each pseudo-crystalline rod contains around twenty molecular chains oriented in parallel and embedded in a matrix of several kinds of protein. In some cases, extensive cross-linking between these proteins produces dramatic cuticle hardening.

The orientation of the chitin rods within the cuticle is controlled and determined by the epidermis; the most common class of which is a helicoidal arrangement, in which the chitin rods are arranged in planes, the planes lying parallel to each other and the cuticular surface. In successive planes, the rod orientation changes by a few degrees in the same direction. This yields the plane-wise rotation of rod axis associated with the helix of a right or left-handed corkscrew; a geometry which, in planar section, gives rise to parabolic patterning and which was first identified in locust and crab cuticle (Neville 1965, Bouligand 1972). It has been identified as the most common form of extracellular architecture in animals, and has also been found in a range of plants (Neville and Levy 1985).

Various measurements have sought to find the refractive index and optical properties of the cuticle of different insect species. Generally, it has been found to be approximately 1.56 at visible wavelengths with very little inherent absorption, except that contributed by additional pigmentation (Land 1972, Vukusic et al. 1999). In a few limited cases, the helicoidal arrangement of chitin rods has been found not only to contribute an optical activity to the reflection from certain insects (Neville and Caveney 1969) but it appears responsible for certain surface iridescence despite the absence of refractive index periodicity. This will be discussed in more depth later in the chapter.

The third category of system comprises plant taxa that exhibit structural colouration brought about by specialised modifications of cell membranes and chloroplasts. The material that underpins such microstructure in these taxa is principally hydrated cellulose fibre. Cellulose itself is a polysaccharide of glucose units that constitutes the chief part of the cell walls of all plants (Preston 1974) and as such it is the most abundant natural fibre in existence (Neville 1975). Several studies have measured the optical properties of this material (Gates 1970, Woolley 1975). In its hydrated form, the refractive index of the cellulose of soybean leaf cell walls measured by an oil infiltration method (Woolley 1971, 1975) was found to be approximately 1.41 (varying slightly with degree of hydration), compared to that of dried cellulose which was approximately 1.55. No data on any inherent optical absorption in this material is available. Plant physiology can cater for differences in refractive index within cell and chloroplast structure either through significant variations in hydration levels of the cellulose or the presence of regions of pure cytoplasm of index approximately 1.33. Studies of the multilayer iridescence associated with certain taxa of algae, suggest that differentiation of refractive index in alternate layers is brought about by neighbouring regions which are alternately high in carbohydrate content and high in protein content (Gerwick and Lang 1977, Pederson et al. 1980). However, the contributory effect of the presence or absence of bromine, a mineral present in this species in high levels, on the refractive index was not discernible.

3 Description and Classification of Natural Optical Systems

Surveys of the natural world reveal an extensive array of optical effects in a broad range of animal, insect and plant taxa. Even with preliminary investigation, it becomes apparent that every instance of natural photonics arises from one, or a combination, of three optical effects: interference associated with multilayering; diffraction associated with surface or bulk periodicities; and wavelength-selective scattering arising from the presence of sub-wavelength sized particles, commonly referred to as Tyndall scattering (Fox 1975, Huxley 1976) in biological systems.

In the vast majority of all species investigated, interference resulting from layers of alternating high and low refractive index has been found to be the most commonly occurring mechanism for strong photonic effects. While examples do exist in which such multilayering comprises systems of broad, relatively flat layers of constant dimensions (Lippert and Gentil 1959, Vukusic and Lawrence unpublished data), it is more often the case that such layering is incorporated within a more specialised design that is best suited for the specific physiology, function or environment of the species in question.

Widespread among multilayer systems are examples wherein the layering is discontinuous, existing in discrete and often compact clusters on or in the surface of the specimen. In some examples, a degree of directional colour invariance is achieved in this way through a spatial averaging of juxtaposing cluster colours (Schultz and Bernard 1989, Vukusic et al. 2000a, 2000b). In other cluster design, a broad-angle visibility is effected by a distribution in orientation of reflecting surfaces (Vukusic et al. 1999).

With certain species, otherwise planar multilayering appears to have additional modulations or periodicities imposed upon it, so that the full multilayer stack takes on a sculpted appearance. This has been shown to be responsible for more complex optical effects including colour stimulus synthesis and even strong wavelength-specific polarisation rotation (Vukusic et al. 2000b). In certain instances it is advantageous for a plant or animal taxon to effect the appearance of silver or gold over a small or large region of its surface (examples include the exocuticle of certain beetles and the scales of certain fish). With few known exceptions variations on a theme of multilayering are responsible for these colour effects; employing one of three types of layered system to produce the appearance of silver or gold. The first of the three of these variations is known as a *chirped* system, in which the dimensions of the high and low refractive index layers within the stack systematically increase or decrease through the system (Neville 1977, Large et al. 2001). The second variation, described as a chaotic reflector, produces a similar effect through random variations in the thicknesses of high and low index material. In both of these systems, the range of layer dimensions developed by the specimen, combined with layer refractive index, determine the position and width of the reflected wavelength band and the quality of the silver or gold-like mirror on its surface (McKenzie et al. 1995). The third type of layered system incorporates a composite of three overlapping but regular multilayer systems, each tuned to a different wavelength band (Denton and Nicol 1965, Denton and Land 1971).

In contrast to the multilayers discussed so far, the reflecting properties of some scarabeid beetles are not based on periodicities in refractive index, but rather result from the helicoidal ordering of chitin microfibrils in planes that make up the insect integument (Neville and Caveney 1969). Such cuticle comprises laminar planes of chitin fibres, the axes of which are systematically rotated by a small amount with each consecutive plane. After a certain number of layers through the system corresponding to a specific distance perpendicular to the planes, the fibre axis has rotated through 180 degrees and is once again parallel to that of the first layer. In this way, two corresponding planes of cuticle occur over a distance that corresponds to half the helical pitch of the system.

The consequent optical effect is that of constructive interference between consecutive planes that have unidirectional chitin axes and which are separated by half the helical pitch. Thus the system may be treated as a multilayer reflector, that reflects a band of wavelength of unpolarised polychromatic light in a manner analogous to cholesteric liquid crystals. As a further consequence of the helical arrangement of chitin fibres, optical reflection from these taxa of beetles is circularly polarised. This is not due to any optical rotation arising at a molecular level due to the L-amino acids of the cuticle protein and the D-amino sugar of the chitin. It instead arises at the supermolecular level and has been termed *form optical rotation*, (Neville 1969) in contrast to the molecular optical rotation exhibited by cholesteric liquid crystals. It should be noted, however, that similar helicoidal structure, found in many other iridescent insect and plant taxa, is not responsible for the strong colouration and does not produce anomalous optical properties such as optical activity. For these species, inherent refractive index periodicity within the species' exocuticle is responsible for the iridescence. In the case of the optically active beetles, it is the sub-wavelength pitch of the helical order that is responsible for the observed form optical activity and iridescent colour.

Diffractive effects in nature arising from surface or complex sub-surface periodicities are less common than those associated with multilayer interference. They are found principally but not exclusively in invertebrates (Parker 2000) and appear principally in Crustacea, Coleoptera and certain Lepidoptera (Stephenson 1969, Hinton and Gibbs 1969, Crowson 1981, Brink and Lee 1999). One may categorise natural diffraction into two components. The first comprises diffractive effects at surfaces that generally correspond to the reflection associated with standard optical diffraction gratings. Individual examples exhibit arrangements of cuticle in series of parallel or near-parallel ridges or striations that are regularly spaced (Parker 1995, Parker 1998, Vukusic unpublished). An observer of a specimen with this structure will observe a monochromatic reflection from a specific region, the hue of which will alter if the specimen, observer or light source changes position.

The second component comprises structures in nature that effect Bragg diffraction at optical wavelengths. This is similar to the Bragg diffraction of X-rays from a crystalline material (Guinier 1963), except that the difference of scale means the effect is seen for ordinary light. These structures are sometimes referred to as volume diffraction gratings or photonic solids. It is, in principle, possible for photonic solids to expel light of certain frequencies for all directions; this implies that light of these frequencies will simply not propagate in the material and that all of it will be reflected from the surface of the solid. When this occurs the material is said to exhibit a *photonic band gap* (see also chapter entitled "Photonic Structures as Interference Devices" by C.G. Ribbing), which is analogous to the familiar electronic band gap of solid state physics except that it applies to photons instead of electrons (Joannopoulos et al. 1995). However, in order for a photonic solid of specific symmetry to exhibit a full photonic band gap, there must be a minimum refractive index contrast between the two media that make up the periodic solid. Natural structures that exhibit these photonic solids are generally composed of cuticle and air, or cuticle and water; both refractive index combinations of which are insufficient to open up a full photonic band gap. However, their combination and structural symmetry can be enough to effect efficient reflectivities over a significant angle range and over a relatively small wavelength band. The selective advantage of colour production using this manner of microstructure rather than a surface grating or a multilayer system, excluding physiological considerations, is that it can sometimes be incorporated into multiple neighbouring grains, each oriented slightly differently, to produce a macroscopic completely angle-independent structural colour.

Other natural examples exist, in which surface periodicity is reduced to sub-wavelength dimensions with a bi- rather than simple mono-grating covering the surface. In such zero-order grating surfaces, the structure resembles and is referred to as a nipple-array. Such nipple-arrays are most commonly found on the outer corneal surfaces of arthropods' ommatidia (Bernhard and Miller 1962, Clapham and Hutley 1973). Effectively, they act as a mechanism for gradually matching the index at the interface of the two media, in most cases cuticle and air. This is in contrast to an abrupt transition between the index of media bounded by a flat interface. In such a way it suppresses the reflection from the corneal surface thus enhancing transmission of light through it for photoreception. By appropriate choice of nipple height, profile, and basal separation, it is possible to facilitate zero reflection over the entire visible spectrum around normal incidence (Bernhard 1967). Mathematical modeling has indicated that such nipple arrays found in the eyes of certain arthropods, produce optimal effect in this respect (Meyer-Rochow and Stringer 1993). At non-zero angles of incidence, where the surface architecture offered by such nipple-arrays no-longer effects zero reflectivity, their profile facilitates identical reflection for both polarisations. Such non-normal-incidence polarisation independent reflectivity has been measured on a flat region of nipple array from the wing of a diurnal moth (Vukusic and Sambles unpublished).

A third category of natural microstructure exists in addition to those that effect interference and diffraction, producing wavelength selective optical scattering. The phenomenon of blue scatter is widespread in nature; the blueness of the sky was long ago investigated by Tyndall (Tyndall 1869). Many other inanimate systems offer strong scattering in the laboratory, such as small concentrations of milk in water, or highly dispersed systems of elementary sulphur or of silver halides in aqueous media. In each case the hue and intensity of the scattered colour depends on the degree of fineness of small particles in the suspension and upon the magnitude of mismatch between the refractive index of the particles and their surrounding medium. Generally, particles that are sufficiently small in size yield deeply violet scattered colour of striking intensity and purity: as particle size is increased the scattered colour passes through deep blue to pale blue. Increasing the particle size beyond approximately one micron results in most visible wavelengths being scattered with sufficient intensity to effect the appearance of white (Fox 1976). Physicists who work with light scattering use several terms to evoke theories associated with different classes of scattering particle. Authors of texts that detail scattering in animate specimens (Fox and Vevers 1960, Byers 1975, Fox 1976) have consistently used the terms Tyndall scattering, Tyndall effect etc., to refer in the broad sense to colours produced through selective scattering of shorter wavelengths by small particles or optical heterogeneities when pigmentation and interference phenomena are not predominant. In certain texts, Tyndall scattering is equated to Rayleigh scattering by particles that are small compared to the wavelength (Longhurst 1967); in others it applies to particles that are the same size or larger than the wavelength (Ditchburn 1963). Tyndall's original experiments embraced both groups of particle size and the expression is usefully retained by many biologists to describe scattering in animate objects.

The most conspicuous examples of Tyndall blues in nature are found in certain mammals, particularly monkeys such as the male baboons and male mandrills. In these creatures hairless patches may be deep violet or blue due to Tyndall scattering associated with strata of discrete black melanocytes arranged in an otherwise transparent layer within the epidermis (Fox 1976).

Purple regions on these same creatures correspond to areas where Tyndall blues colour-add to red pigmentation associated with blood corpuscles close to the surface. This effect is perhaps more commonly observed as port-wine birthmarks in some humans (Verne and Layani 1938).

Whereas Tyndall blue in mammals is effected by disperse systems of solid-in-solid where the melanin granules have a different refractive index to the surrounding epidermis, non-iridescent avian feathers generally achieve blue scatter using air-in-solid systems. In such feathers, air-filled sub-wavelength cavities within the keratin of the feather barbs act as the scattering centres from which Tyndall blues are produced. More often than not, the colour green in avian feathers is the product of the superposition of yellow reflection from additional carotenoid pigmentation in the feathers with the underlying Tyndall blue (Frank 1939, Frank 1976). Green and sometimes purple colouration in many other taxa, e.g., certain snakes, frogs, lizards and fish, are produced by similar effects; namely Tyndall scattering superposed with colour from pigment (Fox and Vevers 1960, Fox 1976).

4 Structure and Function

4.1 Guanine Layered Systems

Guanine and cytoplasm, in the form of layers of alternating high and low refractive index, are commonly found in many fish reflectors, but are also present in certain other vertebrate systems such as tapetal reflectors and iridocytes (Kawaguti and Kamishima 1966, Land 1966).

Many reflecting systems in fish have been the subject of detailed investigation. Among the best understood are those associated with the scales of the sprat *Clupea sprattus* and the herring *C. harengus* (Denton and Nicol 1965, Denton 1970 and Denton and Land 1971). Since their optical systems are very similar to those of most fish reflectors, these will be described in detail. The ventral scales of a young sprat reflect copper hues near normal incidence; complementary hues of green are visible in transmission indicating the absence of significant absorption. A reflecting layer comprises many discrete but co-parallel platelets each comprising a stack of around five guanine crystals of approximate dimensions 6 μm by 24 μm, separated from each other in a medium filled with cytoplasm (Fig. 1). These form the centres from which strong interference occurs. Electron microscope (Kawaguti and Kamishima 1966) and interference microscope (Huxley 1964) measurements on this and other fish scales indicate guanine crystal thicknesses of approximately 100 nm (giving an optical thickness of approximately 180 nm). Were each platelet to act as an ideal quarter-wave reflector (Land 1972) the intervening cytoplasm-filled spaces would also exhibit an optical thickness of 180nm, equating to a physical dimension of approximately 135 nm and reflecting a peak wavelength (λ_{max}) of 720nm (given by $\lambda_{max} = 2(n_a d_a + n_b d_b)$ where n_a, n_b and d_a, d_b represent the refractive indices and thicknesses of the periodic layering (Vašiček 1960). Despite some non-ideal character, namely lower and broader than expected reflectivity, thought to be due to discontinuous platelet coverage across the scale and irregularities in the form and spacing of the guanine crystals, this is found to be the case for the ventral scales described for the sprat. Variations in the colour of reflected light is found on different regions of the same fish, and between different fish, however. These colour differences are thought to be due to be the result of variations in the crystal thickness in younger fish, but to differences in cytoplasm spacing between the crystals in older fish (Denton and Land 1971). While narrow-band colouration exists for these fish scales at the microscopic level, macroscopically fish scales appear rather silvery-white, an indication of rather broad-band reflection. The explanation for this lies in the fact that different coloured regions of each scale overlie (Denton and Nicol 1965) each other (Fig. 1). Optically this is equivalent to differently tuned reflecting filters lying over one another, such that a significant band of wavelength, transmitted by the filter above is reflected by the filter below; the net effect being the reflection of a significant portion of the incident light to produce the appearance of white or silver. Certain other fish, the roach *Rutilus rutilus* (Land 1972) and the *Trichiuridae* ribbonfish (McKenzie 1995) employ either systematic (chirped system) or random variations (chaotic system) in guanine crystal dimensions to cause their silver-white reflection.

The function of silver reflectance in fish has been thoroughly investigated. It appears to serve the purpose of camouflage in mid-water where the distribution of light is symmetric with respect to the plane of their mid-dorsal and mid ventral lines. Specialised orientation and distribution of reflecting platelets within the scales augment the cryptic nature of the silver colouration by offering position-dependent variations in reflectance. For a more complete discussion of the function of structural colour in fish, refer to any of the excellent reviews in literature (Denton 1970, Herring 1994).

Tapetal reflectors, in the form of silver reflecting layers, exist in the eyes of many fish; including rays, dogfish and sharks. As with body-scale reflectors, tapetal reflectors are principally comprised of guanine crystals separated by cytoplasm. In the tapeta of two dog-fish, *Squalus acanthias* and *Scyliorhinus canicula*, reflecting platelets comprise a stack of 12-13 crystals of guanine. Interference microscopy measurements place their thicknesses in the range 60 – 80 nm (Denton and Land 1971) yielding mean reflecting peak values of around 500 nm; however, broader than expected reflection bandwidth, effecting the appearance of silver-white, is produced by either a variation in crystal thickness or spacing, or by the effect of the presence of differently tuned reflecting systems in the same platelet. The function of such tapetal reflectors in fish appears not to be for increasing the total absorption of light via multiple passes through photoreceptors, but rather to reduce the quantity of photopigment required to achieve the same light absorption (Denton 1970).

4.2 Cuticle-Layered Systems

Iridescence originating from layering comprising cuticle as the high refractive index layer is fairly common in nature. The most extensively studied systems are those associated with the scales of Lepidoptera (Mason 1927, Fox 1976, Ghiradella 1993, Vukusic 2000) in which the refractive index contrast between cuticle and air can create dramatic optical effects. Recent research has led to these systems becoming among the best understood in nature. In light of such strength of interest in this area, the cuticular systems associated with Lepidoptera will be the subject of detailed discussion. As items of particular interest, the optical properties of three specific Lepodopteran systems will be described in rather greater depth (see Sects. 4.2.1, 4.2.2, and 4.3.1)

With a few exceptions the seat of colouration in Lepidoptera, whether through pigmentation or structure, lies in the scales that cover the surface of the wings and body (Fig. 2). Each scale is a flattened projection of cuticle from a single epidermal cell within the epithelial layer that makes up the surface of the wing. The arrangement of scales on the wing resembles that of shingles on a roof with, in most species, there being two distinct layers of different scales present. Typical scale dimensions are of the order of 75 μm by 200 μm. While the underside of scales are rather planar and featureless, their interiors and their externally visible top-surfaces exhibit intricate microstructure (Fig. 2). As has been mentioned, the cuticle material itself is a composite of rods of chitin set in a matrix of proteins (Neville 1975). It has been shown to have a refractive index of approximately 1.56-1.58 at visible wavelengths (Land 1972). The presence and concentration of melanin pigment dictates the associated amount of optical absorption (Vukusic et al. 1999).

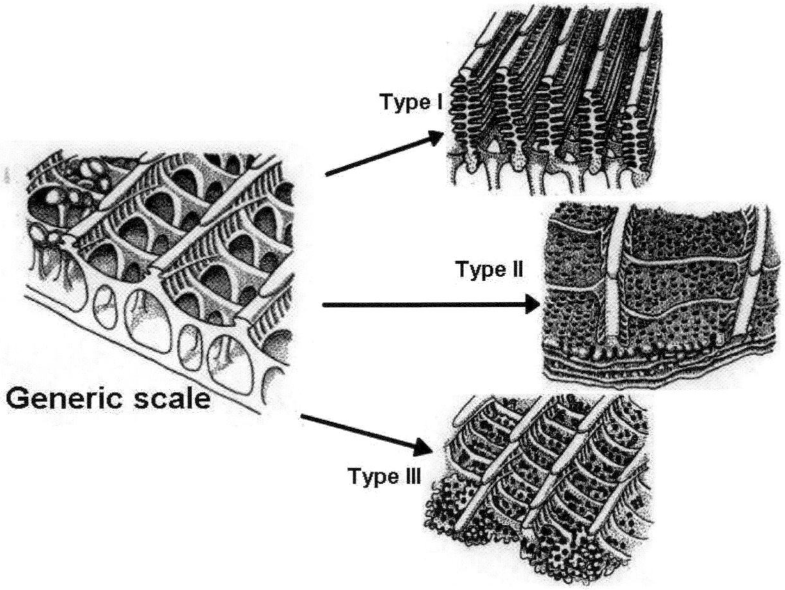

Fig. 2. Schematic diagram to show the classification of microstructure of iridescent butterfly scales, into type I ridge lamellae; type II body lamellae; type III body scattering. (Schematics reprinted with kind permission from H. Ghiradella and Wiley-Liss, © 1998.)

All scales, whether iridescent or not, display one of a variety of forms of ridging extending longitudinally from one end of the scale to the other (Fig. 2). Within the ridging, fine structure may be developed in the form of lamellae or microribbing. Often the ridges may be connected at intervals by a series of arched structures referred to as crossribs. Spacing between ridges on a single scale is often quite uniform, lying in the range 0.5 μm - 5.0 μm depending on species and scale type. In the scales of some species the regions between the ridges and crossribs are hollow, exposing the scale interior with pillar-like trebaculae that run from the base of the ridges to the scale substrate beneath. Pigment other than melanin is laid into pigment granules in this region, while melanin tends to be distributed in the scale structures itself.

Within the category of cuticle-air systems, the different forms of Lepidopteran microstructure have been categorised for the purpose of gaining a valuable overall perspective that itself is partly mirrored in other systems. These microstructures in this system form three principle classes, which themselves may be further subdivided (Vukusic et al. 2000a). Categorisation is based on the nature of structural reflecting and scattering elements and their position in or on the wing-scale. Figure 2 presents schematics of these categories which may be described as: type I microstructure, comprising discrete multilayered systems incorporated into the scale ridging; type II, comprising multilayered systems incorporated into the scale body (colour is generated in both these systems through multilayer interference in the normal way); type III microstructure does not comprise multilayering but other specialised diffraction and scattering systems (dealt with in detail in a later section). The structures associated with all three of these categories are variations on the same general template of scale design. The developmental processes that lead to the formation of each class of scale have been investigated and are described elsewhere (Ghiradella 1976).

Many structurally coloured butterfly species exhibit type I scale microstructure, in which multilayering is incorporated within the ridging of scales (Ghiradella *et al* 1972). In certain species, however, variations exist, so that further categorisations may be made in a way that is associated with the angle subtended by the multilayering with the base of the scale. Type Ia, layering by so-called ridge-lamellae, is defined as a category in which lamellae run parallel or near-parallel to the base of the scale. The brightest structurally coloured butterflies are associated with this category of scale for which there may be as many as twelve ridge-lamellae of cuticle in the multilayer system. Lower brightness from some butterfly wing scales may be attributed to ridges with fewer lamellae (see Sect. 4.2.1).

The visibility of certain *Morpho* butterflies is an excellent example of the effect of type Ia scales (and is described in detail in the case-study for *Morphos*) and can result in absolute reflectivities at blue wavelengths of up to 80% in certain species (Vukusic et al. 1999). Side-to-side tilting of the ridges and of the lamellae surfaces within the ridging provide, in one plane, a broad angle-spread to the reflection. In other *Morpho* species, not only does such ridge tilting provide angle spread in the reflected light, but ridges on the near-transparent scales of a superficial second layer of scales bring about diffraction which further assists broad-angle reflection (Vukusic et al. 1999).

Type Ib scales comprise ridge-lamellae that have moderate inclination to the scale base (Fig. 4). High numbers of ridge-lamellae (greater than five or six) are generally not found within these systems. This limit, together with the steeper angle of inclination facilitates a larger range of structurally reflected colours than is possible with type Ia scales (Vukusic et al. 2001). Additionally, the steeper angle of inclination limits the portion of the observation hemisphere above the scale from which the iridescent colour may be observed even under diffuse illumination. The butterfly *Ancyluris meliboeus* is adorned with patches of type Ib scales and exhibits deep blue to orange structural colouring from the same region at different scale orientations. The colour images in Fig. 4 illustrate how a very small change in wing orientation may extinguish this colour (Vukusic et al. 2002).

Fig. 3. Optical images showing the structural colour and patterning of *Papilio palinurus* and *P. ulysses* (top and bottom left), together with small regions of their iridescent scales (top right: *P. palinurus*; and bottom left *P. ulysses*) taken using a microscope and an unpolarised white light source (Scale bars: left 2 cm; top and bottom right 10 μm). The green colouration of *P. palinurus* is effected through colour mixing of juxtaposed yellow and blue regions, both of which are produced from the same multilayer profile. (Reprinted with permission from *Applied Optics* (**40**, 7, pp1116-1125) © 2001).

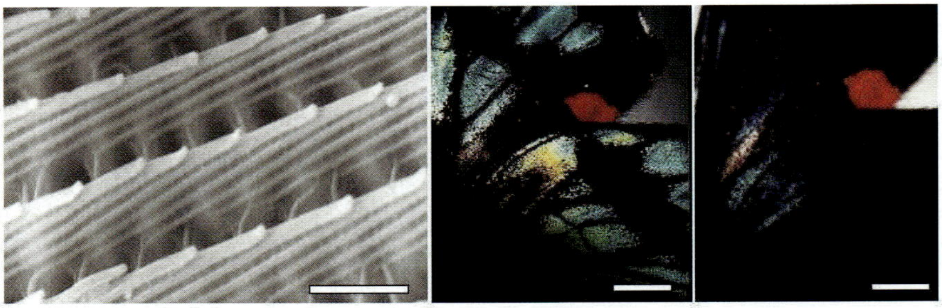

Fig. 4. (Left) SEM image showing type Ib scale microstructure of the butterfly *Ancyluris meliboeus*. The multilayering within the ridging is clearly tilted, with the ridges exhibiting diffracting elements above the multilayering. (Middle and right) optical real-colour photographs of the same regions of *A. meliboeus* wing under identical diffuse illumination, but with a 15 degree difference in wing tilt between them. [Scale bars: left 700 nm; middle and right 5 mm]. (Reprinted with permission from *Nature* (**410**, 36) © 2001).

Type Ic scales comprise cases where this angle of multilayer inclination is extremely steep, to the extent that in some cases it approaches orthogonality with the base of the scale. In such cases, the layering is often formed from an extremely developed system of microribs rather than from the ridge-lamellae. With such steep multilayering, one would expect the portion of the viewing hemisphere above the wing from which the associated structural colour is observable to be extremely limited. Indeed in *Troides magellanus*, a species that exhibits type Ic scales, it is only from near-grazing incidence to the wing that the bright blue-green structural colour becomes visible. This colour appears in addition to the overall yellow colouration derived from its papiliochrome pigmentation, which also exhibits strong UV absorption with some green-fluorescence (Lawrence et al. 2002).

4.2.1 Analysis of Type Ia Scales: M. rhetenor and M. didius

The *Morpho* family of butterflies, indigenous to South and Central America, are generally conspicuous by their brilliant blue iridescent colouration. This blue is so bright that it can reportedly be seen from "a quarter of a mile off" (Bates 1864). Two species of *Morpho* were subjected to detailed study; both exhibited type Ia scale structure.

While one of the species, *M. rhetenor*, exhibits only one layer of developed scales (Fig. 5), the second species, *M. didius* exhibits two distinctly different layers. Cross-sections through the wing scales of each butterfly are shown in Fig. 5 and illustrate the discrete layering associated with the structural colouration of each butterfly. The ridging of the scales of *M. rhetenor* have approximately twelve cuticle layers. There are two types of *M. didius* scales; *ground* scales (Ghiradella 1994) which are akin to the *M. rhetenor* scales, and so-called *glass* scales (Ghiradella 1994) which lie above the ground scales and exhibit high levels of transparency.

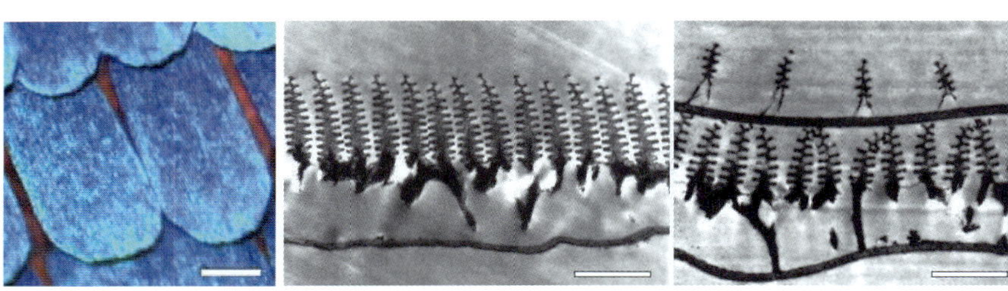

Fig. 5. (Left) Optical image showing neighbouring blue iridescent *M. rhetenor* scales. TEM images of transverse sections through an iridescent scale of *M. rhetenor* (middle) and *M. didius* (right). *M. didius* employs a glass scale above a more reflective ground scale to effect diffraction for a broadening of the angle over-which its blue iridescence is reflected. *M. rhetenor* employs a greater number of cuticular layers to effect higher reflectivity from its single layer of iridescent scales. (Scale bars: left 60 μm; middle and right, 1.4 μm).

To establish the true nature of the optical response from each of the scale structures, it was necessary to examine individual scales. Such examination proved en-

enlightening. Using a method developed for use in the examination of single scales (Vukusic et al. 1999) the reflection, transmission and absorption associated with each scale were measured; their polar-angle dependent reflection and transmission are shown in Fig. 6. The spatial patterns of reflection and transmission from single iridescent ground scales of both butterflies show some similarities; there is a broad angular spread in the reflection from the scale. Reflection from the *M. rhetenor* ground scale appears divided into two distinct wide-angle lobes.

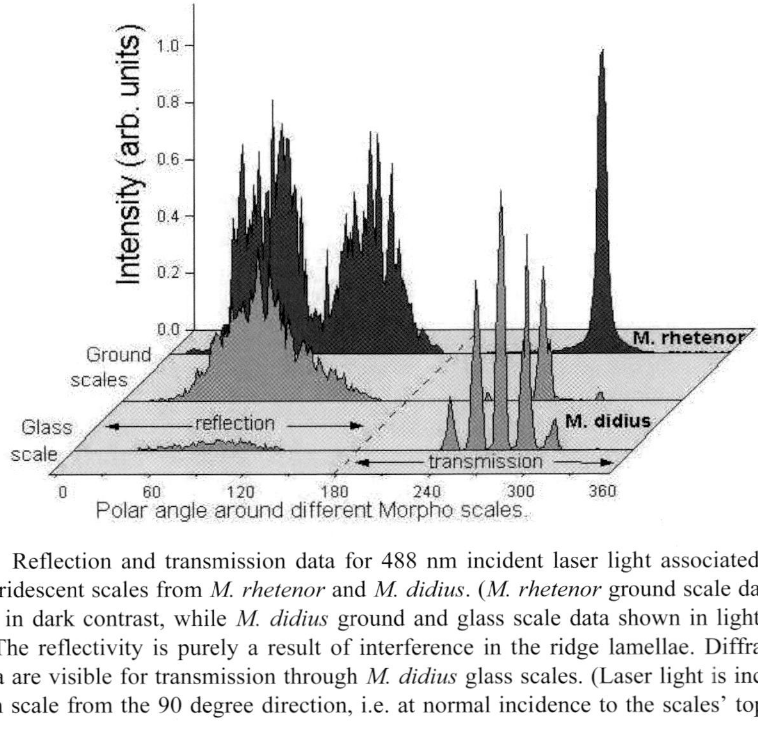

Fig. 6. Reflection and transmission data for 488 nm incident laser light associated with single iridescent scales from *M. rhetenor* and *M. didius*. (*M. rhetenor* ground scale data indicated in dark contrast, while *M. didius* ground and glass scale data shown in light contrast). The reflectivity is purely a result of interference in the ridge lamellae. Diffraction maxima are visible for transmission through *M. didius* glass scales. (Laser light is incident on each scale from the 90 degree direction, i.e. at normal incidence to the scales' top surface).

The reflectivity from the *M. didius* ground scale is not lobed but is nonetheless also spread widely in angle, spanning around 100 degrees in the plane shown (but only approximately 15 degrees in the plane orthogonal to this). Transmission through these ground scales is unremarkable. The transmission through the *M. didius* glass scale is striking; monochromatic collimated light undergoes strong diffraction by the scale ridging as it passes through the scale. The overall reflectivity of both scales is significant. For *M. rhetenor*, the twelve cuticle-layer system reflects a maximum of approximately 80 percent of incident blue light. *M. didius*, reflects less, around 45 percent, a fact associated with the fewer number of cuticle layers in its scale ridging.

Under optical transmission microscopy, both ground scales appear brown, a colour associated with their melanin pigmentation. This pigmentation creates an optical absorption that is an inherent part of the scale optics. In addition to this, it has been associated with cuticle mechanical strengthening through the creation of protein cross-linking (Kalmus 1941, Riley 1997). Additional single scale optical experiments, during which the scale is immersed in liquid bromoform (a fluid whose refractive index matches that of the scale material) indicate that the absorption coefficient (k) at visible wavelengths is approx. 0.060±0.005 (Vukusic et al. 1999).

The purpose of this type Ia scale structure, certainly in the case of *Morpho* sp., seems clear. Not only is it possible to effect high reflectivities of good spectral purity, but the reflectivity may be directed over a broad angle. In the case of *M. rhetenor* the spread in tilt angle of the layers within the ridging and a distribution of tilts of the ridges themselves, creates the broad angle reflection. For *M. didius*, the same effect, coupled with a diffraction-assisted further spread in angle on transmission through the superficial layer of glass scales produces extremely wide-angle high visibility. The description of the full biological significance of this is presented elsewhere (Vukusic et al. 1999).

Type II scales comprise multilayering (described as body-lamellae) that is incorporated into the body of the scale itself (Fig. 2). For this category, ridging is present on the surface of the scales, but it is usually a much simpler non-specialised form of ridging than found on type I scales. As many as ten cuticle layers have been found to exist in some species with iridescent type II scales (Vukusic unpublished). As with type I, type II scales may be categorised in a secondary way. Type IIa comprise the simplest form of such body-lamellae systems in which the lamellae are parallel and flat across the entire area of the scale (allowing for any curvature of the scale itself) (Fig. 7).

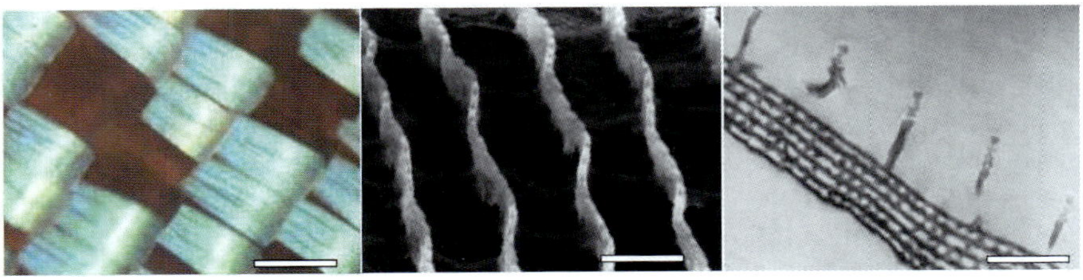

Fig. 7. (Left) Optical micrograph showing iridescent scales on the wing of *Urania leilus*. (Middle) SEM image of the ridges on the surface of the iridescent *U. leilus* scales shown on the left. (Right) TEM image of the transverse section through an iridescent *U. leilus* scale shown left and middle. (Scale bars: left 90 μm; middle 1.2 μm and right 1.0 μm).

Additionally, not only does ridging sit above this system of body-lamellae, but the ridges are often orthogonally interconnected by structures known as crossribs. The spacing of these crossribs is usually quite periodic and is of the order of the spacing of the ridges themselves. It is believed that these and other surface structures associated with type II systems provide one or both of the following effects. They may form impedance matching elements that reduce broadband reflection from the top-most surface of the scale (Vukusic unpublished), and they may constitute diffraction elements that diffuse and spread in angle the selectively reflected colour from the multilayer over which they rest. *Urania leilus* is a species of diurnal moth, the scales of which are of type IIa. Its appearance is less striking than that of *Morpho*; its iridescent regions are less highly reflective and its hue is rather softer in appearance (Fig. 7) (Vukusic and Lawrence unpublished).

Type IIb scales represent a category in which the flat body-lamellae of type IIa have specific modulations imposed upon them. In certain *Papilio* species these modulations create arrays of shallow flat-bottomed concavities across the surface of each scale (Fig. 8).

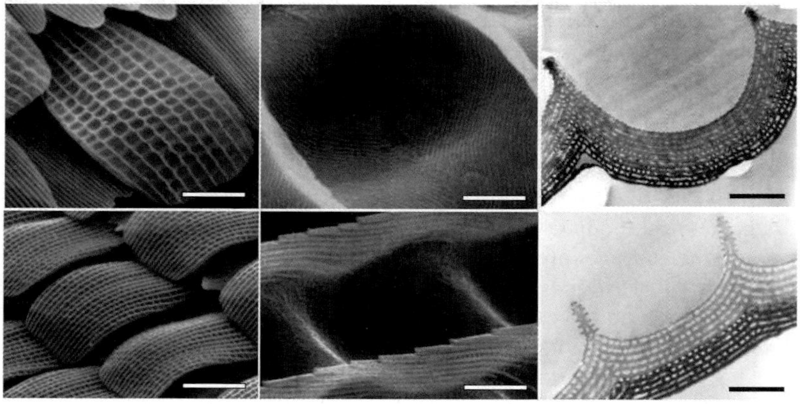

Fig. 8. SEM (left and middle) and TEM (right) micrographs of iridescent scale regions and sections from *P. palinurus* (top row) and *P. ulysses* (bottom row). SEM images of scales from each species show convincing similarity. TEM images of the transverse section through a single concavity of a *P. palinurus* scale (top right) shows a contrastingly deeper profile to that of *P. ulysses* (bottom right). (Scales bars: top left 65 μm; bottom left 120 μm; top middle 1 μm; bottom middle 1.5 μm; top and bottom right 1 μm). (Reprinted with permission from *Applied Optics* (**40**, 7, pp1116-1125) © 2001).

The body-lamellae follow the profile of these concavities. Interestingly, in most type IIb species, the physical dimension of each multilayer remains approximately constant regardless of position across each concavity (Vukusic et al. 2000b). This creates the effect of a sculpted multilayer with uniform layer dimensions, rather than a patterned multilayer of variable dimensions, as exhibited by some tiger beetles (Schultz and Rankin 1985). The particular sculpting associated with some type IIb systems has been shown to produce several unique optical effects, such as the simultaneous production of two structural colours, which combine additively to give the stimulus of a third colour (Vukusic et al. 2000c). Furthermore, through a double reflection from the opposite sides of each concavity, the polarisation of one of the colours may be rotated by 90 degrees (Fig. 9). Additionally with type IIb scales, crossribs are generally absent, ridging appears to play much less of an optical role and the scale surface may exhibit one of a variety of forms of impedance matching surface texture (Fig. 8 middle).

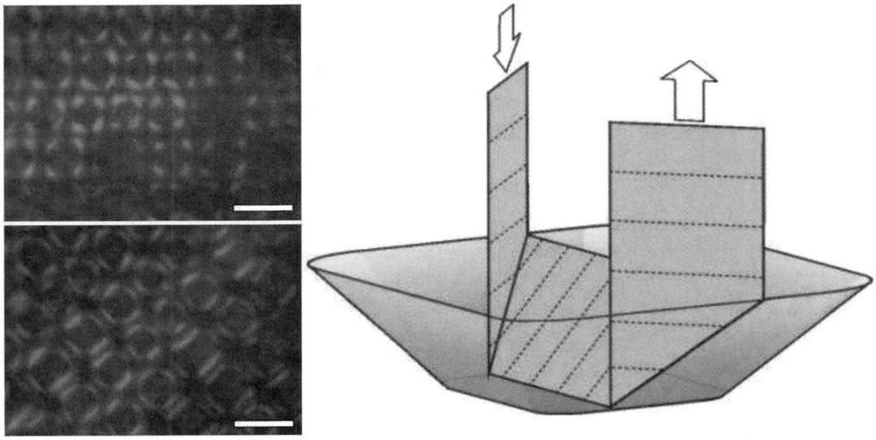

Fig. 9. (Left) Optical micrograph of the same region of the same *P. palinurus* iridescent scale shown figure 1.3 (top right image). In this figure, however, illumination is with linearly polarised white light while the image was captured through a crossed linear analyser. From top to bottom image on the left, the sample has been rotated by 45 degrees (Scale bars: 8 μm). (Right) Schematic diagram to illustrate the mechanism by which the plane of polarisation of light retro-reflected via the sides of each concavity is rotated by 90 degrees (Colour micrographs reprinted with permission from *Applied Optics* (**40**, 7, pp1116-1125) © 2001. Schematic diagram reprinted with permission from *Nature* (**404**, p457) © 2000).

4.2.2 Analysis of Type IIb Scales: P. palinurus and P. ulysses

As has been mentioned, while type I microstructure comprises multilayering sited in scale ridging, type II comprises multilayering within the body of the scale. Among the more interesting of the species studied in detail are those of *P. palinurus* and *P. ulysses* which feature the sculpted multilayering defined as type IIb microstructure. The iridescent scales of these butterflies exhibit modulations in the profiles of their multilayering. These modulations, in orthogonal directions across the scale surface, have the effect of imposing concave structures into the scale. SEM and TEM images of iridescent scales of both butterflies illustrate the geometry of these concavities (Fig. 8).

Close inspection of such images highlights several important features. Firstly, the dimensions of each layer in the direction perpendicular to the local layer surface remain approximately constant, regardless of the position around a concavity. Secondly, the multilayer dimensions are smaller for the blue coloured *P. ulysses* than for the apparently green coloured *P. palinurus*. Finally, there is a distinct difference between the scales of the two species in the depths and profiles of their respective concavities. In *P. palinurus* the concavities are deeper and their sides are steeply inclined with respect to the plane of the scale. For *P. ulysses*, the scale concavities are shallower and their profiles effect much less-inclined side walls.

The effect of the differences in these concavity profiles is only evident on inspection of high magnification optical micrographs (Fig. 3). When illuminated and observed at near-normal incidence, the flat central regions of the concavities reflect yellow in the case of *P. palinurus* and blue for *P. ulysses*. Furthermore, in *P. palinurus*, the inclined sides of each concavity appear as blue annuli around the yellow concavity centres. The inclined sides of *P. ulysses* concavities exhibit no analogous annular reflection pattern (even at ultra-violet wavelengths).

The key to understanding the difference in their optical effects not only lies in appreciating the differences in the thicknesses of their respective multilayering, but also in the differences of their concavity profiles. *P. ulysses*, with smaller layer dimensions than *P. palinurus* reflects a shorter wavelength at normal incidence, i.e. blue in comparison to the yellow reflected by the bottom of *P. palinurus* concavities. However, there is no coloured (or UV) annulus around the central concavity region of *P. ulysses*; this contrasts to the effect observed in *P. palinurus* is caused by differences in concavity wall inclination. Whereas a substantial part of *P. palinurus* concavity sides are inclined at approximately 45 degrees to the plane of the scale, the sides of *P. ulysses* concavities have inclinations of approximately 30 degrees. Consequently, opposite sides of each *P. palinurus* concavity are perpendicular to each other, whereas in *P. ulysses* they are not. Clearly, however, for *P. palinurus*, the single sides of each concavity cannot reflect light back along the incident direction. In fact, each angled side combines with the surface that is orthogonal to itself on the opposite side of each concavity. The blue annulus pattern is effected in this way; normally incident blue light, reflected from one 45 degree inclined surface is directed across the concavity to the opposite orthogonal surface from where it returns parallel to the incident direction (see schematic diagram in Fig. 9). These pairs of inclined surfaces comprise near identical multilayering and are both inclined at approximately 45 degrees to the direction of normally incident light on the scale surface. Accordingly their spectral reflectivity characteristics are closely matched. For *P. ulysses*, in which this surface orthogonality does not exist, such retro-reflection is not possible and the annuli are not seen.

Polarisation conversion of blue light through this double reflection confirms this retro-reflection mechanism in *P. palinurus*. Upon crossing an input linear polariser with an exit analyser (while viewing the sample using normally incident light) all yellow reflected light (reflected directly from the bottom of the concavities) is extinguished while a substantial portion of blue reflected light remains observed (Fig. 9). This necessarily implies that only the blue reflected light has undergone polarisation conversion. Such retro-reflected polarisation conversion is only predicted from orthogonal surfaces when the polarisation vector of the incident light is at 45° to the plane of incidence. It does not occur when the incident polarisation is perpendicular or parallel to the plane of polarisation (Fig. 9 schematic). Rotation of the wing-scale, through 45° in the plane of the wing, changes the regions of the inclined sides of each concavity that exhibit this strong polarisation conversion in the expected manner (Fig. 9 micrographs).

The biological significance of such colour and polarisation effects in *P. palinurus* is easily conjectured. In certain visual systems the close juxtaposition of such small blue and yellow regions may lead to strong colour stimulus synthesis (CSS)

(Burnham et al. 1963). CSS in an observer's visual system is an effect whereby the additive mixture of two or more colours synthesises the stimulus of a different colour to an observer. In the animal kingdom, CSS of green is usually achieved by the additive mixture of pigmentary yellow with structurally effected blue (Fox 1976). Such spatial-averaging CSS is the basis of modern colour television, older systems of colour photography and pointillistic painting (Evans 1943, MacAdam 1956, Agoston 1987). We might assume that CSS of green in the visual systems of predators of *P. palinurus* may provide it with camouflage against foliaceous backgrounds. This, combined with predicted polarisation sensitivity in certain *P. palinurus* photoreceptors, provides an insight to the selection pressures responsible for the development of its microstructure. *P. ulysses* appears to have developed separately under different selection pressures; its colour appearance designed instead to promote long range visibility in a similar fashion to certain *Morpho* sp.

The details of type III Lepidopteran scales, particularly those that exhibit photonic crystal structures, will be discussed in Sect. 4.3 describing non-layered systems.

Structural colour in bird's feathers is found in a variety of species, a very early explanation for which relates to a study of iridescence in male peacock feathers (Newton 1730). Current understanding suggests that the origin of the iridescence is due to the presence of a dense array of rodlets or platelets set into regular lattices beneath the surfaces of feather barbules. In the majority of bird taxa, these rods or platelets have been shown to comprise the pigment melanin, embedded in a generic medium of keratin that makes up the remainder of the feather barbule. In certain taxa the platelets comprise pure melanin (Rutschke 1960), whereas in other species they are filled with pockets of air (Greenwalt et al. 1960). The diameter and separation of the platelets or rodlets varies from species to species. In certain iridescent duck feathers the rodlets are approximately 150 nm in diameter with very small quantities of intermediary keratin (Rutshcke 1966). Conversely, species of cuckoo exhibit smaller rodlets, of the order of 80nm diameter, but with approximately 90 nm of intermediary keratin (Durrer and Villiger 1970). Naturally enough, such dimensions effect visible and in some cases near-UV iridescence in these species. In iridescent feathers of certain other species, the melanin rodlets or platelets are filled with pockets of gas, understood to comprise mostly air (Greenwalt et al. 1960). Barbules from iridescent silver feathers of the pheasant *Lophophorus impeyanus* comprise such gas-filled rodlets: this contrasts to certain hummingbirds (*Crysolampis* and *Clytolaema* sp) which comprise thin gas-filled melanin platelets that are elliptical in transverse section but more planar in longitudinal section. Sections through whole platelets (Fig. 10) exhibit major and minor axis dimensions of approximately 2-3 μm and 1-1.5 μm respectively. Particularly in hummingbirds, the refractive index periodicity afforded by up to fifteen air-filled melanin platelets effects astonishingly bright iridescence (Fig. 10).

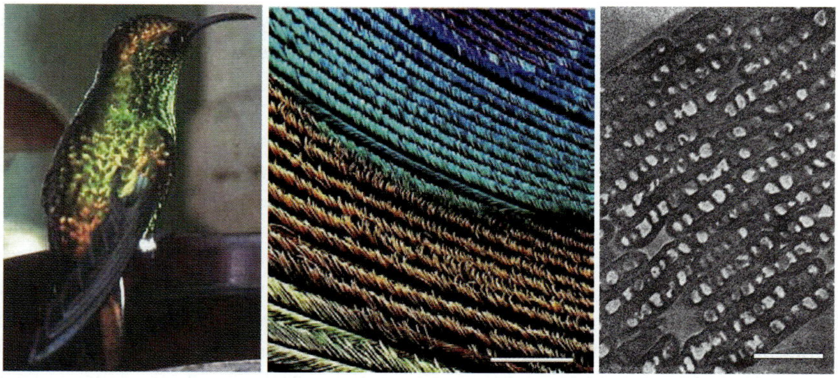

Fig. 10. Iridescent colours on certain hummingbird feathers (left) are produced through interference in microstructures within the feather barbs. Individual barbs (middle) generally comprise a component of layered platelets of melanin filled with gas. A TEM image [reprinted from Greenwalt 1960] of a transverse section through a single barb shows the periodicity which underpins the interference (Scale bar: middle 1 mm; right 120 nm).

4.3 Non-Layered Systems

Certain invertebrate taxa in a very limited range of genera, exhibit microstructure that stems from periodicities of two or three dimensions rather than the often rather standard 1D periodicities of plane multilayered specimens. The colour effects associated with these specimens do not generally appear quite as eye-catching, but the elegance and complexity of their design and function in certain species can be astonishing.

An interesting example of structural colour effected by 2D periodicity, is that exhibited by certain taxa of marine polychaete worm. One of these, *Aphrodita*, a genus of polychaete worm commonly known as the sea-mouse, has a short segmented body that is covered with hair-like structures called setae. These setae, extruded by epidermal cells, are exceptional by their strong iridescence. They comprise bundles of thin-walled cylinders of chitin, held together by sclerotinized protein to form fine and thread-like hairs. The diameter of cylinder centres varies between approximately 100 nm and 360 nm in *Aphrodita* (Parker et al. 2001); in another polychaete species *Pherusa*, optically similar setae exhibit a greater number of cylinder periodicities with a more uniform diameter of approximately 220 nm – 270 nm (Vukusic and Sambles unpublished (Fig. 1.11). The iridescent colour in these setae is effected through diffraction from with the periodicity of the cylinder elements. The detailed optical analysis associated with the structure of *Aphrodita* is presented elsewhere (McPhedran et al. 2001)

A variation of the structure found in iridescent polychaete setae is also found in the tapeta of the eyes of cats. However, whereas polychaete setae comprise bundles of hollow cylinders, the tapeta of cats eyes comprise bundles of solid cylinders or rods made of cuticular material. Within each bundle, rods of diameter 200 nm – 350 nm, are spaced 300 nm – 500nm apart and are arranged in a mutually parallel geometry that yields hexagonal symmetry in transverse cross-section. Strong diffraction arising as a result of the periodicity of these rods produces the eye-shine normally observed with cats' eyes (Pedler 1963). Its physiological function, similar to that of tapeta in certain arthropods, is believed to be for optimising photoreception in the cats' visual system.

In certain Lepidoptera, rather than 2D periodicity such as that just described for polychaete worm setae, a remarkable 3D periodicity exists. Within the previously described context of Lepidopteran microstructure characterisation, such 3D periodicity in scales is categorised as type III microstructure and encompasses species which display colouration that arises from Bragg scattering and Tyndall scattering structures within and on the body of the scale. Unlike colouration arising from multilayering with a characteristic angle-dependence of hue, type III microstructure in these taxa produce broadly constant colour at all angles of observation, often by effecting variations in orientation of microscopic juxtaposed domains in the 3D lattice.

Type IIIa scales are defined as those which comprise an ordered 3D lattice of cuticle within the body of the scale; analogous to an array of spherical holes in a matrix of cuticle. The order imposed by such a system leads to strong scattering of favoured wavelengths in certain directions. Analogous, but inverse, structures lead to the structural colouration in the mineral opal (Sanders 1968, 1971), for which arrays of hydrated silica spheres surrounded by air lead to diffraction colours. The lattices within most type IIIa butterfly scales, however, are usually divided into irregular but distinct domains of a few microns diameter (Fig. 12). By varying the orientation of neighbouring domains, a constant-colour effect is created across the scale and wing through spatial averaging. The diffuse angle-independent nature of the structural colour that is provided by type IIIa scales is believed to help with defense from predation through camouflage.

Type IIIb is rather rarer than type IIIa and is distinctly different. Scales in this category are coloured through Tyndall scattering from sub-wavelength sized scattering structures (Huxley 1976). These structures extend from between the ridges on the surface of the scales down into the body of the scale. In *P. zalmoxis*, the scattering structures are in the form of air-filled alveoli approximately 2 μm long and around 220 nm in internal diameter. The simultaneous presence within the alveoli of a UV-absorbing but blue fluorescing pigment enhances the blue colouration (Huxley 1976).

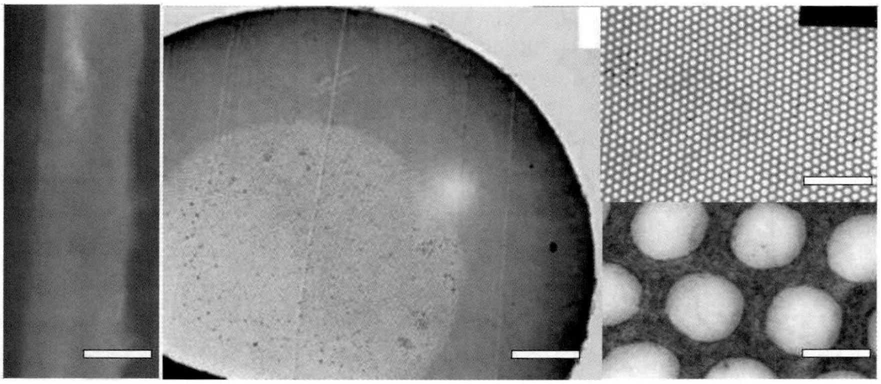

Fig. 11. A single setae from the polychaete worm *Pherusa*, viewed by optical microscopy (left) and by TEM; showing increasing magnifications of transverse sections through the setae. Iridescent effects are visible in the colour image, produced from highly periodic microstructure derived through the longitudinal stacking of hollow cylinders within each seta. The images on the right show magnified regions of the central portion of the middle image (Scale bars: left 0.25 mm; middle 60 μm; top right 4 μm; bottom right 250 nm).

4.3.1 Analysis of Type IIIa Scales: Parides sesostris

P. sesostris exhibits bright green iridescent patches on its ventral wings. Preliminary inspection shows that its colour is angle-independent yet structural in origin; an indication that its microstructure is a form of photonic crystal. SEM micrographs confirm the existence and regularity of its 3D lattice (Fig. 12). The exact function of the advanced form of ridging over the top surface of the scale (Fig. 12 bottom) is unknown, but it is believed to narrow the angle of reflection emerging from the photonic crystal of the scale. It is of interest to ascertain the nature of the photonic crystal microstructure; namely to answer the question "which crystal geometry has nature decided to adopt in the case of this butterfly?". The structure was ascertained in the following way. TEM micrographs of sections through iridescent scales of *P. sesostris* were taken (Fig. 12). These micrographs not only illustrate the domaining of the lattice structure, but show tessellated patterns formed by such thin planes through different orientations of the structure. Using 3D modelling software that permits any orientation of thin section planes to be imaged, the full range of crystal lattice structures were constructed and thin sectional planes taken. The only 3D lattice structure from which it proved possible to extract planar sections that corresponded closely to all those of the TEM micrographs, was the tetrahedral structure i.e. that of the diamond lattice.

It is worth noting out a strong polarisation effect that is observed with the microstructure from this butterfly.

On illuminating the butterfly with linearly polarised light and observing with a crossed linear analyser, strong shorter wavelength colouration, is observed from the domained regions of the scale (Fig. 12 middle). Clearly, owing to the presence of the two crossed polarisers, these colours are only observed through the rotation of the polarisation vector of incident light by 90 degrees on reflection from the structure. Azimuthal rotation of the sample, between such fixed but crossed polarisers, cycles the domain colours from green to shorter wavelengths (ending in the appearance of black where UV is reflected) and back to green.

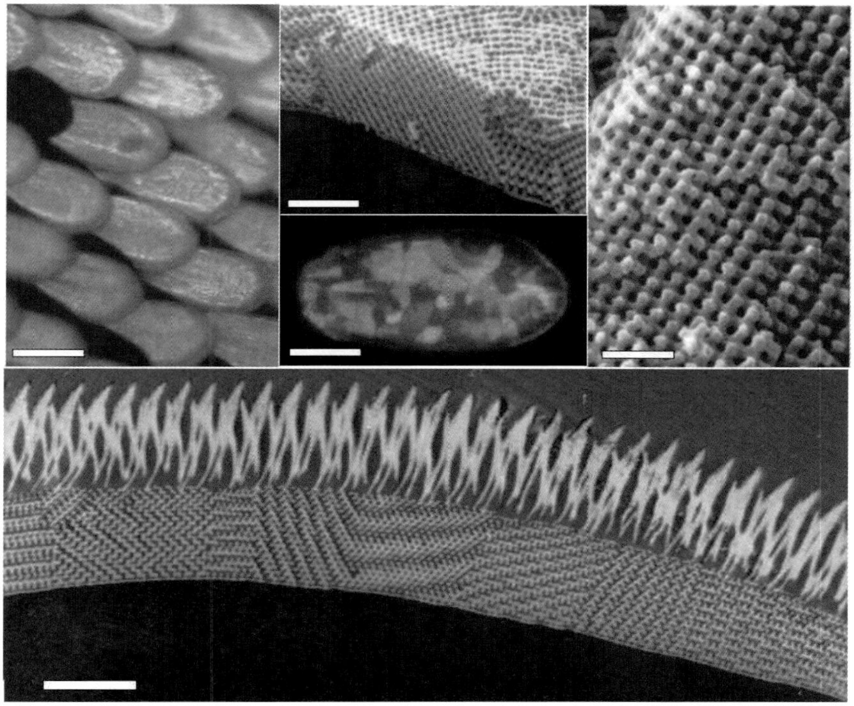

Fig. 12. Optical micrographs; top left shows *P. sesostris* scales in reflection under normal illumination, and (dead centre) a single *P. sesostris* scale viewed in reflection with linearly polarised white light illumination while the image was captured through a crossed linear analyser. Top centre and top right images are SEM micrographs of a fractured transverse section of a *P. sesostris* iridescent scale (the 3D periodic lattice of cuticle is clearly visible). The bottom image is a TEM micrograph of a *P. sesostris* iridescent scale (cuticle material has white contrast, intervening air is black). (Scale bar: top left 90 μm; dead centre 50 μm; top middle 3 μm; top right 1 μm; bottom 6 μm).

This strong polarisation effect is presently attributed to a similar mechanism as that which produces the blue annulus of the *P. palinurus* concavity described in Sect. 4.2.2; namely strong scattering from orthogonally facing and planes within the tetrahedral structure. While the advantage of true tetrahedral microstructure is clear (it achieves optimal scale strength for low mass and photonically it is the structure for which highest reflectivity is achieved over the greatest angle range (Chan et al. 1991)), it is not clear why there appears to be small but evident deviation. Interestingly, a full reciprocal space analysis based on a true tetrahedral lattice of cuticle with ca. 40% occupancy (as calculated from analysis of SEM and TEM micrographs), predicts the characteristic green colouration for this scale structure that is shown in Fig. 12.

4.4 Plant Systems

Understanding of plant iridescence is based on studies of a limited number of taxa. It exists in certain vascular plant leaves, fruit and algae. The mechanisms for their structural colouration have been elucidated in some detail. Interference associated with thin layers of alternating high and low refractive index appears to be the predominant mechanism that underpins the observed iridescence (Fig. 13).

Blue iridescence is prevalent in the genus *Selaginella* (Fig. 13), comprising a range of fern-like understory plants from the rain forests of Malaysia. In the two *Selaginella* species studied in detail, (Hébant and Lee 1984) the iridescence appears to result from the elaboration of two reflective lamellae in the outer wall of the adaxial epidermis. The iridescent juvenile leaves of another neotropical plant *Danaea nodosa*, exhibit a much larger number of layers on the abaxial side of their epidermal cell wall. TEM images of the specimen show 18-30 layers described as alternating electron translucent and electron opaque material (Fig. 13). These layers are responsible for the interference (Graham et al. 1993), the dark and light layer contrast brought about by the staining associated with the TEM sample preparation method. Close inspection of slightly oblique transverse TEM sections through *D. nodosa*, however, indicates the layering is associated with a helicoidal arrangement of cellulose microfibrils. Such structure, results from local strain environments in the plant cells as the layers are being deposited during growth (Gertal and Green 1977), dictating the angle at which the fibres are deposited. It has been reported in the cell walls of a great many other plant taxa (Neville and Levy 1984, 1985) including many in which structural colouration is completely absent. Indeed, it also appears absent from other taxa in which structural colouration is present. Although structurally, the helicoidal patterning appears analogous to that associated with helicoidal chitin rods in the exocuticle of certain scarabeid beetles, it has not been shown to exhibit analogous optical effects.

Modified organelles in the leaves of other taxa have elaborated lamellae for the purpose of effecting structural colour through interference. In the neotropical fern *Trichromanes elegans* optical microscopy shows the origin of the blue-green iridescence as coloured flecks that cover the leaves' surfaces; TEM microscopy indicates that the flecks correspond to elongated and stratified chloroplasts. Within these chloroplasts, referred to as iridiplasts or iridisomes, twenty or more alternating layers of less or more hydrated cellulose, of relatively low refractive index (n=1.35) and high refractive index (n=1.47) respectively, accounts for the colour.

In addition to structural colour in the leaves of certain vascular plants, two species of iridescent fruit have also been identified and investigated (Lee et al. 2000). Fruits from *Delarbrea michieana* and *Elaeocarpus angustifolius* exhibit a brilliant iridescent blue colouration. The origin of the iridescence lies with discrete iridisomes; structures that are in this case produced between cytoplasm and the outer epidermal cell wall.

These iridisomes exhibit structural periodicities comprising wall-like electron-opaque regions surrounding bubble-like electron translucent regions; the differential contrast arising from the TEM sample preparation method. In transverse section the iridisomes are not unlike discrete multilayer systems in appearance, with dimensions that effect blue reflectivity for the refractive indices associated with cellulose of high and low hydration. However, there is a 3D lattice structure associated with these iridisomes, more than the sections alone would suggest. A standard multilayer interference model may well be overly simplistic (Lee 2000).

Fig. 13. (Left) Iridescent blue leaves of the neotropical plant *Selaginella willdenowii*. The colour is caused by interference in thin lamellae in the cell wall of the leaves (leaves are approximately 4 mm long). (Right) TEM images of the outer cell wall of epidermal cells from a juvenile leaf of the plant *Danaea nodosa*. Blue iridescence in the plant is produced by interference in the multilayering. The semi-circular object adjacent to the layering is a single chloroplast. (scale bar: 1.5 μm). (Both images reprinted with kind permission from DW Lee).

Structurally coloured taxa of marine algae usually effect their colour through interference in multiple laminations in cell membranes (Dangeard 1941, Gerwick and Lang 1977, Péderson 1980) or through scattering from rod-shaped or round particles in cytoplasm or vacuoles below their surface (Feldmann 1970). In the Irish moss *Chondrus crispus* blue iridescence results from up to twelve electron opaque layers approximately 50 nm thick, alternating with electron translucent layers approximately 85 nm thick (Péderson 1980). Lower iridescence in certain specimens is attributable to fewer number of laminations, as few as six in certain specimens. Brilliant blue iridescence in the genus *Irideae* is attributable to an even greater number of layers, as many as twenty each of electron opaque and translucent material.

Structural colouration is said to be possibly an anomaly among plants. Unsurprising therefore that the function of such iridescence is not yet fully understood. There are some clues, however. While the iridescent blue optical filter on the leaves of *Selaginella,* for instance, reflects blue wavelengths efficiently, it simultaneously favours efficient transmission of other wavelengths, both longer and shorter than blue. In this way its design purpose may be as a facilitator of anti-reflection over a wavelength range rather than specifically as an efficient blue mirror. Studies on *Selaginella* have shown that prolonged incidence of different intensity ratios of red light (R) to far-red light (FR), bring about strong morphological differences in identical plant species (Hébant and Lee 1984). This suggests phytochrome, a chromoprotein that plays a significant role in initiating developmental processes, is implicated in the selection pressures responsible for the evolution of blue leaf iridescence.

The function of iridescence in blue fruits may be two-fold. Firstly, the structural integrity of epidermal and cellular laminations is much less ephemeral than the chemical species associated with pigmentary colouration after fruits mature or ripen. Subsequently, structural colouration is longer lasting than pigmentary colouration. Additionally, there is higher reflectivity associated with such structural colour over pigmentary colour. Thus a more durable and visible advertisement for dispersal may be effected. Secondly, the transmissivity of such structurally blue fruits to longer wavelengths appears to permit continued carbon dioxide assimilation despite fruit ripeness and even after the fruit has fallen from its tree (Lee 2000).

5 Theoretical Considerations

Natural optical systems are inherently difficult to model accurately. There are several reasons for this. Firstly, they generally comprise very complex structures, the exact dimensions of which are difficult to ascertain with reliability. SEM and TEM microscopy are commonly used to measure these physical dimensions; however, they both necessitate finical sample preparation techniques which, together with the imaging process, may cause a degree of sample deformation. An added complication to this is associated with unsystematic variations in sample dimensions across the region of the specimen of interest. Reliable sample representation may therefore necessitate the modeling of regions of up to several hundreds of micrometers in diameter.

Another difficulty associated with modelling the optics of natural specimens, relates to the accuracy with which the optical parameters of the sample materials are known. Generally, zero or negligible dispersion is assumed, such that a single unvarying refractive index is used to model optical response from near UV across visible to near IR wavelengths. This may well be an oversimplification in many, if not most, cases. Additionally, very little information exists relating the extent of optical absorption present in natural systems, and generally, models assume zero or negligible absorption; again an oversimplification in many cases.

A third area of difficulty relates to extraneous sample regions and macroscopic sample orientational geometry that are not obviously associated with the optics of systems, but which need to be taken into account when trying to model experimentally collected data. A good example of this is exhibited by *Morpho* butterflies for which experimental optical data collected from whole or part wing sections is convoluted with reflection, absorption and scattering effects that are associated with scale alignment and orientation, and to extraneous effects from non-iridescent scales and the wing substrate itself. Either specific study of the isolated optical system (e.g. investigation of single butterfly scales (Vukusic et al. 1999)) or incorporation of the extraneous specimen region into the theoretical model, produces significantly more accurate representations of such natural systems.

The majority of treatments of natural optical systems present a straight-forward theoretical analysis based on fundamental principles of constructive and destructive interference. Several texts explicitly use the method outlined by Huxley (Huxley 1968) for predicting the wavelength and polarisation dependent reflectivities associated with a number of natural systems (Land 1972). Fundamentally of course, this method is a variation on a theme presented in many previous texts and which has been well founded in physics for many years (Abelès 1950, Heavens 1955, Vašiček 1960, Born and Wolf 1964, Mcleod 1969). However, although these methods can only be used to model flat semi-infinite systems of multilayering which do not exist in natural samples, they can in certain cases provide good approximations to real systems. Recent interest in, and optical experimentation with more complex natural optical systems has necessitated much more accurate microstructural models. These are being realised through rigorous application of electromagnetic grating theory, that is described elsewhere (McPhedran et al. 1999, Gralak et al. 2002)

6 Conclusions

Photonics-led selection pressures in nature have led to the development of complex microstructural systems within most major phyla. In the majority of such systems, layers of alternating high and low refractive index are employed to effect surfaces of high reflectance. Certain multilayer systems in particular, achieve high reflectance through optimal layer thicknesses for a limited number of layers; producing "ideal multilayering" through equal optical thicknesses of high and low index layer. Other "non-ideal" multilayer systems effect high reflectance through use of a large number of layers or high contrast between layer refractive index. In certain cases, multilayer systems are designed for broad-band reflectivity, efficiently effecting the appearance of silver or gold surfaces by using chirped systems, chaotic systems or a combination of overlying systems each reflecting different and narrow wavelength bands. Examples exist in which multilayering, incorporated in to spatially discrete bundles, effects strong angle-dependent intensity and hue characteristics dissimilar to those associated with flat semi-infinite multilayer systems. In other examples, additional modulations or specific sculpting imposed on otherwise flat multilayer systems, can effect more intricate optical properties such as colour mixing and polarisation rotation. Certain structurally coloured systems do not exhibit multilayering but instead comprise periodicities of two or of three dimensions that effect strong Bragg diffraction. Optically, such photonic crystal systems exhibit different angle-dependent and polarisation-dependent properties compared to multilayered systems.

It is clear that the gamut of optical microstructure found in nature serves a variety of biological purposes. In individual cases, appreciation for the reasons that particular microstructure has evolved necessitates consideration of the selection pressures associated with such functions as visibility, communication, crypsis, mimicry, and even photoreceptor sensitivity. Sometimes other non-optical functions carry weight in microstructure development; wetability of surfaces, manufacture and mechanical function of material, environmental, ecological and physiological factors each play their part with some degree of significance. However, purely in terms of optics, the lesson is clear. We are spectators in nature's optical arena and would do well to learn from her example.

Acknowledgements

I am very grateful to Prof. Roy Sambles for helpful and inspiring discussions on optics, and to Prof. Helen Ghiradella for her interest and support for the work on Lepidopteran optics. This work was supported by the University of Exeter and by a Biotechnology and Biological Sciences Research Council (BBSRC) advanced research fellowship.

References

Abelès F (1950) Recherches sur las propagation des ondes électromagnétiques sinusoïdales dans les milieux stratifiés. Annls Phys (Series 12) 5:596-640

Agoston GA (1987) Colour Theory and its Applications in Art and Design. Springer-Verlag, New-York

Bates HW (1864) The naturalist on the river Amazons. 2nd ed, London

Bernhard CG (1967) Structural and functional adaptation in a visual system. Endeavor 26: 79-84

Bernhard CG and Miller WH (1962) A corneal nipple pattern in insect compound eyes. Acta Physiol Scandinavica 56: 385-386

Born M and Wolf E (1965) Principles of optics. 3rd Ed, Pergamon, Oxford

Bouligand Y (1972) Twisted fibrous arrangements in biological materials and cholesteric mesophases. Tissue & Cell 4: 189-217

Brink DJ and Lee ME (1999) Confined blue iridescence by a diffraction microstructure: an optical investigation of the *Cynandra opis* butterfly. App Opt 38: 5282-5289

Burnham RW, Hanes RM and Bartleson CJ (1963) Color. Wiley, New York

Byers JR (1975) Tyndall blue and surface white of tent caterpillars *Malacosoma* spp. J Insect Physiol 21:401-415

Chan CT, Ho KM and Soukoulis CM (1991) Photonic band gaps in experimentally realizable periodic dielectric structures. Europhys Lett 16: 563-568

Clapham PB and Hutley MC (1973) Reduction of lens reflection by the "moth eye" principle. Nature 244: 281-282

Crowson RA (1981) The biology of Coleoptera. Glasgow, Academic Press

Dangeard P (1941) Recherches sur les enclaves iridescentes de la cellule des Algues (in French). Le Botaniste 31: 31-57

Denton EJ (1970) On the organisation of reflecting surfaces in some marine animals. Phil Trans Roy Soc B, 258: 285-313

Denton EJ (1971) Reflectors in fishes. Scientific American 224,1: 64-72

Denton EJ and Land MF (1971) Mechanism of reflection in silvery layers of fish and cephalopods. Proc Roy Soc B 178: 43-61

Denton EJ and Nicol JAC (1965) Reflection of light by external surfaces of the herring, *Clupea harengus*. J mar Biol Ass UK 45: 711-738

Ditchburn RW (1963) Light. 2nd ed Blackie, London

Durrer H and Villiger W (1970) Schillerradien des Goldcuckuck (*Crysococcyx cupreus*). Z. Zellforsch 109:407-413

Evans RM (1943) Visual processes and colour photography. J Opt Soc Am 33: 579-614

Feldmann G (1970) Sur l'ultrastructure de l'appareil irisant du *Gastroclonium clavatum* (Roth) Ardissone (Rhodophycées). C R Acad Sci Ser D 270:1244-1246

Fox DL (1976) Animal biochromes and structural colours. Univ California Press, Berkeley

Fox HM and Vevers G (1960) The nature of animal colours. Sidgwick and Jackson, London

Frank F (1939) Die Färbung der Vogelfeder durch Pigment und Struktur (in German). J. Orn. Lpz. **3:** 426-523,

Gates MD (1970) Physical and physiological properties of plants. In: Remote sensing with special reference to agriculture and forestry. Nat Acad Sci pp 224-252

Gertel ET and Green PB (1977) Cell growth pattern and wall micro-fibrillar arrangement. Plant Physiol 60: 247-254

Gerwick WH and Lang NJ (1977) Structural, chemical and ecological studies on iridescence in *Iridaea* (Rhodophyta). J Phycol 13: 121-127

Ghiradella H (1991) Light and colour on the wing: structural colours in butterflies and moths. Appl Opt 30: 3492-3500

Ghiradella H (1994) Structure of butterfly scales: patterning in an insect cuticle. Microsc Res Tech 27: 429-438

Ghiradella H and Radigan W (1976) Development of butterfly scales; II. Struts, lattices and surface tension. J Morph 150: 279-296

Ghiradella H, Aneshansley D, Eisner T, Silbergleid RE and Hinton H E (1972) Ultra-violet reflection of a male butterfly: Interference colour caused by thin layer elaboration of wing scales. Science 178: 1214-1217

Graham RM, Lee DW and Norstog K (1993) Physical and ultrastructural basis of blue leaf iridescence in two neotropical ferns. Amer J Bot 80: 198-203

Gralak B, Tayeb G and Enoch S (2001) *Morpho* butterflies wings color modeled with lamellar grating theory. Opt Express 9: 567-578

Greenstein LM (1966) Nacreous pigments and their properties. Proc Scient Sect Toilet Goods Ass. 45: 20-26

Greenwalt CH, Brandt W and Friel DD (1960) Iridescent colours of hummingbird feathers. J Opt Soc Am 50: 1005-1013

Guinier A (1963) X-ray diffraction. WH Freeman and Company, San Francisco

Heavens OS (1960) Optical properties of thin films. Rep Prog Phys 23:1-65

Hébant C and Lee DW (1984) Ultrastructural basis and developmental control of blue iridescence in *Selaginella* leaves. Amer J Bot 71:216-219

Herring PJ (1994) Reflective systems in aquatic animals. Comp Biochem and Physiol 109A: 513-546

Hinton and Gibbs DF (1969) Diffraction gratings in Phalacrid beetles. Nature 221: 953-954

Huxley AF (1968) A theoretical treatment of the reflection of light by multilayer structures. J Exp Biol 48: 227-245

Huxley J (1976) The coloration of *Papilio zalmoxis* and *P. antimachus* and the discovery of Tyndall blue in butterflies. Proc Roy Soc B 193: 441-453

Huxley J (1976) The coloration of *Papilio zalmoxis* and *P. antimachus* and the discovery of Tyndall blue in butterflies. Proc Roy Soc Lond B 193: 441-453

Joannopoulos JD, Meade RD and Winn J (1995) Photonic Crystals: Molding the Flow of Light. Princeton University Press, Princeton

Kalmus H (1941) Physiology and Ecology of cuticle colour in insects. Nature 148: 428-431.

Kawaguti S and Kamishima Y (1966) Electron microscopy on iridiphores and guanophores of fish. 6th Int Cong for Elect Micros: p419

Land MF (1966) A multilayer interference reflector in the eye of the scallop, *Pecten maximus*. J Exp Biol 45: 433-447

Land MF (1972) The physics and biology of animal reflectors. Progr Biophys Molec Biol 24: 75-106

Land MF (1999) personal communication

Large MCJ, McKenzie DR, Parker AR, Steel BC, Ho Karen, Bosi SG, Nicorovici N and McPhedran RC (2001) The mechanism of light reflectance in silverfish. Proc Roy Soc Lond A 457: 511-518

Lawrence CR, Vukusic P and Sambles JR (2002) Grazing incidence iridescence from a butterfly wing. App Opt 41:437-441

Lee DW (2000) Structural colouration in vascular plants. Photonics Science News 6: 54-60

Lee DW, Taylor GT and Irvine AK (2000) Structural fruit colouration in *Delarbea michieana* (Araliaceae). Int J Plant Sci 161: 297-300

Lippert W and Gentil K (1959) Uber lamellare Feinstrukturen bei den Schillershuppen der Schmetterlinge vom *Urania*- und *Morpho* typ (in German). Z Morphol Okol Tiere 48: 115-122

Longhurst RS (1967) Geometrical and physical optics. 2nd ed Longman, London

MacAdam DL (1956) Perceptions of colour in projected and televised pictures. J Soc Motion Picture Tel Eng 65: 455-466.

Mason CW (1927) Structural colours in insects, II. J Phys Chem 31: 321-354

Mckenzie DR, Yin Y and McFall WD (1995) Silvery fish skin as an example of a chaotic reflector. Proc Roy Soc Lond A 451: 579-584

McLeod HA (1969) Thin film optical filters. Adam Hilger, London

McPhedran RC, Botten LC, Asatryan AA, Nicorovici N, de Sterke CM and Robinson PA (1999) Ordered and disordered photonic band gap materials. Aust J Phys 52: 791 - 809

McPhedran RC, Nicorovici N, McKenzie DL, Botten LC, Parker AR and Rouse GW (2001) The Sea Mouse and the Photonic Crystal. Aust J Chem 54: 241-244

Meyer-Rochow VB and Stringer IA (1993) A system of regular ridges instead of nipples on a compound eye that has to operate near the diffraction limit. Vis Res 33: 2645-2647

Neville AC (1965) Chitin lamellogenesis in locust cuticle. Q J microsc Sci 106: 269-286

Neville AC (1975) Biology of the arthropod cuticle. Springer-Verlag, Berlin

Neville AC (1977) Metallic gold and silver colours in some insect cuticles. J Insect Physiol 23: 1267-1274
Neville AC and Caveney S (1969) Scarabeid beetle exocuticle as an optical analogue of cholesteric liquid crystals. Biol Rev 44: 531-562
Neville AC and Levy S (1984) Helicoidal orientation of cellulose micro-fibrils in *Nitella opaca* internode cells: ultrastructure and computed theoretical effects of strain reorientation during wall growth. Planta 162: 370-384
Neville AC and Levy S (1985) The helicoidal concept in plant cell wall ultrastructure and morphogenesis. In: Brett CT and Hillman JR (eds) The biochemistry of plant cell walls. Cambridge University Press, Cambridge, pp99-124
Newton I (1730) Opticks. 4th ed, reprinted New York, Dover
Parker AR (1995) Discover of functional iridescence and its coevolution with eyes in the phylogeny of *Ostracoda* (Crustacea). Proc Roy Soc Lond B 262: 349-355
Parker AR (1998) Colour in Burgess shale animals and the effect of light on evolution in the Cambrian. Proc Roy Soc Lond B 265: 967-972
Parker AR (1999) The Cambrian light switch. Biologist 46: 26-30
Parker AR (2000) 515 million years of structural colour. J Opt A: Pure Appl Opt 2: R15-R28
Parker AR, McPhedran RC, Botten LC and Nicorovici N (2001) *Aphrodite's* iridescence. Nature 409: 36-37
Pederson M, Roomans GM and Hofsten Av (1980) Blue iridescence and bromine in the cuticle of the red alga *Chondrus crispus* Sackh. Bot Marina 23: 193-196
Pedler C (1963) The fine structure of the tapetum cellulosum. Exp Eye Res 2: 189-195
Phelps CF (1972) Polysaccharides. Oxford biology readers ; 27, Oxford University Press, London
Preston RD (1974) The physical biology of plant cell walls. Chapman and Hall, London
Riley PA (1997) Molecules in focus: Melanin. Int J Biochem Cell Biol 29: 1235-1239
Rutschke E (1960) Die submikroscopische Struktur schillernder Federn von Entenvogeln (in German). Z Zellforsch 73: 432-443
Sanders JV (1968) Diffraction of light by opals. Acta Crst A24: 427-434
Sanders JV and Darragh PJ (1971) The microstructure of precious opal. The Mineralogical Record 2:261-268
Schultz TD and Bernard GD (1989) Pointillistic mixing of interference colours in cryptic tiger beetles. Nature 337: 72-73
Schultz TD and Rankin MA (1985) The ultrastructure of the epicuticular interference reflectors of tiger beetles (*Cicindela*). J Exp Biol 117: 87-110
Stephenson W (1969) The morphology of stridulatory structures in the genus *Ovalipes* Rathbun. Trans R Soc New Zealand 11: 43-71
Tyndall J (1869) On the blue colour of the sky, the polarisation of skylight and on the polarisation of light by cloudy matter generally. Lond Edin Dubl phil Mag 37: 384-394
Vašiček A (1960) Optics of thin films. North-Holland, Amsterdam
Verne J and Leyani F(1938) Les dyschromies (in French). Traité de dermatologie (Paris) 2: 745-811
Vukusic P, Sambles JR, Lawrence CR, and Wootton RJ (1999) Quantified interference and diffraction in single *Morpho* butterfly scales. Proc Roy Soc B 266: 1403-1411
Vukusic P, Sambles JR and Ghiradella H (2000a) Optical classification of microstructure in butterfly wing scales. Photonics Science News 6: 61-66

Vukusic P, Sambles JR and Lawrence CR (2000b) Structural colour: Colour mixing in wing scales of a butterfly. Nature 404: p457

Vukusic P, Sambles JR, Lawrence CR and Wakely G (2000c) Sculpted multilayer optical effects in two species of *Papilio* butterfly. App Opt 40: 1116-1125

Vukusic P, Sambles JR, Lawrence CR and Wootton RJ (2001) Structural colour: Now you see it – now you don't. Nature 410: p36

Vukusic P, Sambles JR, Lawrence CR, and Wootton RJ (2002) Limited-view iridescence in the butterfly *Ancyluris meliboeus*. Proc Roy Soc B 269: 7-14

Woolley JT (1971) Reflectance and transmittance of light by leaves. Plant Physiol 47: 656-662

Woolley JT (1975) Refractive index of soybean leaf cell walls. Plant Physiol 55: 172-174

Photonic Structures as Interference Devices

Carl G. Ribbing

Dept. of Materials Science, Uppsala University and Dept. of Functional Materials, FOI Linköping, Sweden

1 Introduction

Photonic crystals (PC) are based on scattering of electromagnetic waves in an inhomogeneous solid. Both in a basic physics perspective, and as a potential branch of technology it has a very close relationship to optical coatings, based on optical interference. Yet, PC's do not appear to be on the agenda in the community of people involved in the development and production of interference coatings. The concept 'photonic crystals' and the pioneeering works (Yablonovitch 1987, John 1987, Yablonovitch and Gmitter 1989) originate from researchers in the fields of electrical engineering, atomic physics and solid state physics. It is therefore a challenge to write this chapter as an attempt to bridge the gap between two communities that would obviously benefit from lively communication. We shall start with a brief historical introduction.

1.1 Historical Remarks

As it is too early to write the history of photonic crystals, this section will merely sketch a few lines and refer to more complete sources.

As usual the origins of this new and very active field go very far back. Purcell noticed (Purcell 1946) that the relaxation between nuclear spin levels in the radio frequency range corresponds to a relaxation time as long as 10^{21} s. He pointed out that the relaxation rate could be increased by surrounding the nuclei with small metal particles acting as coupled resonant circuits. The relaxation time would then be reduced from the astrophysical to the human range: 10^{3} s. Much later (Hulet et al. 1985) did the opposite to the deexcitation of excited atoms. He demonstrated experimentally that deexcitations of atoms could be quenched. This was done by placing the excited atoms in a transmission line with an upper frequency limit. Above that limit, no photons from deexcitation could propagate. The atoms therefore remained excited. Earlier work on cases with zero transition probability had focussed attention on zero transitions matrix elements from symmetry. These new cases were consequences of the final density of states being equal to zero. The surprising aspect of these early reports was that microscopic, and seemingly isolated systems like nuclei and single atoms, in this direct fashion were sensitive to macroscopic surroundings.

It is characterestic for early work on photonic crystals that those who introduced the concept approached the subject from this atomic physics perspective, i.e. how spontaneneous emission from an atom could be quenched by the surroundings and considerations of localisation of light as an effect of disorder (John 1984).

A second root of photonic energy gaps goes back to early solid state physics and is centered around long range order and crystalline symmetry. It seems an odd coincidence that the same year as Purcell published the note mentioned above, Brillouin wrote, "For any kind of waves (elastic, electromagnetic, or wave mechanical, DeBroglie waves) propagating in a medium with periodic structure, discontinuities are found in the relation between the frequency ν and the wave vector **a** (a vector of length /**a**/=1/λ pointing in the direction of propagation)" (Brillouin 1946). These discontinuites in the dispersion relation are nothing but energy gaps. In a later discussion of wave propagation in crystals, with the dielectric function modulated by a sine wave e.g. from propagating sound, it was demonstrated that forbidden gaps appear, i.e. ω-intervals in the dispersion relation

$$\omega = f(\mathbf{k}) \tag{1}$$

where only imaginary solutions **k** exist (Slater 1958). k is then a measure of the inverse attenuation length. This implies that these particular electromagnetic waves are effectively damped and cannot be transmitted in the bulk of a crystal. In the X-ray region, the gaps are extremely small in comparison with the high frequency of the primary wave. They are treated in detail within the dynamical theory for X-ray diffraction. Mathematically, these various frequency gaps correspond to the famous photonic bandgaps that have now been demonstrated from the visible to the radar range. One important difference is that in the X-ray region the refractive index invariably is very close to 1, while much larger indexvalues appear at lower frequencies from visible to microwaves.

An early, detailed suggestion of a photonic gap, formulated in solid state terminology, considered a one-dimensional (1D) case: A composite layered structure of GaP and GaAs (Bloembergen and Sievers 1970). In Fig. 1 we cite their result with one gap extending from the GaAs TO mode (273 cm^{-1}) to the GaP LO mode (401 cm^{-1}) and another around 800 cm^{-1} that originates from Bragg reflection.

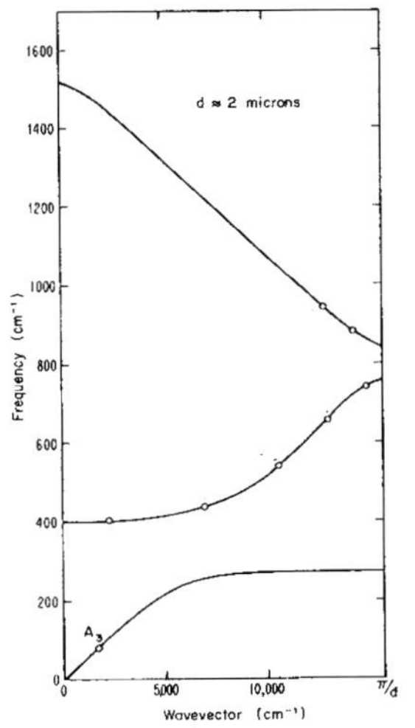

Fig. 1. Dispersion curve for a periodic layered structure of GaP-GaAs with the period d=2 μm. The points mark the occurence of various non-linear effects. Cited with permission (Bloembergen and Sievers 1970).

The authors primarily wanted to study non-linear coupling effects in their structure. However, they commented upon the 800 cm^{-1} gap, obtained as an effect of periodicity resulting in strong Bragg scattering, as well as the coupling to polaritons.

It was not until the late 80ies that these two lines of development were unified into one coherent solid state picture. Analogies with the established description of electronic bands were used and proved to be very powerful. The papers cited above initiated this development, and in particular, when the experimental verification of a photonic gap for all angles of incidence was published, the field was convinced. Some details about degeneracies had created confusion, which were later clarified (Yablonovitch 1993). In Fig. 2 we give the instructive diagram showing how the 'Yablonovite' was made. Photonic crystals for the microwave region have dimensions that permit preparation by ordinary drilling as indicated.

During the 90'ies the field of photonic crystals has seen great expansion. In the early phase, methods for calculation of photonic band structures improved and. They were found to correlate well with experiments. At a later stage of development new methods of preparing photonic crystal structures with a high degree of perfection and with gaps at shorter wavelengths have been in the focus of interest.

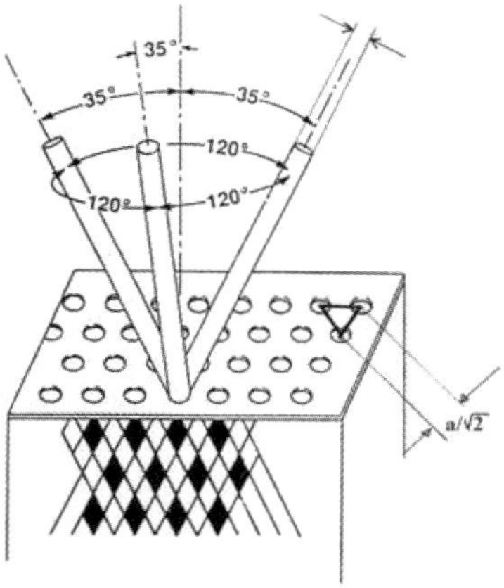

Fig. 2. Preparation of 'Yablonovite', a photonic crystal with a photonic bandgap for all angles of incidence around $\nu \approx 15$ GHz was experimentally demonstrated. For this microwave application the crystal can be prepared as indicated with a cubic lattice constant a of 12.7 mm. Cited with permission (Yablonovitch 1993).

As a new and fast growing field, review articles on the subject were greatly needed. Among others, a special issue of the Journal of Modern Optics was published in 1994 as a response to this interest. At this stage, we cite a selection of reviews (Yablonovitch 1993, 1994; Haus 1994; Pendry 1994; Joannopoulus et al. 1997) and three monographs (Joannopoulus et al 1995; Sakoda 2001, Johnson and Joannopoulus 2002). The MIT-group updates a web-page with international coverage (http://ab-initio.mit.edu/photons/).

2 Light in a Medium

In this section we shall first discuss polarisation and dispersion in a homogenous medium. The ambition is to provide a background to the introduction of photonic crystals, which are inhomogenous. The first part will give an elementary description of the general physical concept of dispersion, which is subtly different from the optical meaning of the same word. The second treats a special case of dispersion: the interaction between an electromagnetic wave and the lattice vibrations in an ionic crystal, usually called the phonon polariton. In the third part, we shall at last cite a photonic bandstructure, as an example of light dispersion in an inhomogenous solid.

2.1 Dispersion of Light in a Homogenous Solid

Solids are not "homogenous" in a microscopic sense. On an atomic scale the charge distribution is in fact quite inhomogenous, and if excited by an electromagnetic wave, microscopic dipoles are created. "Optical homogeneity" is based on tcomparison of atomic dimensions with the wavelength of light. Even UV light has a wavelength λ that is much longer than the distance between atoms. Consequently, a light wave creates an excitation averaged over thousands of atoms. This average is quantified by the complex polarisation, which is included in the dielectric function ε(ω) of the medium.

In this context it is relevant to note that the core problem in effective medium theory (EMT) in optics (Aspnes 1982) is to take one further step in the averaging procedure. The effective medium is inhomogenous on a length scale that is larger than atomic distances, but still small enough to permit averaging over a λ^3 volume (Milton 2002). Photonic crystals represent yet another step towards an inhomogenous medium. In this case the inhomogeniety occurs on a length scale that is comparable to the wavelength. Averaging over a wavelength is therefore not satisfactory, and in their original form EMT models are not applicable. Nevertheless, EMT approaches have been used to simplify calculations by homogenizing a 2 or 3D photonic crystal into a layered structure. This is an approximation of the real situation that simplifies the calculations and is helpful in predicting the opening of gaps (Lalanne 1996). A detailed study compared the results of one EMT model based on PC theory with the conventional Maxwell-Garnett model (García-Vidal et al. 1997) for aligned carbon nanotube films. The results show that, while the ordinary EMT's may be satisfactory for low packing density, more accurate calculations are required to describe structures with high packing density characterized by more complex interactions.

If a plane electromagnetic wave $\mathrm{E} \propto \exp[i(\mathbf{K} \bullet \mathbf{r} - \omega t)]$ is inserted into the electromagnetic wave equation $\varepsilon_1(\omega) \approx 1 - (\frac{\omega_p}{\omega})^2$ and we use the expression for the complex displacement **D**, we obtain a dispersion relation which we write in a form that expresses the frequency as a function of the wavevector **K**:

$$\omega^2 = \frac{(cK)^2}{\varepsilon(\omega)} \tag{2}$$

Noting that both **K** and ε may be complex, we take the square-root of (2), and use a more familiar version for those who work with optical interference coatings:

$$\omega = \frac{c}{N} K \tag{3}$$

In this simple form, dispersion is simply a straight line in a standard ω vs. K diagram. In vacuum N=1, and if we assume a dielectric with a constant N=1.5, typical for glass in the visible, we obtain another straight line. In Fig. 3 we choose to plot (3) in a log-log diagram in order to cover a larger part of the electromagnetic spectrum than just the visible.

In optics "dispersion" implies that the refractive index n varies with wavelength (or frequency). Therefore, the two straight lines in figure 2 when N=n=const are called "non-dispersive". Howwever, no real dielectric solid is truly dielectric over the large frequency range shown on the y-axis in Fig. 3. In general physics, and in particular in wave theory, "dispersion" indicates that the energy and wavevector of the wave are interrelated, e.g. as in (2) above.

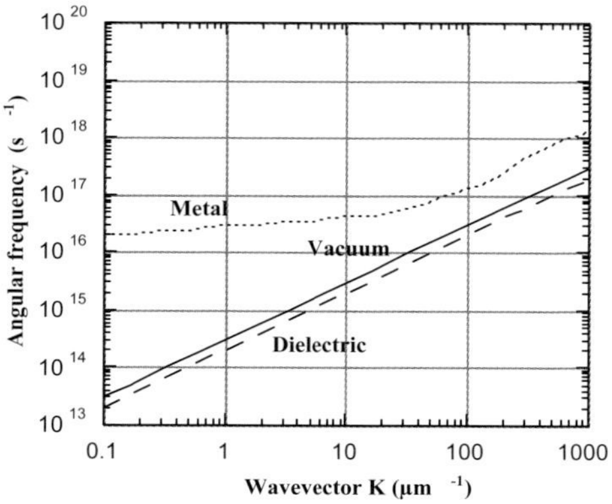

Fig. 3. Dispersion relations for three media as indicated, vacuum n=1, dielectric n=1.5 and for potassium $n=(\varepsilon_1)^{1/2}$ using eq. 4 and $\omega_p= 6.7\ 10^{15}\ s^{-1}$.

In order to illustrate "dispersion" in the optical sense, a third case is included in Fig. 3: The Drude model. For high frequencies, i.e. frequencies well above the plasma frequncy ω_p, the imaginary part of the dielectric function, ε_2, is neglible and we can approximate:

$$\varepsilon_1(\omega) \approx 1 - \left(\frac{\omega_p}{\omega}\right)^2 \tag{4}$$

For a numerical illustration, the value of ω_p is chosen to correspond to the electron density in potassium. The frequency range is exagerated in Fig. 3, so it includes the switch-over from transparent to reflecting behaviour. Primarily, Fig. 3 illustrates that behaviour that in optics is termed non-dispersive corresponds to a straight line in a dispersion diagram, like two of the cases in Fig. 3. 'Optical dispersion', in contrast, is characterized by $dn/d\lambda \neq$ const, as illustrated by the metal.

Furthermore, for low values of K the $\omega(K)$ curve bends over. Considering the situation when $\omega < \omega_p$ we find that (4) and (2) indicate that in this case K is pure imaginary. This is also consistent with (3): an imaginary refractive index N results in an imaginary wavevector K. The imaginary wave vector in turn gives exponential damping of the amplitude and intensity of the electromagnetic wave. Waves excited inside the medium, are extinguished, and for light coming from the outside, the medium exhibits total reflectance.

2.2 The Light Line in the Brillouin Zone

We shall now position the light dispersion curve, often called the *light line*, on the scale of the Brillouin zone. We do this by plotting it together with the dispersion lines for lattice vibrations in Fig. 4 below. Lattice vibrations are quantized and the quanta are called phonons. This kind of diagram is called phonon dispersion, which energy-wise occurs in the infrared. For the sake of distinction, we use **q** for phonon wavevectors.

In this illustration we only sketch the dispersion of photons and phonons from the origin to an unspecified zone boundary. The photon dispersion is given by (3) as in Fig. 3. With the entire zone on the X-axis, the steepness of the photon dispersion becomes manifest. The K-axis in Fig. 3 is thus only a small fraction of the full Brillouin zone. The slope of a dispersion curve gives the velocity of the corresponding wave. The photon dispersion is therefore as much steeper than that of the phonon, as the velocity of light is higher than that of sound. The dash-dotted branches in Fig. 4 are the lower and upper polariton branches. A polariton is a quasi-particle, i.e. a "mixture" of the electromagnetic wave and the phonon that proceeds through the crystal, once the phonon has been excited. The diagram shows that it is only in the intermediate range that the dispersion of a polariton is really different from either of its origins. For small q it is close to a photon and for large q it is close to a phonon. The almost vertical behaviour of the light line illustrates that even for a high energy photon, higher energy than that of the valence electrons in a medium, the photon momentum, $\hbar$ is small. If on scale, the light line would not even be discernible from the y-axis. Photon excitation alone can strongly increase the energy of a valence or conduction band electron, but not significantly change the momentum. With Fig.4 as a background, we can better understand the interaction between transverse electromagnetic waves and lattice vibrations, i.e. the photon-phonon interaction. To first order, a photon only interacts with optical phonons because the E-field of the wave will accelerate the ion cores with positive and negative net charges in opposite directions. The general condition for an interaction is that both wavevector and energy match.

The left part of Fig. 4, shows the so called Reststrahlen Band on the common frequency axis. It is characteristic for a large group of compounds with some ionicity that they exhibit at least one band of high normal bulk reflectance in the infrared.

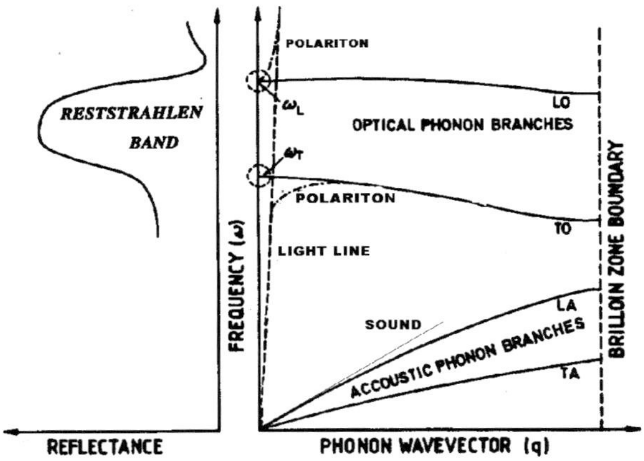

Fig. 4. Dispersion relations for light and phonons in a compound. The slope of the LA-branch close to the origin (fine line) is proportional to the velocity of sound in the medium. The slope of the light line (dashed) is proportional to the speed of light, $^c/n$. The dash-dotted branches are the polariton dispersion curves. The left part of the diagram is the bulk normal reflectance. (From Ribbing and Wäckelgård 1991).

As seen in the diagram this band is located between the two phonon frequency parameters ω_T and ω_L. The Lorentz oscillator model, gives a negative real part of dielectric function in this frequency interval. Just as in the metal case, the negative dielectric function corresponds to an imaginary wavevector **K**, which in turn causes high reflectance. E.g. a sodium chloride crystal is a dielectric with high transparency in the visible. Yet, in the far infrared (wavelength range 40-60 μm) it is optically metal-like, i.e. $\varepsilon_1=\mathrm{Re}[\varepsilon]<<0$ because of ionic excitation. The metal acts as an optical high pass filter. Above the plasma frequency the metal is highly transmitting until interband transitions from lower electronic levels cause absorption as well as increasing reflectance. The Reststrahlen Band, in contrast, is an optical stop band. ω_T and ω_L are the lower and higher frequency limits of this stop band. There is wide variation in such stop bands and the wavelengths corresponding to the limits have been tabulated (Dobrowolski 1978).

2.3 The Photonic Bandstructure – Dispersion for a Small Zone

Figures 3 and 4 together constitute a basis for a photonic band structure. Each branch, or "band" in a band structure, is a dispersion relation for electromagnetic waves in the medium. It gives the frequency, or energy, of an electromagnetic wave travelling in a certain direction in wavevector space, as a function of wavevector **K**.

Under fairly general circumstances both electronic and photonic bands are unaffected by the lattice close to the origin, called Γ. Near Γ photon dispersion is linear, $\omega(K) \propto K$, like the light line in a homogenous medium. This is in contrast to electron bands that are parabolic close to Γ.

It is only when the electron or photon approaches a zone boundary that the dispersion markedly departs from these simple behaviours. The electrons scatter against the lattice potential so the strength and symmetry of that potential become important. Conversely, it is not surprising that the light lines in Fig. 3 for an ordinary bulk solid are unaffected by the symmetry of an ordinary crystal. These photons have too small wave vector values to interact with any zone boundary. To create such an interaction we must *shrink the Brillouin zone* by a factor 10^2-10^4. This is what is needed for long range order to play a role. Shrinking the BZ is done by *increasing the lattice constant* a. It should be about 1000 times bigger than in an ordinary atomic crystal. We have thus arrived at a receipe for a photonic crystal: *Prepare a periodic structure of varying dielectric function with a lattice constant that is of the same order of magnitude as the wavelength of the light you wish to manipulate.* We expect visible light to interact with the lattice of a photonic crystal, the lattice constants of which are of the order 500 nm. The details will depend upon the choice of materials and the symmetry of the structure.

Ordinary crystals scatter X-rays in a pattern that is characteristic of the crystalline structure. Short wavelength X-rays have K-values that are large enough to interact with the BZ boundaries. In this sense, X-ray diffraction is an extreme case of photonic crystal diffraction in which the scattering power of ionic cores replaces the refractive index variation in a photonic crystal. In essence this is the same argument that is used to understand the limit of resolution in microscopy. An EM wave cannot resolve much finer details than the wavelength.

We are now in a position to discuss a realistic photonic bandstructure and start by citing results for the Yablonovite structure in Fig. 2 (Yablonovitch 1993). The cylindrical holes form a triangular 3D lattice as indicated, with a diameter d such that d/a =0.469. This gives a 78 % volume fraction of air with a remaining long-range ordered "skeleton". The material used in the microwave experiment was a Stycast with $n^2 = 12$. Drilling the inclined holes clear through the material results in a distorted fcc-structure. The Brillouin zone in the insert of Fig. 5 accounts for this. It rests on the hexagonal face through which the drill exits. The hexagonal faces where the drill enters and exits have L_3 points, i.e. 120° rotational symmetry. This is not true for the other hexagonal faces that have L_1 points i.e. only 360° rotational symmetry. This reduced symmetry is essential. It has been shown that the fcc symmetry results in band degeneracy at the W-point, even for an index ratio as large as 4 (Leung and Liu 1990). It was also found that the reduced symmetry of the diamond structure was sufficient to lift this degeneracy already for index ratio as low as 2 (Leung1997).

An important difference between electronic and photonic band structures is that EM waves have a vectorial character in their polarization. This difference from the calculations of electronic bands caused some of the early confusion about the existence of photonic gaps. We know that s- and p-polarized light interact

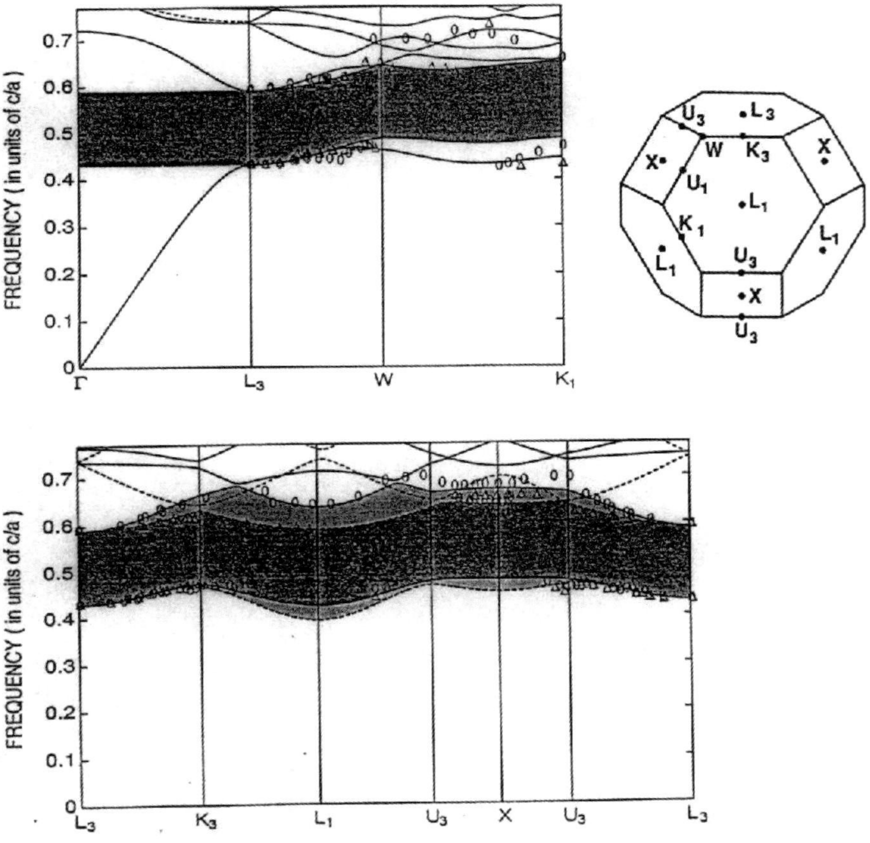

Fig. 5. Photonic bandstructure for Yablonovite as shown in fig. 2 with d/a=0.469. Solid curves are for s-polarized (TE-mode) radiation and dashed for p-polarized (TM-mode). The black area is the gap for both polarizations, the shaded for one of them only. The points are experimental from transmission measurements in the microwave region. Oval points for s-, and triangles for p-polarized radiation. The Brillouin zone is a modified fcc-zone. Cited with permission (Yablonovitch 1993).

differently with solids, except at normal incidence. We therefore expect differences between s- and p-bands, except at the Γ-point, where they should be degenerate. This difference is seen in fig 5 as two sets of lines, full and dashed respectively, that coincide at Γ.

The bands specify waves that can travel through the crystal: $\omega=\omega(\mathbf{K})$ as a function of the **K**. The points on the x-axis, are the symmetry symbols on the zone. A photonic bandstructure is generic, i. e. it is applicable on any length scale. For a given symmetry and material properties, it scales with the ratio of the wavelength to the lattice constant: λ/a.

At this stage a general comment can be made about "gaps" in band structures. In Fig. 5 the black band covers a frequency range, where the waves are evanescent, i.e strongly damped. Yet, a band structure includes only the high symmetry parts of all possible K values in the first BZ. It may therefore be difficult to accept that this selection of all possible states proves that there are no allowed states in this gap range. General arguments for electronic as well as photonic states can be formulated showing that both band minima and maxima should appear in high symmetry K directions. Therefore, the gap shown in Fig. 5 is in fact valid for all vectors **K** in the first BZ. Such a gap is said to be complete. The concept of complete photonic gaps has become a major issue in the study of photonic crystals, primarily for technical applications like photonic fibers, supression of de-excitations and omnidirectional mirrors, but also as a key scientific point.

Looking at the dispersion curves in Fig. 3 and considering equation (3) alone, we cannot explain that a dispersion curve for a medium exhibits a break or "gap". There must be a mechanism, triggered by long range order, that creates a gap in the bulk crystal, be it electronic or photonic. In both cases the explanation lies in scattering and interaction between the scattered waves within the structure. The crystalline order is a crucial condition for obtaining gaps. It is familiar that light interferes after scattering against a structure. If conditions on periodicity and coherence are fulfilled, the scattered waves can combine in a way that e.g. all forward scattering is effectively extinguished, and the light intensity is totally reflected. Secondly a certain "contrast" between the two dielectric functions is required to create strong enough scattring to give a photonic crystal a complete gap. In passing it is interesting to notice that the German and Scandinavian languages use the same word "Gitter" for "lattice" and "grating".

Conversely, in a dielectric photonic crystal, photons in the first band i=1 close to the zone center Γ do not "feel" any influence from of zone edges. The bands close to the origin in Fig. 5 are as linear as the vacuum line in Fig. 3 – they just have different slopes.

3 Theoretical Overview

The aim of this section is to present, only slightly more formally, an overwiev of the theory underlying the description and analysis of photonic crystals. The ambition is not to give a complete and independent treatment, but to present and make understandable. In particular this is for a reader who is familiar with conventional optical filters but not quite at ease with the Schrödinger equation and the basics of solid state physics. With this in mind, we shall start with with an object that should be a bridge to mutual understanding: the optical multilayer.

3.1 The Optical Multilayer as a One-Dimensional Photonic Crystal

Considering the remarks about photonic crystals and imaginary K-vectors that have already been made above, the correspondence between the stop bands of a multilayer and photonic gaps of a 1 D photonic crystal seems obvious. Indeed, the connections are there, however, slightly more involved.

It is known that optical multilayers, given a sufficient number of layers, can exhibit strong, omnidirectional stop bands even with moderate index contrast n_H/n_L. Because of the influence of polarisation, this stopband is at shorter wavelengths than the design wavelength λ_0 for normal incidence. A detailed analysis showed that the particular case of a 23 QWOT stack of silica and titanium dioxide, with $n_H/n_L \approx 1.7$ has a reflectance above 98% for all angles of incidence and both polarisations in the wavelength range $\lambda_0/\lambda = 1.1 - 1.17$ (Carniglia 1999).

We now switch language to that of photonic bandstructures and assume that it has a *complete* photonic bandgap. *Complete gap* implies that it covers the entire Brillouine zone, i.e. for all directions of incidence and for both polarisations. The presence of such a gap makes all electromagnetic waves inside the structure with frequencies in that gap evanescent. Waves incident at *any* angle and with either polarisation, are totally reflected. Consequently, the photonic crystal with a complete bandgap will be an ideal mirror for the frequency range of the gap. A complete photonic bandgap is thus a *sufficient condition* for a mirror to be ideal. However, the periodic dielectric multilayer with a a small index ratio does not have a complete photonic bandgap. Obviously such a gap is *not a neccessary condition* on an ideal mirror.

In a conventional optical description this is easy to understand as an effect of Snell's law and refraction at the front surface. If light is incident from air, i.e. $n_0=1$, then the refracted light inside the second medium, A, will be within a cone, the apex angle of which is given by the refractive index of that medium: $\phi_A \leq \arcsin(1/n_A)$. In e.g. ordinary glass we do not need to make waves more than 42° from the normal extinct, since higher angles do not occur.

Within the photonic band formalism, one phrases same reduction for necessity somewhat differently. Only states inside the photonic crystal, that can couple to states at the outside, need to be made evanescent by the presence of a gap. In other words, the gap does not need to be complete! I.e. a complete photonic gap is a sufficient, but not a neccessary, condition on an ideal mirror.

A closer photonic band analysis shows that for *normal* incidence parallel to the z-axis, a 1-D photonic crystal exhibits a gap for arbitrarily small index contrast, n_H/n_L,. However, for off-normal incidence in the xz-plane, the wavevector **K** has a component parallell to the surface of the multilayer. This destroys the symmetry of the K_x and K_y directions. For a particular K_x, any K_y value in the plane is permitted, when this symmetry restriction is lifted. Therefore the gap is filled with allowed states. The bands also split, according to polarisation. In other words, the gaps of 1-D crystals is not complete, even for a large index contrast (Winn et al. 1998). This has to be reconciled with the fact that dielectric multilayer mirrors can in fact be made omnidirectional.

The detailed explanation requires that both the states inside the multilayer – i.e. the bandstructure along K_x and K_y for the multilayer – and those in the medium of incidence are considered. If the surrounding medium is vacuum/air the dispersion relation for a wave is

$$\omega = cK = c\sqrt{K_x^2 + K_y^2} \tag{5}$$

simply the light line for vacuum shown in Fig. 3. Inside the multilayer K_y is again a free variable, so the projected bands cover an area. As usual for non-normal incidence, the s and p branches are different. We show this combined analysis (Winn et al.1998) in Fig. 6. The dispersions for p and s waves are shown separately in the two halves of the diagram as marked, in both cases as a function of K_x which

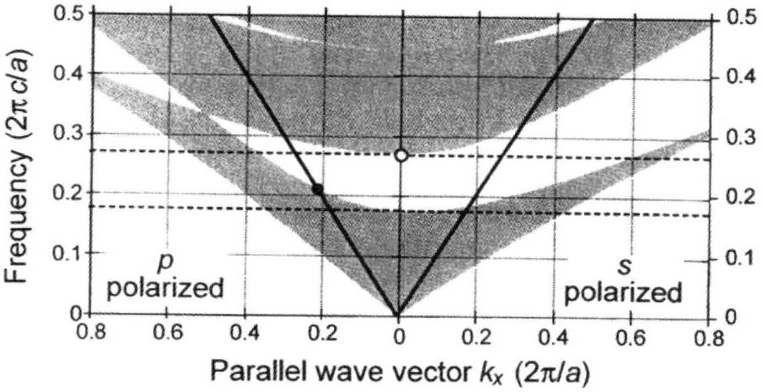

Fig. 6. Gray areas are the projected bandstructure for a stack with n_H=3.4 and n_L=1.7. The heavy, full lines are the light lines above which the vacuum states are found. The circles show the ω-range of the omnidirectional mirror. The dashed horizontal lines indicate the stop band for normal incidence. (Cited from Winn et al. 1998)

is symmetry restricted. The vacuum waves are in the triangular area above the heavy straight lines. The waves that are permitted in the multilayer are found in the gray areas, obtained when the waves with different K_y values are projected onto one plane.

Focussing first on the permitted modes in the multilayer, the diagram confirms that there is no complete gap, although n_H/n_L=2. However, if we consider only waves that are above the heavy lines, we find an ω interval between the two small circles for which there are no permitted modes in the multilayer. In this interval the multilayer is totally reflecting for both polarisations and all angles of incidence. This is in agreement with the results of the ordinary optical multilayer calculations just cited (Carniglia 1999), that the frequency interval with omnidirectional reflectance is above that of the stopband at normal incidence.

The detailed analysis of optical dielectric multilayers within the conceptual framework of photonic crystals has thus not given new end results. This would have been very surprising considering the long history, well-established theory and accurate empirical verification of conventional multilayer designs. It has also been pointed out that the perfectly periodic stack is not the optimum mirror and that in particular, the front layer thickness should be different. The common assumptions about an infinite number of layers and neglect of absorption represent idealisations that are not fulfilled. It is therefore unavoidable that a real multilayer exhibits a drop in reflectance at some angle of incidence (Baumeister 1999). A more detailed mathematical analysis of the angular variation of reflectance from optical multilayers is given in elsewhere in this volume (Tikhonravov 2002).

3.2 Some Features of Optical Bandstructure Calculations

This volume is not the right place for complete derivations, or even complete descriptions, of the various methods that have been used to calculate photonic bandstructures. We shall only give short acounts in order to bring out the distinction between two major schemes: plane wave (PW) and on-shell methods.

The PW methods rely on the Fourier theorem that any wave package can be obtained as a linear combination of plane waves. It is therefore sufficient to study the behaviour of one such wave and then write the final result as a linear combinaton of such PW's. This is not different from the conventional derivation of the wave equation for a single plane wave (Born and Wolf 1980). The complex dielectric function should have an explicit space dependence in anticipation of the periodic variation in a photonic crystal. In the following step, the derivatives with respect to time are carried out which gives the explicit frequency dependence. Then, in contrast to the conventional wave equation in E, the equation

$\Theta H(\mathbf{r}) = (\frac{\omega}{c})^2 H(\mathbf{r})$ in **H** is obtained by using the operator:

$$\Theta = \nabla \times \left(\frac{1}{\varepsilon(\mathbf{r})} \nabla \times \right) \tag{6}$$

It is deliberately introduced to cast this magnetic field waveequation into the form of an eigenvalue equation, like the Schrödinger equation. The interaction between the EM wave and the medium is dominated by the E-field. The reason to use the operator Θ is that it is Hermitean (Joannopoulus et al. 1995). Once the theoretical treatment is made, the E-field can be obtained.

The formal similarity between this new form of the waveequation and the time-independent Schrödinger equation $H\Psi = E\Psi$ is sufficiently convincing to accept that it can by solved by a linear combination of magnetic plane waves, $\mathbf{H}(\mathbf{r},t) = \mathbf{H}(\mathrm{r})e^{i\omega t}$ describing any mode pattern. Therefore, if we have a solution it will give us connected values of the wavevector **K** and the eigenvalues $(\omega/c)^2$. If $\varepsilon(\mathbf{r})$ is periodic in space, the magnetic plane waves satisfy the Bloch theorem. Because of the symmetry of the reciprocal lattice, all solutions can then be translated to a corresponding state in the first Brillouin zone**.** This is the same procedure as used for electronic bandstructures and phonon dispersion relations. The result is that for each K value in the first zone, we collect a series of solutions $\omega_n(\mathbf{K})$ where n is a *band-index* starting at lower ω-values and increasing for higher frequencies. One band is the function $\omega=\omega_n(\mathbf{K})$ for a constant n – a surface in a 4-dimensional space. By convention, the photonic band structure is the cross-sections that are obtained from several of these surfaces, when K is varied along the symmetry directions in the first BZ.

As already mentioned, a complication with photonic bandstructures, in contrast to dispersion relations of electrons and phonons, is the vector nature of light. We have to include that s and p polarized light interact differently with a medium. In the photonic crystal literature, radar and microwave terminology is frequently used. s polarized light is often called ”TE” for Transverse Electric and p polarized ”TM” for Transverse Magnetic. The well-known difference in angular dependence for p and s polarized light reflectance shows up also in the 1-D photonic crystal as mentioned above. It is characteristic that the gap between the s branches (white) for normal incidence between the dashed lines in Fig. 6 grows in size when the angle of incidence is increased. It is the rapid decrease of the gap for p polarized light that closes the gap beyond an angle of incidence corresponding to $k_x \approx 0.37$.

The disappearance of the gap because the angular dependence of the p polarized light agrees with the traditional bulk reflectance of a solid. Yet, it is not universal. In 2-D and 3-D bandstructures the bands for s and p polarized waves are usually given together on the same K-axis. It then turns out that the topology determines whether it is the TE or the TM modes that close the gap. The TM bandgaps survive in crystals where the high ε component is non-connected, and TE bandgaps where it is connected (Joannopoulus et al. 1997).

From this schematic description of the plane wave method, an important limitation emerges. The solutions are obtained in the form of a multivalued function ω_n at one selected **K.** In order to obtain this solution, the numerical values of the periodic function $\varepsilon(\mathbf{r})$ must be used. In other words *the dielectric function* (or the refractive index) *must be known before the frequency*. In essence this means that the plane wave method cannot directly be used to calculate bands in a frequency region where one or both of the material components exhibit significant variation with frequency, i.e. ”dispersion” in the optical sense. For model calculations, to demonstrate the effects and appearance of photonic gaps, this is not a serious limitation. This also applies to real cases, when the variation of the refractive index is weak, as for many oxides in the visible or semiconductors in parts of the infrared. It may also be possible to model the real variation in $\varepsilon(\omega)$ or $n(\lambda)$ with a step function, used in solving the problem once for each constant value. However, in cases with strong dispersion, e.g. metals around their plasma edges, this constitutes a serious barrier. We conclude this part with two web adresses where plane wave program packages for the calculation of photonic band structures are available: The MIT group: http://ab-initio.mit.edu/mpb/ and the Amsterdam group: http://www.science.uva.nl/research/scm/sprik/pbs.htm

The characteristic feature of the on-shell method is that the frequency ω_i (or energy) is fixed first, and then the corresponding solution(s) **K** are calculated. The "on-shell" method is related to tight-binding calculations that are used to compute electronic bands in solids. The starting point is then an atomic level α with a free atom eigenenergy value. The corresponding band $E_n(\mathbf{K})$ is then derived by parametrizing the interaction with the neighbours, shell by shell. The wavefunction is written so as to satisfy the Bloch condition. On might say that the hypersurface $\omega_n(\mathbf{K})$ is cut horisontally. This is in contrast to the plane wave method, where the cut is vertical. Consequently, if $\varepsilon(\omega)$ is known, from a model or a table of experimental results, the value can be inserted into $\varepsilon(\mathbf{r}, \omega)$. For the selected ω value, the corresponding **K** values are calculated. Calculations of photonic bands with the on-shell method have been performed and described (Pendry 1994) and a detailed account of the program package has also been published (Bell et al. 1995). A compiled version with graphics included is available on the web (http://www.elec.gla.ac.uk/~areynolds).

The starting point is again Maxwell's two curlequations, expressed in ω and **K**, once the explicit time dependence has been eliminated. The components x, y, z of **K** are approximated by truncation of their expansions and then inserted into the Maxwell equations. One then has the $\mathbf{B}_x$ component expressed as a function of $\mathbf{E}_y$ and $\mathbf{E}_z$ etc. These equations for the **B**- and **D**-components are Fouriertransformed back into real space. One then has a finite element discretization that expresses each component of **B** and **D** in the other **E** and **H** components at the site and at a neighbouring site. The H and the E fields are defined on two interpenetrating simple cubic cells that are displaced half a body diagonal from each other. The result of the derivation relates the E_x, E_y, H_x and H_y values in one simple cubic cell to the next cell along the direction of propagation z. If all values in one cell are known, the corresponding values in the next cell can be calculated by multiplication by the transfer matrix **T**,

$$\mathbf{F}(z + \mathbf{a}_3) = \mathbf{T}(z)\mathbf{F}(z) \tag{7}$$

where F is a vector that contains the xy components of **E** and **H**.

As mentioned the on-shell method is applicable in cases where $\varepsilon(\mathbf{r}, \omega)$ is a strongly varying function of ω. The solutions are **K** values that are obtained for a fixed ω. It is also useful when there are abrupt changes in $\varepsilon(\mathbf{r},\omega)$ on a short length scale, e.g. if metal particles are embedded in a dielectric and $\omega < \omega_p$. In the opposite case - if $\varepsilon(\mathbf{r}, \omega)$ varies slowly with r on the length scale of a wavelength - it may be neccessary to use free space solutions to take advantage of this.

The Korringa-Kohn-Rostoker method (KKR), originally developped for electronic band calculations (Ziman 1964), has been used also for the calculation of photonic bands (Moroz A. 1994), originally as a plane wave method, but at a later stage also as an on-shell method (Modinos et al. 2001) and applied to the cases of small metal spheres in gelatine and silicon spheres in a non-absorbing Drude metal.

4 Structural and Material Aspects

In this section we shall discuss the structure and choice of materials for different kinds of PC.

4.1 Photonic Crystals Made from Oxides and Semiconductors

Just like in the case of optical multilayers, the purely dielectric case is the prototype device. In both cases the index contrast is an important parameter. The ratio n_H/n_L is of primary importance when deciding how many layers are needed to satisfy the target criteria. In photonic crystal design, the same ratio, or the square of it: $\varepsilon_A/\varepsilon_B$, determines whether the crystal has a complete gap or not.

In the 3-D case photonic crystal case, the choice of structure enters as an extra parameter. Different structures require different minimum ratios to give a complete gap. To obtain a complete gap, one should avoid letting a band in one part of the Brillouin zone overlap a gap in another part. Therefore complete gap formation is favoured if the center-to-boundary distances are close to constant. The zone for the fcc structure is the one closest to spherical.

The fcc structure was selected for some of the early attempts to demonstrate complete gaps. It failed to show a complete gap for an index ratio as high as 4 because of degeneracy at the W point. The non-primitive diamond structure, that also has a fcc space lattice, has lower symmetry and the index ratio 2 is sufficient (Leung 1997).

The need for high index materials for demonstration of complete gaps became obvious. The selected material should also be suitable for micromechanical preparation of periodic structures. It should be no surprise that silicon emerged as a favourite. Si is practically dielectric for wavelengths longer than 700 nm through the communication and thermal IR wavelengths, and has a high and almost constant n-value greater than 3.4 in that region. Secondly, Si is one of the most used materials for micromechanical processing, with excellent mechanical properties and suited for litography, including dry and wet etching (Madou 1997).

Preparation of photonic crystals for control of visible light requires a precision better than 100 nm, which has been a difficult challenge to meet. The communication NIR wavelenghts around 1.5 μm are in focus for many potential applications. The demands on absolute precision are then reduced and several micromechanical techniques have been used sucessfully. In this context we cite the "wood-pile" structure that has been prepared in Si. It consists of Si rods with cylindrical or square cross-sections that are placed in layers (Özbay et al. 1994). The resulting structure is face-centered tetragonal (fct) which proved favourable for complete gap formation.

A very interesting theoretical study of the interaction between the structural photonic gap and the phonon-polariton gap was done by the Ames group in Iowa (Sigalas et al. 1994). They calculated the IR transmittance vs. frequency for a 2-D square lattice of GaAs cylinders in air with a filling ratio of 0.449. The key point was that the PC was designed to give a structural gap in the vincinity of the polariton gap. This entails calculations in a range with strong dispersion and resonance absorption at ω_T. The Reststrahlen band shows up as a strong gap interacting with the structural gap that is split into two.

We conclude this subsection with reference to a feature issue of MRS Bulletin that was recently published (Polman and Wiltzius 2001) that contains eight reports on material issues and preparation methods of interest for the realization of photonic crystals.

4.2 Metallodielectric Photonic Crystals

Just like in the traditional field of optical filters, so has the introduction of metals into PC's created new possibilities. The characteristic optical behaviour of metals is formally due to the large negative value of the dielectric function for frequencies below the plasma frequency ω_p, the range with negative ε and imaginary K. The physical origin is the electron gas, the behaviour of which can be reasonably described by the classical Drude model. For these same basic reasons, metals in PC's have been shown to give rise to new and striking effects. They are commonly referred to as metallodielectric photonic crystals (MDPC). The MDPC's are different from what had already been explored in earlier effective medium studies of cermets, which typically contained metal particles of size $<<\lambda$ in a dielectric host (Aspnes 1982).

At least two different aspects of periodic, connected metal patterns in a dielectric host have been reported. In the first case a copper wire mesh was used with a structure analogous to that of diamond. The "bonds" were 1 cm copper wires with 1.25 mm square cross section, corresponding to a unit cube length of 2.3 cm. The transmission was measured in the microwave range 1-15 GHz. The results showed that a metallic crystal has photonic gaps as expected from symmetry. It also exhibited a "metallic" gap from 0 to 6.5 GHz which is halfway to the photonic gap defined by the structure. Finally it was observed that for p polarized incident radiation, the crystal transmitted *longitudinal* modes in a range around 13 GHz. It was pointed out that the observations represented a new regime of photonic crystals in which $a<<\lambda$ (Sievenpiper et al. 1996).

In the second case, a simple cubic crystal of fine aluminium wires, ≈ 1 μm radius, was studied theoretically (Pendry et al. 1996). The numerical calculations were made with the transfer matrix method using somewhat modified boundary conditions. Adapted results are cited below in Fig. 7. This bandstructure also shows the dispersion for the imaginary values of **K** in the left half of the diagram, as in Fig. 6. The light line is included for comparison in the right hand part. There are no permitted transverse modes below the frequency marked ν_p = 8.2 GHz. This frequency has the same function as the plasma frequency of a bulk metal.

$$\omega_p = \sqrt{\frac{n_{eff} e^2}{\varepsilon_0 m_{eff}}} \tag{8}$$

The authors show that this reduction of the bulk plasma frequency of Al by a factor of almost 10^6 is due to two factors. The wire mesh represents a "dilute" metal, i.e. the volume effective density of conduction electrons n_{eff} is reduced in proportion to the metal volume fraction. Secondly, the conduction electron effective mass m_{eff} is *increased* from inertia because of self-induction in the wires. The oscillations of the conduction electrons correspond to a current that generates a magnetic field. This field contributes to the total momentum of the electrons via the magnetic vector potential. Again, the finer the wires, the larger is this field at the surface of the wire. The authors show that this self-induction corresponds to an increase of m_{eff} in (8) by a factor larger than 10^4.

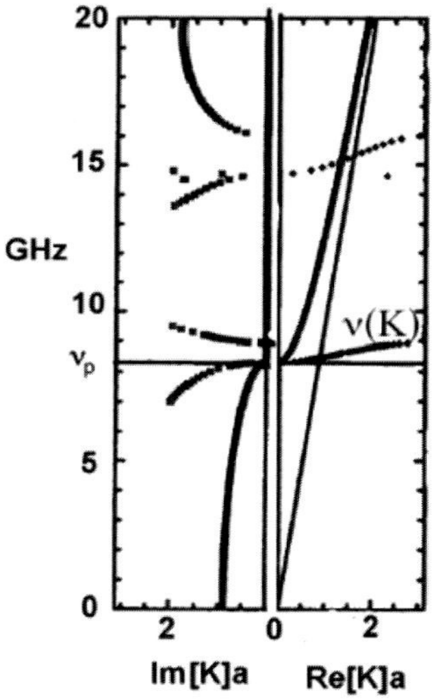

Fig. 7. Calculated photonic bandstructure for a sc structure of Al wires with radius 1 μm, and unit cell length 5 mm. The straight line in the real part of the diagram is the light line. ν_π indicates the reduced plasma frequency. A longitudinal plasmon mode ν(K) originates from this value. Adapted with permission (Pendry et al. 1996).

The combined effect of decreased n_{eff} and increased m_{eff} gives the reduction of the plasma frequency by the factor 10^6. A remarkable observation from this analysis is that the original electron density of Al cancels, so this numerical value is only a result of the *macroscopic* parameters r and a. The plasmon modes are seen as the line starting at ν_p in the right half of Fig. 7. Just like the ordinary volume plasmons, the MDPC plasmons have weak dispersion. Likewise, the MDPC plasmons are shown to have a lifetime that depends on the dissipation of the metal in about the same fashion as the their bulk counterparts.

The allowed transverse modes for frequencies above ν_p, in the absence of dissipation, were shown to follow the modified dispersion relation.

$$K = \frac{\sqrt{\omega^2 - \omega_p^2}}{c_0} \tag{9}$$

It is pointed out that while this mesh, with a metal content of only $1{:}10^6$, is quite opaque for waves with lower frequencies than ν_p, it will be effectively transparent for visible light. The small effective volume needed for such en element suggests new improved device application in the radar and microwave ranges. A fully transparent, aluminium microwave oven appears physically feasible!

A recent review of metallic PC's describes in particular the possibilities to develop compact microwave and radar components (Lourtioz and Lustrac, 2002).

5 Potential Applications

Space does not permit an extensive discussion of potential applications of PC's so this will be little more than a listing of references to some recent works on different methods. The first subsection, however, will be used for a common feature of several potential applications.

5.1 Defects in a Photonic Crystals

It has been shown that controlled introduction of defects in PC's has effects that are closely analogous to the doping of semiconductors. Typically, a small extra piece of the high index dielectric in a PC with a complete gap manifests itself as a defect state in that gap. This is a state that is split off the conduction band, just like a donor state in a semiconductor. It is referred to as a donor mode. Conversely, if the defect is created be the removal of a small piece of the high index dielectric, the defect state in the gap originates from the valence band and is called an acceptor mode. It was shown that the volume of the defect controls how far from the band edges the defect states are removed (Yablonovitch et al. 1991).

Phenomenologically, the important aspect of the doped photonic crystal is that the defect, with the associated mode, functions as a cavity of utmost perfection. The bound state is surrounded by the forbidden gap that stops transmission to the outer world and quenches deexcitations, e.g. by interactions with phonons. Q-values greater than10^9 have been predicted (Yablonovitch 1994) for high frequency dielectric cavities.

From an application perspective, the *linear defect* is even more promising. An extra, or a missing, line of the dielectric component in a 2-D or 3-D PC with a complete gap creates a defect mode in the gap. This defect mode functions as a high efficiency communication channel.

5.2 Specific Applications

Conventional optical fibre with a refractive index gradient has limited capacity when it comes to sending signals through sharp bends. The possibility to use a linear defect in a PC with a complete gap has been discussed. A production method, such that an extra air hole in the middle was developed and the Bath group demonstrated mode confinement to that central air hole defect (Cregan et al. 1999). The Sandia and MIT groups in collaboration have demonstrated the feasibility of sending signals through sharp bends – with a radius of curvature smaller than one wavelength (Lin et al. 1998). There are spin-off companies producing such holey fibres (e.g. i.maxwell@redfernpolymer.com. or http://optics.org/article/ole/feature/1/90/2?year=2001&mont=12).

Improved light sources and detectors based on PC technology have been suggested (Schnitzer et al. 1993). This implies that the spontaneous emission is concentrated to the mode of the cavity. In this arrangement the diode emission is still thresholdless.

The new possibility offered by PC's to engineer a surface layer such that the thermal emission is selectively reduced is of potential technical interest. For instance for military platforms there are reasons to search for a *selectively* low emitting surface. This leaves most of the radiative cooling unimpaired and blocks only the radiation that would otherwise be a detection hazard. For IR suppression the thickness required is comparable to that of a paint layer (Rung and Ribbing 2002), but to achieve sufficient long-range order over large areas at a reasonable cost represents a formidable challenge.

6 Summary and Conclusion

In conclusion it remains to be seen whether the advent of PC's will revolutionize the field of optoelectronics or if it is merely a reform of language to describe optical multilayers in 3-D. It appears that holey fibers, truly PC's, or just improved total reflectance, will find market segments in the near future. There is also good hope, perhaps primarily for special 2-D PC's, for a range of applications, also in the fields of microwave and radar. In the latter cases the need for improved production methods is less of a problem – which it still is in the optical range.

Acknowledgements

The authors that have given their permission to cite their diagrams are acknowledged, as well as the Swedish Natural Science Research Council and the Swedish Defence Research Agency for supporting our work with photonic crystals.

References

Aspnes DE (1982) Optical properties of thin films. Thin Solid Films 89:249-262

Baumeister P (1999) Dependence of the reflectance of a multilayer reflector on the thickness of the outer layer. Appl Optics 38:6034-6035.

Bell PM, Pendry JB, Martín Moreno L, Ward AJ (1995) A program for calculating photonic band structures and transmission coefficients of complex structures. Comp Phys Commun 85:306-322

Bloembergen N, Sievers AJ (1970) Nonlinear optical properties of periodic laminar structures. Appl. Phys. Lett. 17:483-485

Born M, Wolf E (1980) Principles of optics. Pergamon Press, Oxford, Sec. 1.3.4

Brillouin L (1946) Wave propagation in periodic structures. McGraw Hill, Ch. VII

Carniglia CK (1999) Hot or Hype? Reflections on the 'Perfect Mirror'. Photonics Spectra June: 148-153

Cornelius CM, Dowling JP (1999) Modification of Planck Blackbody Radiation by Photonic Band-Gap Structures Phys. Rev. A59:4736-4746

Cregan RF, Mangan BJ, Knight JC, Birks TA, Russel PSJ, Roberts PJ, Allan DC (1999) Single-Mode Photonic Band Gap Guidance of Light in Air. Science 285:1537-1539

Dobrowolski J (1978) Reststrahlen Filters. In: Driscoll W, Vaughan S (eds) Handbook of optics. McGraw-Hill, New York, Ch 8, sec 99

García-Vidal FJ, Pitarke JM, Pendry JB (1997) Effective Medium Theory of the Optical the Optical Properties of Aligned carbon nanotubes. Phys Rev Lett 78:4289-4292

Haus JW (1994) A brief review of theoretical results for photonic band structures. J Mod Optics 41: 195-207

Hulet RG, Hilfer ES, Kleppner D (1985) Inhibited Spontaneous Emission by a Rydberg Atom. Phys. Rev. Lett. 55: 2137-2140

Joannopoulus JD, Meade RD, Winn JN (1995) Photonic crystals: Molding the Flow of Light. Princeton Univ. Press, Singapore

Joannopoulus JD, Villeneuve PR, Fan S (1997) Photonic Crystals: putting a new twist on light. Nature 386:143-149

John S (1984) Electromagnetic absorption in a disordered medium near a photon mobility edge. Phys Rev Lett 53: 2169-2173

John S (1987) Strong localization of photons in certain disordered dielectric superlattices. Phys Rev Lett 58:2486-2489

Johnson SG, Joannopoulos JD (2002) Photonic Crystals The Road from theory to Practice, Kluwer Academic Publ.

Lalanne P (1996) Effective medium theory applied to photonic crystals composed of cubic or square cylinders. Appl Opt 35:5369-5380

Leung KM, Liu YF (1990) Full vector wave calculation of photonic band structures in face-centered-cubic dielectric media. Phys Rev Lett 65:2646-2649

Leung KM (1997) Diamondlike photonic band-gap crystal with a sizable gap. Phys Rev B56:3517-3519

Lin SY, Chow E, Hietala V, Villeneuve PR, Joannopoulos JD (1998) Experimental demonstration of guiding and bending of electromagnetic waves in a photonic crystal. Science 282:274-276

Lourtioz J-M, Lustrac A. (2002) Metallic photonic crystals C. R. Physique 3:79-88

Milton GW (2002) The theory of composites Cambridge University Press, Cambridge

Modinos A, Stefanou N, Yannopapas V (2001) Applications of the layer-KKR method to photonic crystals. Optics Express 8:197-202

Moroz A (1994) Inward and outward integral equations and the KKR method for photons. J. Phys. Condens. Matter 6:171-202

Oswald JA, Wu BI, McIntosh KA, Mahoney LJ, Verghese S (2000) Dual-band infrared metallodielectric photonic crystal filters. Appl. Phys. Lett 77:2098-2100

Özbay E, Abeyta A, Tuttle G, Tringides M, Biswas R, Chan CT, Soukoulis CM, Ho KM (1994) Measurement of a three-dimensional photonic band gap in a crystal structure made of dielectric rods. Phys Rev B50:1945-1948

Pendry JB (1994) Photonic band structures. J Mod Optics 41: 209

Pendry JB, Holden AJ, Stewart WJ, Youngs I (1996) Extremely Low Frequency Plasmons in Metallic Mesostructures. Phys Rev Lett 76:4773-4776

Petersen KE (1982) Silicon as a Mechanical Material. Proc IEEE 70:58-95

Polman A, Wiltzius P (Guest Ed.)(2001) Materials Science Aspects of Photonic Crystals. MRS Bulletin 26: Issue 8

Purcell EM (1946) Spontaneous emission probabilities at Radio Frequencies. Phys Rev 69:681

Ribbing CG, Wäckelgård E (1991) Reststrahlen band as property indicators for materials in dielectric coatings. Thin Solid Films 206:312-317

Rung A, Ribbing CG (2002) Calculated photonic strucures for infrared emittance control. Appl. Optics 41:3327-3331

Sakoda K (2001) Optical Properties of Photonic Crystals. Springer, Berlin, Heidelberg

Schnitzer I, Yablonovitch E, Scherer A, Gmitter TJ (1993) The single-mode light-emitting-diode. In Photonic Band Gaps and Localization (Ed Soukoulis. Plenum, New York) 369-378.

Sievenpiper DF, Sickmiller ME, Yablonovitch E (1996) 3D wire mesh photonic crystals. Phys Rev Lett 76:2480-2483

Sigalas MM, Soukoulis CM, Chan CT, Ho KM (1994) Electromagnetic-wave propagation through dispersive and absorptive photonic-band-gap materials. Phys Rev B49:11080-11087

Slater JC (1958) Interaction of waves in crystals. Rev. Mod Phys. 30:197-222

Tikhonravov A, Design of Optical Coatings, this volume. Moscow State University, Moscow, Russia

Wijnhoven JEGJ, Bechger L, Vos WL (2001) Fabrication and Characterization of Large Macroporous Photonic Crystals in Titania Chem. Mater. 13:4486-4499

Winn JN, Fink Y, Fan S, Joannopoulus JD (1998) Omnidirectional reflection from a one-dimensional photonic crystal. Opt Lett 23: 1573-1575

Yablonovitch E, Gmitter TJ (1989) Photonic band structure: the face-centerd-cubic case. Phys Rev Lett 63: 1950-1953

Yablonovitch E (1987) Inhibited spontaneous emission in solid state physics and electronics. Phys Rev Lett 58:2059-2062
Yablonovitch E, Gmitter TJ, Meade RD, Rappe AM, Brommer KD, Joannopoulos JD (1991) Donor and acceptor modes in photonic band structure. Phys Rev Lett 67:3380-3383
Yablonovitch E (1993) Photonic band-gap structures. J Opt Soc B10: 283-295
Yablonovitch E (1994) Photonic crystals. J Mod Optics 41:173-194
Ziman JM (1964) Principles of the Theory of Solids. Cambridge Univ. Press, sec. 3.7

Abbrevations and Symbols

EMT		Effective medium theory	
D		(Complex) Dielectric displacement	As/Vm
$\varepsilon(\omega)= \varepsilon_1(\omega)+i\varepsilon_2(\omega)$		Complex dielectric function	-
ε_0		Permittivity of vacuum	As/Vm
K, k, q		Wave vectors of EM waves, electrons and phonons respectively	m^{-1}
λ		Wavelength	m^{-1}
LO	ω_L	Longitudinal optical (phonon frequency)	s^{-1}
P C		Photonic Crystal	
T E		Transverse electric (mode)	
T M		Transverse magnetic (mode)	
T O	ω_T	Transverse optical (phonon frequency)	s^{-1}
ω		Angular frequency	s^{-1}
$\omega_{\square}$		Plasma frequency	
ν		Frequency: $\omega/2\pi$	$Hz = s^{-1}$

Some Fundamentals of Optical Thin Film Growth

Norbert Kaiser

Fraunhofer Institute for Applied Optics and Precision Engineering, Jena, Germany

1 Introduction

Optical phenomena occur at surfaces and in thin films. However, unlike in microelectronics and opto-electronics, surface and thin film science is often hardly of any help in describing the basic processes taking place on optical surfaces. Optical surfaces are never of a well-defined single-crystalline or amorphous type, and the growth of optical films takes place under rather unclean and consequently undefined conditions. Our final interest is always optical, and there may also be an interest in the mechanical, thermal and electrical properties. The dependence of optical properties on thin film growth conditions is a severe constraint. There is a significant conflict between models based on ideal- and such based on real-structure films. This will be explained in Sect. 2. What we need is models that relate real film structure to desired optical properties. Such knowledge would allow the design and manufacture of coatings to be based on more realistic production conditions. Section 3 deals with aspects of thin film growth of variable dimensions, from isolated nanoclusters to continuous macroscopic films. The focus is on dielectric (Sect. 4) and metallic films (Sect. 5) and their optical properties.

Investigations into process parameters, real structure and film properties have been made already for many years, as is evident from the classical work by (Hass 1938) "Struktur und Optik aufgedampfter Metallschichten" (Structure and Optics of Evaporated Metal Films). Early work by (Heitmann 1968) on dielectrics shows that optical film properties are sensitive to deposition conditions, especially substrate temperature. For further reading, the following literature is recommended as an introduction to the field of structure-related optical properties of thin films: (Aspnes 1982, Guenther et al. 1986, Macleod 1981, 1982, Ohring 1992, Pulker 1999). As a sourcebook for the experimental and theoretical understanding of surface and thin film processes, (Venables 2000) is very useful. It contains also the resources of the World Wide Web.

2 Real Structure

The ideal structure of solids either shows an ideal single-crystalline order or is ideally amorphous. But there is no ideal crystalline or amorphous structure in the real world. The real structure of single crystals always has a certain amount of intrinsic and extrinsic defects. Intrinsic defects are those which are caused by atoms that do not change the stoichiometry of the crystal. Defects that produce non-stoichiometry are called extrinsic defects. Defects may generally have a major influence on the physical properties of the solid, for instance on light transmission, electrical conductivity, etc.

The growth of thin films by physical vapour deposition is an extremely non-equilibrium process, which takes place at very high supersaturations, at comparatively high concentrations of impurity atoms, and on undefined substrates. Consequently, the number of extrinsic and intrinsic defects in thin films may exceed that in bulk solids by about five orders of magnitude. Due to the usual deposition conditions in optics and due to the surface structure of optical components, optical films are mostly polycrystalline. It means that the real film consists of defect-rich crystalline grains, which are separated by impurity-enriched grain boundaries. This is the reason why film properties differ significantly from bulk properties. Figure 1 shows this fact schematically. Many transmission electron microscopical (TEM) studies (Lissberger and Pearson 1976, Guenther and Pulker 1976) have shown that optical thin films condense with a columnar microstructure, as for example the 300 nm thick MgF_2 film shown in Fig. 2.

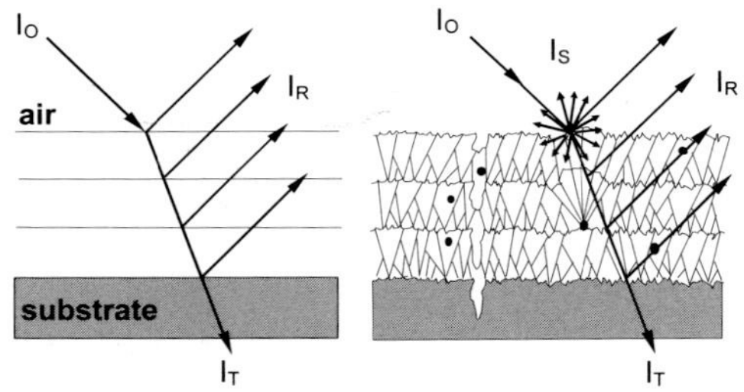

Fig. 1. Left: Ideal single-crystalline substrate coated with ideal single-crystalline films. The incoming intensity I_o is split into reflected and transmitted parts I_R and I_T, respectively. The conservation of energy is given by $Io = I_R + I_T$. Right: Real substrate with real coatings (highly exaggerated). Part of the incoming intensity I_o is absorbed (I_A) or scattered (I_S). The conservation of energy is given by $Io = I_R + I_T + I_A + I_S$.

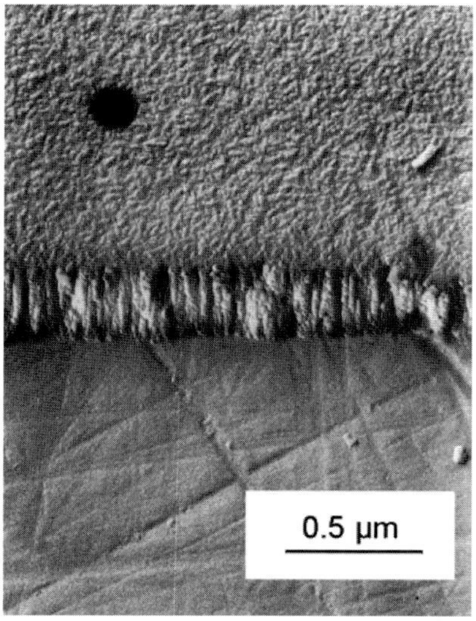

Fig. 2. TEM micrograph of 300 nm MgF_2 on BK7 substrate. The micrograph shows a carbon-platinum replica of a fracture edge. At the bottom the BK7 surface with polishing scratches can be seen. The fracture edge in the middle shows the columnar volume structure of MgF_2, and at the top the rough surface of the film is visible. Note that some polishing scratches are transmitted on the film surface. Micrograph by Heinz Müller, Jena.

The real structure of thin films is a metastable non-equilibrium state far away from its energetic minimum. It tends to change whenever enough energy is available. This makes things difficult. Figure 3 shows an interesting example of non-equilibrium growth of LiF. LiF is an extremely important coating material for the vacuum ultraviolet spectral range. However, under certain conditions it grows in the form of so-called whiskers. These are quasi-one-dimensional crystals which grow unidirectionally. The lengths of such crystals can be several centimeters and can be explained by a vapour – liquid - solid phase transformation mechanism. Whisker crystals are feared by everyone dealing with thin films. Whisker growth causes enormous failures in electronic assemblies. In optical coatings, whiskers are highly undesirable as well. On the other hand, film growth can be adjusted so that films become continuous at a thickness of only a few nanometers. Figure 4 shows a cross-section transmission electron micrograph (XTEM) of a Cr/Sc magnetron-sputtered multilayer. Such layers are superior mirrors for extreme ultraviolet radiation (see Chapt. "Coatings for EUV/X-ray mirrors" by Sergey Yulin).

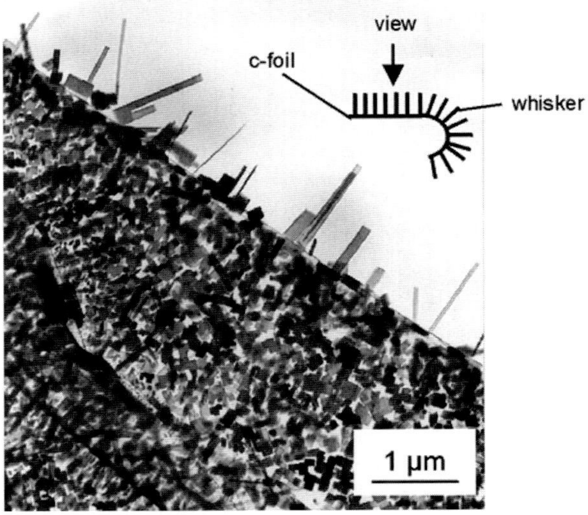

Fig. 3. TEM micrograph of a 100 nm thick LiF-film on carbon foil (substrate temperature 300°C, deposition rate 5 nm/s). Note that despite a high average film thickness of 100 nm, there is unidirectional whisker growth only; the film is completely discontinuous and useless for optics. Whiskers are visible where the supporting carbon foil is rolled as shown in the included schematic diagram. Micrograph by Heinz Müller, Jena.

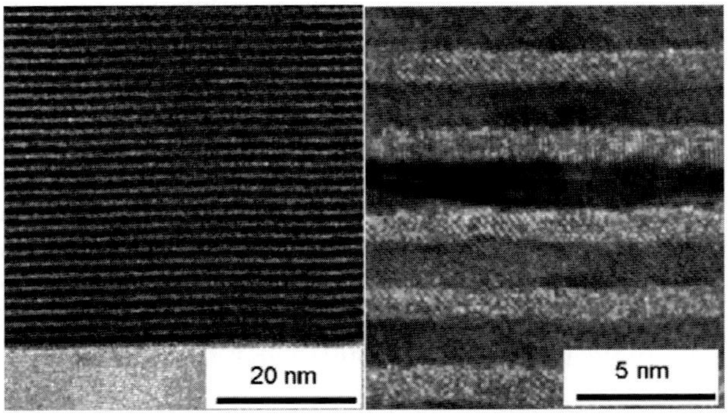

Fig. 4. XTEM micrograph of Cr/Sc-multilayer mirrors with period spacing d = 1.6 nm (left) and with d = 3.2 nm (right). Mirror after (Kuhlmann et al. 2002). Micrograph by Ute Kaiser, Jena.

Real structure is the link between thin film deposition parameters and thin film properties. For engineering electronic and optical properties, it is vital to control stoichiometry, mean grain size, and grain size distribution of the thin-film structure.

To engineer film properties through real structure, models are indispensable. In the following, thin film formation will be described as a kinetic adsorption and diffusion process starting with nucleation, coalescence and subsequent thickness growth, whereby all states can be influenced by deposition parameters.

3 Growth of Thin Films

3.1 Nucleation

The production of thin films by vapour deposition is a classical case of heterogeneous nucleation, i.e. the condensation of adatoms or admolecules at a substrate which is composed of atoms different from those of the vapor. Figure 5 shows the various stages of thin film growth.

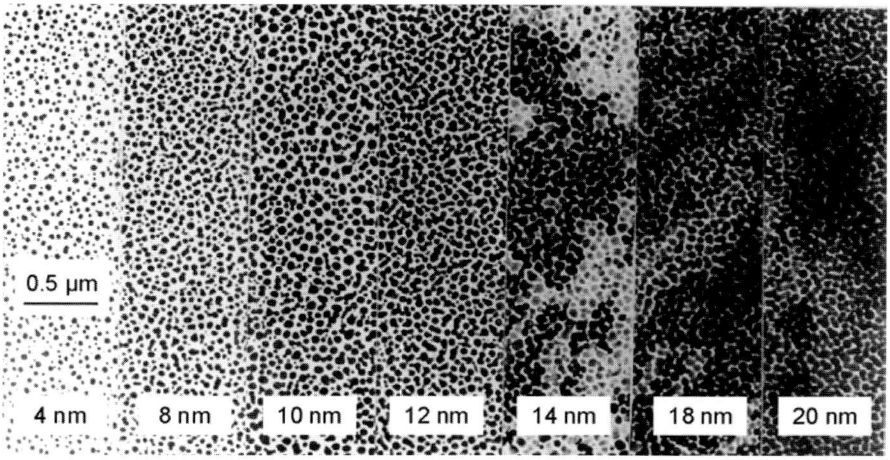

Fig. 5. TEM micrographs of antimony film formation (nucleation, growth of nuclei, coalescence, channels, holes, homogeneous film). It is special for Sb that at percolation an amorphous-crystalline phase transition (change of contrast!) occurs between 12 nm and 14 nm average film thickness (after Kaiser 1984).

Nucleation takes place at very high supersaturations S, defined as

$$S = p_V \,/\, p_e \tag{1}$$

p_V - vapour pressure of deposit material evaporated from the source at temperature T, p_e - equilibrium vapour pressure of the substrate material at temperature Ts. The deposition rate R is related to p as

$$R = p_V / \sqrt{2\pi mKT} \tag{2}$$

K – Boltzmann's constant, m – molecular weight of deposit material, T – temperature of the evaporation source. Since S is extremely high (S = 10^5 – 10^{50}, compared to only about 4 in a Wilson chamber!), the vapour cools down at the substrate surface very rapidly (splat-quenching), and consequently there is no time for rearrangement of the atoms into stable configurations at its energetic minimum. Adatoms stick where they land and become buried under the next arriving adatoms. Both the production of the vapour and its condensation are processes that are very far from equilibrium. As visible in Fig. 5, the isolated Sb-clusters are frozen in a metastable amorphous phase. However, at the very moment when the clusters come into contact (percolation), explosive crystallization occurs (Kaiser 1984).

Like all phase transitions, the formation of thin films is characterized by the formation of nuclei and their growth. Depending upon interaction energies of substrate atoms and film atoms, three different growth modes (Fig. 6) may occur.

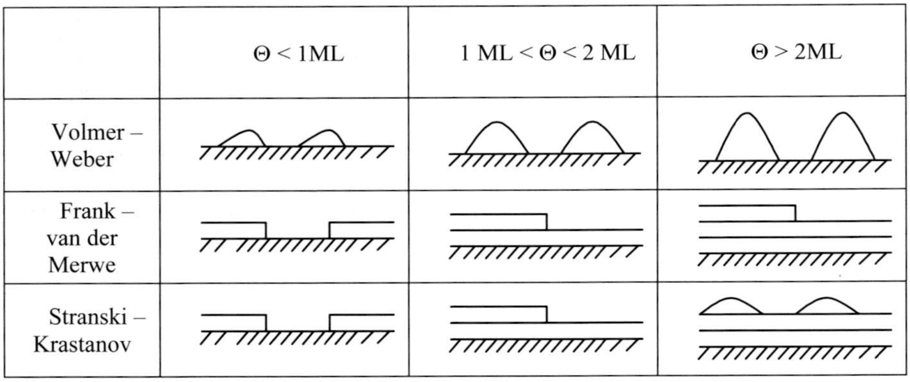

Fig. 6. Initial states of film growth after (Bauer 1958). Θ is the substrate surface coverage in monolayers.

- *Layer by Layer:* In the two-dimensional (2D) Frank-van der Merwe mode, layers of material grow on top of each other. Interaction between substrate and film atoms is larger than between adjacent film atoms.
- *Island:* In the Volmer-Weber mode, separate three-dimensional (3D) islands form on the substrate. Interaction between film atoms is larger than between adjacent film and substrate atoms.
- *Layer-plus-Island:* In the Stranski-Krastanov mode, one or two monolayers form first, followed by individual islands.

Growth modes can be systematically classified in terms of surface energies considering Young's equation.

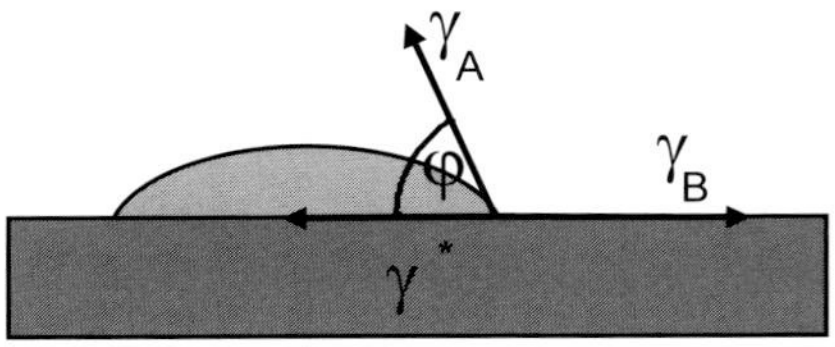

Fig. 7. Wetting angle φ of a liquid nucleus on a substrate is described by Young's equation: $\gamma_B = \gamma^* + \gamma_A \cos\varphi$, γ_B - surface energy of substrate; γ_A - surface energy of film material; γ^* - interface energy film – substrate.

Island growth ($\varphi > 0$) requires $\gamma_B < \gamma_A + \gamma^*$, whereas layer growth ($\varphi = 0$) requires $\gamma_B > \gamma_A + \gamma^*$. Layer-plus-island growth occurs because the interface energy increases with film thickness; typically the layer on top of the substrate is strained to fit the substrate. The growth mode is not only controlled by interface energies but also by supersaturation. Generally, growth tends to shift from island to layer as supersaturation increases .

If the growing thin film has a particular crystallographic orientation relative to the substrate, the term epitaxial growth is used. In microelectronics the substrate is normally a single crystalline semiconducting material, and films are required to be single crystalline with low defect density. In optics, the substrate surface is always polished or solidified from a melt and is far from being single crystalline. Such a configuration has a low surface energy γ_B and a low film-substrate interface energy γ^* compared to the surface energy γ_A of the film material. Consequently, nucleation of optical films mostly follows the Volmer Weber island growth. There is no crystallographic orientation of the growing island film to the substrate during nucleation. This holds also for single crystalline optical substrates like CaF_2 and MgF_2. The surface is always polished and undefined in the sense of crystallography. Non-epitaxial growth of optical thin films is also supported by the low substrate temperature. Epitaxial growth requires at least 500°C, whereas in optics substrates are not allowed to be heated above 300°C or in the case of CaF_2 even not above 150°C. Despite of such non-epitaxial growth conditions, the very first seconds of the birth of a growing film remain enormously important for tailoring the final optical properties. Substrate temperature, deposition rate, energy of admolecules hitting on the substrate, and residual gas are the factors influencing the number, size and orientation of the growing nuclei. Rules of thumb are:

- The number of nuclei per unit area increases with increasing deposition rate. At higher rates there are more but smaller nuclei per unit area than at lower rates. Consequently, substrate coverage is larger at higher deposition rates.
- The number of nuclei per unit area decreases with increasing substrate temperature. At higher substrate temperatures there are fewer but larger nuclei per unit area than at lower temperatures.
- Nuclei tend to become crystalline instead of amorphous with increasing substrate temperature and decreasing residual gas pressure.
- Substrate defect-induced nucleation leads to seeds of rough defects.

- Additional momentum transfer to the growing admolecules increases the number of nuclei per unit area and supports amorphous structures.
- Any type of extrinsic defects leads to undefined amorphous structures. The number of nuclei per unit area increases with increasing defect concentration.
- Interdiffusion of nuclei with the substrate surface increases with increasing substrate temperature and increasing energy of the admolecules.
- Coevaporation of different materials changes nucleation compared to the evaporation of one single material.
- Nucleation can be stopped and started again by barrier layers (stratification).

Quantitative investigations into atomistic nucleation and growth are known for metals on dielectric and metallic substrates. These examples are summarized by (Venables 2000). The influence of ion assistance have also been taken into consideration (Sillanpää et al. 2001). Driven by micro-electronics and opto-electronics, nucleation processes on semiconductor surfaces have been investigated very intensively. In the context of MOS (metal-oxide-semiconductor) structures, the growth of oxide layers on metals and semiconductors has been investigated also; however, our final interest is optical. There is a lack of investigations into the nucleation of dielectric layers on optical substrates.

3.2 Coalescence

The next step of three-dimensional film formation is the growth of islands until they touch each other to form a continuous network. This so-called coalescence process is extremely important for the final film properties. After solid-like or incomplete coalescence of two islands there may remain a grain boundary between them. This is called coalescence-induced boundary formation. Solid-like coalescence leads to mechanical stress in island films and influences further film growth. By the way, this can be used to adjust mechanical stress in thin films.

During liquid-like or complete coalescence, islands may fuse together in a liquid-like fashion to form a new, larger and boundary-free island. This new island covers less substrate area than the two islands together before coalescence. Hence, new substrate area for secondary nucleation is available. Such processes have been investigated for many metals, e.g. for Al by (Barna et al. 1988).

Surface energies and supersaturation are the factors controlling these processes through material transport by surface and bulk diffusion. The transition from isolated islands to a continuous macroscopic network can be characterized by a percolation threshold thickness. Percolate, from Latin "percolare", means to pass or cause to pass through a porous material. This can be understood, for example, when we look at measurements of electrical conductivity between two macroscopically separated contacts on an insulating substrate during the growth of a metallic layer between them. At percolation thickness, coalescence creates a continuous path for electrons through the network, and conductivity rises steeply by several orders of magnitude. Obviously, a phase transition takes place at the percolation threshold, with optical film properties changing exponentially (see also

Fig. 5). It will be shown that this phenomenon is crucial when it comes to engineering very thin semitransparent metal films.

3.3 Thickness Growth

The mean percolation thickness for many films is between 1 nm and 20 nm. The thickness required for producing metal mirror layers or dielectric quarterwave systems is, however, larger than that value. Further material deposition does not simply increase thickness; moreover, the real structure is determined by four processes.

- *Shadowing* – a geometric interaction between the arriving admolecules and the roughness of the growing surface. This effect is dominant at low substrate temperatures T_S. It occurs because the vapour beam is directed. Arriving material travels in an essentially straight line from the source, in the same way as illumination emanates from a light source.
- *Surface diffusion* – mobility of admolecules at surfaces and interfaces such as grain boundaries; dominant at medium substrate temperatures T_S.
- *Bulk diffusion* – mobility of admolecules in the volume of grains; dominant at high T_S.
- *Recrystallization* – phase transition as a complete change of crystal orientation; dominant at percolation thickness, large film thickness and high T_S.

By these processes the growing film develops a particular crystallographic orientation relative to the substrate. The films are polycrystalline, and the crystallites often show a preferred orientation, which is called growth texture. Texture influences the properties of films through orientation-dependent physical properties such as refractive index and mechanical stress.

For most metals and dielectrics, activation energies for diffusion are related to the melting temperature T_m of the material and the substrate temperature T_s. Thus, different basic processes can be expected to dominate different ranges of T_s/T_m and determine the real structure. This is the concept of structure zone models introduced by (Movchan and Demchishin 1969). In the simplest case, real structure exists in three different zones:

- Zone I: $T_s/T_m < 0.3$ (low mobility; admolecules stick where they land; the result is a fine-grained porous real structure).
- Zone II: $0.3 < T_s/T_m < 0.5$ (surface diffusion occurs with activation energies between 0.1 eV and 0.3 eV; a columnar real structure is obtained.
- Zone III: $T_s/T_m > 0.5$ (bulk diffusion occurs with activation energies above 0.3 eV, resulting in a rough, equiaxed, grained real structure).

Such a generalization of homologous structure zone models in terms of physical processes is very important for understanding the engineering of film properties. Structure zone models have been deduced for sputtered films (Thornton

1986) and for ion-assisted deposition (Messier et al. 1984). Additionally, (Grovenor et al. 1984) takes the influence of substrate morphology into consideration.

Up to now it has only been mentioned briefly that the growth of thin films for optics by physical vapor deposition is a rather contamination-affected process. In fact, under high-vacuum conditions in a normal production environment, the rate of residual gas molecules (water, hydrocarbons, etc.) striking the substrate surface is of the same order of magnitude as the deposition rate of the film material. At a total pressure of 10^{-4} Pa it takes only about one second to form one monolayer of residual gas molecules on the substrate. Consequently, extrinsic impurity defects, whether present on the substrate surface or resulting from the evaporation source or from the residual gas, have a tremendous influence on film growth. The effect is comparable to that of the substrate temperature. Impurities markedly change the surface energies at all stages of film growth. Impurities which are not soluble in the lattice are segregated into the grain boundaries and onto the growing film surface. At a critical impurity concentration, a fixed passivation layer forms on which secondary nucleation may occur. Therefore, nondiluted impurities are concentrated at interfaces.

Concerning real structure, large impurity concentrations act in the same way as low substrate temperatures. Structure zones are shifted along the T_s/T_m axis towards a fine crystalline, porous structure. (Barna and Adamik 1995) introduced the „basic structure zone model", which starts from the basic case of impurity-free films (Fig. 8) and then takes the increasing influence of impurities into consideration

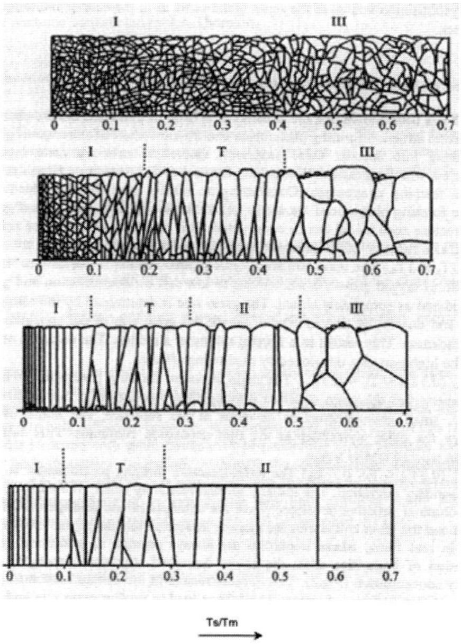

Fig. 8. Basic (a) and real structure zone models for low (bottom), medium (middle), and high (top) impurity concentration after (Barna and Adamik 1995).

4 Dielectric Films

4.1 Optical Properties

Refractive index

The relation between refractive index n and density ρ is given by the Lorentz-Lorenz equation:

$$\frac{n^2-1}{n^2+2}=C\rho \tag{3}$$

The factor C gives the relation between the macroscopic thin-film properties - refractive index and density ρ - and the microscopic polarizability α of an isotropic medium.

$$C=\frac{4\pi}{3}\alpha\frac{N_A}{M} \tag{4}$$

N_A is Avogadro´s number ($6.02\text{x}10^{23}$ particles/mol) and M is the molecular weight. By plotting the macroscopic refractive index expression $(n^2 - 1)/(n^2 + 2)$ versus density ρ (Lorentz-Lorenz plot), we can determine C and then e.g. the polarizability α. At optical frequencies, the refractive index of a dielectric is a function only of the electronic polarizability. Typical bulk values are in the order of $\alpha \approx 5\text{x}10^{-24}\ \text{cm}^3$ for TiO_2. In TiO_2 thin films, $\alpha \approx 30\text{x}10^{-24}\ \text{cm}^3$ has been measured by (Mergel et al. 2000).

To express the differences between compact material and void-rich thin films, the term *packing density p* (more correctly, packing coefficient) is used. ρ_f is the density of the thin film (solid + voids), and ρ_b is the density of the bulk (solid) material.

$$p=\frac{\rho_f}{\rho_b} \tag{5}$$

With p and n_f as mean refractive index of the film and n_b as mean refractive index of the bulk material, the Lorentz-Lorenz equation is:

$$p=\frac{(n_f^{\,2}-1)(n_b^{\,2}+2)}{(n_f^{\,2}+2)(n_b^{\,2}-1)} \tag{6}$$

If $p < 1$, the film consists at least of two materials. In the simplest cases these are solid film material and voids. The voids can be filled with air and/or with other materials like water and hydrocarbons. A reasonable model which relates the composite film refractive index n_f with the refractive index n_b of the bulk material and the refractive index n_v of the voids in the films is that of (Kinosita und Nishibori 1969).

$$n_f = (1-p)n_v + pn_b \tag{7}$$

Another important effective medium model for the refractive index of an array of separated columns is that of (Bragg and Pippard 1953).

$$n_f^2 = \frac{(1-p)n_v^4 + (1+p)n_v^2 n_b^2}{(1+p)n_v^2 + (1-p)n_b^2} \tag{8}$$

where n_b is used for the index of the bulk part and n_v for the index of the void material surrounding the bulk columns. If the columns are very densely packed, the following expression can be used. Equations 7–9 satisfactorily cover the range of refractive indices of composite films.

$$n_f^2 = pn_s^2 + (1-p)n_v^2 \tag{9}$$

The knowledge about optical anisotropy and optical inhomogeneity is vital for the calculation and reproducible production of optical thin films.

Absorption

Ideal transmitting optical elements should have zero absorption between the short wavelength transmission and infrared transmission cut-offs. In real materials, absorption is caused by defects which may be intrinsic or extrinsic. Absorption can be strongly influenced by external conditions. These are temperature, pressure, electrical and magnetic fields, and irradiation with energetic particles. Dynamic effects occur during energy deposition if optical constants and defect concentrations are dependent from temperature. This is typical for phase transitions (melting or recrystallization). Intrinsic absorption in a homogenous solid is described by Lambert´s law:

$$I(z) = I_0 e^{-\beta z} \tag{10}$$

with I_o being the incident intensity in direction z normal to the coating substrate interface and β the single photon absorption coefficient. Scatter losses are neglected. 1/β is the optical penetration depth at which the intensity I is reduced to the part 1/e. β relates to the extinction coefficient k and the wavelength by:

$$\beta = \frac{4\pi k}{\lambda} \tag{11}$$

Real solids have typical absorption coefficients of about $\beta = 10^{-3}$ cm^{-1} in their transparence region, and $\beta = 10^{6}$ cm^{-1} in the region of fundamental absorption outside the transparence band. The transition from fundamental absorption into the transparence band at the short UV cutoff can be described by Urbach´s rule. This is important in the UV region.

Homogeneously distributed sub-wavelength-sized defects can cause impurity levels of donor and acceptor type in the band gap. Refractive index becomes complex. Local defects with sizes in the order of film thickness are very strong absorption centers. They are known to be starting points for laser induced damage (Shaw-Klein et al. 1993, Kozlowski and Chow 1994, Reichling et al. 1998).

Scattering

Furthermore, real structure causes scatter losses. All types of real-structure phenomena such as grains, pores, defects, and rough interfaces are directly connected to scattering (Duparré 1995).

4.2 Fluorides and Oxides

Classical fluoride optical interference film systems are now applied down to the vacuum ultraviolet excimer-laser wavelength range. Here, the thicknesses of the individual quarterwave fluoride films can be as low as about 10 nm. As has been known for a long time (Koch 1965), optical coatings contain considerable amounts of water and hydrocarbons. Structure zones of MgF_2 (melting temperature T_m = 1538 K; we are talking in K rather than in °C when calculating the T_s/T_m –zones) are in the following substrate temperature T_S regions:

Zone I:	$T_S < 188°C$
Zone II:	$188°C < T_S < 496°C$
Zone III:	$T_S > 496°C$

The real structure of some important fluorides is shown in Fig. 9.

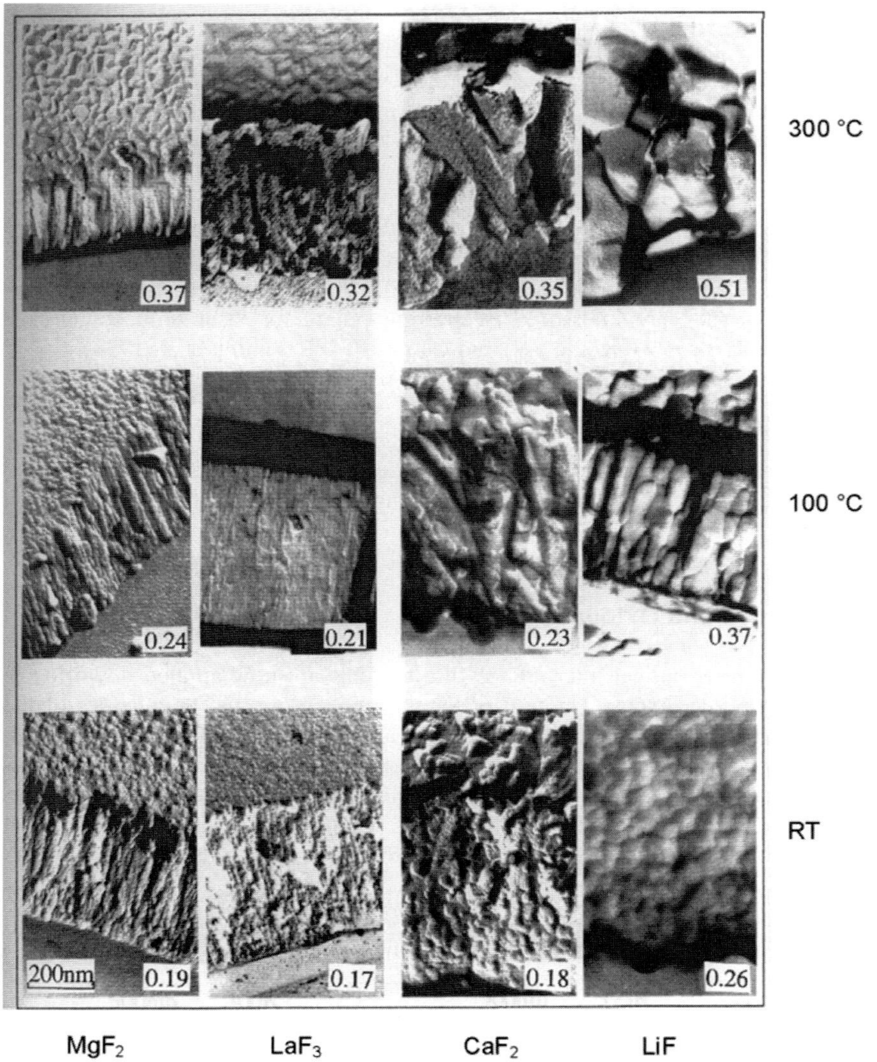

Fig. 9. TEM micrographs of Carbon-Platinum replicas from MgF_2, LaF_3, CaF_2, and LiF fracture edges (after Kaiser et al. 1992).

Optical effects are inhomogeneities and anisotropies, i.e. complex refractive index changes with variable film thickness and angle of light incidence. Optical anisotropies caused by columnar film growth can be used for special applications, such as polarizers for normal incidence (Hodgkinson 1999, 2000). This is the most prominent example of how microstructure can be tailored via growth phenomena.

Fig. 10. Helical MgF_2 columns (after Hodgkinson and Wu 1999, and Robbie et al. 1995).

The connection between optical properties and the real structure of oxide films has been investigated very intensively as well. Recent examples are the application of effective medium theories on the basis of the classical Lorentz-Lorenz model to TiO_2 films (Mergel 2000, 2001).

5 Metal Films

The growth of metal films starts with islands corresponding to the Volmer - Weber mechanism. Island films are moderately transparent to visible and infrared radiation. At percolation thickness the film becomes continuous and is transparent to the visible spectral region while reflecting infrared. Finally, at larger thicknesses, all radiation is reflected. This is one of the most exiting phenomenon in optics: properties change with film real structure only by the addition of a few nanometers of metallic material. Schematically this is shown in Fig. 11.

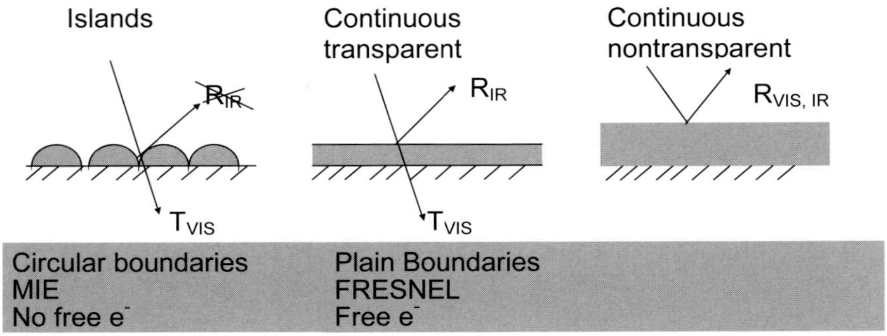

Fig. 11. Growth of a metal, starting from islands over continuous transparent to nontransparent films (after Kaiser et al. 2001).

5.1 Metal Island Films

As has been known for many years, the optical behaviour of metal island films (Ag, Au, Cu, ...) is not similar to that of the corresponding bulk material (Fragstein 1958). 50 years ago this effect was called "Anomalien der optischen Konstanten " (Anomalism of optical constants). Today, these properties are of outstanding importance in optics (overview in Kreibig 1995). Metal islands may interact with light over surface plasmons. These are strong collective oscillations of free electrons in the island in phase with the incoming light. The Mie-resonance frequency of the oscillation depends upon three factors:

- the metal´s electronic properties,
- size, shape, arrangement, and orientation of the islands,
- properties of the host material in which the islands are embedded or on whose surface they are fixed.

Since metals are highly absorbing, only very small changes in cluster size and distribution can be used to engineer the optical properties of the ensemble. Optical constants can be calculated on the basis of Maxwell-Garnett models and the Drude theory of free electrons (Monard 1997, Stenzel et al. 2000).

Practical applications of ultra-thin films require a macroscopic matrix in which the sensitive nanometer-sized metal clusters are embedded. In an optical film design, metal islands can be embedded in barrier layers such as SiO_x, Al_2O_3 or NiCr to prevent oxidation, or in other matrices such as transparent conductive oxides. Metal island films have become important elements of present-day optics and optoelectronics. Intense research is being conducted into active optoelectronic elements consisting of island films, including optically active individual molecules. As an example, the Stranski-Krastanov layer-plus-island growth of Ge islands on Si substrates for the production of self-assembled quantum-dots may be mentioned

(Thanh et al. 2000). Such deposition can only be performed under ultra-high vacuum conditions and on clean single-crystalline surfaces. In optics, deposition conditions are far from being as well defined. However, there are many industrial applications making use of the optical, electronic and mechanical properties of metal island films. Classical examples are the red and yellow colours of medieval church windows, of photographic systems or solar absorbers. Recent applications in the field of nanooptics are based on ultra-fast light switching phenomena. They can be used for novel components in integrated optics, for biosensors or optical tweezers. Metal island films are also useful as intermediate layers intended to change the surface energy and thus to tailor adhesion, bonding, wetting, or mechanical stress. We are just beginning to use these possibilities for the engineering of novel optical films of extraordinary properties.

5.2 Continuous Transparent Metal Films

Above percolation thickness, the optical behaviour of metal films is optically similar to that of the corresponding bulk materials, and they are transparent to visible and reflective for infrared radiation as long as thickness is below about 20 nm. In this thickness interval, transparent metal films can be used as neutral beamsplitters, induced transmission filters, solar control coatings, thermal insulating coatings, transparent electrodes, and heating layers. For example, silver is used for neutral-colour highly transparent and thermal insulation coatings (low E), variably coloured solar control coatings, and low-E sun coatings (Hill and Nadel 1999, Gläser 1999). to reduce absorption, the percolation thickness should be as low as possible. Maximum wetting of the substrate can be achieved in the two-dimensional Frank-van der Merwe growth mode. Unfortunately, as already mentioned, optical film growth starts with three-dimensional islands. In that case, the percolation thickness can be decreased by increasing the number of islands per unit area. At the same mean film thickness, many small islands cover a larger substrate area than do fewer but large islands. Tailoring percolation thickness to minimum values by increasing the deposition rate and reducing the substrate temperature is not possible, due to severe technological restrictions typical for large-area architectural coatings. However, the concept of changing surface energies (Fig.7) can be applied successfully. It can be implemented by plasma pretreatment of the substrate or by using ultra-thin bonding layers such as Cr, Pd or NiCr. Also, ion assistance can be used (Ensinger 1997). At larger thicknesses, however, abnormal grain growth occurs (Dannenberg et al. 2000) as predicted by Grovenor's structure zone model. Wetting can also be improved by the use of semiconductive transparent oxides such as In_2O_3, SnO_2, ZnO_2, and In_2O_3 doped with 2% to 10% Sn. They are frequently used instead of, or in mixtures with, metals. Finding out the best engineering technology is currently a field of keen competition between large-area coaters. Figure 12 shows the real structure of a thin Ag film embedded in two dielectric layers. To find stable production conditions, we need models for calculating the optical properties of such configurations.

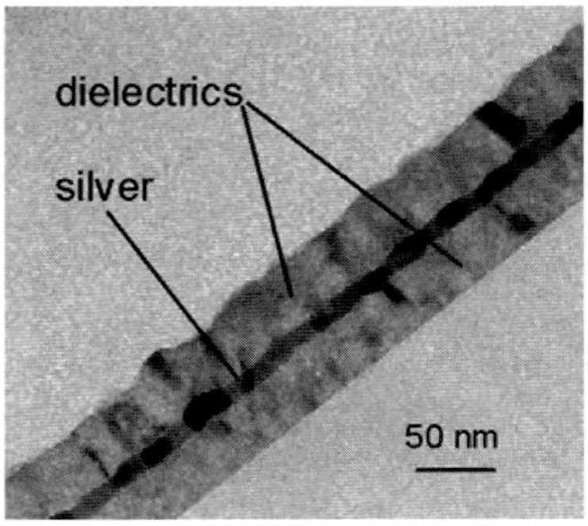

Fig. 12. Fig. XTEM micrograph of a semitransparent silver film between two dielectric layers. Micrograph by Ute Kaiser, Jena.

Another huge field of application of transparent continuous metal films is multilayer mirror coating for soft X-radiation (especially in the "water window": 2.3 nm ... 4.4 nm) and for extreme ultraviolet radiation (10...100 nm). This is outlined in the Chapter on "Coatings for EUV/X-ray mirrors", Sergey Yulin.

5.3 Continuous Nontransparent Metal Films

Continuous nontransparent metal films have been used as mirrors for years. It has been shown that, at high deposition rates and low substrate temperatures, nucleation density is very high, and consequently, thin films have a fine polycrystalline microstructure and maximum reflectivity. This has been known for many years in the industrial production of metal mirrors as the "quick-and-cold rule". Under these conditions (see rules of thumb for nucleation) there are many small nuclei, and the metal films feature smoothness, low contamination and homogeneously high reflectance.

6 Summary

The growth of optical thin films starts with the adsorption and diffusion of individual atoms. The structure of polished optical surfaces is rather less defined in terms of surface energy and can only be characterized by integral values. The maximum permissible substrate temperature in optics is about 300 °C, and deposition rates are in the range of several nm/s. At the same rate, foreign molecules from the residual gas impinge on the substrate surface. Due to these circumstances, nucleation starts statistically with the formation of three-dimensional islands (Volmer-Weber). Next, coalescence leads to a macroscopic network, and at percolation thickness, the optical and also all other properties change abruptly. Further thickness growth often leads to a polycrystalline real structure. Films consist of grains ranging from a few nm to several μm in extension. Grain boundary regions can have volume fractions comparable to those of the grains. Consequently, the properties of thin films differ markedly from the bulk properties of the same material. Thin film properties are directly related to the film's real structure. The study of structure–property relationships is therefore indispensable for the engineering of thin films. Substrate temperature and deposition rate can only be varied within a limited range. Film growth can be influenced more effectively by changes of the substrate surface energy and an increase in the energy of adatoms.

Substrate surfaces can be pretreated with different kinds of plasmas and ions, and by pre-nucleation with sub-nanometer bonding islands. The energy of adatoms can be increased by all types of energetic deposition, especially by using ion assistance. Where applicable, sputter technology leads to much better defined film properties, since the energy of adatoms is in the region of several eV compared to only several 0.1 eV with electron beam or boat evaporation.

In contrast to electronics, film growth in optics is rather undefined and poorly understood. Despite the fact that optical coatings are used with an ever increasing leverage effect on optics, we are only just beginning to understand film growth in detail. For controlling the optical properties of films nowadays it is of utmost importance to have a thorough knowledge of the real structure on the atomic scale.

References

Aspnes DE (1982) Optical Properties of Thin Films. Thin Solid Films 89:249–262

Barna A, Barna PB, Pocza FP (1988) Surface chemical phenomena influencing the growth of thin fims. J Vac Sc Tec 6:472–485

Barna PB, Adamik M (1995) Growth mechanisms of polycrystalline thin films, In: Matacotta, Ottaviani G (Eds) Science and Technology of Thin Films. World Scientific Publishing, Singapore, pp 1–28

Bauer E (1958) Wachstum dünner Schichten. Zeitschrift für Kristallographie 110:372–394

Bodemann A, Kaiser N, Raupach L, Weissbrodt P, Hacker E (1996) 248 nm laser interaction studies on LaF_3/MgF_2 optical coatings by mass spectroscopy and X-ray photoelectron spectroscopy. Proc SPIE 2714:405–415

Czigany Z, Adamik M, Kaiser N (1998) 248 nm laser interaction studies on LaF_3/MgF_2 optical coatings by crosssectional transmission electron microscopy. Thin Solid Films 312:182

Dannenberg R, Stach EA, Groza JR, Dresser BJ (2000) In-situ TEM observations of abnormal grain growth, coarsening, and substrate de-wetting in nanocrystalline Ag thin films.Thin Solid Films 370:54

Duparré A (1995) Light Scattering of Thin Dielectric Films. In Hummel RE, Guenther KH (eds.) Handbook of Optical Properties, Volume I, Thin Films for Optical Coatings. CRC Press Boca Raton, London, Tokyo

Ensinger W (1997) Low energy ion assist during deposition – an effective tool for controlling thin film microstructure. Nuclear Instruments and Methods in Physics Research B 127/128:796–808

Eva E, Mann K, Kaiser N, Anton B, Henking R, Ristau D, Weissbrodt P, Mademann D, Raupach L, Hacker E. (1996) Laser conditioning of LaF_3/MgF_2 dielectric coatings at 248 nm. Applied Optics 35:5613–5619

Fragstein C, Römer H (1958) Über die Anomalie der optischen Konstanten. Zeitschrift für Physik 151:54

Gläser HJ (1999) Dünnfilmtechnologie auf Flachglas. Verlag Karl Hofmann, Schorndorf

Grovenor CRM, Hentzell HTG, Smith DA (1984) The development of grain structure during growth of metallic films. Acta metal 32:773–781

Guenther KH, Pulker HK (1976) Electron microscopical investigations of cross sections of optical thin films. Applied Optics 15:2992–2997

Guenther KH, Smith DJ, Bangjun (1986) Structure and related properties of thin film optical coatings. Proc SPIE 678:2

Hass G (1938) Struktur und Optik aufgedampfter Metallschichten. Annalen der Physik 31:245

Heitmann W (1968) The influence of various parameters on the refractive index of evaporated dielectric thin films. Applied Optics 7:1541–1543

Hill RJ, Nadel SJ (1999) Coated Glass Applications and Markets. BOC Coating Technology (1999)

Hodgkinson I, Wu Q (1999) Birefringent Thin Film Polarizing Elements. World Scientific Singapore, New Jersey, London Hong Kong

Hodgkinson I, Lakhtakia A, Wu Q (2000) Experimental realization of sculptured-thin-film polarization-dicriminatory light-handedness inverters. Optical Engineering 39:2831–2834

Kaiser N (1984) Crystallization of amorphous antimony films. Thin Solid Films 116:259–265

Kaiser U, Kaiser N, Weissbrodt P, Mademann D, Hacker E, Müller H (1992) Structure of thin fluoride films deposited on amorphous substrates. Thin Solid Films 217:7–16

Kaiser U, Kaiser N (1994) C-adsorption behaviour of thin fluoride films. Thin Solid Films 237:250–254

Kaiser N, Yulin S, Feigl T, Kuhlmann T (2000) Si-based multilayers with high thermal stability. Proc. SPIE 4146:91–100

Kaiser N (2002) Review of the fundamentals of thin-film growth. Applied Optics 41:3053–3060

Kinosita K, Nishibori M (1969) Porosity of MgF_2 films evaluation based on changes in the refractive index due to adsorption of vapour. J Vac Sc Techn 6:730–733

Koch H (1965) Optische Untersuchungen zur Wasserdampfsorption in Aufdampfschichten (insbesondere MgF_2-Schichten). Phys Stat Sol 12:533

Kozlowski MR, Chow R (1994) The role of defects in laser damage of multilayer coatings. Proc SPIE 2114:640–649

Kreibig U, Vollmer M (1995) Optical Properties of Metal Clusters. Springer, Berlin Heidelberg New York

Kuhlmann T, Yulin S, Feigl T, Kaiser N, Gorelik T, Kaiser U, Richter W (2002) Chromium-scandium multilayer mirrors for the nitrogen K_α line in the water window region. Applied Optics 41:2048–2052

Lewis B, Anderson JC (1978) Nucleation and Growth of Thin Films. Academic Press (1978)

Lissberger PH, Pearson JM (1976) The performance and structural properties of multilayer optical filters. Thin Solid Films 34:349–355

Macleod HA (1981) Performance-limiting factors in optical coatings. Proc SPIE 288:580
Macleod HA (1982) Microstructure of optical thin films. Proc SPIE 325:21
Mademann D, Raupach L, Weissbrodt P, Hacker E, Kaiser U, Kaiser N (1993) Investigations of thin fluoride films for optical application by surface analytical methods and electron microscopy. Fresenius J of Anal Chem 346:173–176
Mergel D, Buschendorf D, Eggert S, Grammes R, Samset B (2000) Density and refractive index of TiO_2 films prepared by reactive evaporation. Thin Solid Films 371:218–224
Mergel D (2001) Modelling TiO_2 films of various densities as an effective optical medium. Thin Solid Films 397:216–222
Messier R, Giri AP, Roy RA (1984) Revised structure zone model for thin film physical structure. J Vac Sci Technol A2(2):500–503
Monard H (1997) Optical properties of silver, gold and aluminium ultra thin granular films evaporated on oxidized aluminium. Thin Solid Films 310:265–273
Movchan BA, Demchishin AV (1969), Phys Metal Metallogr 28:83
Ohring M (1992) The Material Science of Thin Films. Academic Press, Inc., Harcourt Brace Jovanovich, Publishers
Pulker HK (1999) Coatings on Glass. Elsevier, Amsterdam
Reichling M, Bodemann A, Kaiser N (1998) Defect induced laser damage in oxide multilayer coatings for 248 nm. Thin Solid Films 320:264–279
Robbie K, Friedrich LJ, Dew SK, Smy T, Brett MJ (1995) J Vac Sci Technol A13:1032
Shaw-Klein LJ, Burns SJ, Jacobs SD (1993) Model for laser damage dependence on thin-film morphology. Applied Optics 32:3925
Sillanpää J, Koponen IT, Grønbech-Jensen N (2001) A rate-equation model for the growth of metallic thin films in ion beam assisted deposition. Nuclear Instruments and Methods in Physics Research B 184:523–530
Stenzel O, Lebedev AN, Schreiber M, Zahn DRT (2000) Simulation of linear optical losses of absorbing heterogeneous thin solid films. Thin Solid Films 372:200–208
Thanh VL, Yam VY, Zeng Y, Bouchier D (2000) Nucleation and growth of self-assembled Ge/Si (001) quantum dots in single and stacked layers. Thin Solid Films 380:2–9
Thornton JA (1986) The microstructure of sputter-deposited coatings. J Vac Sci Technol A4(6):3059
Venables JA (2000) Introduction to Surface and Thin Film Processes. Cambridge University Press

Abbrevations

EUV	extreme ultraviolet spectral region
IR	infrared spectral region
TEM	transmission electron microscopy
UV	ultraviolet spectral region
VIS	visible spectral region
XTEM	cross section transmission electron microscopy

Symbols

α	polarizability
β	single photon absorption coefficient
d	thickness (period spacing)
Θ	substrate surface coverage
φ	wetting angle
γ_B	surface energy of substrate
γ_A	surface energy of film material
γ^*	interface energy film – substrate
I_o	incoming intensity
I_A	absorbed intensity
I_S	scattered intensity
I_R	reflected intensity (reflectance)
I_T	transmitted intensity (transmittance)
k	extinction coefficient
K	Boltzmann´s constant
m	molecular weight
n	refractive index
n_f	refractive index of the film material
n_b	refractive index of the bulk material
n_v	refractive index of the void material
N_A	Avogadro´s number
p	packing density
p_e	equilibrium vapour pressure
p_V	vapour pressure of deposit material
R	deposition rate
ρ	density
ρ_f	density of the thin film
ρ_b	density of the bulk material.
S	supersaturation
T	temperature
T_m	melting temperature
Ts	substrate temperature
z	direction normal to the coating substrate interface

Design of Optical Coatings

Alexander V. Tikhonravov

Research Computing Center, Moscow State University, Moscow, Russia

1 Introduction

The first optical coating designs were obtained by direct analysis of spectral properties of single layers and multilayers. In particular, the first anti-reflection (AR) and high-reflection (HR) coatings were obtained by this approach. Soon, however, it was recognized that such design by direct analysis has limited potentialities and only a few types of optical coatings can be found by this approach.

Intensive development of design methods was started in the early fifties in the last century. It is not surprising that methods from microwave theory were widely borrowed because the development of design methods in that area had been started nearly two decades earlier. In particular, the theory of Chebyshev filters was successfully applied to the design of narrow-band interference optical filters (Baumeister 2000; Delano and Pegis 1969). In parallel the design approach based on the theory of equivalent layers was elaborated. In the following years various types of optical coatings were successfully designed by this approach or its combination with other design techniques. These results were summarized in the book written by A. Thelen (Thelen 1989).

A revolutionary step in optical coating design was taken in 1958 by P. Baumeister when he introduced the merit function estimating the closeness between designed and target spectral characteristics and proposed considering the design of optical coatings as an optimization problem (Baumeister 1958). At first only the optimization of a merit function from a properly chosen starting design was discussed. As a rule all optimization routines converge to the local minimum of the merit function which is close to the starting design. This means that the final design is usually close to the starting one. For this reason the corresponding design procedure is commonly called refinement.

Unfortunately, good starting designs are not available for many important design problems and in the seventies and eighties a lot of effort was applied to the development of nonlocal design techniques. One of the most universal techniques of this type is based on the Fourier transform. The application of the Fourier transform for optical coating design was first proposed by the Estonian scientist L. Sossi (Sossi 1974) but only after a series of works by J. Dobrowolski, P. Verly, and their colleagues (Dobrowolski and Lowe 1978; Verly et al. 1989; Verly et al. 1992) has the Fourier transform technique become a really universal design tool.

In 1982 the needle optimization technique was invented (Tikhonravov 1982 -2). The outstanding potentialities of this technique have become evident only a few years later with the wide implementation of personal computers. In 1993 the needle optimization technique was first introduced in the OptiLayer design software. Now practically all commercial thin film software uses some versions of this technique.

As late as 1994 J. Dobrowolski wrote, "The design of filters with any but the simplest spectral characteristics remains a complicated problem ..." (Dobrowolski 1994, p 42.14). The existence of a universal and extremely effective design technique (needle optimization technique) has changed the situation fundamentally. It is no longer a problem to obtain theoretical designs with the required spectral properties. The designer can concentrate attention on feasibility issues, on finding the best design from a manufacturing point of view, on more effective and less time-consuming organization of his own labor. The content of this chapter reflects this new situation.

Design examples of this chapter have been calculated using the OptiLayer design software (Tikhonravov and Trubetskov 2002 - 2). To write down multilayer designs in a formula form standard notations are used (Thelen 1989). Dispersion of refractive indices are not taken into account to shorten the description of design examples.

2 Modern Approaches to the Design of Optical Coatings

Descriptions of various design methods and many useful multilayer designs can be found in several texts (Dobrowolski 1994; Macleod 1986; Thelen 1989). This section presents the modern state of the art in general optical coating design techniques. The question of design optimality and the key role of the design total optical thickness are discussed.

2.1 Formulation of the Design Problem

Consider a multilayer coating consisting of m homogeneous and isotropic layers (Fig. 1). The z-axis is oriented in the direction from the substrate to the ambient medium. Layers are numbered in the same direction.

Let us denote the physical thicknesses of layers by $d_1, d_2, ..., d_m$ and their refractive indices by $n_1, n_2, ..., n_m$. The refractive indices of the substrate and of the ambient medium are denoted n_s and n_a, respectively. For the sake of simplicity we assume that all layers and outer media are non-absorbing and consider the normal incidence case.

The Abeles matrix method (Abeles 1950) associates the characteristic matrix of the layer with each layer. At the normal light incidence the characteristic matrix of the j-th layer is

$$\mathrm{M}_j = \begin{pmatrix} \cos\varphi_j & (i/n_j)\sin\varphi_j \\ in_j \sin\varphi_j & \cos\varphi_j \end{pmatrix} \tag{1}$$

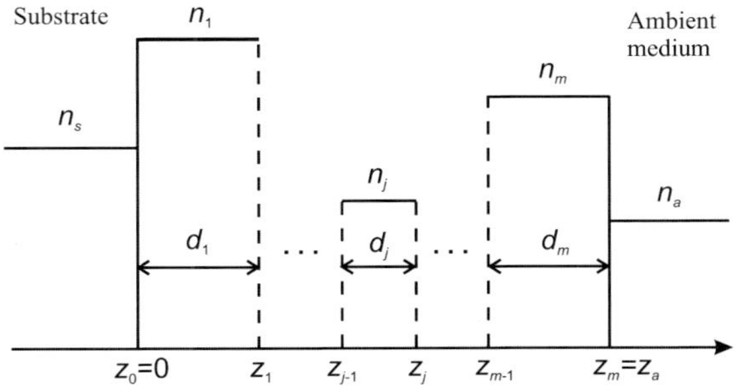

Fig. 1. Schematic of multilayer coating.

where

$$\varphi_j = \frac{2\pi}{\lambda} n_j d_j \tag{2}$$

is the phase thickness of the j-th layer and λ is the wavelength of the incident light in vacuum.

Along with the wavelength we shall also use the wavenumber

$$k = 2\pi / \lambda \tag{3}$$

as a spectral parameter.

The product of characteristic matrices of all layers

$$M = M_m M_{m-1} \ldots M_1 \tag{4}$$

is known as the characteristic matrix of a multilayer coating.

Note: In some texts, for example, in the book "Design of Optical Interference Coatings" by Alfred Thelen (Thelen 1989), the characteristic matrix of a coating is calculated by multiplying layer characteristic matrices in the reverse order. This is connected with numeration of the layers in the reverse order, i.e., in the direction from the ambient medium to the substrate.

Let us denote the elements of the coating characteristic matrix by m_{ij}:

$$M = \begin{pmatrix} m_{11} & m_{12} \\ m_{21} & m_{22} \end{pmatrix}. \tag{5}$$

The amplitude transmission and reflection coefficients, t and r respectively, are expressed through these elements by the formulas

$$t = \frac{2n_a}{n_a m_{11} + n_s m_{22} + n_a n_s m_{12} + m_{21}}, \quad r = \frac{n_a m_{11} - n_s m_{22} + n_a n_s m_{12} - m_{21}}{n_a m_{11} + n_s m_{22} + n_a n_s m_{12} + m_{21}}. \tag{6}$$

Equations (6) are the basic formulas for calculating all spectral characteristics of a multilayer coating. The intensity transmission and reflection coefficients (transmittance T and reflectance R) are expressed through the amplitude coefficients by the formulas

$$T = \frac{n_s}{n_a}|t|^2, \quad R = |r|^2. \tag{7}$$

When solving a design problem we ought to find the number of layers of a multilayer coating and the refractive indices and thicknesses of its layers that provide the target spectral characteristics. In the following we shall call the parameters m (the number of layers) and n_j, d_j ($j = 1, \ldots, m$) design parameters.

Let us assume that the total number of layers m and the refractive indices of all layers are set. Thus at the moment only the layer thicknesses can be varied.

Let the target spectral characteristic be the reflectance $\tilde{R}(\lambda)$. In the general case there is no multilayer coating featuring exactly $\tilde{R}(\lambda)$ (Thikhonravov 1993). Thus we can only speak about designing a multilayer coating with a reflectance that is in some respect close to $\tilde{R}(\lambda)$.

Let us introduce the vector X with the coordinates being layer physical thicknesses:

$$X=\{d_1, d_2, \ldots, d_m\}. \tag{8}$$

We can consider this vector as an element of the m-dimensional parametric space.

Because the number of layers and the layer refractive indices are fixed the reflectance depends only on the layer thicknesses, i.e., on the vector X. We shall denote the reflectance $R(X,\lambda)$.

Obviously, it is impossible to calculate $R(X,\lambda)$ at all λ in the spectral region of interest. Thus let us introduce a wavelength grid for this region. Let the total number of grid points be equal to L. Denote these points by $\lambda_1, \lambda_2, \ldots, \lambda_L$. There are many ways to estimate the deviation of the actual reflectance from the target reflectance. The most common one is to estimate it as a weighted mean square deviation on the introduced wavelength grid:

$$F(X) = \frac{1}{L}\sum_{j=1}^{L}\left[\frac{R(X,\lambda_j) - \tilde{R}(\lambda_j)}{\Delta R_j}\right]^2. \tag{9}$$

Values ΔR_j in Eq. (9) are called design tolerances. The function $F(X)$ is commonly called the merit function. Evidently its value is entirely determined by the vector of design parameters X. Quite often it is convenient to set all tolerances to be equal to 0.01. In this case $F^{1/2}(X)$ represents the mean square deviation of the actual reflectance from the target reflectance in percent.

The introduction of the merit function allows one to formulate the design problem as an optimization problem. Optimization of the merit function is currently the most common approach to the design of multilayer optical coatings.

2.2 Refinement

A procedure, first proposed by Philip Baumeister (Baumeister1958) and commonly known as refinement, is a key element of modern design techniques. The essence of refinement is the optimization of the merit function starting from a given point in parametric space. This point is known as a starting design for the refinement procedure.

The optimization of functions is one of the well elaborated topics in modern applied mathematics. Various optimization routines are described in the well-known books (Press et al. 1986). With modern state of art of design software it is not necessary to consider the details of various refinement techniques. Nevertheless, acquaintance with some basic ideas is useful for the designer of optical coatings.

The main problem related to refinement is non-convexity of the merit function. In general, the merit function possesses a great number of local minima. Due to this fact one can anticipate only convergence to the local minimum which is closest to the starting design.

Various first order optimization methods, which make use of the merit function gradient, are most widely used for the refinement. The gradient is a vector composed of merit function partial derivatives with respect to the design parameters. A crucial issue for the effectiveness of all first order optimization methods is the accuracy of the gradient calculation. An up-to-date approach to this calculation is based on precise analytical expressions. The details of this approach can be found in (Furman and Tikhonravov 1992).

Along with the first order methods more sophisticated high order optimization methods have been also applied to the design of optical coatings (Tikhonravov and Trubetskov 1992; Tikhonov et al. 1993). Unfortunately there is no universal optimization method that is most suitable in all situations. For this reason all modern commercial design codes have in disposal several optimization routines. When solving practical design problems it is worthwhile to empirically find the most suitable one.

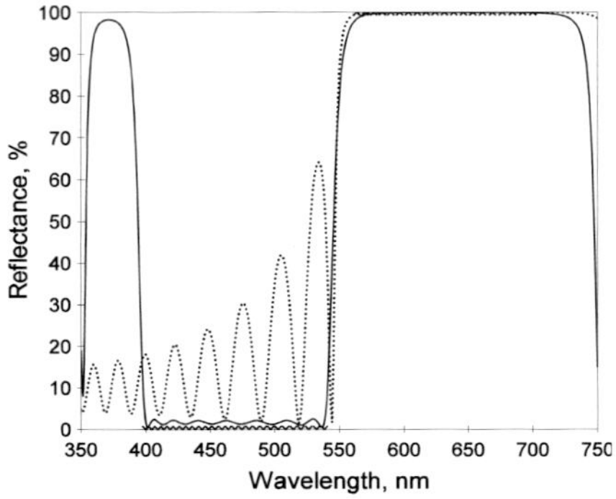

Fig. 2. Design of the short-wave-pass filter by refinement: dashed line - reflectance of the starting design, solid line - reflectance of the final design.

Figure 2 illustrates an application of the refinement procedure to the design of short-wave-pass filter. The target reflectance is zero in the spectral region from 400 *nm* to 540 *nm* and 100 % in the spectral region from 560 *nm* to 700 *nm*. We design a two-component multilayer filter with alternating high and low refractive indices, $n_H = 2.35$ and $n_L = 1.45$. The refractive indices of the substrate and ambient media are $n_s = 1.52$ and $n_a = 1.00$.

The twenty one layer quarter wave mirror with the central wavelength of 640 *nm* is used as a starting design. Its reflectance is shown by dashed line in Fig. 2. Target reflectance values are indicated by crosses.

The reflectance of the final 21-layer design is shown by solid line in Fig. 2. The refinement was accomplished by the damped-least-square optimization method with analytically calculated merit function gradient. The refinement procedure takes only a few seconds on modern PC.

The above example shows that the refinement is an extremely efficient design procedure if a proper starting design is known. A high performance of modern personal computers enables also a successful application of refinement procedure to some design problems with unknown starting designs. If the total number of design layers is relatively small (up to 10-20) then the efficient design approach is to start refinement procedure many times with randomly set starting designs and then to select the best final design.

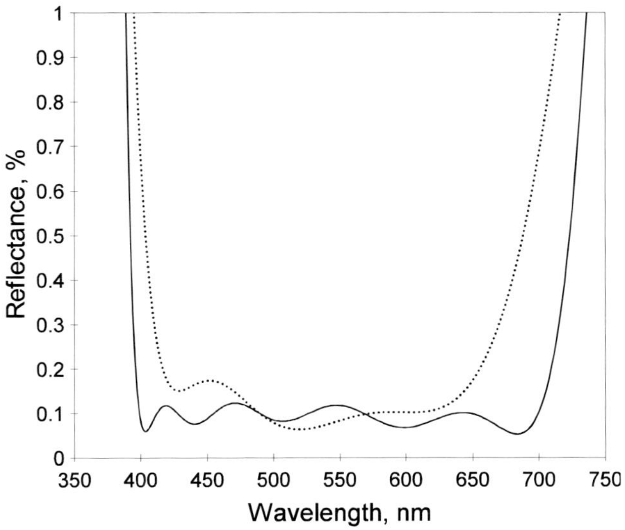

Fig. 3. Reflectances of the 4-layer (dashed line) and 10-layer (solid line) AR coatings obtained by refinement with randomly set starting designs.

Figure 3 shows the reflectances of 4-layer and 10-layer AR coatings obtained using this approach. The refractive indices of the layers and outer media are the same as before. The design formulas of the obtained design are 0.215*H* 0.361*L* 1.972*H* 0.937*L* and 0.211*H* 0.364*L* 2.038*H* 0.577*L* 0.117*H* 1.090*L* 0.579*H* 0.113*L* 1.057*H* 0.942*L* with the reference wavelength 550 *nm*.

2.3 Needle Optimization Technique

Unfortunately, good starting designs are not readily available for many design problems. The efficiency of the approach based on refinement with randomly set starting designs decreases rapidly with growing number of design layers. To solve complicated design problems requiring large numbers of design layers, one ought to use non-local design techniques.

A powerful non-local design technique, the needle optimization technique, was first proposed in 1982 (Tikhonravov 1982 - 2). From a physical point of view, the main idea of the needle optimization technique can be described as follows. A multilayer optical coating is an interference structure with spectral performance determined by interference effects among the light waves reflected from various layer boundaries and transmitted through these boundaries. The interference effects are determined by the phases and amplitudes of the above waves. When the merit function is optimized with respect to layer thicknesses, the phase relationships among all reflected and transmitted waves are changed. The optimization process is finished when it becomes impossible to minimize the merit function using interference effects produced by the existing layer boundaries.

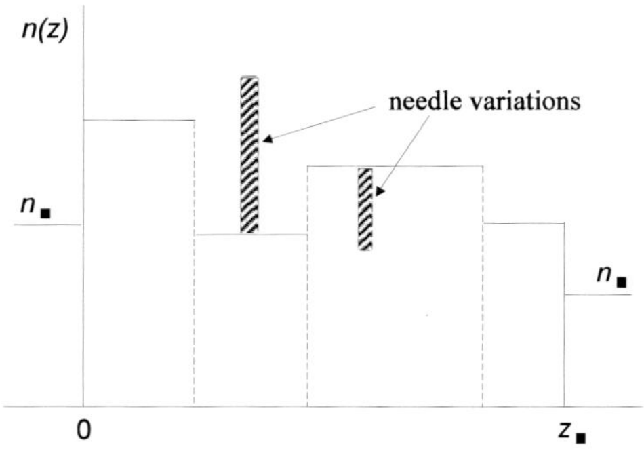

Fig. 4. Needle variations of the refractive index profile.

Clearly, to proceed with optimization, one ought to introduce into consideration some new physical effects. This can be done by the insertion of new layers into a multilayer structure. Such insertions are equivalent to variations of the refractive index profile (see Fig. 4). These variations are referred to as needle variations. If new layers are inserted in proper places, then new interference effects associated with these layers improve the correspondence between actual and target spectral characteristics.

The way to find proper places for the insertion of new layers is provided by rigorous mathematical considerations (Furman and Tikhonravov 1992; Tikhonravov 1985 -1). The following formula presents the merit function variation when the needle variation of the refractive index profile is done at the point z inside a multilayer structure:

$$\delta F = Q(z)\left[n_k^{\,2} - n^2(z)\right]\delta z. \tag{10}$$

Here δz is the thickness of a new layer, n_k is its refractive index and $Q(z)$ is a special function that is calculated by the effective mathematical algorithm *without actually carring out layer insertions.*

The emphasized words in the last sentence reflect the essence of the needle optimization technique and explain its outstanding efficiency. To find proper insertion places it is not necessary to make actual insertions in various places inside a multilayer structure. Function $Q(z)$ is calculated only once and immediately gives all necessary information for the optimization procedure.

Suppose that we can use several materials with refractive indices n_k for the design. Let us introduce the perturbation function

$$P(n_k, z) = Q(z)\left[\hat{n}_k^{\,2} - n^2(z)\right]. \tag{11}$$

Suppose that $P(n_k, z)$ is negative at some point z. It follows from Eq. (10) and (11) that needle variation of the refractive index profile at this point with sufficiently small δz will necessarily cause the merit function to decrease.

In general, all places where the function $P(n_k, z)$ is negative may be considered as appropriate places for the insertion of new layers. At the same time, better results are usually achieved when the insertions are done at the places where $P(n_k, z)$ has its most negative values.

In the case of two-material optical coatings, the choice of refractive indices of new layers is obvious. In the case when more than two materials are available for the design, it is necessary to calculate $P(n_k, z)$ for all available refractive index values. Generally, better results will be obtained with those n_k values that provide the most negative values of $P(n_k, z)$.

In practice needle variations are not necessarily thin. The thicknesses of new layers are determined so as to cause the maximum decrease of the merit function, and often they turn out to be big enough.

The insertion of new layers leads to a new multilayer structure with a higher number of layers. It is quite possible that the phase shifts between the waves reflected and transmitted through these boundaries are not optimally adjusted. Thus the optimization of the new structure with respect to the thicknesses of its layers may effect a further decrease of the merit function.

In general, the design procedure based on the needle variation technique consists of a sequence of insertions of new layers followed by a corresponding sequence of optimizations of the new structures with respect to the thicknesses of its layers.

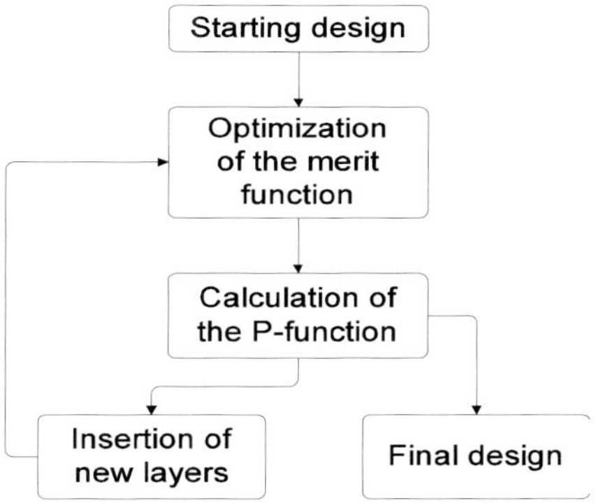

Fig. 5. General schematic of the needle optimization procedure.

A general schematic of the needle optimization procedure is shown in Fig. 5. To start the procedure, a starting design is specified. The choice of a starting design is rather arbitrary. As discussed in the next section the most essential parameter is its total optical thickness. The procedure runs through the loop shown in Fig. 5 until the *P*-function becomes nonnegative for all z values. This means that an optimal design is obtained and it is impossible to improve the merit function further either by the optimization with respect to layer thicknesses or by insertions of new layers.

2.4 Question of Design Optimality

To discuss the question of design optimality consider as an example the design of a bandpass filter with the following target transmittance: 100% in the ranges from 400 *nm* to 500 *nm* and from 600 *nm* to 700 *nm*, 0% in the range from 500 *nm* to 600 *nm*. Substrate, ambient medium, and layer materials are the same as before.

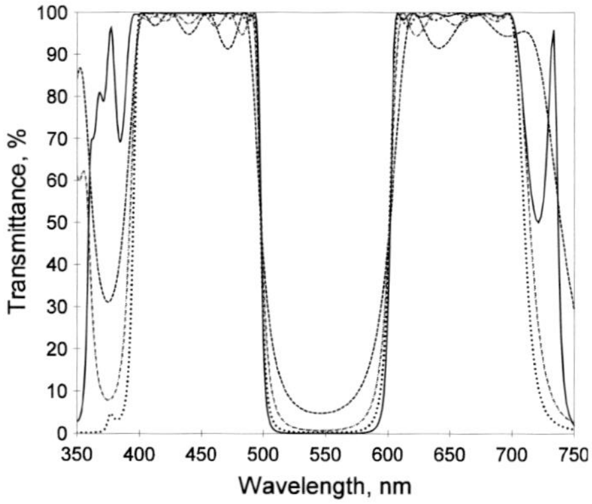

Fig. 6. Transmittances of four bandpass filter designs with the design parameters listed in Table 1.

Figure 6 shows the transmittances of four different designs with different numbers of design layers and different values of the merit function estimating the closeness between actual transmittances shown in Fig. 6 and the target transmittance.

Table 1 summarizes several major parameters related to the obtained results. Consider the first two columns of this table. The merit function value is successively decreased as the fitting of target transmittance is improved. It is not surprising that for the first three designs the improvement is achieved due to increasing the total number of design layers. But in the last line, corresponding to the design with the lowest merit function value, the total number of design layers is smaller than in the third line! It seems to be strange that a nearly three times decrease of the merit function is not associated with an increase of such an important design parameter as the total number of design layers.

This contradiction is explained if one looks at the third column of Table 1. It becomes evident that the essential decrease of the merit function is achieved due to the increase of the design total optical thickness.

The considered example shows that the total optical thickness is an extremely important design parameter which should be included in the list of the main design parameters along with the total number of design layers.

Table 1. Summary of results for the bandpass filter.

Number of design layers	Merit function value	Design total optical thickness, μm
13	9.10	2.35
23	3.85	3.45
41	1.65	4.49
30	0.62	5.98

Let us ask ourselves what design from Table 1 is an optimal one. Obviously, it is difficult to answer this question without knowing specific manufacturing demands, potentialities of deposition equipment, specific accuracy of deposition monitoring, and so on. We see that it is difficult to formalize the concept of design optimality. It is sensible to talk only about practically optimal designs and it is likely that such designs will be different for different designers.

2.5 The Role of the Design Total Optical Thickness and Synthesis by Gradual Evolution

To select a practically optimal design it is worthwhile to obtain a set of solutions to the design problem with various combinations of major design parameters. In this respect it is important to recognize a key role of the design total optical thickness.

It was rigorously proved in (Tikhonravov 1982 - 1) that the ratio of reflectance R and transmittance T can be expressed in the form of a Fourier transform integral of the type

$$R(k)/T(k) = \int_{-2x_a}^{2x_a} F(t)\exp(ikt)dt \tag{12}$$

where k is the wavenumber and x_a is the design total optical thickness.

The most important fact here is that the Fourier transform of *R/T* differs from zero only in the interval $(-2x_a, 2x_a)$ specified by the design total optical thickness. If x_a is small then all Fourier harmonics of *R/T* are slowly varying functions of k. This means that it is impossible to achieve a good approximation of rapidly varying spectral characteristics if the design total optical thickness is small.

Quite often target spectral characteristics have abrupt variations. This is a typical feature of all edge filters, bandpass filters, etc. The example of the previous section (Fig. 6) confirm the above conclusion about the role of the design total optical thickness in the approximation of rapidly varying target characteristics.

The design total optical thickness plays a key role in the achievement of a high quality approximation of target spectral characteristics of all types, not only those with rapidly varying target dependencies. Realization of this fact results in the expansion of the needle optimization technique that will be referred to as synthesis by gradual evolution.

The term "gradual evolution" is borrowed from the paper published by J. Dobrowolski in 1965 (Dobrowolski 1965). The process of gradual evolution as described by J. Dobrowolski consists of adding several new layers to the existing multilayer structure and optimization of a new structure with respect to the design parameters of new layers.

Synthesis by gradual evolution based on the needle optimization technique is organized as follows. Suppose that the needle optimization procedure is finished, i.e. it is impossible to further decrease the merit function by insertions of new layers and optimizations with respect to layer thicknesses. In this case to proceed with optimization it is required to increase the total optical thickness of design. This can be done in different ways. For example, it is possible just to increase a thickness of one of the coating layers. After this the needle optimization procedure is started again.

Usually increase of a layer thickness causes an increase of the merit function. But the following needle optimization procedure entirely compensates this increase and results in a further decrease of the merit function value.

Synthesis by gradual evolution is easily automated. It has been implemented in commercial software codes (Tikhonravov and Trubetskov 2002 - 2). Results presented in Fig. 6 and Table 1 were obtained by gradual evolution synthesis (see Table 1 for the design thicknesses and corresponding merit function values). It should be noted that in the process of gradual evolution the total number of design layers is usually increased. A relatively small number of design layers in the last line of Table 1 is the result of some additional operations described in Sect. 3.3.

3 Practical Aspects of Optical Coating Design

The goal of this section is to provide guidelines for more effective and less time consuming design. How to improve the feasibility properties of a theoretical design is also discussed. The design of a narrow band reflector is used to illustrate the content of this section.

3.1 Target Specification

Correct specification of design targets, in the first place correct choice of the wavelength grid where design targets are specified, is essential for the effectiveness of the design procedure. The choice of a too sparse wavelength grid may slow down or entirely destroy the convergence of optimization routines. It is also possible that a sparse wavelength grid would not allow reproduction of essential features of the spectral characteristics of theoretical designs. On the other hand, too dense wavelength grids also slow down the design procedure and may provoke too strict demands on the quality of target approximation, especially in the spectral regions of abrupt target variations.

As shown in Sect. 2.5, with increasing total optical thickness of the design its spectral characteristics oscillate more rapidly. Thus the density of the wavelength grid ought to be dependent on the design total optical thickness because more dense grids are required to reliably reproduce spectral characteristics of thicker designs. The following formula is useful for estimating the number of wavelength grid points:

$$L = 20\frac{x_a\left(\lambda_u - \lambda_l\right)}{\lambda_u \lambda_l}. \tag{13}$$

Here L is the number of wavelength grid points, x_a is the total optical thickness of the design and λ_l and λ_u are the lower and upper bounds of the wavelength region of interest.

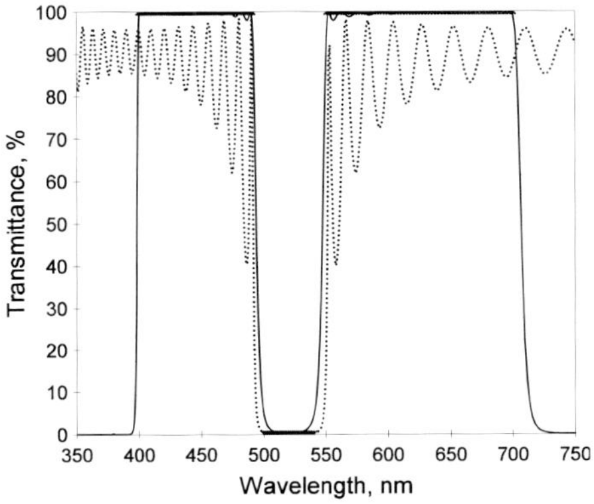

Fig. 7. Design of a narrow band reflector: dashed line - transmittance of the starting design (see Sect. 3.2 for its choice), solid line - transmittance of the final design.

Consider the choice of wavelength grid for the design of a narrow band reflector with the high-reflection region from 500 *nm* to 540 *nm* and with a high transmittance in the remaining parts of the spectral region from 400 *nm* to 700 *nm*. Let us estimate the total optical thickness of the sought for design as 8 μm. The basis for this estimation is provided in the next section. Equation (13) gives

$$L = 20\frac{8000 \cdot 300}{700 \cdot 400} \approx 170. \tag{14}$$

If a spectral region of interest is narrow, then homogeneous wavelength grids are appropriate. In the case of wide spectral regions it is better to use inhomogeneous wavelength grids where the density grows with decreasing wavelength. From a theoretical point of view the most appropriate choice is a wavelength grid with equidistant points in the inverse wavelength scale (wavenumber scale). Such a grid is used to specify the design targets in Fig. 7.

When target spectral characteristics have abrupt variations, special precautions should be taken in the spectral regions of these variations. Otherwise, the discrepancy between target and actual characteristics in these regions will dominate in the merit function value and it will be difficult to achieve good target approximation in other spectral regions.

A simple and effective decision is just to omit several grid points in the areas of abrupt variations. This is done in Fig. 7 where 5 wavelength grid points are omitted in the region from 490 *nm* to 500 *nm* and another 5 points are omitted in the region from 540 *nm* to 550 *nm*.

Another possible decision is to increase target tolerances in the vicinities of abrupt variations. The relative contribution of these vicinities to the merit function value will be decreased.

Tolerances are also useful to control the level of approximation of target characteristics in various spectral regions. It may be worthwhile to play with tolerances as the design process proceeds if it becomes evident that the approximation accuracy is not sufficient in some essential spectral regions.

3.2 Starting Design Choice

The needle optimization technique does not require special starting designs. To achieve the desired accuracy of target function approximation one can vary the total optical thickness of a starting design or use the process of gradual evolution. Nevertheless, in some cases the choice of a proper starting design may facilitate solution of the design problem.

Undoubtedly, the designer's personal experience is of utmost importantance for selection of a starting design because only a few general rules can be recommended. Those rules concern application of quarter wave mirrors and their modifications as starting designs.

Quarter-wave mirrors may serve as good starting designs for edge filters and band pass filters of various types. In Sect. 1.5 of the book (Furman and Tikhonravov 1992) it is shown that variations of layer thicknesses have little influence on spectral characteristics in the high reflection regions compared to their influence in the other spectral regions. Due to this fact varying layer thicknesses can lead to an increase of transmittance beyond the mirror high reflection zones without noticeable disturbance of the reflective properties. Design of the edge filter in Sect. 2.2 (see Fig. 2) serves as the corresponding example.

Quarter-wave mirrors have been thoroughly studied and there are several formulas that are particularly important for proper choice of a starting design.

The lower λ_l and upper λ_u boundaries of high reflection zones of a quarter-wave mirror at normal light incidence are defined by the equations (Furman and Tikhonravov 1992)

$$\lambda_l = \frac{\lambda_0}{2m+1}\left[1+\frac{2}{\pi(2m+1)}\arcsin\frac{n_H-n_L}{n_H+n_L}\right]^{-1} \quad \text{and} \tag{15}$$

$$\lambda_u = \frac{\lambda_0}{2m+1}\left[1-\frac{2}{\pi(2m+1)}\arcsin\frac{n_H-n_L}{n_H+n_L}\right]^{-1}, \tag{16}$$

where λ_0 is the central wavelength of the main high reflection zone, n_H and n_L are the refractive indices of high and low index materials and m is an integer.

Formulas for the main high reflection zone centered at λ_0 are obtained when *m=0*. Other *m* values correspond to more narrow high reflection zones centered at $\lambda_0/3$, $\lambda_0/5$, … .

The starting design for the refinement procedure illustrated by Fig. 2 in Sect. 2.2 was chosen according to the above formulas. Assuming $\lambda_l = 550nm$ as a low wavelength bound of the main high reflection zone we get from (15) $\lambda_0 \approx 640nm$.

Consider the choice of a starting design for the narrow band reflector for which the target spectral characteristic was specified in the previous section. As before materials with n_H=2.35 and n_L=1.45 are used for the design. Equations (15) and (16) enable estimating the width of the second high reflection zone of a quarter-wave mirror with central wavelength $\lambda_0 \approx 1560nm$. The width of this zone, which corresponds to m=1 in Eqs. (15) and (16), is close to 60 *nm* and it is centered at 520 *nm*. We shall use the 19-layer quarter wave mirror with $\lambda_0 \approx 1560nm$ as a starting design for the needle optimization procedure. The starting design transmittance is the dashed line in Fig. 7.

The transmittance of the final design obtained by the needle optimization procedure is the solid line in Fig. 7. The structure of the final design is discussed in the next section.

The number of layers of a starting quarter-wave mirror mainly affects the steepness of the transmittance slope between the stopband and the passband regions. As shown in Sect. 2.5 thicker designs feature more rapid variations of spectral characteristics. Thus one should increase the number of starting design layers if higher steepness of transmittance slopes is required. This is connected with the well established fact that the design total optical thickness is a rather conservative parameter that does not vary considerably in the course of the needle optimization procedure (Tikhonravov et al. 1996).

The total optical thickness of the starting 19-layer design is 7.41 μm. The slightly higher value of 8 μm was used for the estimation of the required number of spectral grid points (see Eq. (14)).

It may happen that the target high reflection zone is wider than that of a quarter-wave mirror. To set starting designs in such situations it is worthwhile to use combinations of several quarter wave mirrors with shifted central wavelength values or to use mirror-like structures with successively increasing or decreasing layer thicknesses.

3.3 Meeting Feasibility Demands

Feasibility demands on optical coating designs depend on the specific manufacturing and monitoring environment. At the same time there are some general requirements that nearly all optical coating designers encounter. As a rule, thin coating layers with physical thicknesses less than 10 *nm* are difficult to manufacture (Thelen 1995) and it is desirable to exclude such layers from the final design. In general it is also desirable to minimize the total number of design layers. There is a simple procedure allowing accounting for both reguirements. We use the design of a narrow band reflector to illustrate this procedure.

Needle optimization gives a 71-layer final design with spectral transmittance given by the solid line in Fig. 7. The achieved merit function value is equal to 0.841 (the merit function is analogous to that in Eq. (10) with all tolerances set to 1 %).

It turns out that the final design has several thin layers with physical thicknesses less that 10 *nm*. The following procedure enables excluding thin layers without essential impact on the quality of target approximation. Consider the design formula of the obtained 71-layer coating. The beginning of this formula is 1.876*H* 0.750*L* 0.110*H* 0.939*L* 0.951*H* ... with the reference wavelengh 550 *nm*. Here 0.110*H* corresponds to the third layer having high refractive index and physical thickness equal to 6.46 *nm*. Let us substitute 0.110*H* for 0.110*L*. This operation excludes the thin high index layer and forms a new low index layer from the former layers 2, 3, and 4. This new layer has optical thickness 1.799*L*.

Table 2. Parameters of the initial and modified narrow band reflector designs (see the text for details).

Number of design Layers	Merit function value	Thin layers (thinner than 10 nm)
71	0.841	yes
61	0.854	no
51	0.877	no

The above operation decreases the number of design layers by two and increases the merit function value. Note that the design total optical thickness is kept unchanged. Due to this fact the refinement of the modified design compensates for the merit function increase almost entirely.

The same operations with subsequent refinement of modified designs are applied to all thin layers. As a result a 61-layer design with the merit function value of 0.854 is obtained. The thinest layer of this design is 14.43 *nm* thick. We see that the described procedure of thin layer removal affects the merit function but insignificantly. We have only a 1.5 % increase of the merit function value while all thin layers have been excluded and the total number of design layers has been decreased by 10. The spectral transmittance of the new 61-layer design is not distinguishable from that of the former 71-layer design.

In many cases it is possible to noticebly decrease the total number of design layers without much impact on the merit function value even if there are no thin layers. The same procedure as before is useful for this purpose. If we allow a merit function increase of not more than 5 % from the initial value of 0.841, then the number of design layers can be futher reduced to 51. The merit function value corresponding to this design is 0.877 and it has practically the same transmittance as the initial 71-layer design. Table 2 summarises the parameters related to the obtained results.

3.4 Material Selection

The choice of materials for optical coatings is mainly determined by various manufacturing and operating requirements. These requirements are discussed in Chapter "Film Deposition Methods" of this book by H. K. Pulker. In this section we only consider questions related to the design process itself.

Suppose that it is possible to use a set of materials with different refractive indices. A theoretical insight to proper selection of materials is provided by the maximum principle in thin film optics (Tikhonravov 1985 - 2; Tikhonravov 1993). For normal light incidence and non-absorbing layer materials there are two fundamental conclusions following from this principle.

1. *The minimum merit function value is provided by the design with two alternating refractive indices, i.e. by the design comprising just two materials.*
2. *If more than two materials are available, the best results are achieved with the pair of materials having the lowest and the highest refractive indices.*

These conclusions are valid for all types of optical coatings: antireflection coatings, beam splitters, edge filters, high reflectors, etc. The first conclusion means that from a theoretical point of view two-component multilayers enable achieving better results than any rugate structure. Practical examples illustrating this conclusion can be found in (Tikhonravov et al. 1996).

As soon as one works with absorbing layer materials the above conclusions are not valid in general. At the same time, they are usually still true for dielectric materials with small extinction coefficients (about 10^{-4} or less).

To highlight the difference between oblique and normal light incidence and also the difference between the *s* and *p* polarized cases, it is worthwhile to consider effective refractive indices, which are introduced by the equations (Furman and Tikhonravov 1992)

$$n^s = \sqrt{n^2 - n_a^2 \sin^2\theta_a}, \qquad n^p = n^2 / \sqrt{n^2 - n_a^2 \sin^2\theta_a}, \tag{17}$$

where θ_a is the angle of light incidence.

Effective refractive indices in the *s* case are always less than the corresponding *n* values, whereas effective refractive indices in the *p* case are always more than the *n* values. Essential parameters, however, are not the absolute values of the effective refractive indices but the correlation factors between the effective indices of layers and external media. This is a consequence of the fact that all spectral characteristics are invariant with respect to simultaneous multiplication of all indices by a constant value (Thelen 1989).

The concept of effective refractive indices allows one to reduce the design problem for the *s* or *p* polarized light to the same form as in the case of normal light incidence (see, for example, Furman and Tikhonravov (1992)). So the above conclusions are valid for the effective refractive indices when the *s* and *p* cases are considered separately.

In the *s* case the dependence of n^s on n is always monotonic and the minimum and maximum effective indices correspond to the minimum and maximum values of *n*. Thus, in the *s* case it is also better to use materials with the lowest and highest refractive indices. In the p case the dependence of n^p on n is not monotonic. For high incidence angles the minimum n^p value may correspond to the intermediate refractive indices from the available set of indices. In this case using materials with intermediate refractive index values will provide better design results.

The most complicated design problems at oblique light incidence are those with targets specified simultaneously for both s and p polarizations, especially if these targets are specified at several large incidence angles. In this case using intermediate refractive indices turns out to be quite necessary.

4 Optical Coatings in Telecommunications: Specific Design Approaches

Successful manufacturing of WDM filters is only possible due to the error self-compensation provided by turning point optical monitoring (Macleod 1972). This type of monitoring requires that the optical thicknesses of filter layers are multiples of a quarter-wave at the filter central wavelength. Thus layer thicknesses of WDM filters cannot be varied continuously and the general design approaches considered in Sect. 2 are not well suited for WDM filter design. This section presents effective techniques elaborated specifically for the design of telecommunication filters.

4.1 Effect of Error Self-Compensation

The effect of thickness error self-compensation plays a key role in the successful manufacturing of WDM filters. This effect is provided by turning point optical monitoring, i.e. by the monitoring of transmittance extrema at the filter central wavelength (Macleod 1972). For successful design and manufacture of WDM filters it is essential to understand the main reason for filter property degradation and the essence of the error self–compensation mechanism.

From a physical point of view any WDM multi-cavity filter is a typical resonance structure (see Fig. 8) with spectral transmittance defined by the reflectivities of filter mirrors and the phase properties of filter cavities. Errors in thicknesses of filter layers affect mirror reflectivities and the phase properties of cavities in different ways. It is a well-known fact that thickness errors have a small impact on the intensity reflection coefficient of a quarter-wave mirror in the vicinity of the mirror central wavelength (Furman and Tikhonravov 1992, p 85). At the same time thickness errors in the outermost mirror layers and thickness errors in spacer layers essentially affect the phase properties of filter cavities. It is therefore understandable that the main reason for filter property degradation is the disturbance of the phase properties of its cavities.

Consider one of the spacer layers and suppose that its physical thickness is d and its refractive index is n. Ideally, the optical thickness of this layer nd should be equal to $s\lambda_0/2$, where λ_0 is the filter central wavelength and s is a spacer order (integer value). Denote phase shifts on reflection from the left and right mirrors by φ_r and φ_l, respectively. The filter possesses the required spectral properties if the condition

$$\varphi_l(\lambda_0)+\varphi_r(\lambda_0)-4\pi nd/\lambda_0=0 \quad (\mathrm{mod}\, 2\pi) \tag{18}$$

is maintained for all its cavities at the central wavelength λ_0.

For a theoretical WDM filter design $nd=s\lambda_0/2$, $\varphi_l(\lambda_0)$ and $\varphi_r(\lambda_0)$ are both equal to *0* or π and condition (18) is fulfilled. Errors in the thicknesses of mirror layers and spacer layers disturb this condition and, as a sequence, degrade filter properties.

The degradation of filter properties is caused by the combined effect in several filter layers. Individual thickness errors are compensated by errors in other layer thicknesses as soon as the condition (18) is maintained. In principle, the self-compensation effect can be achieved due to the variation of the thickness of a layer that is deposited several steps later after a layer with noticeable thickness error or due to the variations of thicknesses of several subsequent layers (Tikhonravov and Trubetskov 2002 - 1).

Thickness variations in outer mirror layers have higher impacts on the phase shift on reflection than thickness variations in those layers that are far from the mirror interface. It is therefore understandable that thickness errors in internal mirror layers are not as dangerous for WDM filter properties as thickness errors in spacer layers and nearby mirror layers. This conclusion is entirely confirmed by the direct estimation of WDM filter sensitivity to errors in thicknesses of individual filter layers (Tikhonravov and Trubetskov 2002 -1).

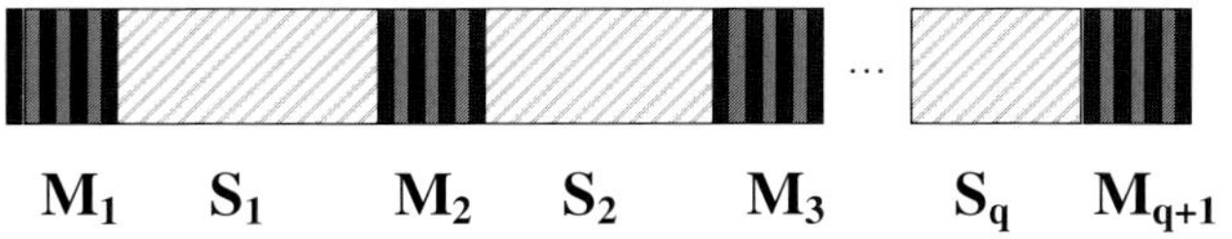

Fig. 8. General schematic of WDM filter: M_1, M_2, ... are multilayer mirrors: S_1, S_2, ... are spacer layers forming filter cavities.

For better understanding of the effects connected with thickness errors it is necessary to stress that the self-compensation mechanism provided by turning point optical monitoring is nothing pther than the mechanism of maintenance of the phase properties of filter cavities (Tikhonravov and Trubetskov 2002 -1). It is therefore clear that the most important role is played by the effects of error self-compensation in several mirror layers deposited just after the deposition of each spacer layer. Indeed, the deposition of these layers starts the final stage of cavity formation. Thickness variations in the first layers of the right mirror are able to compensate for errors in the left mirror and in the spacer layer. With increasingg number of deposited layers of the right mirror compensation for previous errors becomes more and more difficult because of a small impact of the corresponding thickness variations on the resulting spectral transmittance.

It is also clear that the self-compensation mechanism has a local character, i.e., it works independently for each filter cavity. If the phase properties of one of the filter cavities are not properly maintained during the formation of this cavity nothing can be improved later when layers corresponding to other cavities are deposited.

4.2 Automatic Design of Telecommunication Filters

The application of turning point optical monitoring requires that the optical thicknesses of WDM filter layers are multiples of a quarter-wave at the filter central wavelength. For this reason the design of WDM filters is a specific optimization problem where specific design approaches are required rather than general approaches like those based on the needle optimization technique.

The number of layers of WDM filters may readily exceed a hundred. However, this does not mean that the number of design parameters is that big. In fact the number of essential design parameters is much smaller. As mentioned above any WDM filter is a typical resonance structure (see Fig. 8) with spectral transmittance entirely determined by the number of the filter cavities, the reflectances of the filter mirrors and the phase properties of the filter cavities. When designing WDM filters one actually varies these parameters.

The reflectance of a quarter-wave filter mirror is determined by the number of layers of this mirror. This means that its value cannot be varied continuously, but takes only specific values defined by the integer parameter, namely the number of mirror layers m. The phase properties of a filter cavity are defined by the optical thickness of the corresponding spacer layer and the phase shifts on reflection from mirrors surrounding this layer. Thus, these properties are also defined by the integer parameter, in this case by the spacer order s.

Let the total number of filter cavities be equal to q. Denote $m_1, \ldots, m_{q+1}$ the numbers of layers in the mirrors, and $s_1, \ldots, s_q$ the orders of spacers. The above discussion shows that $q, m_1, \ldots, m_{q+1}, s_1, \ldots, s_q$ are the main design parameters of a WDM filter. Thus the design of WDM filters is a specific integer optimization problem. In principle, one can try to design WDM filters by direct computation of spectral transmittances for various combinations of the integer parameters q, m_j, s_j (trial and error method). However, the number of possible combinations is so big that a reliable design using this approach is impossible even on modern supercomputers. A key to elaboration of an extremely efficient automated design technique is the combination of ideas from classical design approaches with advanced integer optimization techniques.

Classical design ideas enable construction of a filter prototype that adheres to the spectral requirements of any particular WDM filter. It is a well-known fact that Chebyshev filters provide maximum steepness of the transmittance slope between a passband and a stopband. Due to this fact the theory of Chebyshev filters enables one to determine the minimum number of filter cavities required to achieve a specified steepness of the transmittance slope (Tikhonravov and Trubetskov 2002 - 1).

With the number of cavities determined it is possible to find a filter prototype using Thelen's design approach (Thelen 1966). It is convenient to design a filter prototype and then a filter itself for the case when the incidence medium has the same refractive index as the substrate. Later the filter design can be adjusted to the actual incidence medium by a standard AR coating.

A general schematic of Thelen's prototype has the form

n_s | Eq. Layer 1 | Eq. Layer 2 | … | Eq. Layer 2 | Eq. Layer 1 | n_s

Equivalent layers 1 and 2 are composed by several quarter wave layers with alternating high and low refractive incices. The numbers of quarter wave layers are chosen so that to provide maximum transmittance at the filter central wavelength (Thelen 1966). In order to obtain the required number of filter cavities the equivalent layer 2 is repeated *q - 1* times where *q* is the estimated number of cavities.

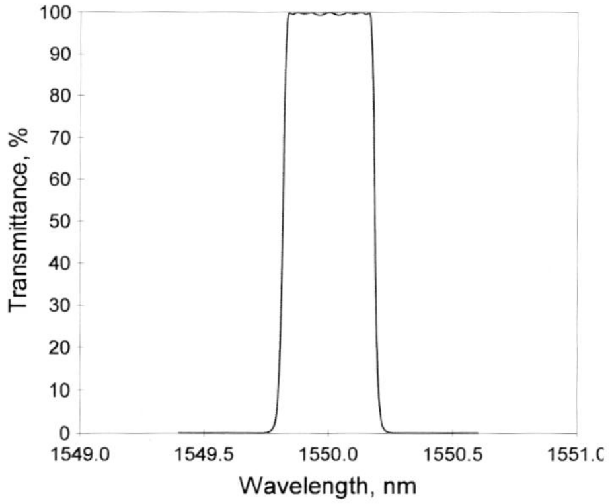

Fig. 9. Transmittances of two different WDM filter designs (transmittances are very close and almost indistinguishable in the figure) with the following specifications: λ_0 = 1550 *nm*, 0.35 *nm* bandwidth at -0.5 dB transmittance level, 0.5 nm bandwidth at -30 *dB* transmittance level.

It is always possible to specify a set of filter prototypes with nearly equivalent spectral properties. This is because there are two ways to control the width of the filter passband. The first one is to change the number of layers of the filter mirrors and the second one is to change the orders of the filter spacers. Increasing the number of mirror layers or increasing the spacer order narrows the passband region while decreasing these parameters broadens it. Due to this fact starting with the Thelen prototype it is possible to find several nearly equivalent filter prototypes that have smaller numbers of mirror layers but higher spacer orders.

As soon as one of the filter prototypes is chosen it is possible to start integer optimization using this prototype as a starting design. Rapid convergence of the optimization procedure is assured by modern discrete optimization methods (Tikhonravov and Trubetskov 2002 - 1). In this way immersed WDM filter designs are obtained. They are adjusted to air incident medium by AR coating at the top of the immersed design.

The described approach turns out to be extremely efficient from a computational point of view (Tikhonravov and Trubetskov 2002 - 1). It enables the designer to obtain a set of structurally different filter designs with nearly equivalent spectral properties. Fig. 9 presents transmittances of two 7-cavity WDM designs obtained for the following design specifications: λ_0 = 1550 *nm*, 0.35 *nm* bandwidth at – 0.5 dB transmittance level, 0.5 nm bandwidth at -30 dB transmittance level. The layer materials have refractive indices 2.065 and 1.465, the substrate has refractive index 1.52 and the incident medium is air. The designs are essentially different: the first one has 272 layers while the second one has 256 layers and thicker spacer layers than the first design. It is worthwhile to note that it took only a few minutes to obtain both designs.

The ability to easily obtain a set of filter designs with nearly equivalent spectral properties provides the designer with an additional opportunity to choose one that is most manufacturable. Pre-production sensitivity analysis is useful for this purpose (Tikhonravov and Trubetskov 2002 - 1).

5 Summary and Conclusions

Modern design techniques allow one to design optical coatings with quite complicated spectral characteristics. The main problem now is not to obtain a design with the required spectral properties but to find the one which is most manufacturable. In this respect the recognition of the key role of design total optical thickness is extremely important. Varying this parameter it is possible to obtain a set of solutions to the design problem and to select the one which is optimal from a practical point of view.

Further development of design techniques requires concentrating attention on various feasibility issues. Specific design approaches are required for some types of optical coatings because of specific demands on their parameters connected with coating manufacturability. An extremely effective technique of this type has been elaborated for the design of WDM filters. Specific design approaches for coatings for ultrashort pulse laser applications are considered in Chapter "Optical Interference Coatings for Ultrafast Optics" by G. Tempea.

There is continually increasing demand for a general design technique that will allow one only to construct coatings with desired spectral characteristics but also to achieve the minimum design sensitivity to manufacturing errors. The development of such a technique will raise the production yield of complicated coatings required for modern high-technology applications.

References

Abeles F (1950) Recherches sur la propagation des ondes electromagnetique sinusoidales dans les miliex stratifies. Application aux couches minces. Ann de Physique 5:596-640, 706-782

Baumeister P (1958) Design of multilayer filters by successive approximations. J Opt Soc Am 48:955-958

Baumeister P (2000) Optical coatings technology. Book used in a short course Engineering 823.17. U of California, Los Angeles

Delano E, Pegis RJ (1969) Methods of synthesis for dielectric multilayer filters. In: Wolf E (ed) Progress in Optics, vol. 7. North-Holland, Amsterdam, pp 67-137

Dobrowolski JA (1965) Completely automatic synthesis of optical thin film systems. Appl Opt 4:937-946

Dobrowolski JA, Lowe D (1978) Optical thin film synthesis program based on the use of Fourier transforms. Appl Opt 17:3039-3050

Dobrowolski JA (1994) Optical properties of films and coatings. In: Bass M (ed) Handbook of Optics, 2-nd ed, McGraw-Hill, New York

Furman Sh, Tikhonravov AT (1992) Basics of optics of multilayer systems. Editions Frontieres, Gif-sur-Yvette

Macleod HA (1972) Turning value monitoring of narrow-band all-dielectric thin film optical filters. Opt acta 19:1-28

Macleod HA (1986) Thin film optical filters. McGraw-Hill, New York

Press WH, Flannery BT, Teukoksky SA, Vetterling WT (1986) Numerical recipes. Cambridge U Press, Cambridge

Sossi L (1974) On the theory of synthesis of multilayer dielectric light filter (in Estonian). Eesti NSV Tead Akad Toim Fuus Mat 23:229-237

Thelen A (1966) Equivalent layers in multilayer filters. J Opt Soc Am 56:1533-1538

Thelen A (1989) Design of Optical Interference Coatings. McGraw-Hill, New York

Thelen A (1995) Design of a hot mirror - contest results. In: Optical Interference Coatings, Optical Society of America, Washington, D.C. pp 2-10

Tikhonov AN, Tikhonravov AV, Trubetskov MK (1993) Second order optimization methods in the synthesis of multilayer coatings. Comp Math Math Phys 33:1339-1352

Tikhonravov AV (1982) The accuracy obtainable in principle when solving synthesis problem. USSR Comp Math Math Phys 22:143-157

Tikhonravov AV (1982) Synthesis of optical coatings using optimality conditions (in Russian). Vestnik MGU, physics and astronomy series 23:91-93

Tikhonravov AV (1985) Mathematical methods for the analysis and synthesis of layered media (in Russian). Doctor of Sci thesis, Moscow University

Tikhonravov AV (1985) Optimal control problems related to the synthesis of stratified media. Diff Equat 21:1025-1031

Tikhonravov AV, Trubetskov MK (1992) Thin film coating design using second order optimization methods. In: SPIE Proceedings 1782:156-164

Tikhonravov AV (1993) Some theoretical aspects of thin film optics and their applications. Appl Opt 32:5417-5426

Tikhonravov AV, Trubetskov MK, DeBell GW (1996) Application of the needle optimization technique to the design of optical coatings. Appl Opt 35:5493-5508

Tikhonravov AV, Trubetskov MK (2002) Automated design and sensitivity analysis of WDM filters. Appl Opt 41:3176-3182

Tikhonravov AV, Trubetskov MK (2002) OptiLayer design software, http://www.optilayer.com

Verly PG, Dobrowolski JA, Wild WJ, Burton RL (1989) sinthesis of high rejection filters with the Fourier transform method. Appl Opt 28:2864-2875

Verly PG, Dobrowolski JA, Willey RR (1992) Fourier transform method for the design of wideband antireflection coatings. Appl Opt 31:3836-3846

Abbreviations

n_s	- substrate refractive index
n_a	- ambient refractive index
n_H	- refractive index of high index material
n_L	- refractive index of low index material
n_j	- refractive index of the *j*-th layer from the substrate
n^s	- effective refractive index in the *s*-case
n^p	- effective refractive index in the *p*-case
d_j	- physical thickness of the j-th layer from the substrate
φ_j	- phase thickness of the j-th layer from the substrate
M_j	- characteristic matrix of the j-th layer from the substrate
M	- coating characteristic matrix
m_{ik}	- elements of coating characteristic matrix
r	- amplitude reflection coefficient
t	- amplitude transmission coefficient
R	- intensity reflection coefficient (reflectance)
T	- intensity transmission coefficient (transmittance)
λ	- wavelength
λ_0	- central wavelength of quarter wave mirror or filter, reference wavelength for design formula
L	- total number of spectral points
X	- vector of design parameters
F	- merit function
P	- perturbation function of the needle optimization procedure
x_a	- design total optical thickness
θ_a	- angle of light incidence
φ_l, φ_r	- phase shifts on reflection for the left and right mirrors of WDM filter cavity.
q	- total number of cavities (spacer layers) of WDM filter
s	- order of spacer layer of WDM filter

Coating Materials

Martin Friz[1] and Friedrich Waibel[2]

[1]Merck KGaA, D-64579 Gernsheim, Germany
[2]Umicore Materials AG, FL-9496 Balzers, Liechtenstein

1 Coating Materials for Different Deposition Techniques

A large number of different deposition techniques are used for the production of thin films for optical applications, as outlined in chapter 'Thin Film Deposition Techniques' by H. K. Pulker. The two most important categories are physical vapour deposition (PVD), namely thermal vaporisation and sputtering, and chemical vapour deposition (CVD). It is obvious, that for each deposition technique suitable coating materials are required.

The PVD processes normally use inorganic elements or compounds and gases, whereas the CVD processes, dip coating and spinning, use liquid inorganic and organic compounds and gases.

Liquid compounds and gases are normally purchased directly from the producer, because they need no special preparation.

Solid materials have to be compact and in the appropriate form or shape, free of gas inclusions or even be prepared according to a special recipe. Targets must also fulfil structure requirements (grain size, texture, precipitation). These operations are the task of companies specialised in the production of coating materials and targets.

This chapter focuses on solid coating and sputtering materials. The requirements on these materials are discussed, their properties listed and their production described.

Earlier summary articles or chapters in monographs on coating materials have been written by H. A. Macleod (Macleod 1986) , H.K. Pulker (Pulker 1979, 1999) and E. Ritter (Ritter 1975).

2 Requirements for Coating Materials

2.1 Evaporation Materials

Apart from the elements only a relatively small number of inorganic compounds can be evaporated to produce a coating with the same composition as the starting material. Suitable compounds are found mostly among the chalcogenides and halides. In some cases it is necessary to change the evaporation process (e.g. reactive evaporation, evaporation from several sources or flash evaporation) in order to ensure that the resultant coating has the desired stoichiometry.

Evaporation materials have to fulfil a series of requirements in order to meet the demands of the coating process and to achieve the required film properties.

2.1.1 Chemical Purity

The chemical purity of vapour deposition materials influences not only the coating properties but also the way the material behaves during evaporation. As will be shown, a purity of least 99.99 %, normal in metallurgy and often specified by optical coating companies, is indeed a step in the right direction. However, it does not indicate whether the material will be suitable, nor how the material compares to materials of a different origin.

Even minute concentrations of transition elements can have a marked effect on the transmission properties of dielectric layers. Unfortunately, there are only few publications that address this topic. Nowhere is there a description, for instance, of the relationships between levels of contaminants and optical absorption in thin films. The most detailed articles containing quantitative accounts of impurity-specific losses can be found in the literature on optical waveguides made of silicon dioxide (SiO_2) (Newns et al. 1973; Schultz 1974). Particularly troublesome in the visible range are minute quantities of cobalt (Co), chromium (Cr), copper (Cu), iron (Fe), nickel (Ni) and vanadium (V). Cobalt, chromium and vanadium exhibit the greatest molar absorption or extinction coefficient k in the order of 100 litres·mol^{-1}·cm^{-1}. Using Beer's law of absorption

$$A = 1 - e^{-k \cdot c \cdot d}$$

enables an absorptance A of 10^{-3} % to be estimated for a layer thickness d = 100 nm and an impurity level c = 100 ppm (equivalent to about 0.003 mol/litre). Compared to this value, other causes of optical losses in the layer, e.g. scattering or absorption caused by variations in stoichiometry, often have a much greater effect. This means the aforementioned purity of 99.99 % is almost invariably adequate. For many important applications (e.g. AR coatings for spectacles and photographic lenses, filters, cold-light mirrors) levels of impurities even totalling 1000 ppm may be tolerated.

Prerequisite is, of course, that the strongly absorbing impurities mentioned are actually detected during chemical analysis of the evaporation material. It is therefore necessary to define limits for the interfering impurities – usually the elements listed above. The methods of analysis most widely used nowadays for detecting impurities are flame atomic absorption spectrometry (FAAS) and optical emission spectrometry with plasma excitation (ICP-OES). For substances with light cations, e.g. for magnesium fluoride (MgF_2), aluminium oxide (Al_2O_3), silicon monoxide (SiO) and silicon dioxide (SiO_2), x-ray fluorescence may also be used with excellent limits of detection for heavier elements.

Chemical analysis must also be able to detect foreign substances, which, for instance, might have a detrimental effect on deposition behaviour or the structure of the coating. A well known effect is that of a high oxide content in fluorides, most noticeable perhaps in MgF_2, as it can result in splashing during evaporation. Many oxides have lower vapour pressure than the corresponding fluorides, and form an oxide layer over the melt during evaporation. This causes the melt to overheat and the oxide layer to tear, the actual cause of splashing. Determination of non-evaporable residue is one possible way of analysing the oxide content. Another option is inert gas extraction.

Evaporation behaviour and structure of the layer may be influenced by high residual gas content in the evaporation material. In such cases it is best to determine the residual gas content by heating the substance in a high-vacuum recipient while monitoring pressure versus time. A mass spectrometer should be used to test the liberated gases, to be able to find the precise cause.

2.1.2 Physical Properties

Testing of dimensions, particle sizes and weights is a vital part of quality control. For tablets it is quite easy to deduce their density, which in the case of substances with poor melting properties, e.g. zirconium oxide (ZrO_2), is a critical measure of the stability of the tablet. Extreme differences in temperature within a tablet during electron beam evaporation lead to significant heat stress. In such cases cracks in the tablet can not be completely avoided, and disintegration can only be prevented by keeping the porosity of the tablet sufficiently high; yet the porosity must not be so high as to make it impossible to maintain a low adsorbed gas content.

Some substances, e.g. some titanium suboxides, are relatively brittle and produce large amounts of abraded particles during storage and transport. The problem can be lessened by optimising the packaging with padding or vacuum wrapping. An abrasion test in a running wheel is needed to allow various production batches to be assessed and compared.

Various vapour deposition materials with favourable properties are prepared from substance mixtures, whereby new compounds are formed by solid state reactions during the production process. The melting and evaporation properties of these materials depend to a very large extent on the completeness of the chemical reaction. X-ray diffraction (powder diffraction) is used to check the phase purity of the end product. Also, differences in the stoichiometry of a compound may negatively affect evaporation characteristics and coating properties. A good example of this are the various titanium sub-oxides, where determination of stoichiometry may well be an appropriate part of quality control.

2.1.3 Process Suitability

Decades of experience in the manufacture and use of evaporation materials have shown that factors affecting process suitability are so complex that they do not allow quality to be defined simply by physical and chemical properties. Therefore, substance testing under actual conditions of use continues to be a major feature of quality control. The assessment criteria differ from substance to substance. Common criteria are, however, the way the material behaves during the heating or melting phase, the pressure profile for the system and the heating power needed for a given rate of evaporation. In the case of oxides, in particular during reactive evaporation, it is also necessary to record the evaporation data (pressure and rate fluctuations) over a prolonged period. In the case of mixed oxides, the reflectance of the coating is measured as it builds up, and the refractive index determined as an additional control for correct stoichiometry. In certain cases, such as when hafnium oxide (HfO_2) is used in UV applications, spectrophotometric measurements are taken to confirm the suitability of the material for this spectral range.

2.2 Sputtering Targets

Sputtering of metallic layers is straightforward. For the sputtering of dielectric layers there are two different approaches. The first is to use ceramic target materials of the desired layer composition. The second is a reactive sputtering process, where a metallic target is sputtered in a reactive atmosphere of oxygen or nitrogen. Most sputtering processes for interference coatings use metallic target materials. In reactive sputtering processes there is not only a reaction on the substrate but also on the surface of the metallic sputter target. A stable sputtering process needs optimised management of reactive gas and cathode power. Most processing problems (such as arcing) result from dielectric layers grown on the target surface.
Ceramic targets are seldom used because of two severe disadvantages. One disadvantage is that most of the ceramic materials are non-conductive and cannot be operated using Direct Current (DC) power. To install RF (Radio Frequency) power is expensive, it induces a low sputter rate and is complicated. The second disadvantage is the high production cost of ceramic target materials.

Ceramic targets are only used where an extremely precise stoichiometry of the dielectric layer is required . The best example is the production of ITO-layers with In_2O_3/SnO_2 targets. The use of conductive TiO_2 targets is also increasing (Weigert 2001).

Materials for sputtering targets have to fulfil specific requirements: The most important characteristics include:

- purity
- chemical homogeneity in case of alloys (stoichiometry)
- homogeneity of the structure (grain size, texture, precipitates, crystallinity)
- target density
- no oxide or other impurity inclusions (metals and alloys)
- no pores and voids

- good electrical conductivity (ceramics and semiconductors)
- no surface contamination
- dimension, target surface roughness
- mechanical stability of the compound target

The overall purity of sputtering targets for interference coating applications is 99.99%. It allows a sum of metallic impurities of 100 ppm. This is not very high compared with microelectronic applications (99.9999%, sum of impurities of 1 ppm in maximum).

Several of the most common analytical techniques to control impurity values and contents of alloying elements are XRF, ICP, GDS, GDMS, and CGHE. (Venzago and Weigert 1994; Wilhartitz et al.1990).

The optimised performance (purity, structure, homogeneity, etc.) for each target, from target-to-target, and from lot-to-lot is very important for good sputter results. The critical elements that influence the absorption levels are Co, Cr, Cu, Fe, Mn, Ni, V, etc.

Studies show a direct correlation between grain size and target performance. A small and homogeneous grain size induces a constant sputter rate over the entire target surface and produces consistent film thickness uniformity. It also affects the control of second phase precipitates (Marx and Murphy 1990). Precipitates larger than 10 µm cause particle generation during the sputtering process.

These quality factors are strongly influenced by the way the material is produced (see Chap. 4).

Nearly all modern sputtering processes are based on magnetically enhanced cathode processes. For the planar magnetron the sputtering target is just a sheet of metal or ceramic. Very often it is produced as a compound target bonded on a cooling or a special backing plate, to produce a mechanically stable system (weak or brittle materials) or to save costs of expensive target materials. Rotating cathodes are increasingly used for large area coatings (tube as target geometry).

The criteria for the choice of technology include target lifetime, target utilisation, material recycling and uptime of the sputtering process or stability of the process during target lifetime.

3 Survey on Materials, Properties and Applications

Properties of thin films including optical, mechanical, electrical, thermal, and chemical properties are influenced by deposition parameters. Therefore refractive indices and transmittance range given in this chapter are only meant as guidance values. Detailed information on the influence of deposition parameters on film properties is given in chapters 'Some Fundamentals of Optical Thin Film Growth' by N. Kaiser and 'Film Deposition Methods' by H.K. Pulker.

3.1 Dielectrics

3.1.1 Fluorides

Ultraviolet Spectral Range

The development of ultraviolet light sources, especially the Excimer laser, has had a significant impact on new applications in material processing and medicine (Stamm et al. 1997). Several fluorides with band gap energies around 10 eV are suitable for such applications in the wavelength range even below 200 nm. Extensive data on refractive index and extinction of fluoride films in the UV range using different coating techniques have been published between 1987 and 1991 (Lingg et al. 1987; Zukic et al. 1990; Kolbe et al. 1991).

Table 1. UV properties of fluoride films

Material	Transmittance Data	Refractive index
Aluminium fluoride (AlF_3)	k ~10^{-4} (λ=200 nm) k ~10^{-3} (λ=200 nm), IAD $T_{cut\text{-}off}$ = 120 nm	n = 1.4 (λ=200 nm) [a] n = 1.5 (λ=200 nm), IBS [a] n = 1.38 (λ=193 nm) [b]
Barium fluoride (BaF_2)	k ~10^{-2} (λ=140-230 nm)	n ~1.7 (λ=160-210 nm) [c]
Calcium fluoride (CaF_2)	k ~10^{-2} (λ≥180 nm)	n < 1.3 (λ≥180 nm) [c]
Cryolite (Na_3AlF_6)	$T_{cut\text{-}off}$ = 120 nm	n = 1.37 (λ=193 nm) [b] n = 1.42 (λ=193 nm) [d]
Dysprosium fluoride (DyF_3)	$T_{cut\text{-}off}$ < 190 nm	n = 1.55 (λ=193 nm) [b]
Gadolinium fluoride (GdF_3)	k ~10^{-3} (λ=200 nm) $T_{cut\text{-}off}$ =123 nm	n = 1.68 (λ=200 nm) [a] n = 1.58 (λ=193 nm) [b]
Lanthanum fluoride (LaF_3)	k ~10^{-3} (λ=200 nm) k ~10^{-3} (λ=200 nm), IAD k <$2.2\cdot10^{-2}$ (λ≥145 nm)	n = 1.7 (λ=193 nm) [a] n = 1.82 (λ=200 nm), IBS [a] n = 1.71 (λ=193 nm) [d] n ≥ 1.85 (λ=135-180 nm) [c]
Magnesium fluoride (MgF_2)	k ~10^{-4} (λ=200 nm) k ~10^{-3} (λ=200 nm), IAD k <10^{-4} (λ=140-230 nm)	n = 1.42 (λ=200 nm) [a] n = 1.46 (λ=200 nm), IBS [a] n ~ 1.5 (λ=140-220 nm) [c]
Neodymium fluoride (NdF_3)	k ~ 10^{-3} (λ=200 nm)	n = 1.75 (λ=200 nm) [b] n = 1.75 (λ=193 nm) [d]
Terbium fluoride (TbF_3)	T = 50% (λ=140 nm, d=100 nm) [e]	
Ytterbium fluoride (YbF_3)	T = 50% (λ=142 nm, d=100 nm) [e]	
Yttrium fluoride (YF_3)	k = 0.0025 (λ=193 nm)	n = 1.61 (λ=193 nm) [f]

[a] Kolbe et al. 1991 [b] Cowell 1996[d] [c] Zukic et al. 1990
[d] Izawa et al. 1994 [e] Lingg et al. 1987 [f] Kolbe et al. 1990

Today research and development in ultraviolet coatings focuses on applications that meet the stringent requirements for optical components used in the new generations of wafer-steppers for the semiconductor industry working at 193 nm (ArF-Excimer) and 157 nm wavelength (F_2-Excimer).

The materials of choice for this application are magnesium fluoride (MgF_2) for the low refractive index layer and lanthanum fluoride (LaF_3) for the high refractive index layer (Bernitzki et al. 1999; Ristau et al. 2001). However, a series of problems still exists before the technological demands of chip production are met:

- small refractive index difference between the two materials used
- optical inhomogeneity of LaF_3 films
- optical losses
- laser induced damage thresholds
- mechanical stress of the thin films

Therefore alternative materials such as aluminium fluoride (AlF_3), cryolite (Na_3AlF_6) and various rare-earth fluorides are being evaluated. The influence of ion-assisted deposition (IAD) techniques on thin film properties is also being researched.

A detailled description on DUV/VUV coatings is given in chapter 'Optical Coatings for the DUV/VUV' by R. Thielsch.

Visible Spectral Range

Magnesium fluoride (MgF_2) is still the dielectric material with the highest consumption. This is based on a combination of material properties like low refractive index and film durability. A large area of applications is the refinement of mineral glass lenses with antireflection coatings for consumer optics (spectacle lenses, camera objectives, and binoculars). However this is by far outnumbered by the manufacture of the so-called optically variable pigment (OVP). This pigment is a multilayer coating made from metals and magnesium fluoride and was originally developed for optical document security. It is now applied to banknotes and valuable documents worldwide and also for a variety of decorative applications (Phillips and Bleikolm 1996; Phillips and Nofi 1999). Other fluorides still being used for applications in the visible range are cerium fluoride (CeF_3) for antireflection coatings and cryolite (Na_3AlF_6) respectively chiolite ($Na_5Al_3F_{14}$) for filters.

Infrared Spectral Range

Thanks to its outstanding mechanical properties for μm film thickness thorium fluoride (ThF_4) has been the material of choice for low index films in the infrared region up to wavelengths around 15 μm (Heitmann and Ritter 1968). ThF_4-films did not show the problem of cracking or peeling-off, encountered with other fluorides. But the main drawback of this material – its radioactivity – caused more and more users to look for a less hazardous replacement and ignoring it in new designs. Therefore, at the end of the 1980s an intense search for alternative materials and processes began for non-radioactive coatings with low losses and high damage thresholds especially for use with the CO_2-laser at 10.6 µm. The most promising candidates are barium fluoride (BaF_2), yttrium fluoride (YF_3) and ytterbium fluoride (YbF_3). It could be shown that films deposited with ion assistance (e.g. BaF_2 or YbF_3) show absorption data and damage thresholds comparable to those obtained from conventionally deposited ThF_4 (Rahe et al. 1990; Schnellbügel et al. 1994).

3.1.2 Oxides, Oxynitrides, Nitrides and Carbides

Film Preparation

Oxides are an important class of coating materials, because they generally form hard, abrasion resistant, chemically and environmentally stable films with a good variety of refractive indices and spectral ranges of high transmission.

Many oxides react at high temperatures with refractory metal boats. Therefore the electron beam evaporation with water-cooled copper crucibles or molybdenum liners is an adequate method for non-activated thermal evaporation.

Since quite a few oxides dissociate at high temperatures, at least to a certain degree, reactive evaporation in an oxygen atmosphere is advisable in most cases, but the level of oxygen has to be adjusted to the actual needs. Reactive evaporation of suboxides is in certain cases, like titanium oxide, a preferred choice.

The metal is used as starting material for the deposition of oxides by reactive arc evaporation and reactive DC-sputtering. If RF sputtering is applied, the oxide can be used as the starting material. For RLVIP the metal or a slightly reduced and hence conducting oxide is the starting material.

In general, all activated evaporation processes, such as IBAD, RLVIP, arc etc. increase and stabilise the refractive index and the transparency of most oxide films, especially the ones with high refractive index.

Nitride, oxynitride and carbide films are normally produced by reactive magnetron sputtering, ion beam sputtering, RLVIP, reactive arc evaporation or by ion-assisted reactive electron beam evaporation.

Important Materials

SiO_2: It can be deposited by electron beam evaporation with IAD (Alvisi et al. 1999) or without IAD (Scherer et al. 1996), by RLVIP (Lorenz et al. 1991), and by ion beam (Tabata et al. 1996), by reactive (Edlou et al. 1993) or RF sputtering. A summary of properties is given by Anderson and Ottermann (Anderson and Ottermann 1997).

SiO: Before the availability of electron beam evaporators for SiO_2 evaporation SiO_x films were prepared by the reactive evaporation of SiO (Ritter 1971). Their composition corresponds to Si_2O_3 and the refractive index is around 1.55. Today SiO is used mostly for multilayer coatings in the infrared range.

Al_2O_3: For optical applications it is mostly deposited by electron beam evaporation, although all the methods mentioned with SiO_2 can be applied (Edlou et al. 1993; Martin et al. 1999).

HfO_2: Used in the UV region and at certain laser wavelengths as a high index material. The films are produced by electron beam evaporation of HfO_2 (Reicher et al. 2000; Andre et al. 2000) or by reactive sputtering. (Edlou et al. 1993). Reactive evaporation of Hf reduces the number of film defects and enhances the laser-induced damage threshold at 1060 nm (Andre et al. 2000).
ZrO_2: Films of ZrO_2 prepared by electron beam evaporation show strong inhomogeneity of the refractive index (Klinger and Carniglia 1985). The refractive index at the film/air interface is smaller by about 0.1 than at the film-substrate interface, called negative inhomogeneity (Bennett et al.1989). This is due to the film structure and the lower packing density at the film-air interface. By using IAD the inhomogeneity changes to positive and under certain conditions homogeneous films can be obtained (Cho and Hwangbo 1996).
Ta_2O_5: Tantalum oxide shows a wide spectral range of high transparency, low losses and high stability. Besides TiO_2 it is probably the most used high index material. It is produced by ion assisted reactive electron beam evaporation (Zoeller et al. 1996), by ion plating, ion beam sputtering (Dods et al. 1999) and reactive DC magnetron (Edlou et al. 1993) or RF sputtering . A summary of properties is given by Bange (Bange 1997).
Nb_2O_5: Niobium oxide exhibits a somewhat higher refractive index than Ta_2O_5. But the UV transparency is not as good as with Ta_2O_5. Films can be produced by RLVIP (Edlinger et al.1989) and by reactive RF magnetron or dual AC magnetron sputtering (Sullivan et al. 2000)
TiO_2: Suboxides are often used for the production of TiO_2 films by evaporation (mostly electron beam), namely TiO, Ti_2O_3 and Ti_3O_5 (The paper by Selhofer et al. 2002 contains a large list of references). The use of TiO2 as starting material in reactive electron beam evaporation is also reported (Chen et al. 1997).
Other deposition methods are reactive arc evaporation (Bendavid et al. 2000), radical beam assisted reactive evaporation (Yamada et al. 1999), RLVIP, ion beam sputtering (Albertinetti and Minden 1996), reactive magnetron sputtering (Edlou et al. 1993) and helicon plasma sputtering (Wang et al. 1999).

A summary of properties is given by Bennett et al. (Bennett et al. 1989) and Anderson et al. (Anderson et al. 1997)
In_2O_3: Indium oxide doped with 5-10% Sn (ITO) is widely used as a transparent conductive film. It is deposited by reactive evaporation (Luff et al. 1997), by ion-assisted reactive evaporation, by DC magnetron sputtering (Lehan 1996), or RF sputtering (Kulkarni et al. 1998).
WO_3: Tungsten oxide films are prepared by thermal evaporation (Hussain 1999) or DC reactive magnetron sputtering. A summary of properties is given by Bange (Bange 1997).
SiON: Produced by dual ion beam sputtering (Brunet-Bruneau et al. 1996) or reactive magnetron sputtering of Si in an atmosphere of oxygen and nitrogen (Pinard and Mackowski 1997).
Si_3N_4: Produced by reactive magnetron sputtering in nitrogen (Martin et al. 1997), RLVIP (Lorenz et al. 1991) or ion-assisted electron beam evaporation of Si (Lee et al. 1999).
SiC: Produced by ion-beam sputtering from a SiC target (Larruquert and Keski-

Kuha 2000) or by ion-assisted electron beam evaporation of SiC.
B_4C: Produced by ion-beam sputtering from a B_4C target (Larruquert and Keski-Kuha 1999)

Table 2 summarises important properties of oxide, oxynitride, nitride and carbide films.

Table 2. Properties of oxide, oxynitride, nitride and carbide films

Coating material	Film composition	Deposition method	Range of transparency µm	Refractive index at 550 nm	Structure and packing density (p)	Mechanical properties
Si	SiO_2	R DC sputtering, RLVIP	0.2-9	1.45-1.5	A, p=0.98	Hard, CS
SiO_2	SiO_2	EB	0.2-9	1.45-1.46	A, p=0.9-0.98	Hard, CS
SiO	SiO	BE	0.7-9	2.0 (0.7 µm)	A	Hard, TS
Al_2O_3	Al_2O_3	EB	0.2-7	1.63 (T=300)	A, p=1	Hard, CS
HfO_2 or Hf	HfO_2	R EB	0.25-12	1.95 (T=250)	A/C	Hard
Y_2O_3	Y_2O_3	EB	0.25-12	1.87 (T=250)	A/C	Hard
ZrO_2	ZrO_2	EB	0.34-12	1.95-2.05 (T=200)	C, p=0.67-0.82	Hard, CS
Ta	Ta_2O_5	R DC sputtering, RLVIP	0.3-10	2.03-2.09	A	Hard, CS
Ta_2O_5	Ta_2O_5	R EB, RLVIP	0.3-10	2.1 (T=250)	A	Hard, CS
Nb	Nb_2O_5	R DC sputtering, RLVIP	0.38-8	2.35 –2.4	A	Hard, CS
Nb_2O_5	Nb_2O_5	R EB, RLVIP	0.38-8	2.2-2.4	A (T <200)	CS
Ti	TiO_2	R DC sputtering, RLVIP	0.4-3	2.2-2.4	A (T <200)	Hard, CS
TiO	TiO_2	R EB	0.4-3	2.2-2.4	A (T <200)	Hard, TS
Ti_2O_3	TiO_2	R EB	0.4-3	2.1-2.4	A (T <200)	Hard, TS
Ti_3O_5	TiO_2	R EB	0.4-3	2.1-2.4	A (T <200)	Hard, TS
In-Sn alloy or oxide	ITO	R DC sputtering	0.4-1.5	1.95-2.0	A (T <150)	
Zn, ZnO	ZnO	R DC sputtering		2.0		Not very stable
Si	SiO_xN_y	R magnetron or IBS	0.4-9	1.5-2.0		Hard
Si	Si_3N_4	R magnetron sputtering or IAD EB	0.25-9	2.0	A	Hard, CS
Si	SiC	IBS or IAD EB	1.5-11	2.95	Polycrystalline	

R reactive, A amorphous, CS compressive stress, EB electron beam evaporation, BE boat evaporation, TS tensile stress, C crystalline, T substrate temperature [°C], IBS ion beam sputtering

Mixtures of Oxides

Some refractive indices cannot be realized with pure evaporation materials. There is a gap between 1.63 (Al_2O_3) and 1.85 (Y_2O_3). There are also no pure oxides with indices of about 2.1, needed for three-layer AR coatings on glass. Some materials have disadvantages: ZrO_2 and HfO_2 do not melt and therefore form uneven surfaces during evaporation, which often causes inhomogeneous thickness distribution and poor thickness reproducibility. Ta_2O_5 normally needs postbaking to form transparent films if conventional electron beam evaporation without ion assistance is used. The grain structure, density, and stress of thin films can also be modified by using mixtures. Table 3 lists some of the commercially available mixtures.

Table 3: List of some commercially available mixtures

Substance	Supplier (Trade name)	Refractive index at ~ 550 nm	Transmittance range [μm]	Evaporation
SiO_2/Al_2O_3	Umicore (Lima) Merck (L5)	1.48-1.50	0.20-7	EB, 1600 °C
Al_2O_3/Pr_6O_{11}	Merck (M1), Umicore (Paso)	1.7-1.9	0.38-9	EB, 2100-2200°C
Al_2O_3/La_2O_3	Merck (M2), Merck (M3)	1.7-1.8	0.21-10	EB, ~2100°C
ZrO_2/Ta_2O_5	Umicore (Pisa)	2.05	0.35-10	EB, 2100°C
ZrO_2/TiO_2	Cerac, Merck (H1), Optron (OH-5)	2.1	0.40-7	EB, 2200-2400°C
ZrO_2/MgO	Cerac	2.1	0.35-10	EB, ~2200°C
ZrO_2/Al_2O_3	Optron (OM-4, OM-6)	1.68-1.75	0.35-10	EB, 1850-2200°C
TiO_2/Pr_6O_{11}	Merck (H2) Umicore (Ida)	2.1	0.40-7	EB, 2200-2400°C
TiO_2/Al_2O_3	Umicore (Dralo)	2.2	0.40-7	EB, 1700-1800°C
TiO_2/La_2O_3	Merck (H4)	2.1-2.2	0.40-7	EB, 2200-2300°C

Description of Mixed Materials in Detail

Electron beam evaporation is recommended for all of the materials described below. Oxygen partial pressure is about $2*10^{-4}$ mbar (except Lima), deposition rate 0.2 to 0.8 nm/s (if not otherwise noted). Substrate temperature 30 to 300°C (if not otherwise noted).

SiO_2/Al_2O_3: A mixture of SiO_2 and Al_2O_3 lowers the stress in SiO_2 films and enhances chemical stability (Selhofer and Müller 1999).

Al_2O_3/Pr_6O_{11} melts completely before evaporation starts. The melt forms an even surface during evaporation. Refilling of fresh material is possible. Durable films on glass and plastic substrates (Friz et al. 1992).

Al_2O_3/La_2O_3 melts partially before evaporation starts. Refilling possible. Durable films on glass and plastic substrates (Merck 1993, Koenig 1994, Friz et al. 1997). Application besides AR coatings are multilayer coatings, beamsplitters, and polarisers.
ZrO_2/Ta_2O_5: In contrast to pure ZrO_2 films this mixture has a homogeneous refractive index. In addition the ZrO_2 facilitates the oxidation of Ta_2O_5 (Wille 1974).
ZrO_2/TiO_2 forms homogeneous films with index 2.1. The material forms only a thin melted layer on top of the tablet or disc. Careful control of e-beam power distribution is needed to maintain an even surface, refilling not recommended. Deposition rate 0.2 to 1 nm/s. Very low absorption in transmittance range, hard and durable films on glass and plastic substrates. Application mainly AR coatings.
TiO_2/Pr_6O_{11} melts before evaporation starts. Even surface during evaporation, refilling of material possible. Boat evaporation also possible. Rate 0.2 to 1 nm/s. Stable index, absorption in NUV below 400 nm, very weak absorption band at about 450 nm. Application for AR-coatings on glass and plastic lenses (Kraus and Rheinberger 1962; Selhofer and Müller 1999).
TiO_2/Al_2O_3: The addition of Al_2O_3 lowers the stress and the refractive index.
TiO_2/La_2O_3 forms a melt before evaporation starts, refilling possible. Stable index, low absorption in transmittance range (Friz et al.1992, 1997; Merck 1993; Niederwald et al.1999). Application for AR coatings on glass and for multilayer coatings, substrate temperature 250 to 300°C. Also for coatings on plastic substrates at low substrate temperatures.
Ta_2O_5/Al_2O_3: Mixtures with different compositions can be used to deposit medium index films with refractive indices in the range of 1.66 to 1.9 and a range of transparency from 0.35-7 μm. Application for AR coatings on glass substrates with different refractive indices (Ganner 1986).
ZrO_2/Al_2O_3: Mixtures with different compositions can be used to deposit medium index films with refractive indices in the range of 1.63 to 1.75. Refractive index of films varies with composition of the mixture but is almost independent from evaporation temperature, evaporation temperature range 1850-2200°C. Application for three-layer AR coatings on glass substrates with different refractive indices (Canon 1976).

Applications of Oxide, Oxynitride, Nitride and Carbide Films

AR-Coatings

Oxides are used as high and medium refractive index films in most AR coatings. These include: titanium oxides, tantalum oxide, zirconium oxide, yttrium oxide, aluminium oxide and proprietary oxide mixtures, such as Pisa, Ida, Dralo, H1, H2, H4, OH-5, Paso, M1, M2, M3, OM-4, OM-6 etc. An alternative to high index oxide films are nitride films, e.g. silicon nitride.

SiO_2, Lima or Substance L5 is used as low index film in AR coatings, where MgF_2 cannot be used because of stability (soft due to low substrate temperature, e.g. on organic eyeglass lenses) or compatibility with the deposition technique used (large area coating by sputtering or dipping).

Multilayer Systems
Oxide films are also widely used in multilayer systems, such as cold light mirrors, heat reflecting filters, colour separators, narrow band interference filters, e.g. DWDM filters, and laser coatings.

Normally SiO_2 is the low index material. TiO_2-SiO_2 multilayers yield broad reflectance bands because of the large difference in refractive index. They are therefore applied for many mass products. TiO_2 is often replaced by tantalum oxide for multilayer systems with very high requirements in respect to accuracy, thermal stability and optical losses. Tantalum oxide is, like hafnium oxide, also a good high index material for multilayer systems used in the ultraviolet spectral region.

Transparent Conductive Coatings
A further important application of oxide films is the use as transparent conductive coatings in displays or in transparent heated windows. Transparent conductive coatings are formed by indium tin oxide (ITO), Al-doped zinc oxide (Gläser 1990) and F-doped tin oxide (Gläser 1990).

Films for Electrochromic Displays
Electrochromic displays use also oxide films such as tungsten or nickel oxide.

3.1.3 Sulfides, Selenides and Tellurides

Sulfides, selenides and tellurides used in vacuum evaporation include Sb_2S_3, Sb_2Se_3, ZnS, ZnSe, ZnTe, CdS, CdSe, CdTe, and PbTe. All materials can be evaporated from resistively heated boats made of Ta or Mo. Indirect heated crucibles made from vitreous carbon or alumina are also recommended. Use of electron beam is recommended in literature especially for ZnS and ZnSe.

All of these materials sublime. Except Sb_2S_3 they all decompose during evaporation, but recombine on the substrate surface. Recombination is controlled by substrate temperature during deposition. Optimum substrate temperature is about 150-180°C.

Absorption of films depends strongly on stoichiometry and impurities. High purity 5N materials are recommended with impurity levels of about 1 ppm. Materials sintered or sublimed in vacuum are most suitable. Film stoichiometry has to be controlled very carefully by control of substrate temperature and deposition rate. Absorption can also be influenced by the oxygen partial pressure (Hawkins et al. 2000, Seeley et al. 1979, 1980, Bin et al. 1997).

Sb_2S_3 and **Sb_2Se_3** are mainly of historical interest. Recently Sb_2S_3 was used for making AR coatings on silver halide optical substrates (Matsushita Electric Ind. Co. Ltd. 1994).

ZnS in the past was a very important high index material because it can be evaporated easily from resistively heated boats. It was used e.g. for interference filters in the visible range (ZnS/Cryolite) and in large quantities for cold light mirrors (ZnS/SiO, ZnS/MgF_2). Since ZnS films are rather soft as compared with oxide films, ZnS was replaced by TiO_2 and other oxides in many applications. Today ZnS is used mainly for IR applications (e.g. AR coatings on Ge), for holographic applications and for decorative coatings. ZnS is also an interesting material for production of electroluminescent thin film devices.

Optimum substrate temperature is about 150-180°C. At this temperature the sticking coefficient at the substrate surface is zero for Zn and S atoms. Only ZnS molecules after recombination stick to the substrate (Goldfinger et al. 1963; Ritter and Hoffmann 1969).
ZnSe is used mainly for IR applications, e.g. for AR coatings on Ge or ZnSe (Rudisill et al. 1974; De et al. 1995).
ZnTe, CdS, CdSe, CdTe. These materials are of interest mainly for electrooptical applications (e.g. CdS, CdTe and CuInSe for solar cells or photoresistive films).
PbTe is used for filters in the 5 to 40 µm region in combination with ZnS. PbTe shows the highest refractive index of all evaporation materials (Hawkins et al. 2000; Seeley et al. 1979, 1980). Also (Ge,Pb)Te can be used (Zhang 1998).

Table 4. Properties of some sulfides, selenides and tellurides

Material	Density [g/cm³]	Refractive index	Transmittance range [µm]	Sublimation temperature [°C]
Sb_2S_3	4.12	3.2 at 550 nm	0.5 - 10	400 – 550
ZnS	3.98	2.3 at 550 nm 2 at 10 µm	0.4 - 15	1000 - 1100
ZnSe	5.42	2.6 at 633 nm 2.2 at 10 µm	0.6 - >15	800 - 1000
ZnTe	6.34	2.8 at 1 µm	0.7 - 20	1100
CdS	4.82	2.5 at 550 nm 2.26 at 10 µm	0.5 - >14	~680
CdSe	5.81	2.5 at 550 nm 2.4 at 10µm	0.7 - 25	600- 700
CdTe	5.85	3.1 at 3 µm 2.6 at 10µm	0.9 - 25	1000 - 1100
PbTe	8.16	5.6 at 5 µm	3.5 - >40	1100 - 1200

Toxicity
All of the materials mentioned are toxic, some are highly toxic, carcinogenic, mutagenic and dangerous for the environment. Therefore special precautions for handling and special disposal regulations have to be observed. It is recommended to carefully avoid any contact with the skin, to avoid inhalation of dust, and to remove gases (H_2S, H_2Se) which might exit the evaporation systems after opening.

3.2 Metals, Alloys, Semiconductors and Carbon

The evaporation and sputtering of metals, even of refractory metals, poses no big problems due to the development of high power e-beam and sputtering equipment. The metal can also be used as starting material in reactive processes, such as RLVIP, arc evaporation, reactive sputtering or IBS, for the production of oxide, nitride or carbide films.

For very pure metal films, attention has to be paid to the influence of the residual gas, since many metals react very easily with oxygen and water vapour and also with nitrogen or hydrocarbons. UHV technology may be necessary in certain cases.

Metal films show predominantly tensile stress. By doping with gases the stress can be lowered.

Silver, aluminum and gold films are soft. Chromium films are much harder, but still do not reach the hardness of oxide films.

Metal films are used as reflectors, as semitransparent beam splitters, as transparent conductive films, as absorbers, and in multilayer reflector stacks in the EUV/X-ray region.

Aluminum is widely used as reflector in the IR, visible and UV region (Méndez et al. 2000).

Protected silver films yield very high reflecting mirrors for the visible and near IR region. Semitransparent silver films can be used as transparent heat reflectors (Lee et al. 1996) and transparent conductive films. Semitransparent gold films serve also as transparent conductive films.

Figures with the reflectance of Ag, Al, and Au films from the UV to the IR can be found on pages 394-395 in Pulker's book (Pulker 1999).

The metals W, Mo, Cr, and Sc serve in combination with Si, C, or B_4C in reflector stacks for the EUV/X-ray region (Chapter 'Multilayer Coatings for EUV/X-ray Mirrors' by S. Yulin).

Flash evaporation and especially sputtering are the best deposition methods for alloys.

The semiconductors Si and Ge can be thermally evaporated (Rafla-Yuan et al. 1997) or sputtered and are used as high index films in the IR-region. Si is also used in multilayer stacks in the EUV/X-ray region. It readily forms oxide or nitride phases, so the background pressure has to be kept very low for very pure films.

Table 5 shows properties of Si- and Ge-films.

Table 5. Properties of silicon and germanium films

Coating material	Film composition	Deposition method	Range of transparency in µm	Refractive index at 2 µm	Structure	Mechanical properties
Si	Si	EB	1.0-9	3.4	A (T<300)	Hard
Ge	Ge	EB	1.7-25	4.4	A (T<300)	

EB electron beam evaporation, A amorphous, T substrate temperature [°C]

Carbon films, diamond like carbon films (DLC) and diamond films are deposited by various PECVD and PVD methods (Matthews and Bachmann 1991). They have found applications in optics as hard coatings in the infrared and as coatings in the EUV/X-ray range (Chapter 'Multilayer Coatings for EUV/X-ray Mirrors' by S. Yulin).

3.3 Cermets

Cermet films are combinations of metals and dielectric materials. Metal nanoparticles are embedded in a dielectric matrix. Typical examples are coatings that absorb light at a low reflectance level, e.g. for sunglasses.

A material used for this purpose is a mixture of SiO_2 and molybdenum (Balzers Patent- und Beteiligungs-AG 1967), which is evaporated by electron beam. It yields hard, environmentally stable coatings with a brown transmission color and moderate reflection.

3.4 Organic Materials

Although there are manifold examples of deposition methods and applications of organic thin film materials only a few are used for optical interference coatings.

The manufacture of multicolour interference coatings consisting of metallic layers and transparent acrylate polymer is described (Shaw et al. 1995). Acrylate monomer as starting material is evaporated in vacuum to form the dielectric film.

Enormous efforts in research of organic thin films for optoelectronic applications has been stimulated by C.W. Tang's observation of electroluminescence in a stack of low molecular weight organic semiconductors in 1987 (Tang and Van-Slyke 1987). Strictly speaking the conducting and light emitting films are no optical interference coatings but the final devices might well need such stacks for improved performance. The basic materials involved act as charge transporting layers (holes, electrons) and as emitter materials. Spin coating and evaporation in vacuum are the preferred processes to deposit the layers.

The section on optical interference coatings prepared from organic thin film materials would not be complete without mentioning multilayer films prepared by coextrusion of polymers to build stacks of up to 1000 individual layers (Schrenk et al. 1991). Various devices like polarizers, mirrors and filters have been demonstrated and are commercially available (Weber et al. 2000).

4 Industrial Production of High Quality Coating Materials

4.1 Evaporation Materials

High quality coating materials have to fulfil many requirements as outlined in Chap. 2.1. Industrial production of evaporation materials has to take into account all these demands by using optimised production steps. Raw materials and intermediates have to be controlled by several analytical methods. The finished products have to be analysed according to specifications including chemical analysis, weight and dimensions and stability. Also suitability for vacuum evaporation has to be tested (applications test). The most important production steps will be demonstrated in some examples below.

Production of Zirconium Oxide Discs

The raw materials are analysed for impurity levels. Grain size distribution is also analysed since it influences the sintering behaviour. Forming of discs needs organic binding materials to be added to the raw material, a technique commonly used in production of ceramic parts. In many cases the raw material is spray dried to form a free flowing granulate. Discs are then formed using hydraulic or mechanical presses. In this step it is very important to avoid contamination e.g. by tool abrasion. In the following step the discs are sintered in air to remove the organic additives, which is very important since even a very low carbon content causes spitting during heating and evaporation. In the last step the discs are sintered to reach the desired density and structure.

Similar processes are used to produce tablets of e.g. Al_2O_3, HfO_2, Nb_2O_5, Ta_2O_5, and TiO_2.

It is also necessary to control the density of the materials. In some cases high densities are required, but in general discs with too high densities are undesirable since they easily crack during heating up to the melting or evaporation temperature. Also spitting can occur during release of oxygen if densities are too high.

Some oxide materials, e.g. zirconium oxide, titanium oxide and hafnium oxide, are sintered in vacuum to produce oxygen deficiency. This step needs accurate control of pressure and temperature.

Production of Titanium(III)-Oxide

Ti_2O_3 is produced by sintering of discs or granules made from a mixture of TiO_2 and Ti. The raw materials are mixed and formed to discs or granules, which are then sintered in high vacuum to form Ti_2O_3. In this process the grain size distribution of raw materials is critical, as is homogeneity of the mixture and temperature profile during sintering. The reaction has to be complete since even a very low residue of TiO_2 causes degassing and spitting during melting and evaporation.

Other suboxides of titanium, namely TiO and Ti_3O_5 are produced in similar processes.

Production of Magnesium Fluoride Granules

MgF_2 granules are mainly produced by melting in vacuum. Oxygen content is very critical with MgF_2 because too high oxygen levels cause spitting during evaporation and absorption of the coatings in UV. Therefore the raw material and all production steps have to be optimised to maintain low oxide levels. The raw material is melted in vacuum furnaces to yield polycrystalline ingots. The ingots are crushed and the desired grain sizes are separated by sieving processes.

Similar processes are used to produce other fluoride materials like BaF_2, CaF_2, LaF_3, PbF_2, YbF_3, YF_3, etc. Sublimation in vacuum can be used to further reduce impurity levels, especially oxygen.

Production of Silicon Monoxide

Silicon monoxide is produced by synthesis in a vacuum sublimation process. A mixture of SiO_2 and Si powders is heated to about 1400°C in high vacuum. SiO is formed in gaseous state and is sublimed out of the crucible. The SiO vapour is condensed at surfaces at lower temperatures to form thick layers of black coke like SiO. The physical appearance of the material is controlled by the condensation rate and temperature. In this process very careful temperature control is absolutely necessary to get the desired quality. In a second step the material is crushed and separated into the desired grain sizes.

In case of SiO granules the density, form, grain size, gas content and the SiO_2 residue is critical for many applications.

Production of Mixtures – Substance H4 Patinal®

The raw materials TiO_2, Ti and La_2O_3 are mixed and formed to granules and sintered in high vacuum, where a solid state reaction takes place and a new chemical compound is formed. Therefore the temperature profile in the vacuum process has to be controlled very carefully. Otherwise spitting and even disintegration of granules may occur. The mixture homogeneity with respect to the constituents including oxygen is very important since refractive index and absorption depend on composition.

Similar processes are used for production of other materials, e.g. Substance H1, Substance H2, Ida, Dralo, and Pisa.

Silicon Dioxide Granules

Silicon dioxide granules are made from fused silica by crushing and sieving. For special applications in coating plants with automatic feeding systems it is necessary to avoid needle like particles which may block the feeding system. Therefore special equipment must be used to remove the needle like particles

Table 6. Typical specifications for evaporation materials

Material	Co [%]	Cr [%]	Cu [%]	Fe [%]	Ti [%]	V [%]	Zr [%]	Oxide [%]
ZrO_2 [1)]	≤ 0.0005	≤ 0.002	≤ 0.001	≤ 0.01	≤ 0.2	≤ 0.005		
HfO_2 [2)]	≤ 0.0005	≤ 0.005	≤ 0.0005	≤ 0.005	≤ 0.005	≤ 0.005	≤ 1	
MgF_2 [3)]	≤ 0.0005	≤ 0.0005	≤ 0.0005	≤ 0.001		≤ 0.001		≤ 0.15
SiO [4)]	≤ 0.0005	≤ 0.001	≤ 0.002	≤ 0.005		≤ 0.0005		
Ti_2O_3 [5)]	≤ 0.0005	≤ 0.005	≤ 0.001	≤ 0.02		≤ 0.01		

[1)] Merck # 108904 Zirconium(IV) oxide discs about 6 g Patinal®.
[2)] Merck # 111626 Hafnium(IV) oxide UV granules (grey) about 1-4 mm Patinal®.
[3)] Merck # 108545 Magnesium fluoride granules about 1-4 mm Patinal®.
[4)] Merck # 107720 Silicon monoxide granules about 2-8 mm Patinal®.
[5)] Merck # 110203 Titanium(III) oxide tablets about 1 g Patinal®.

4.2 Sputtering Targets

Sputtering targets are produced by a melting and casting process or by powder metallurgy. Thermal spraying and electroplating are only used for special geometry and applications. The selection of the production process depends on desired material characteristics (metal or ceramic, melting point, kind of alloy, purity levels), target size and geometry, e.g. flat, tube, bonded, etc. Table 7 gives an overview of the production processes for sputtering targets.

Table 7. Production procedures for sputtering targets

Casting Metallurgy

Vacuum Induction Melting (VIM)
Continuous Casting
Electron Beam Melting (EB)

Powder Metallurgy

Powder Preparation

Cold Isostatic Pressing (CIP)
Sintering
Hot Pressing (HP)
Hot Isostatic Pressing (HIP)

Heat Treatment
Homogenisation, Recrystallization

Transformation Technique
Forging, Hot and Cold Rolling, Extrusion Pressing

Mechanical Tooling
Machining, Sizing

Bonding
Soldering, Diffusion Bonding, Roll Plating, Explosion Welding

Cleaning / Packing

Metallurgical melting processes are normally the most economic manufacturing methods. Metals such as Zn, Sn, In and their alloys can be melted in resistively heated open air furnaces. Materials with higher melting temperature up to 2000°C are melted in vacuum induction furnaces. The vacuum ensures high purity and low levels of volatile impurities (gases, Cl, P, K, Li, etc.). Any geometry of raw material and a recycling of used sputtering targets are possible. Chromium, titanium and niobium can be melted by electron beam melting. Most cast ingots are post-treated by forging, extrusion, rolling and high temperature annealing processes to improve the homogeneity and to reach a fine grain size structure. Also, special materials like silicon, germanium, SiAl alloys are produced as cast. Si and Ge single crystal targets are used when very accurate film distributions are required.

Melting processes are not applicable in the case of high melting temperatures (W, Ta, Mo, ceramics), of extremely brittle and stress loaded castings (ceramics), of sublimation before melting (Cr, ITO) and in the case of alloys with a miscibility gap in the liquid phase (Ag-Ni). In these cases powder metallurgical production techniques have to be used. After powder mixing and blending a press and heat treatment step is needed (cold pressing plus sintering, hot pressing or hot isostatic pressing). The advantages are adjustable target compositions, controlled microstructure, homogeneous grain size, adjustable amount of intermetallic phases and a possible near net shape production. The disadvantage is reduced density and a higher gas content in the finished product.

Another production method is thermal spraying, e.g. of silicon tube targets. The raw material is liquefied by a plasma and the molten droplets are directed towards the target.

Most targets have to be bonded on a backing or cooling plate to ensure good thermal contact to the cooling system of the cathode and to accommodate thermal and mechanical stresses. Common methods are: solder seal with indium, roll plating, diffusion bonding, explosion welding and epoxy gluing.

5 Summary and Outlook

The different deposition technologies for thin films for optical applications require a large number of evaporation materials and sputtering targets, which have to fulfil specific requirements. Their development and production is the task of specialised companies.

Coating materials for optical applications comprise dielectric materials, especially fluorides, oxides, nitrides and carbides, elementary and compound semiconductors, metals and their alloys.

Fluoride films are still a domain of classical evaporation technology. Ion assisted evaporation and sputtering are of increasing importance for oxide, nitride, carbide and compound semiconductor films. Ion assisted processes yield films of highest quality and sputter processes are well suited for mass production. The trend to these technologies will continue and pose new challenges for coating materials in future.

Acknowledgements

The preparation of this chapter within a relatively short period was only possible thanks to the cooperation of several people. The authors would like to thank their colleagues Friedrich Koenig, Klaus Leitner and Elmar Ritter for valuable contributions.

References

Albertinetti N, Minden HAT (1996) Granularity in ion-beam-sputtered TiO_2 films. Appl. Opt 35: 5620-5625

Alvisi M, De Nuncio G, Di Gulio M, Ferrara MC, Perrone MR, Protopapa L, Vasanelli L (1999) Deposition of SiO_2 films with high laser damage thresholds by ion-assisted electron-beam evaporation. Appl. Opt. 38: 1237-1243

Anderson O, Ottermann C (1997) Silicon oxides. In: Bach H, Krause D (eds) Thin films on glass. Springer, Berlin Heidelberg, pp 161-173

Anderson O, Bange K, Ottermann C (1997) Titanium oxide. In Bach H, Krause D (eds) Thin films on glass. Springer, Berlin Heidelberg, pp 137-161

Andre P, Poupinet L, Ravel G (2000) Evaporation and ion assisted deposition of HfO_2 coatings: Some key points for high power applications. J. Vac. Sci. Technol. A 18: 2372-2377

Balzers Patent- und Beteiligungs-AG (1967) Lichtabsorbierendes Sonnenbrillenglas. Schweizer Patent 447 476

Bange K (1997) Properties and characterisation of dielectric thin films. In: Bach H, Krause D (eds) Thin films on glass. Springer, Berlin Heidelberg, pp 101-225

Bendavid A, Martin PJ, Takikawa H (2000) Deposition and modification of titanium dioxide films by filtered arc deposition. Thin Solid films 360: 241-249

Bennett JM, Pelletier E, Albrand G, Borgogno JP, Lazarides B, Carniglia CK, Schmell RA, Allen TH, Tuttle-Hart T, Guenther KH, Saxer A (1989) Comparison of the properties of titanium dioxide films prepared by various techniques. Appl. Opt. 28: 3303-3317

Bernitzki H, Lauth H, Thielsch R, Blaschke H, Kaiser N, Mann K (1999) Current status of radiation resistance of dielectric mirrors in the DUV. SPIE 3578

Bin F, Zhang F (1997) Study on optical properties of PbTe and ZnSe at low temperature. J. of Shanghai Jiaotong Univ. 31: 69-71

Brunet-Bruneau A, Vuye G, Frigerio JM, Abelès F, Rivory J, Berger M, Chaton P (1996) Infrared ellipsometry investigation of SiO_xN_y thin films on silicon. Appl. Opt. 35: 4998-5004

Canon Kabushiki Kaisha (1976) Three layer antireflection film US Pat. 3 934 961

Chen JS, Chao S, Kao JS, Lai GR, Wang WH (1997) Substrate-dependent optical absorption characteristics of titanium dioxide thin films. Appl. Opt. 36: 4403-4408

Cho HJ, Hwangbo CK (1996) Optical inhomogeneity and microstructure of ZrO_2 thin films prepared by ion-assisted deposition. Appl. Opt. 35: 5545-5552

Cowell R (1996) Durable optical coatings for energetic UV laser sources. SPIE 2703:369-374

De CK, Misra NK, Gosh T B (1995) Preparation and structural characterisation of ZnSe thin films by x-ray diffraction technique. Indian J. Phys.,A 69A: 261-266

Dods SRA, Zhang Z, Ogura M (1999) Highly dispersive mirror in Ta_2O_5/SiO_2 for femtosecond lasers designed by inverse spectral theory. Appl. Opt. 38: 4711-4719

Edlinger J, Ramm J, Pulker HK (1989) Properties of ion-plated Nb_2O_5 films. Thin Solid Films175: 207-212

Edlou SM, Smajjkiewicz A, Al-Jumaily GA (1993) Optical properties and environmental stability of oxide coatings deposited by reactive sputtering. Appl. Opt. 32: 5601-5605

Friz M, Koenig F, and Feiman S. (1992) New materials for production of optical coatings. Proc. SVC 35th Ann. Techn. Conf 1992, 143-149

Friz M, Schallenberg UB, Laux S. (1997) Plasma ion assisted deposition of medium and high refractive index thin films. Proc. SVC 40th Ann. Techn. Conf. 1997: 280-292

Ganner P (1986) Medium-Index mixed-oxide layers for use in AR-coatings. Proc. SPIE 652: 69 –76

Gläser HJ (1990) Dünnfilmtechnologie auf Flachglas. Verlag Karl Hoffmann, Schorndorf

Goldfinger P, Jeunehomme M (1963) Mass spectrometric and Knudsen cell vaporization studies of 2B-6A compounds. Trans. Far. Soc. 59: 2851-2867

Hawkins GJ, Hunneman R, Sherwood R, Barrett BM (2000) Infrared filters and coatings for the High Resolution Dynamics Limb Sounder (6-18 μm) Appl. Optics 39: 5221-5230

Heitmann W, Ritter E (1968) Production and properties of vacuum evaporated films of Thorium fluoride. Appl. Optics 7:307-309

Hussain Z, (1999) Vacuum temperature-dependent ellipsometric studies on WO_3 thin films. Appl. Opt. 38: 7112-7127

Izawa T, Yamamura N, Uchimura R, Yakuoh T (1994) Damage thresholds and optical stabilities of fluoride HR coatings for 193 nm. SPIE 2114:297–308

Klinger RE, Carniglia CK (1985) Optical and crystalline inhomogeneity in evaporated zirconia films. Appl. Opt. 24: 3184-3187

Koenig F, Friz M (1994) Development of medium and high refractive index coating materials for the visible and UV spectral range. Proc. SVC 37th Ann. Techn. Conf. 1994: 118 -121

Kolbe J, Müller H, Schink H, Welling H, Ebert J (1990) Laser induced damage thresholds of dielectric coatings at 193 nm and correlations to optical constants and process parameters. NIST Spec. Publ. 801:404–416

Kolbe J, Kessler H, Hofmann T, Meyer F, Schink H, Ristau D (1991) Optical properties and damage thresholds of dielectric UV/VUV coatings deposited by conventional evaporation, IAD and IBS. SPIE 1624:221–235

Kraus T, Rheinberger P (1962) Use of rare earth metal in vaporizing metals and metal oxides. U.S. Patent 3 034 924

Kulkarni AK, Lim T, Khan M, Schulz KH, (1998) Electrical, optical and structural properties of indium-tin-oxide thin films deposited on polyethylene terephtalate substrates by rf sputtering. J. Vac. Sci. Technol. A 16: 1636-1640

Larruquert JI, Keski-Kuha RAM (1999) Multilayer coatings with high reflectance in the extreme-ultraviolet spectral range of 50 to 121.6 nm. Appl. Opt. 38: 1231-1236

Larruquert JI, Keski-Kuha RAM (2000) Reflectance measurements and optical constants in the extreme ultraviolet for thin films of ion-beam-deposited SiC, Mo, Mg_2Si, and InSb and of evaporated Cr. Appl. Opt. 39: 2772-2781

Lee CC, Chen SH, Jaing CC (1996) Optical monitoring of silver-based transparent heat mirrors. Appl. Opt. 35: 5698-5703

Lee CC, Chen HL, Hsu JC, Tien CL (1999) Interference coatings based on synthesized silicon nitride. Appl. Opt. 38: 2078-2082

Lehan JP (1996) Determination of grain size in indium tin oxide films from transmission measurements. Appl. Opt. 35: 5048-5051

Lingg LJ, Targove JD, Lehan JP, Macleod HA (1987) Ion-assisted deposition of lanthanide trifluorides for VUV applications. SPIE 818:86–92

Lorenz H, Eisele I, Ramm J, Edlinger J, Bühler M (1991) Characterization of low temperature SiO_2 and Si_3N_4 films deposited by plasma enhanced evaporation. J. Vac. Sci. Technol. B 9: 208-214

Luff BJ, Wilkinson JS, Perrone G (1997) Indium tin oxide overlayered waveguides for sensor applications. Appl. Opt. 36: 7066-7072

Macleod HA (1986) Thin-film optical filters. Second edn. Adam Hilger LTD, Bristol, pp 398, 504-511

Martin PJ, Bendavid A, Takikawa H (1999) Ionized plasma vapor deposition and filtered arc deposition; processes, properties and applications. J. Vac. Sci. Technol., A 17: 2351-2359

Martin PM, Stewart DC, Bennett WD, Affinito JD, Gross ME (1997) Multifunctional multilayer optical coatings. J. Vac. Sci. Technol. A 15: 1098-1102

Marx D, Murphy R (1990) Sputtering targets: Challenges for the 1990s. Solid State Technology, March: 11-14

Matsushita Electric Ind Co Ltd. (1994) Manufacture of optical parts. JP 05188204

Matthews A, Bachmann PK (eds) (1991) Diamond and diamond-like carbon coatings 1990. Elsevier Sequoia S.A., Lausanne

Méndez JA, Larruquert JI, Aznarez JA (2000) Preservation of far UV aluminum reflectance by means of overcoating with C_{60} films. Appl. Opt. 39: 149-156

Merck KGaA (1993) Process for producing evaporation material for preparation of medium refractive index optical films. EP 0 574 785

Merck KGaA (1993) Material for vapour deposition for the production of high refractive optical layers. EP 0 561 289

Newns GR, Pantelis P, Wilson JL, Uffen RWJ, Worthington R (1973) Absorption losses in glasses and glass fibre waveguides. Opto-electronics 5:289-296

Niederwald H, Laux S, Kennedy M, Schallenberg U, Duparré A, Mertin M, Kaiser N, Ristau D (1999) Ion-assisted deposition of oxide materials at room temperature by use of different ion sources. Appl. Opt. 38: 3610-3613

Phillips RW, Bleikolm AF (1996) Optical coatings for document security. Appl. Optics 35:5529-5534

Phillips RW, Nofi M (1999) Colors by chemistry or by physics. Proceedings SVC:494-499

Pinard L, Mackowski JM (1997) Synthesis and physicochemical characterization of silicon oxynitride thin films prepared by rf magnetron sputtering. Appl. Opt. 36: 5451-5459

Pulker HK (1979) Characterization of optical thin films. Appl. Opt. 18: 1969-1977

Pulker HK (1999) Coatings on Glass. Second, revised edn. Elsevier, Amsterdam, pp 219-223, 394-428

Rafla-Yuan H, Rancourt JD, Cumbo MJ (1997) Ellipsometric study of thermally evaporated germanium thin film. Appl. Opt. 36: 6360-6363

Rahe M, Oertel E, Reinhardt L, Ristau D, Welling H (1990) Absorption calorimetry and laser induced damage threshold measurements of AR-coated ZnSe and metal mirrors at 10.6 μm. SPIE 1441:113–126

Reicher D, Black P, Jungling K (2000) Defect formation in hafnium dioxide thin films. Appl. Opt. 39: 1589-1599

Ristau D, Günster S, Bosch S, Duparré A, Masetti E, Ferré-Borrull J, Kiriakidis G, Peiró F, Quesnel E, Tikhonravov A, (2001) UV-optical and microstructural properties of MgF_2- and LaF_3-coatings deposited by IBS and PVD processes. Optical Interference Coatings Banff, OSA Technical Digest: ThA5-1–ThA5-3

Ritter E, Hoffmann R (1969) Influence of substrate temperature on the condensation of vacuum evaporated films of MgF2 and ZnS. J. Vac. Sci. Technol. 6: 733-736

Ritter E (1971) Die Eigenschaften verschiedener Siliciumoxid-Phasen in dünnen Schichten. Vakuum-Technik 21: 42-48

Ritter E (1975) Dielectric film materials for optical applications. In: Hass G, Francombe MH, Hoffman RW (eds) Physics of thin films, vol 8. Academic Press, New York, pp 1-49

Rudisill, JE, Braunstein, M, Braunstein AI (1974) Optical coatings for high energy ZnSe laser windows. Appl. Optics 13: 2075-2080

Scherer K, Nouvelot L, Lacan P, Bosmans R (1996) Optical and mechanical characterization of evaporated SiO_2 layers. Long term evolution. Appl. Opt. 35: 5067-5072

Schnellbuegel A, Hagedorn H, Anton R (1994) Ion assisted deposition of non-toxic coatings for high power CO_2 laser optics. Proc. SPIE 2253: 839-848

Schrenk WJ, Wheatley JA, Lewis RA, Arends CB (1991) Nanolayer polymeric optical films. Tappi 1991 Polymers, laminations and coatings conference

Schultz PC (1974) Optical absorption of the transition elements in vitreous silica. J.Am.Ceram.Soc. 75:309-313

Seeley JS, Hunneman R, Whatley A (1979) Infrared multilayer interference filter manufacture: supposed longwave limit. Appl. Optics 18: 3368-3370

Seeley JS, Hunneman R, Whatley A (1980) Far infrared filters for the Galileo-Jupiter and other missions. Appl. Optics 20: 31-39

Selhofer H, Müller R (1999) Comparison of pure and mixed coating materials for AR coatings for use by reactive evaporation on glass and plastic lenses. Thin Solid Films 351: 180-183

Selhofer H, Ritter E, Linsbod R (2002) Properties of titanium dioxide films prepared by reactive electron beam evaporation from various starting materials. Appl. Opt. 41: 756-762

Shaw DG, Cline DS, Dawson EP, Langlois M (1995) Multicolor interference coating. US-Patent 5877895

Stamm U, Pätzel R, Bragin I, Kleinschmidt J, Voß F, Basting D (1997) Recent developments in industrial excimer laser technology. SPIE 3092:485–492

Sullivan BT, Clarke GA, Akiyama T, Osborne N, Ranger M, Dobrowolski JA, Howe L, Matsumoto A, Song Y, Kikuchi K (2000) High-rate automated deposition system for the manufacture of complex multilayer coatings. Appl. Opt. 39: 157-167

Tabata A, Matsuno N, Suzuoki Y, Mizutani T (1996) Optical properties and structure of SiO_2 films prepared by ion-beam sputtering. Thin Solid Films 289: 84-89

Tang CW, VanSlyke SA (1987) Organic electroluminescent diodes. Appl. Phys. Lett. 51:913-915

Venzago C, Weigert M. (1994) Application of the glow discharge mass spectrometry (GDMS) for the multi-element trace and ultratrace analysis of sputtering targets. Fresenius J Anal Chem 350: 303-309

Wang X, Masumoto H, Someno Y, Hirai T (1999) Design and experimental approach of optical reflection filters with graded refractive index profiles. J. Vac. Sci. Technol. A 17: 206-211

Weber MF, Stover CA, Gilbert LR, Nevitt TJ, Ouderkirk AJ (2000) Giant birefringent optics in multilayer polymer mirrors. Science 287:2451-2456

Weigert M (2001) New developments in materials for sputtering on glass. International Glass Review Issue 3: 72-76

Wilhartitz P, Ortner HM, Krismer R, Krabichler H (1990) Classical analysis including trace-matrix separation versus solid state mass spectrometry: A comparative study for the analysis of high purity Mo, W and Cr. Microchim. Acta (Wien) II: 259-271

Wille B (1974) Refracting light permeable oxide layer and method for its manufacture. U. S. Patent 3 783 010

Yamada Y, Uyama H, Watanabe S, Nozoye H (1999) Deposition at low substrate temperatures of high quality TiO_2 films by radical beam-assisted evaporation. Appl.Opt. 38: 6638-6641

Zhang, S et al (1998) The effects of processing conditions on PbGeTe film performances. Proc SPIE 3175: 429-432

Zöller A, Götzelmann R, Matl K, Cushing D (1996) Temperature-stable bandpass filters deposited with plasma ion-assisted deposition. Appl. Opt. 35: 5609-5619

Zukic M, Torr DG, Spann JF, Torr MR (1990) Vacuum ultraviolet thin films. 1: Optical constants of BaF_2, CaF_2, LaF_3, MgF_2, Al_2O_3, HfO_2, SiO_2 thin films. Appl. Opt. 29:4284–4292

Abbreviations

AC	Alternating Current
AR	Antireflection
CGHE	Carrier Gas Heat–Extraction
CIP	Cold Isostatic Pressing
CVD	Chemical Vapour Deposition
DC	Direct Current
DLC	Diamond Like Carbon
DUV	Deep Ultra-Violet
DWDM	Dense Wavelength Division Multiplexing
EB	Electron-Beam
EUV	Extreme Ultra-Violet
FAAS	Flame Atomic Absorption Spectrometry
GDMS	Glow Discharge Mass Spectroscopy
GDS	Glow Discharge Spectroscopy
HIP	Hot Isostatic Pressing
HP	Hot Pressing
IAD	Ion Assisted Deposition
IBS	Ion Beam Sputtering
IBAD	Ion Beam Assisted Deposition
ICP	Inductively Coupled Plasma
ICP-OES	Inductively Coupled Plasma Optical Emission Spectroscopy
IR	Infrared
NUV	Near Ultra-Violet
OVP	Optically Variable Pigment
PECVD	Plasma Enhanced Chemical Vapour Deposition
PVD	Physical Vapour Deposition
R	Reactive
RF	Radio Frequency
RLVIP	Reactive Low Voltage Ion Plating
UHV	Ultra High Vacuum

UV	Ultra-Violet
VIM	Vacuum Induction Melting
VUV	Vacuum Ultra-Violet
XRF	X-Ray Fluorescence

Symbols

A	Absorptance
c	Concentration
d	Thickness
k	Extinction Coefficient
n	Refractive Index
λ	Wavelength
T	Transmittance

Film Deposition Methods

Hans K. Pulker

Thin Film Technology Group, Institute of Ion Physics, University of Innsbruck, Austria

Thin films of elements, alloys and chemical compounds can be formed on solid substrates by various wet and dry chemical and physical deposition methods. Depending on the applied process, depositions are performed in air or in environmentally controlled atmosphere, under reduced pressure or in vacuum.

In this chapter mainly physical vapour deposition (PVD) processes are treated because they are the preferred technologies for film deposition in interference optics. Only in few variants chemical vapour deposition (CVD) processes are important for this purpose. PVD processes are performed under vacuum and are based on purely physical effects occasionally also associated by unwanted chemical reactions. Intentionally forced chemical reactions by adding reactive gas to the coating chamber are used to deposit stoichiometric chemical compound films in the reactive deposition process. In these processes the reactivity is generally positively influenced by the presence of a gas discharge plasma.

All the processes are used to deposit coatings in the thickness range between a few nanometers up to some microns. Single films or multilayers can be deposited homogeneously or with graded composition.

1 Physical Vapour Deposition

In physical vapour deposition (PVD) there are two fundamental transfer processes of the coating material into the vapour phase: evaporation and sputtering.

A growing number of coating process variants based on these two transfer principles are in technical use. In all cases the vapourized material is transported through a reduced atmosphere and condensed on a non-heated or heated and generally rotated substrate where the coating is formed.

In many cases technical vacua between 10^{-2} and 10^{-4} mbar are sufficient to perform a coating. For high quality films, however, low pressures, lower than 10^{-6} mbar, and precisely controlled gas compositions are required.

Although sputtering is performed in the working gas pressure range of 10^{-2} to 10^{-3} mbar, a much lower starting pressure is required to obtain clean and reproducible conditions in the actual working pressure range.

There are very different kinetic energies of the volatilised coating material species in the various PVD variants. Simple evaporation yields typically 0.05 to 0.1 eV particle energy whereas sputtering may yield energies between 1 and 40 eV or higher values. Ion and plasma assisted or operated processes may reach even more than 100 eV.

Higher particle energies generally have a positive influence on adherence, density, structure and microstructure of the films and improve the optical film properties and their environmental stability. Very high energy values may cause increased residual optical absorption in chemical compound films and high compressive mechanical stress values.

PVD coating processes are carried out in sealed cylindrical or cubic stainless steel chambers, which are equipped with all required installations, e.g. vapour sources and substrate holders as well as control and measurement devices for optimum deposition.

Different pumping systems are used to evacuate a coating chamber. Most common are mechanical displacement pumps combined with baffled synthetic oil operated diffusion pumps or turbomolecular pumps. Cleaner vacua are achieved with refrigerator cryo pumps. They can pump all gases and have the highest pumping speed of all high vacuum pumps.

Modern coating systems automated to a high degree and generally installed in humidity controlled and dust free clean rooms.

Optical coatings are mainly produced in batch processes but increasingly also in continuously operated coating units or in roll to roll coaters.

1.1 Thermal Vaporization

Coating substrates with thin films by evaporation or sublimation and subsequent condensation was probably first made by Faraday in 1857 with exploding metal wires in inert atmosphere. He was later followed by Nahrwold in 1887 who performed evaporation by resistance heating of platinum wires under vacuum. After that this technique was used by Kundt 1888 to deposit metal films for refractive index determinations of metals. Under much better developed technological conditions metal films were later deposited by Pohl and Pringsheim in1912.

Processes of film deposition by evaporation or sublimation and vapour condensation on solid surfaces at low residual gas pressures have been developed and improved in parallel with the development of vacuum technology (Dushman and Lafferty 1962, Wutz et al. 1982, Yarwood 1982). If only physical effects are involved film formation is generally performed at pressures between 10^{-6} and 10^{-8} mbar and down to the lowest values of modern vacuum technology.

If in the film forming process also chemical reactions are involved this special case is termed reactive evaporation or better reactive deposition. Reactive techniques are very important for the deposition of stoichiometric chemical compound films. For instance, most metal oxides decompose during heating and vaporisation. By controlled addition of oxygen gas to the residual atmosphere up to a total pressure of about $1 - 2\ 10^{-4}$ mbar and at a medium deposition rate of about 0.5 - 1 nms^{-1} stoichiometric films can be obtained if these conditions are strictly maintained.

The reactions are sometimes also activated by uv radiation or by ionic or electronic bombardment and cold plasmas. To improve mainly mechanical film quality substrates are heated to 150° and 350°C, respectively, prior to deposition .

There are different methods of heating a material for evaporation. Most materials melt before evaporation starts. Vaporised atoms and molecules have molecular velocities of the order of 10^5 cm s^{-1} and their mean energies are below 0.1 eV.

Evaporation is a very fast process. The number of evaporating particles depends strongly on the temperature. Evaporation of material from a crucible does not occur uniformly in all directions, but is dependent on $\cos^b \alpha$ where α is the angle from the normal to the source surface. The exponent b determines the exact shape of the vapour cloud. An increase of the b value decreases the emitted amount at larger emission angles. Directions perpendicular to the source are therefore favoured because $\cos \alpha$ is then a maximum. This is a very important fact in considerations of proper lateral thickness uniformity on a plane horizontal or curved substrate surface.

Expressions have been calculated for different shaped small area, ring and disc sources, etc. However, real evaporation sources do not precisely obey the cosine law. It could be shown that improved correlation with effective thickness distributions can be achieved by using a virtual source concept (Graper 1973, James and Matthews 1990). In many cases, however, the experience of a skilled operator is extremely helpful in finding a fast solution of such problems.

1.2 Evaporation of Metals and Elemental Semiconductors

Careful investigations of the evaporation rate in vacuum by Hertz, Knudsen and Langmuir (Bach and Krause 1997, Frey and Kienel 1987, Leyland and James 1991, Maissel and Glang 1970) led to the formulation of the following equation:

$$dN / A\,dt = \alpha_e [(p_e - p_h) / \sqrt{(2\pi\ mkT)}] \qquad cm^{-2}s^{-1} \qquad (1)$$

dN	:	number of evaporating atoms	α_e	:	evaporation coefficient
A	:	surface	p_e	:	equilibrium vapour pressure
t	:	time	p_h	:	hydrostatic pressure
m	:	atomic mass	k	:	Boltzmann´s constant
			T	:	temperature (Kelvin)

With $\alpha_e = 1$ and $p_h = 0$ the maximum evaporation rate is obtained.

Two important statements are the consequence follow from such considerations: The maximum evaporation rate is mainly determined by the material's surface temperature and this rate cannot be greater than the value established by the equilibrium vapour pressure.

For calculation of evaporation rates empirically obtained vapour pressure values must be applied. Useful vapour pressures for evaporation under vacuum are in the range of 10^{-2} mbar.

Vapour pressure data can be found in the literature (Dushman and Lafferty 1962, Pulker 1999).

By multiplying equation 1 by the mass of an individual atom the mass evaporation rate G per unit area is obtained:

$$G = m\, dN / A dt = (m / 2\pi kT)^{1/2}\, p_e \tag{2}$$

or

$$G = 4.4 \cdot 10^{-2} \,(M / T)^{1/2}\, p_e . \qquad \mathrm{g\,cm^{-2}\,s^{-1}} \tag{3}$$

M = molar mass and p_e is given in mbar.

The double integral yields finally the total amount of evaporated material:

$$m_{total} = \int_t \int_A G \; dA\, dt . \qquad \mathrm{g} \tag{4}$$

For most metals mass evaporation rates are in the range of 10^{-4} g $cm^{-2}s^{-1}$. Sometimes, however, higher rates may be desirable. In Table 1 mass evaporation rates of various metals with the corresponding evaporation temperatures are shown (Frey and Kienel 1987).

Table 1. Mass evaporation rate versus evaporation source temperature of some metals

Material		Zn	Ag	Al	Au	Ti	Pt	Ta
Mass evaporation rate		Temperature °C						
$g\,cm^{-2}s^{-1}$	10^{-4}	325	1020	1190	1405	1570	2080	3020
	10^{-3}	400	1160	1310	1550	1780	2270	3300
	10^{-2}	480	1300	1490	1710	1980	2490	3650
	10^{-1}	600	1500	1640	1900	2210	2740	3800

For optical applications the most common metal and elemental semiconductor films are Al, Ag, Cr, Au, Ni, Rh, Si and Ge.

The film properties are influenced by the production method. Careful evaluations are required to select the proper crucible material, heating method, deposition rate and residual atmosphere to prevent film contamination with possibly co-evaporating source material and incorporation of reactive gas.

This can be demonstrated by the different optical constants of evaporated Al-films, shown in Table 2, which were obtained by Hass (Hass and Waylonis 1961) in the period 1945 to 1961.

Because of the higher and cleaner vacua and faster deposition rate the newer films, possibly containing less oxygen, clearly show better optical values particularly in the blue and near ultraviolet region. Many other metal films are also very sensitive to oxidation and even traces of oxide phases cause degradation of their optical properties.

Today electron beam heating mostly performed with the 270° deflection beam e-gun is the preferred evaporation technology for nearly all metals. High rates and fast rate changes can be achieved with it. To obtain less- or even non-contaminated metal films a high deposition rate is required.

Table 2. Optical constants of vacuum-evaporated aluminium films prepared and measured by Hass in 1946 and 1961

λ [nm]	n (1946)	n (1961)	k' (1946)	k' (1961)
644	1.12	-	6.26	-
650	-	1.30	-	7.11
578	0.89	0.93	5.68	6.33
564	0.76	0.82	5.50	6.99
491	0.57	-	5.20	-
492	-	0.64	-	5.50
435	0.40	-	4.16	-
436	-	0.47	-	4.84
380	-	0.37	-	4.25

λ wavelength, n refractive index, k' extinction coefficient.

With exception of special cases as for instance mirror coating, evaporation from resistance heated refractory metal boats and wires is less often used. Selected less frequently applied techniques in metal evaporation are heating the coating material with electric arcs (Martin and Bendavid 2001) and heating with lasers (Chrisey and Hubler 1994). Fast deposition of not too extended areas with particles of relatively high kinetic energy is characteristic of these processes. Both techniques are operated in a pulsed mode to decrease the formation of unwanted droplets. However, special droplet separators must be applied to obtain high quality films.

1.2.1 Evaporation of Alloys and Composite Materials

Because of the different vapour pressures of their individual components evaporation of alloys is much more complicated than evaporation of single metals. It is difficult to obtain a homogeneous film of a certain composition by evaporating from one single source. The more volatile components of the alloy are depleted in the evaporation source after a time. Therefore it is often problematic to maintain a desired alloy composition in thicker films and over larger substrate surfaces. Similar problems exist with multi-component mixtures and cermets.

One method to overcome this difficulty is to use controlled multi-source evaporation with mixing the vapours of the single constituents on their way to the substrate. With this method all known source-heating techniques are applicable., However, the method is expensive and it is difficult to coat larger areas uniformly (Rickerby and Matthews 1991).

Other methods are single source evaporations. In the first mode electron beam evaporation from one molten pool of a rod fed crucible with continuous alloy material supply is performed.

In the second mode satisfactory results have been obtained by flash evaporation where the relevant material combination is supplied continuously to an evaporation source in small portions, the temperature of which is considerably higher than the evaporation temperature of all components of the combination. The varying vapour pressures can be neglected in explosive evaporation and a homogeneous film is obtained after condensation of the composite vapour (Maissel and Glang 1970). Homogeneous starting material of uniform grain size and constant supply of the material mixture to the source are extremely important for a perfect layer. Vibrating feeders are often used for the supply (Pulker 1999).

A third possibility of using the single-source mode consists in the evaporation of only 10 to 15% of a crucible filling with a material mixture composed of materials not too different in their vapour pressures. Thin films with fairly good and uniform composition can be obtained in this way.

1.2.2 Evaporation of Chemical Compounds

In contrast to metals most chemical compounds decompose at least partially if not completely during evaporation. Stoichiometric films can therefore often not be obtained by direct evaporation. However, there are exceptions.

Some simple metal halides, for instance, can be evaporated practically without dissociation and form stoichiometric compound films on the substrates. Technically important metal fluorides are LiF, MgF_2, CaF_2, AlF_3, LaF_3, NdF_3, PbF_2 and only in special cases ThF_4 (Pulker 1999).

Compounds with complex anions like $Na_3(AlF_6)$ dissociate almost without exception and show fractionation phenomena of their components. In the case of cryolithe NaF appears first in the vapour followed by AlF_3. Consequently NaF is found in the deposited film and also the desired $Na_3(AlF_6)$ as well as $Na(AlF_4)$ both formed by recombination processes (Pulker and Zaminer 1970).

A relatively high number of metal oxides is used for optical films. Such oxides are e.g. SiO_2, Al_2O_3, MgO, Nd_2O_3, Gd_2O_3, Y_2O_3, La_2O_3, ZrO_2, HfO_2, Ta_2O_5, Nb_2O_5, GeO_2, and SnO_2. Only few suboxide molecules can be evaporated having unchanged composition of the starting material. Examples are SiO, GeO and SnO. Evaporation of suboxides can be performed from resistance heated boats.

Oxides like SiO_2, Al_2O_3, ZrO_2, Ta_2O_5, etc., are generally evaporated – with pronounced dissociation – by electron beam heating. The dissociated materials are partly recombined during condensation.

Using the method of reactive deposition (Auwärter 1952, Frey and Kienel 1987, Ritter 1966) completely oxidised, i.e. stoichiometric films can be obtained after evaporation and subsequent condensation of oxidic starting materials.

$$SiO_2 \xrightarrow{\Delta T} SiO + \tfrac{1}{2} O_2 \qquad \text{evaporation}$$

$$SiO + \tfrac{1}{2} O_2 \rightarrow SiO_2 \qquad \text{condensation}$$

Since most of the oxygen liberated during evaporation disappears in the vacuum pumps, oxygen gas must be admitted into the coating chamber during condensation and film formation.

Metal sulfides, selenides and tellurides such as ZnS, CdSe, ZnTe, as well as III/V compounds like InAs, GaAs, and others also show decomposition. However, most of them can also be deposited as stoichiometric films by reactive techniques or by substrate temperature controlled multi-source evaporations (Pulker 1999).

1.2.3 Evaporation Sources

Various types of evaporation sources are available to heat and to vaporize a starting material for film deposition. Reviews concerning this topic can be found in nearly all books dealing with the production of PVD-films (Pulker 1999). The choice of a proper evaporation source depends on the material to be deposited, the thickness, the number and the required purity of the films, and also on the special deposition process used and on the existing coating plant with its installations.

1.2.3.1 Resistively Heated Sources

Some conducting materials like C, Fe, Ti, Rh,.., in the form of wires or rods can be sublimated by direct current flow. However, this technique is only rarely applied.

Most commonly used is the evaporation or sublimation of the coating material from a resistively heated container in the shape of a boat, a crucible, a coil or a strip made of refractory metal (Mo, W, Ta) or C, in some cases with a ceramic (Al_2O_3, BN, TiB_2) insert.

Unwanted film contamination can occur through undesirable chemical reactions between film material/hot crucible wall or/and reactive gases/hot crucible, forming volatile compounds. For evaluation of the best source material and its shape and for evaporation of a given coating material information can be obtained in the coating material catalogues of suitable companies.

1.2.3.2 Inductively Heated Sources

A coating material can be heated and evaporated inductively in direct or in indirect manner. This might be of interest when larger amounts of material, that becomes volatile at lower temperatures have to be evaporated. However, since the costs of the apparatus are high, application of this technique is limited.

1.2.3.3 Electron Beam Heated Sources

Electron beam heating is today the universal method to evaporate practically all starting materials used in optical film deposition. The coating material in a water-cooled crucible heated and molten by high-energy electron bombardment has no chance for unwanted chemical reactions with the cold walls. Thus, film contamination can be excluded.

Beam shaping devices enable influencing the area energy density so that an optimum power concentration for a maximally uniform evaporation for each material can be obtained. Various crucible sizes and shapes are available and to minimize the heat-loss to the coolant water ceramic inserts can be used in special cases (Pulker 1999).

Two technical solutions of the various possibilities to realize electron beam heating are of practical importance: the beam deflection gun and the tele-focus gun.

With the deflection gun, electrons emitted from a hidden hot-filament cathode and accelerated in an electric field are deflected magnetically by about 270° and focused on the coating material in the, generally rotating, anodic crucible. In this design traces of evaporated material from the electron emitter cannot cause film contamination. The acceleration voltage ranges between 6 and 10 kV. The focal spot size can be modified by variation of a Wehnelt potential and there is also an xy-sweep. For metals area power densities up to 10 $kWcm^{-2}$ are required, whereas dielectrics generally need only 1 – 2 $kWcm^{-2}$.

With the tele-focus design, the electron emitter and electron acceleration system are usually at a greater distance from the crucible. Tele-focus e-guns are often operated with higher power and used with sweeping beam, e.g. to heat large rectangular crucibles in roll to roll plastic foil metallization.

1.2.4 Flash Evaporation

Not only alloys are often problematic to evaporate homogeneously in an economic manner, but also various mixtures of e.g. different refracting metal oxides.

In film system production it is sometimes difficult or impossible to find a single metal oxide with a certain refractive index value required for the realisation of the design of a special optical film system.

In this case it may be possible to mix appropriate metal oxides to obtain this desired refractive index in the mixture. The material mixtures are compressed and sintered and then the compacted material is brought to a proper, uniform and small grain size.

Flash evaporation is a useful technology for evaporation of such material granulates. It is performed with a continuous supply of the relevant granulated material to a hot evaporation source with a temperature considerably higher than the evaporation temperatures of all constituents of the mixed coating material (Frey and Kienel 1987, Maissel and Glang 1970). For the continuous material supply a vibrating feeder can be used (Pulker 1999). As evaporation sources both resistance heated boats and electron beam guns can be employed. A homogeneous film is generally obtained after condensation.

1.2.5 Reactive Evaporation

As already mentioned above, heating of chemical compounds for evaporation generally results in a more or less strong material decomposition independent of the applied heating method.

Since in the case of metal oxides part of the liberated easily volatile oxygen disappears in the pumping system non-stoichiometric and optically strongly absorbing films are obtained.

In order to obtain a stoichiometric non-absorbing oxide film a relatively high oxygen partial pressure of 1 to 2.10^{-4} mbar and a slow coating material deposition rate of 0.5 to 1 nm sec^{-1} are required.

Metal oxide films are generally produced by using sub-oxides or metallic starting materials. Basically also sulphides and nitrides or other compounds can be produced in a similar manner.

In reactive evaporation it is advantageous to activate the reaction partners and care must be taken that the gas components consumed during the chemical reaction are continuously replenished.

Reactive evaporation was invented by M. Auwärter in 1952. Concerning the place of the reaction there are principally three possible sites: on the substrate or film surface, in the vapour phase or on the evaporant surface.

Many considerations and theoretical treatments concerning reactive evaporation have been undertaken and are partly reviewed e.g. in (Pulker 1999).

There are collisions between coating material species and reactive gas molecules in the gas space but the reaction probability is very low. The reaction takes place to a far greater extent on the substrate and film surface exposed to the incident vapour and gas molecules.

Chemical film formation is decreased at elevated substrate temperature. The reason is very likely a reduced residence time of the gas molecules on the substrate surface important for the progress of the chemical reaction. Nevertheless, in most reactive evaporations substrates are heated up to 200° and 300°C to improve the mechanical film properties.

Summarizing, it can be stated that oxide film formation with reactive evaporation is influenced by the rate of incidence of the coating material atoms or molecules and the reactive gas molecules, the residence time of the relevant reaction partners and the substrate temperature. Concerning the influence of the starting compound on the properties of reactively evaporated TiO_2 films, results from careful investigations have been published (Pulker et al. 1976, Selhofer et al. 2002).

1.2.6 Ion and Plasma Assisted Evaporation

For interference optical applications high quality low-loss films are required that have good adherence and abrasion resistance with temperature and environmentally stable optical and mechanical characteristics.

It is not possible to obtain chemical compound films with such characteristics in conventional reactive evaporation by deposition onto unheated or heated substrates.

However, energy input directly into the growing film, e.g. by interactions with activated states and bombardment with energetic particles, in order to eliminate the deficiencies inherent in classical reactive evaporation improve both the optical and the mechanical film properties (Konuma 1992).

Coating experiments with activated and ionized oxygen molecules basically proved that the presence of a cold plasma enhances chemical reactivity, improves stoichiometry and so decreases residual optical absorption in the films. Bombardment of a growing film from an ion gun with argon and oxygen ions with energies up to several hundred electron volts improves not only stoichiometry and transparency, but also the morphology and density of films deposited even on unheated substrates.

High density bombardment-grown amorphous films result not only in high bulk-like and stable refractive indices but also in high intrinsic compressive mechanical stress.

Although compressive stress improves film hardness and abrasion resistance, too high values are disadvantageous in multilayer applications.

1.2.6.1 Activated Reactive Evaporation ARE

The reaction probability in collisions between the reactants in reactive film formation can be enhanced by activating one or even both of them (Auwärter 152, Bunshah and Raghuram 1972, Heitmann 1971). This means it is advantageous for the occurrence and for the yield of a reaction when the reactants are activated, e.g. in electronically excited states.

Activation can be performed by uv radiation, electron impact or all together in the presence of a gas discharge plasma.

In order to generate mainly activated states of oxygen, glow-discharge sources with a hollow cathode inside a quartz tube have been used successfully (Heitmann 1971, Pulker 1999).

With TiO as the starting material residual absorption of TiO_2 films is decreased by a factor of 10 and with Ti_2O_3 or Ti_3O_5 the factor is even 100 compared with TiO_2 films obtained with ordinary oxygen (Pulker 1999).

Although the resulting absorption is slightly higher, substrate heating is required for highly adherent and hard films, since film formation occurs at mainly thermal energies.

1.2.6.2 Ion Beam Assisted Deposition IBAD

Most properties of evaporated films are mainly determined by the lack of mobility of the condensing atoms and molecules due to their low thermal energies lower than 0.1 eV.

Controlled bombardment of a growing film by argon ions or in the reactive mode also by oxygen ions with energies from around 10 to up to several hundred electron volts improves density and stoichiometry and therefore the optical properties of oxide coatings deposited on unheated substrates (Allen 1983, Martin and Netterfield 1986, Martin et al. 1983). Adding energy to the film growth process also appears to improve adhesion (Hirsch and Varga 1978) and to modify stress (Herrmann and McNeil 1982). Ion beam techniques can be applied also before and after film deposition. Substrates can be ion beam cleaned with inert or reactive gas ions, depending upon substrates involved, and freshly deposited films can be bombarded to further improve their properties.

Hot-cathode Kaufman-type (Kaufman et al. 1982, Kaufman 1986) ion guns are often used in IBAD, but because of erosion of the filament and possible film contamination also other cold-cathode installations have been applied (Rickerby and Matthews 1991). It is impossible to discuss and compare all the variants of this technique here. However, it should be mentioned, that larger substrate surfaces can hardly be bombarded uniformly. Therefore intermittent treatment of rotating substrates is generally applied.

Considerable efforts have been directed to the analysis of plasma conditions and their correlation with the properties of the deposited films.

The mechanism responsible for densification is momentum transfer, as was shown by molecular-dynamic calculations (Müller 1986).

1.2.6.3 Ion and Plasma Plating

Plating processes are different ion and plasma technologies for supplying energy uniformly and simultaneously over extended areas to films growing on unheated substrates. The technologies are characterized by the use of very dense cold plasmas. The energy input during film formation changes film morphology and film density and with it all related properties. In such processes, where all molecular and atomic partners in the coating process are ionised and activated to varying degrees, the effective surface temperature of the substrate is raised. However, the total thermal load can be kept low enough so that most plastic substrates can be coated without damage.

It is possible to perform such depositions in reactive and non-reactive modes. For interference coatings the reactive modes are preferred. Besides oxides also other compounds, e.g. oxynitrides and nitrides, can be synthesized (Pulker 1999).

Reactive Low Voltage Ion Plating RLVIP

Reactive ion plating is a process for production of chemical compound films mainly by direct synthesis from the elements, e.g. $2\ Ta + 2.5\ O_2 \rightarrow Ta_2O_5$, or from suboxides, e.g. $2\ TaO + 1.5\ O_2 \rightarrow Ta_2O_5$.

The reactive ion plating process (Pulker et al. 1985) is performed in the specially designed automatic box-type plating system BALZERS BAP 800 shown schematically in Fig. 1. It can be used for deposition of single layer and multilayer oxide coatings onto unheated glass and other unheated substrates. The evaporations are made by two special 270° deflection type electron beam evaporators. The starting materials, metals or suboxides, form electrically conducting melts. Very effective ionisation and activation of the evaporating coating material atoms and the admitted reactive gas molecules results mainly from a low-voltage (typically 60 V) high-current (typically 60 A) electron beam of a non-self-sustaining thermionic arc discharge plasma directed to the anodic crucible of the e-beam evaporators. The cathode of the non-self-sustaining arc discharge is a resistance heated Ta filament in a separate chamber kept under a relatively high argon pressure. The substrate holder is electrically insulated. In contact with the formed plasma cloud the substrates receive a relatively high negative self-biasing potential of -15 to -25 V with respect to the plasma, which together with the repulsive forces of the positive ions from the anodic crucible (potential typically between 70 and 40 eV) mainly determines the impact energy of the film forming ions.

The total pressure in the plant is in the low 10^{-3} mbar range. Film deposition is started and stopped by opening or closing moveable shutters in front of the electron beam evaporators. Film thickness and deposition rate are controlled by optical thickness monitoring and oscillating quartz crystal monitoring (Edlinger et al. 1987).

Quarter-wave films in the visible are deposited within 3 to 5 min. A uniformity in thickness distribution of 61% over the whole substrate holder of 800 mm can be achieved with a simple static correction shield (Pulker 1999).

Reactive Advanced Plasma Source Ion Plating RAPSIP

Another ion plating technology is based on the use of the advanced plasma source APS from LEYBOLD. The high power plasma source is a relatively new development (Matl et al. 1991, Beisswenger et al. 1994) that is applied in a batch box-coater system. The principle of the RAPSIP is shown in Fig. 2.

Fig.1

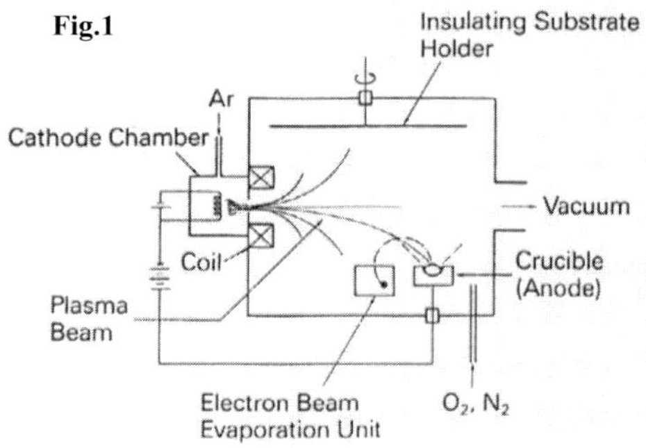

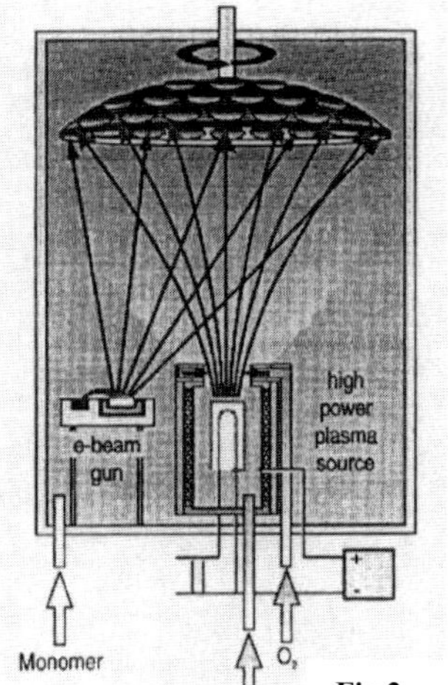

Fig. 1. Schematic of Reactive Low Voltage Ion Plating (RLVIP)

Fig. 2. Schematic of Reactive Advanced Plasma Source Ion Plating (RAPSIP)

The plasma source, which consists of a large area LaB_6 cathode, a cylindrical anode tube and a solenoid magnet, is placed in the center of the plating chamber bottom next to the electron beam evaporators. The large cylindrical LaB_6 cathode is heated indirectly by a graphite filament heater. A dc glow discharge plasma in Ar is switched on. The solenoid magnet surrounding the source forces the electrons in axial directions. Due to the magnetic field the electrons move in the direction of the substrate holder. They spiral along the magnetic field lines thus bringing the plasma into the coating chamber. The reactive gas inlet in the form of a ring shower is located on the top of the anode tube. Blown directly into the dense plasma the reactive gas is ionized and activated. Since the dense plasma fills the total volume of the chamber, the evaporant also becomes partly ionized (Pulker 1999). Improved chemical reactions typical for IAD occur, but also ion plating takes place because of a negative self-biasing potential of the substrates and the chamber walls. This potential can be higher than −100 V and is mainly determined by the discharge voltage, the magnetic field strength of the solenoid and the gas flow. The generation of the plasma is completely separated from evaporation. Therefore the process is compatible with any evaporation material like metals, semiconductors, suboxides and oxides.

With the APS 8 kW plasma power is possible and a total ion current of 5 A with excellent uniformity over a large substrate holder surface of about 1 m^2 can be achieved (Zöller et al. 1996). It is further possible to use the APS for a plasma enhanced chemical vapour deposition PECVD process. There is an additional gas inlet for a monomer gas to be added to the plasma leading to a deposition process of e.g. a hydrophobic top-coat on the surface of an antireflectance coating of ophthalmic lenses (Zöller 1997). Film properties of oxide coatings obtained by different reactive deposition technologies are shown in Table 3.

Table 3. Refractive indices (n_{550} films, n_{589} bulk materials) of various metal oxide films obtained by reactive ion plating techniques and conventional reactive evaporation

Film material	SiO_2	Al_2O_3	ZrO_2	Ta_2O_5	TiO_2
Reactive evaporation Ts = 300°C	1.46	1.61	1.96	2.10	2.34 – 2.44
RAPSIP Ts = 50°C	1.47	1.63	2.10	2.12	2.38
RLVIP Ts = 50-80°C	1.48	1.66	2.18	2.24	2.48 – 2.55
Bulk material	1.46 (fused silica) 1.54 (quartz)	1.75 – 1.77 (corundum)	2.17 – 2.20 (baddeleyite)	-	2.60 – 2.90 (rutile) 2.49 – 2.56 (anastase)

Ts: substrate temperature.

1.2.7 Laser Induced Evaporation / Ablation

An energetic laser beam directed to a coating material target situated in a vacuum chamber causes vaporization if the energy is absorbed by the material and the vapour condenses on a substrate and forms a coating. The advantage of such a process is the extraordinary high cleanliness in the heating technique. A light beam can also easily be brought into a coating plant and focused on a material target where it produces the desired vaporization. Therefore such processes have been studied extensively (Brinsmaid et al. 1957, Miller 1994, Pulker 1999, Reisse et al. 1994, Sankur and Cheung 1988, Sankur 1992).

CO_2-lasers, Nd-YAG –lasers and excimer lasers are generally used for this purpose. Power densities of 10^7 to 10^{10} W cm^{-2} are possible with highly focused laser beams. As a consequence of the very high energy densities, the ablation process is different to thermal evaporation.

A plasma is formed on the target spot and the large amount of material ions ejected from the plasma plume have high kinetic energies of some ten electron volts (Reisse et al. 1994). Unfortunately, in this vaporization process the ablated species contains droplets, which drastically decrease the quality of optical films.

This problem can be partially overcome by using pulsed irradiation with pulse durations between 20 and 200 ns and repetition rates between 40 and 150 Hz. In this fast explosive-like vaporization process the ablated species might contain fewer and smaller droplets than in the continuous mode of operation. With, e.g. excimer laser pulses it is possible to ablate a target and obtain a film without changes in the composition. This congruent vaporization is important for the deposition of alloy films.

For wider applications of laser ablation the energetical efficiency should be increased by means of stronger laser light absorption in the starting material.

However, the high costs, the mentioned problems with splashing of the eroded material and other difficulties still limit the use of laser ablation in coating technology (Chrisey and Hubler 1994, Dielemann et al. 1995).

1.2.8 Arc Source Vaporisation

Cathodic arc source vaporization is not a new technique to obtain coatings (Boxman 2000), but it has more recently undergone considerable development to a reliable and powerful technology for tribological coatings (Boxman et al. 1996). A vacuum arc is a type of electric discharge where the electron/atom interactions and ionizing processes occur in a metal vapour released from the cathode by the arc.

The metallic cathode, with the grounded vacuum coating chamber as anode, works in the low-voltage high-current mode. Unlike the case of the abnormal discharge where the entire cathode is emissive, 1 to 10 μm sized cathode spots are formed in the discharge with a duration of about 10^{-6} s. The current density in the spots, moving fast and randomly across the cathode surface, is between 10^6 and 10^8 A cm^{-2} (Boxman et al. 1996).

The arc source plasma contains a high concentration of positive metal ions with considerable energy of about 5 to 50 eV. Reactive deposition is possible in the presence of a reactive gas like oxygen or nitrogen. Highly adherent and dense films of high hardness, e.g. TiN, have been achieved. The problem is droplets, commonly referred to as macroparticles. Their size ranges between 0.1 and 100 μm. They appear independent of the mode of operation, dc or pulsed.

Not a problem in most tribological coatings, droplets in films prevent the use of arc vaporization for applications in precision optics. Recently coating experiments with different filtered arc configurations have been performed. They resulted in a very good film quality, but also in a marked decrease of coating material flux (Martin and Bendavid 2001). The filter devices are also not inexpensive.

1.3 Sputter Ablation

Cathodic sputter ablation or sputtering determines the ejection of atoms from a solid cathodic target as a consequence of the momentum transfer action of bombarding high energy positive gas ions or atoms out of a gas discharge plasma.

At low argon working gas pressure between 10^{-2} and 10^{-3} mbar and target-substrate distances of 5 to 15 cm, the high energy of the sputtered atoms result in good film quality even when the films are deposited on unheated substrates. Besides metal-sputtering also reactive processes, e.g. for oxide or nitride formation, are possible.

Sputtering was discovered 150 years ago in England by Grove (Grove 1852) and in Germany by Plücker (Plücker 1858). More about the history of the oldest PVD-technology is found in the monograph Coatings on Glass (Pulker 1999).

Since about 1955 gas-discharge sputter deposition has become an important film production technology. The invention of magnetron cathodes improved the sputter rate. High frequency sputtering can be applied for depositing insulating chemical compounds from insulating targets, but low deposition rates and target cooling problems often restrict the use of this technique in practice. State of the art in gas discharge sputtering is the use of twin or dual magnetron sputter cathodes operated in dc or pulsed dc mode (Schiller et al. 1995, Szczyrbowski and Teschner 1995). The technology is successfully used for mirror coating particularly of Al without and with reflectance enhancement with dielectric films for antireflective and antistatic layer deposition, for production of transparent conductive films, and for coatings on ophthalmic lenses. For precision optical film deposition newly designed sputter machines are known which enable production of laser mirrors, broad and narrow band interference filters, beam splitters and colour filters. Also film systems for EUV exist (Kaiser 2002). Magnetron sputtering is very important in large area glass coating for architectural and display applications (Bräuer 2002).

1.3.1 Reactive Gas Discharge Sputtering

Reactive sputtering with simple metal dc-magnetron cathodes may cause problems because of insulating compound formation on the target surface and its well known consequences. Although most of the problems can be overcome by the use of the bipolar twin or dual magnetron technology (Schiller et al. 1995, Szczyrbowski and Teschner 1995) mentioned above, there exists another interesting technical solution in the form of a cyclic process: metal deposition followed immediately by oxidation (Scobey et al. 1989). In practice alternating dc-magnetron sputtering of two to three monolayer metal films is carried out with subsequent oxidation in a high intensity reactive plasma zone. The reactive plasma zone is separated physically from the metal deposition zone by a region of relatively low pressure. The substrates are mounted on a cylindrical drum which rotates fast in a differentially pumped vacuum chamber. Baffled dc-magnetrons sputter the metal films and in a long narrow zone reactive plasma sources with high-intensity ion flux perform effective oxidation. The use of multiple stations enables a considerable increase in deposition speed. Rates comparable with those in reactive evaporation can be achieved. Not only metal oxides but also other materials can be deposited in straightforward synthesis by this technology. Curved substrates can be coated with single films or multilayers. However, it is impossible to coat larger flat surfaces with this method. Another technology uses metallic twin magnetrons combined with an additional reactive microwave plasma (Boling et al. 1995, Ebert and Ebert 2002).

1.3.2 Ion Beam Sputtering IBS

Energetic ion beams can be used for substrate cleaning, for deposition and for treatment of films during and after deposition. Modern broad-beam ion sources of the Kaufman type (Kaufman and Robinson 1987) operating in clean vacua between 10^{-4} and 10^{-5} mbar in dc or high frequency mode are used with 45° impact angle to erode the target.

In case of insulating targets, the positive argon ions are neutralised by electrons coupled into the beam.

In dual ion beam sputtering a second reactively operated ion gun is used, with a beam of lower energy, directed at normal incidence to the growing film, as in ion beam assisted deposition (Wei et al. 1995).

The sputtered atoms and molecules arrive at the unheated substrate with energies between 20 and 100 eV. The energy input into the growing film in clean, reactive, low-pressure environments results in dense, pure, highly-adherent, amorphous and stoichiometric layers with excellent optical properties and high stability (Wei 1989).

Multilayers can be produced by changing the target. However, the substrate area that can be coated uniformly is only 10 to 12 cm in diameter and the deposition time for a one quarter-wave (visible range) thick film is about 20 min (Pulker 1999).

2 Chemical Vapour Deposition CVD

Chemical vapour deposition is a material synthesis method in which the constituents of a vapour phase react in heterogeneous reaction sequences to form a solid film on a solid substrate. The search for suitable reactants and reactions is still in progress. Most used reactions can be put into one of the following four categories: decomposition, reduction and oxidation, hydrolysis and polymerisation. Activation of the reactions is possible by various means, e.g. temperature, radiation, plasma, etc.

Film formation on the substrate can take place in air at atmospheric pressure or at low pressure and in controlled residual gas, at high or at low temperatures and with or without additional activation. This clearly shows that CVD encompasses an interdisciplinary range of chemistry and physics (Gordon 1997, Kern 1988).

For coating glass, low temperature processes are preferred. Many of them require substrate temperatures between 300 and 600°C. This is normally too high for plastics and there are only a few processes which run below 100°C.

Many low temperature reactions are known for the deposition of metal oxide, nitride and oxinitride films on glass. Starting compounds are metal alkoxides, alkyls, halides or acetyl-acetonates. Most frequently pyrolysis and pyrolytic oxidation reactions are used to form metal oxides like TiO_2, ZrO_2, HfO_2, SiO_2, Al_2O_3, Ta_2O_5, and Nb_2O_5 (Ree 1996, Pulker 1999). The use of plasmas allows lowering the surface temperature, but elevated temperatures up to nearly 200°C are still required to avoid incorporation of unwanted reaction products (H_2,HCl,CH_x) in the films. Operating in a pulsed mode improves the film quality and its use is therefore preferred.

CVD technologies are very important for film production in the electronics industry (Kern 1989) and in the deposition of hard and wear resistant coatings on tools (Yee 1978). In optics CVD is used less often. Only plasma polymerisation for the deposition of a base or top coat in ophthalmics and in car head lights is used. However, a modified CVD plasma impulse technology originally used in ultra low loss optical fiber production (Lydtin 1986, Bauch et al. 1987) became important for high reflectance optical multilayer fabrication.

2.1 Plasma Impulse Chemical Vapour Deposition PICVD

Plasma impulse CVD is a new very economic coating technology developed by SCHOTT (Segner 1991). Starting with high purity precursers low-loss and stable metal oxide layers, like SiO_2 and TiO_2, can be obtained in a pressure range of 0.5 to 5 mbar with high deposition rates of several 100 nm min^{-1} on flat or dome-shaped glass substrates smaller than 20 cm in diameter with temperatures between 40 and 150°C. A microwave impulse plasma initiates the typical reaction sequence of gas phase and surface reactions, e.g. :

$SiCl_4 \xrightarrow{\frac{1}{2}O_2,\,Ar,\,plasma} SiO + 2Cl_2$ homogeneous gas phase reaction

$SiO \longrightarrow$ diffusion to substrate

$\frac{1}{2}O_2 + SiO \longrightarrow SiO_2$ heterogeneous surface reaction

Gas filling, reaction time and exhaust of the by-products occur very fast in the millisecond range.

a) 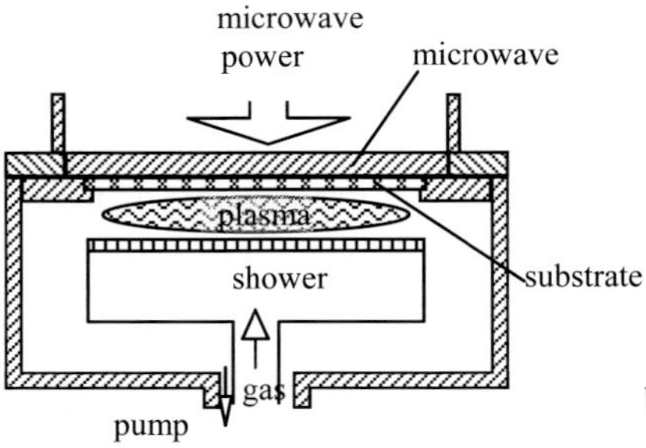

b) 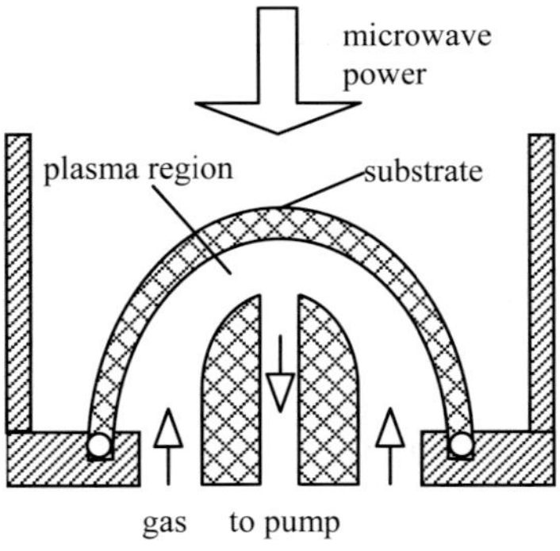

Fig. 3. Schematic of plasma impulse chemical vapour deposition recipients
a) for flat substrates, **b)** for dome-shaped substrates

The film is deposited pulse per pulse with one or several monolayers. Since the reactions run with complete efficiency, the deposition rate is controlled only by mass flow at a constant microwave power and pulse repetition time. The reactor is shown schematically in Fig. 3.

In 1993 SCHOTT-AUER successfully introduced the plasma impulse CVD process into industrial production of 50 mm cold light mirrors for halogen lamps (Bewig et al. 1997). A special multistage PICVD coater has been developed in which each reflector forms its own mini-vacuum-recipient. Evacuation, gas supply and exhaust are performed from the bottom. A moveable microwave antenna for the deposition phase is on the top. The coating yield and equipment up-time of this coater have been reported to be excellent (Krause 1997).

3 Summary and Outlook

The possibility to improve film performance in conventional reactive and even in activated reactive processes is rather limited. However, it is very effective in higher energy coating processes particularly with continuous or pulsed plasma assistance. Substituting substrate pre-heating by direct energy input into the growing film via energetic particle bombardment and irradiation improves film morphology and the stoichiometry of chemical compound films. Avoidance of substrate heating is of great importance because of the enormous increase in coating plastics. A tendency became apparent that sputtering is more commonly used also for precision optical coatings.

Problems with reactive gas processes, formerly limiting applications, could be overcome by bipolar twin magnetron technology or by other means with multiple plasmas.

For interference optical film production PVD technologies are dominant. CVD is only rarely used. An exception is the plasma impulse technology PICVD.

In general thickness uniformity, in situ measurement technology, the speed of the processes and the yield per run are important topics besides improvements in automatic vacuum and deposition control and the gradual replacement of batch processes by continuous deposition.

References

Allen T (1983) Proc. International Ion Engineering Congress, ISIAT 83 and IPAT 83, Kyoto, Japan Vol. 2 pp 1305

Auwärter M (1952) BALZERS AG, Austrian Pat.Nr.192650), US Pat. No.2.920.002 (1960), FRG Pat. Nr. 1.104.283 (1968)

Bach H, Krause D (1997) Thin Films on Glass. Springer, Berlin

Bauch H, Krause D, Kersten R (1987) CVD in microwave produced plasmas for fiber preforms. J. Opt. Commun. 8 (4):130

Beisswenger S, Götzlmann R, Matl K, Zöller A (1994) Proc. Soc. Vac. Coaters, 37 th Ann. Techn. Conf. 1-878068 – 13 – X

Bewig L, Küpper T, Langfeld R (1997) Cold Light Reflectors. Chapter 6.4 In: Bach H, Krause D. (eds.) Thin Films on Glass. Springer, Berlin pp. 346-350

Bräuer G (2002) Chapter 7 of this book

Brinsmaid D, Koch G, Keenan W, Parson W (1957) KODAK US Pat. No.2.784.115, Application made May 4 (1953)
Bunshah R, Raghuram A (1972) J. Vac. Science Technology. 9:1385
Boling N (1995) A high rate reactive sputtering process for batch, in-line, or roll coaters. Proc. Soc. Vac. Coaters 38th Ann. Techn. Conf. pp286-289
Boxman R (2000) Vacuum arc deposition: early history and recent developments Proc. ISDEJV Xian, China
Boxman R, Martin P, Sanders D (eds.) (1996) Handbook of Vacuum Arc Science and Technology. Noyes, New York
Chrisey D, Hubler G (eds.) (1994) Pulsed Laser Deposition of Thin Films, Wiley, New York
Dieleman J, Biermann U, Hess P (eds.) (1995) Photon Assisted Processing of Surfaces and Thin Films. Elsevier, Amsterdam
Dushman S, Lafferty J.M (1962) Scientific Foundation of Vacuum Technique, 2nd. ed. Wiley, New York
Ebert M, Ebert J (2002) Magnetron Sputtern im Mikrowellenplasma im Vergleich zu konventionellen Aufdampfverfahren. Photonik Vol. 34 No. 1: 34-37
Edlinger J, Buehler M, Pulker HK (1987) Proc. Internat. Sympos. on Trends and New Applications in Thin Films, Strasbourg (F), March, Vol. II, p. 735, Supplem. a la Revue: Le Vide, les Couches Minces, No. 235 (Janv./Fev. 1987)
Faraday M (1857) Phil. Trans. 147, pp 145
Flory F (ed.) (1995) Thin Films for Optical Systems. Marcel Dekker, NY
Frey H, Kienel G (1987) Dünnschicht-Technologie. VDI-Verlag, Düsseldorf
Gordon R (1997) Chemical vapour deposition on glass, J. Non-Cryst. Solids 218 :81- 91
Grove WR (1852) Phil. Trans. Roy. Soc. London 142. pp 87
Graper EB (1973) J.Vac. Sci. Technol. 10 pp 100
Hass G (1946) Optik 1. pp 8
Hass G, Waylonis M (1961) J. Opt. Soc. Am. 51 pp 719
Heitmann W (1971) Appl. Optics. 10:2414
Herrmann WC, McNeil J.R (1982) Proc. SPIE, Vol. 325:101
Hirsch EH, Varga I.K (1978) Thin Solid Films 52:445
James AS, Matthews A (1990) Surface Coating Technology 41 pp 305
Kaiser N (2002) Chapter 3 of this book
Kaufman HR, Cuomo J.J, Harper J.M.E (1982) J. Vac. Sci. Technology 21:725
Kaufman HR (1984) Fundamentals of Ion Source Operation. Commonwealth Scientific Corp., Virginia
Kaufman HR (1986) J. Vac. Sci. Technol. A4:764
Kaufman HR, Robinson R S (1987) Operation of Broad Beam Sources. Commonwealth Scientific Corp., Alexandria pp 108-110, pp 177-179
Kern W (1988) Deposition Technologies and Applications. In: Schuegraf K.K (ed.), Handbook of Thin Film Deposition Processes and Techniques. Noyes, New York, pp 1-26
Kern W (1989) Chemical Vapour Deposition. In: Levy R.A (ed.) Microelectronic Materials and Processes. Klüwer Academie , Dortrecht pp 203-247
Konuma M (1992) Film Deposition by Plasma Techniques. Springer, Berlin
Kundt A (1888) Ann. Physik. 34, pp 473
Krause D (1997) PICVD Chapter 5.3 In: Bach H, Krause D (eds.) Thin Films on Glass. Springer, Berlin pp 244-261

Leyland A, James A.S (1991) Chapt.4 in Rickerby D.S, Matthews A (eds.) Advanced Surface Coating. Blackie, Glasgow
Lydtin H (1986) Lightwave Technology LT 4(8) pp 1034
Maissel LI, Glang R (1970) Handbook of Thin Film Technology. McGraw-Hill, New York
Martin PS, Bendavid A (2001) Review of the Filtered Vacuum Arc Process and Materials Deposition. Thin Solid Films 394:1-15
Martin PJ, Bendavid A (2001) Thin Solid Films, 394: 1-15
Martin PJ, Macleod H.A, Netterfield R.P, Pacey C.G, Sainty W.G (1983)
Applied Optics. 22:178
Martin PJ, Netterfield R.P (1986) Progress in Optics. 23:115
Matl K, Klug W, Zöller (1991) Mat. Sci. and Engng. A140:532
Miller JC (1994) Laser Ablation. Springer, Berlin
Müller KH (1986) J.Vac.Sci.Technol. A4 461 and Physical Rev. B, Vol. 35, No. 15 (1987-11) 7906
Nahrwold R (1887) Ann. Physik 31 pp 467
Plücker J (1858) Pogg. Ann 103 pp 88
Pulker HK (1999) Coatings on Glass, 2nd.rev. ed. Elsevier, Amsterdam
Pulker HK (1999) Coatings on Glass. 2nd revised ed. Elsevier, Amsterdam, Chapter 6 pp 141-144
Pulker HK, Paesold G, Ritter E (1976) Refractive indices of TiO_2 films produced by reactive evaporation of various titanium-oxygen phases. Appl. Optics 15:2986-2991
Pulker HK, Ritter E (1971) In: Auwärter M (ed.) Ergebnisse der Vakuumtechnik und der Physik Dünner Schichten. Bd.2, Wiss. Verlagsges. GmbH Stuttgart pp 244
Pulker HK, Zaminer Ch (1970) Thin Solid Films 5:421
Pulker HK, Haag W, M. Buehler, Moll E (1985) Proc. Munich, IPAT 85:299, CEP Consultants Ltd., Edinburgh
Pohl R, Pringsheim P (1912) deut. Phys. Ges. 14. pp 506
Rees W.S (1996) CVD of Non – Metals. VCH – Verlag, Weinheim
Reisse G, Weissmantel S, Keiper B, Steiger B (1994) Thin Films. Proc. DGM Informationsgesellschaft Verlag, TATF-94 pp 320
Reisse G, Weissmantel S, Keiper B, Johansen H, Martini T, Scholz R (1994) Proc. TATF-94 285:89
Rickerby DS, Matthews A (1991) Advanced Surface Coatings. Blackie, Glasgow and London
Ritter E (1966) Journal Vac. Science Technology. 3:225
Sankur H, Cheung T.J (1988) Appl. Phys. A 47:271
Sankur H (1992) Thin Solid Films. 218:161
Scobey M, Seddon R, Seeser J, Austin R, LeFebvre (1989) US-Patent No 4,851,095
Schiller S, Geodicke K, Kirchhoff V, Kopte T (1995) Pulsed Technology – a New Era of Magnetron Sputtering (Twin Magn.) In: Proc. SVC 38th Ann Techn. Conference, Chicago
Segner J, (1991) Plasma Impulse Chemical Vapour Deposition – a Novel Technique for the Production of High – Power Laser Mirrors. Mat. Sci. Engng. A140:733
Selhofer H, Ritter E, Linsbod R (2002) Properties of titanium dioxide films prepared by reactive electron beam evaporation from various starting materials. Appl. Opt. 41(4):756-762

Szczyrbowski J, Teschner G (1995) Reactive Sputtering of SiO_2-Layers onto Large Scale Substrates using an AC Twin Magnetron Cathode. In: Proc. SVC 38 th Ann. Techn. Conf. Chicago
Wutz M, Adam H, Walcher W (1982) Theorie und Praxis der Vakuumtechnik, revised and enlarged ed., Vieweg, Braunschweig
Wei DT, Kaufman HR, Lee Cheng-Chung (1995) Chapter 6. In: Flory FR (ed.) Thin Films for Optical Systems. Marcel Dekker, NY
Wei DT (1989) Ion beam interference coating for ultra – low optical loss. Appl. Opt. 28:2813-2817
Yarwood J (1982) High Vacuum Technique, 4th.edn. Chapman & Hall, London
Yee KK (1978) Protective Coatings for Metals by CVD. Int. Metals Rev. 1:19-42
Zöller A, Götzlmann R, Matl K, Cushing D (1996) Appl. Opt. 35:5609
Zöller A (1997) Vakuum in Forschung und Praxis. Wiley VCH Weinheim 1:32

Abbreviations

PVD	Physical Vapour Deposition
CVD	Chemical Vapour Deposition
ARE	Activated Reactive Evaporation
IBAD	Ion Beam Assisted Deposition
IAD	Ion Assisted Deposition
RLVIP	Reactive Low Voltage Ion Plating
APS	Advanced Plasma Source
RAPSIP	Reactive Advanced Plasma Source Ion Plating
PECVD	Plasma Enhanced Chemical Vapour Deposition
EUV	Extreme Ultra Violet
IBS	Ion Beam Sputtering
PICVD	Plasma Impulse Chemical Vapour Deposition
e-gun	Electron Beam Evaporation Device

Symbols

N	Number of Atoms
A	Surface
t	Time
m	Atomic Mass
α_e	Evaporation Coefficient
P_e	Equilibrium Vapour Pressure
P_h	Hydrostatic Pressure
k	Boltzmann´s Constant
T	Temperature
T_s	Substrate Temperature
M	Molar Mass

Γ	Mass Evaporation Rate
λ	Wavelength
n	Refractive Index
k	Extinction or Absorption Coefficient
$\cos^b\alpha$	Spatial Vapour Distribution
b	Exponent e.g. 1-5

Large Area Deposition

Günter Bräuer

Fraunhofer Institute for Electron Beam and Plasma Technology, Dresden, D
Fraunhofer Institute for Surface Engineering and Thin Films, Braunschweig, D

1 Magnetron Sputtering

1.1 Introduction

The preferred process for deposition of high precision thin films on large areas (glass panes of up to 3.2 m in width, transparent plastic foils of up to 2 m in width or large metal sheets) is magnetron sputtering.

"Sputtering" means the ejection of atoms from a solid surface (a so-called target) by impact of energetic ions. It was first observed more than 100 years ago as a disturbing effect eroding the cathodes in glow discharges.

The sputter process takes place in a glow discharge plasma at low pressures. The physics and characteristics of glow discharges are described in many books, for example in (Frey 1995, Haefer 1987, Pulker 1999). Figure 1 shows the principles of a simple diode sputter process. A vacuum chamber is pumped down to a typical pressure of 10^{-3} Pa and then backfilled with argon to a pressure of about 10 Pa. If a voltage of several kV is applied between the two planar electrodes, a self-sustaining glow discharge is ignited. Positive argon ions are generated by collisions with energetic electrons. They are accelerated towards the cathode, hit the target and transfer their momentum to the target atoms which are emitted as a consequence of collision cascades being generated in the first several tens of atomic layers below the surface. These sputtered particles form a thin film on the substrate (anode) located on the opposite side of the cathode. Argon ions striking the target also create secondary electrons which are accelerated towards the anode and gain enough energy to produce further argon ions.

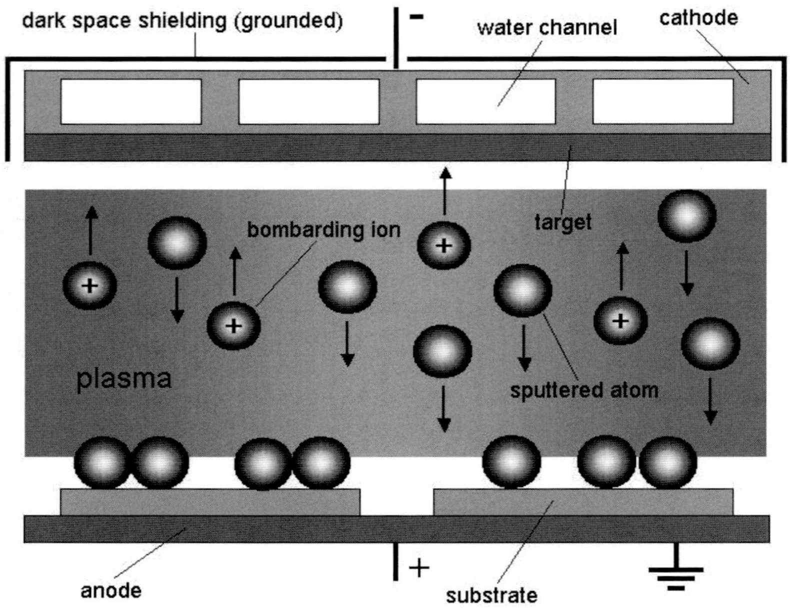

Fig. 1. The principle of diode sputtering

An exhaustive description of the physics of cathode sputtering and the various dependencies is given for example in (Behrisch 1981) and (Oechsner 1975). The most important number is the sputter yield Y giving the mean number of emitted target atoms per incident ion. Y depends on several parameters like target and ion mass, ion energy and angle of incidence. Of further importance is the energy distribution of sputtered particles. It has a maximum at a value corresponding to half of the surface binding energy of the target material. The mean energy is around 5 eV und thus around ten times higher than the energy of evaporated particles (around 0.5 eV). This difference is of great importance for the density of the growing film.

Diode discharges as shown in Fig. 1 are not very effective with regard to industrial applications of sputtering. The working pressure is rather high and thus emitted particles loose their energy on the way to the substrate by various collisions with the species in the plasma. The high plasma impedance requires voltages in the kV range. Bombardment of the substrate with fast electrons results in temperatures of up to 500 °C making the process unsuitable to generate films on temperature sensitive substrates like plastic materials. The most severe disadvantage, however, is the very low deposition rate D describing the speed of film growth. Its value usually is given in nm/s or µm/h. For diode sputtering, D typically is around 0.1 nm/s.

The disadvantages described above could be overcome by the use of additional magnetic fields to reduce the plasma impedance or to increase the plasma density. In a combination of electric and magnetic fields electrons move on rather complicated cycloidal paths governed by the Lorentz force. Thereby the probability for electron-argon collisions is drastically increased, resulting in a strong decrease of plasma impedance and allowing operation pressures in the range of 0.1 – 1 Pa.

The principle set-up of a planar magnetron is shown in Fig. 2. The cathode in general consists of a copper body to which the target plate is fixed. Permanent magnets inserted into the copper body create a magnetic field in front of the target. The regions of high electron density and thus high ion density are located in those areas where the angle between the magnetic and electric field lines is larger than 45°. The plasma density has a maximum at 90°.

Intensive cooling of the target plate is necessary since approximately 85% of the applied electrical power is transformed to heat. Only 1 – 2% is used for the emission of target atoms, meaning that sputtering in principle is a process with an extremely low efficiency.

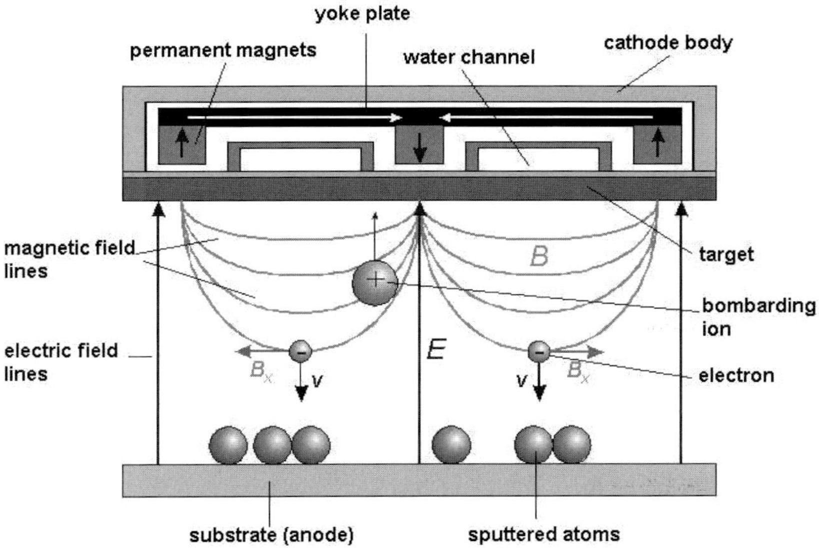

Fig. 2. The magnetron cathode

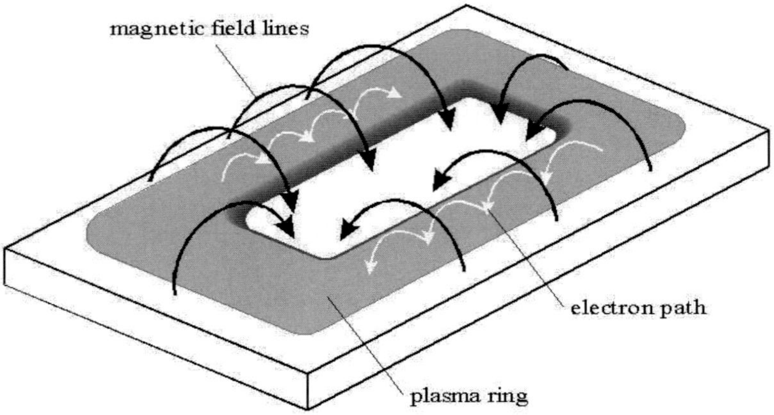

Fig. 3. Perspective view of a rectangular magnetron

Depending on the application, magnetron cathodes can be constructed with circular or rectangular shape using different magnet arrangements. However, at least one closed path with magnetic field lines perpendicular to the electrical field has to be formed in front of the target. Figure 3 shows a perspective view of the plasma ring ("race track") above a rectangular magnetron cathode.

While ion bombardment of the target surface in diode sputtering is uniform, the variation of ion density across the target of a magnetron cathode causes an inhomogeneous material erosion. As a consequence, used targets show a V-shaped erosion pattern. The target utilization of a standard planar magnetron is limited to 25 – 30%, which is the most severe disadvantage of magnetron sputtering. Many attempts have been made to increase the utilization by use of specially designed magnet arrays. One of these solutions enabling values up to 60% is described in (Kukla 1990). Another solution is the "C-MAG" described below.

Magnetron cathodes typically operate at discharge voltages around 500 V and power densities of up to 40 W/cm^2 related to the full target area. The achievable deposition rates for metals are in the range of 5 – 20 nm/s.

1.2 Reactive Sputtering of Compound Layers

Optical thin film stacks on large area substrates require fast and stable deposition of metals and compounds with high reproducibility and excellent uniformity. Depending on the application, thickness deviations have to be kept within tolerances ranging from ± 1% to ± 5% across a coating width up to 3 m. Table 1 summarizes oxides and nitrides that are of special importance for large area coating. They are arranged with regard to increasing index of refraction. Most of them are highly insulating.

Table 1: The most important compounds for optical coating on large areas

Material	Index of Refraction n (550 nm)*)
SiO_2	1.46
Si_3N_4	1.95 – 2.05
SnO_2	1.95 – 2.05
ZnO	1.95 – 2.05
$In_2O_3 – SnO_2$	1.95 – 2.05
Ta_2O_5	2.14 – 2.20
TiO_2	2.35 – 2.55

*) In general, the index of refraction depends on deposition conditions.

Due to their uniform plasma distribution in the longitudinal direction, sputter magnetrons provide very good coating uniformities up to widths of nearly 4 m. This is the main reason why magnetron sputtering has become the leading production process for large area thin film deposition. However, its applicability for manufacturing optical coatings has suffered from some serious drawbacks for a long time.

There are basically two approaches to generating a plasma for a sputter process: the use of DC power or the use of RF power with a frequency of 13.56 MHz. Both processes run very stably if a metallic film is deposited. To produce an oxide or nitride layer with a DC sputter process, the target must be conductive since DC power cannot be coupled through any insulating material. The desired compound is then formed by sputtering the metallic target in a gas mixture of argon and the related reactive component (oxygen or nitrogen). This process is called reactive sputtering. If RF power is used, any insulating compound target may be used since RF can be coupled even through any nonconductive material.

RF sputtering of dielectrics like SiO_2, Si_3N_4 or Al_2O_3 on small scales is a very common process, e.g. in the microelectronics industry. However, for economic mass production of such films on large areas the use of RF is prohibitive because of low deposition rates, high equipment costs (power supply and matching network) and uniformity problems for substrate widths larger than 1 m.

DC powered magnetrons enable fairly high deposition speeds and power supplies up to 200 kW are available at reasonable costs. The main disadvantage of the reactive process is the fact that the desired reaction cannot be limited to the substrate, but also takes place on the target and all other surfaces in the vicinity of the magnetron. As a consequence of insulating layers being built up on these surfaces, process stability is not sufficient and achievable deposition rates are limited by the formation of arc discharges. In particular, this holds for SiO_2, the most important low refractive index material and the only one which can be deposited by sputtering.

"Arcing" may limit the process stability in a drastic manner. A typical magnetron discharge operates at 500 V and a power density of 10 – 20 W/cm^2. Under certain process conditions the discharge may collapse into a small thread of ions and electrons called an arc. Such an arc discharge is characterized by low voltage and, since the total power is focused on a very small target spot, by a power density in the 10^5 W/cm^2 range. Arcs generate particulates which may contaminate the growing layer. Severe arcing may even give rise to local melting of the target or local destruction of thin film and substrate.

Besides arcing the reactive DC sputter deposition of insulating materials is accompanied by another substantial problem often described by the term "disappearing anode" (Scholl 1993). As mentioned above, the anode is formed by metallic surfaces in the vicinity of the cathode like shieldings, chamber walls, etc. During deposition of an insulator, all these surfaces are covered with the nonconductive material and after a process time of several hours any effective anode is absent. As a conesquence, the impedance of the discharge increases and an undefined glow distribution, drifting potentials, unstable process parameters and undefined film properties are observed.

1.3 The Rotatable Magnetron Cathode

During the last 15 years many attempts have been made to eliminate arcing in reactive DC sputtering of insulating materials. Arcs are caused by the thin dielectric layers accumulated in the noneroding areas of the conductive targets which steadily charge up due to the bombardment with positive ions. One possible solution of the arcing problem is therefore the minimization or elimination of these areas. It can be realized by plasma scanning of the entire target surface. The so-called C-MAG or rotatable magnetron cathode (McBride 1990, Wallace 1999) outlined in Fig. 4 employs a tubular target which is continuously rotated around a fixed magnet array. The rotation of the target through a confined plasma results in elimination of deposits and thus permits arc free operation at high deposition rates together with an excellent material utilization of up to 80%. For a system used in production coaters the target tube outer diameter is around 15 cm. For self-supporting materials (e.g. Ti, Al) the tube is the target itself. Brittle materials must be supported by a backing tube.

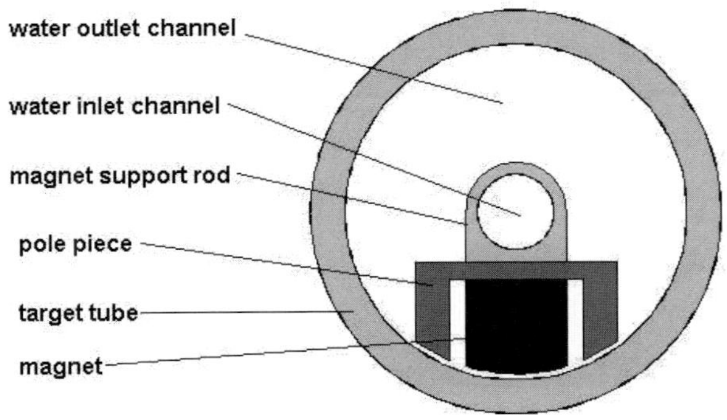

Fig. 4. The rotatable magnetron cathode

1.4 Pulse Magnetron Sputtering

Theoretical considerations published in (Szczyrbowski 1997) indicate that arcing should be effectively suppressed if the magnetron is driven by AC power at frequencies above 10 kHz. Figure 5 outlines the setup of a mid-frequency powered dual magnetron system. Two identical planar magnetron cathodes are mounted close to each other in a sputter chamber and connected to the output of a mid-frequency (MF) power supply operating typically at a frequency in the 50 kHz range. At any time, one of the magnetrons is at negative potential and acts as a sputter cathode, while the second one acts as an anode. The momentary cathode generates secondary electrons which are accelerated towards the anode and neutra-

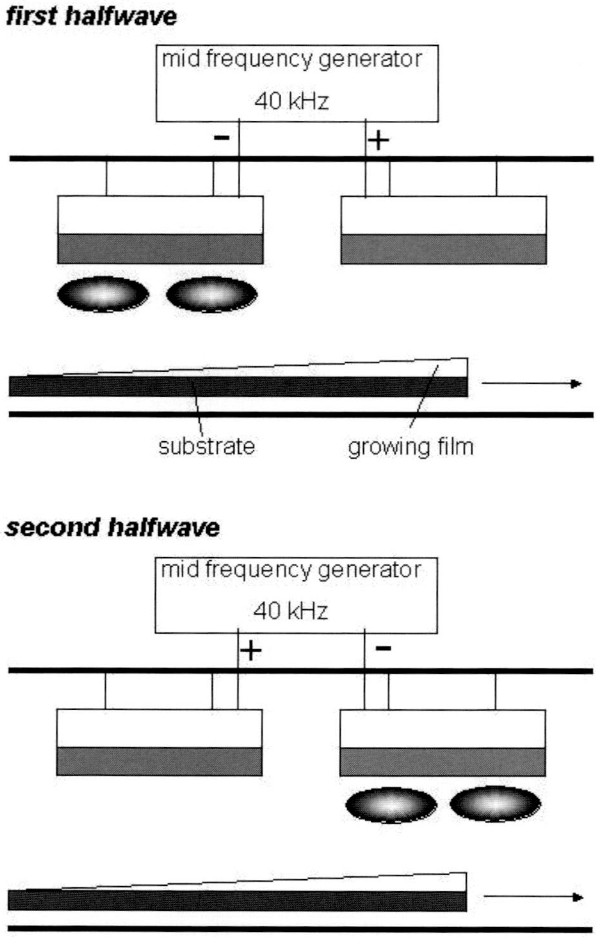

Fig. 5. Operation principle of a mid-frequency powered double magnetron system

lize positive surface charges having built up in insulating areas during the negative half cycle.

Such sputter arrangements can be operated at very low arcing frequencies for process times of several hundred hours. Furthermore, they solve the disappearing anode problem since independent of the actual conductivity of the environment the conductive erosion zone of the momentary anode is always able to collect the electrons from the plasma.

MF powered dual magnetrons with lengths up to 3.75 m permit deposition rates up to 5 times higher than those obtained by conventional reactive DC sputtering. Table 2 summarizes some deposition data for the most important optical materials. Deposition rates are given in the unit nm · m/min. This unit is often used instead of nm/s for large area coating since the substrate is always moving at a certain speed during deposition. The number represents the film thickness (in nm) at a substrate speed of 1 m/min.

Table 2: Dynamic deposition rates for different compound layers

(Dynamic deposition rate: film thickness in nm at a substrate speed of 1 m/min)

Material	DC single magnetron 3750 mm	MF double magnetron > 3000 mm
SiO_2	5 - 10	70
Si_3N_4	20	40
SnO_2	40	80
TiO_2	5 – 10	40

The quality of the deposited films is very high (Jäger 1998, Ruske 1999, Szczyrbowski 1997). For TiO_2, the very stable rutile structure could be realized with pulse magnetron sputtering (Szczyrbowski 1997). In general, compared to those produced by DC sputtering, the films exhibit higher density, increased hardness, a smoother surface and improved optical properties. Thus, they have opened the door for new opportunities in large area coating.

The film thickness uniformities fulfill the highest requirements on optical interference coatings on large areas. Figure 6 gives an example for SiO_2 deposition on a 2500 mm glass substrate using a 3200 mm long magnetron system. Thickness deviations are less than $\pm$ 1%.

1.5 Process Control

An inherent property of any reactive sputter process is the hysteresis behaviour of target voltage (at constant current), deposition rate and reactive gas partial pressure as a function of the reactive gas flow (Schiller 1979 Schiller 1984). It is caused by the fact that with increasing flow rate the target surface is continuously covered with reaction products. The hysteresis is more or less pronounced for different target materials. Figure 7 gives an example for AlN deposition. Depending on the reactive gas flow, two stable operating regimes exist: the metallic mode and the reactive mode.

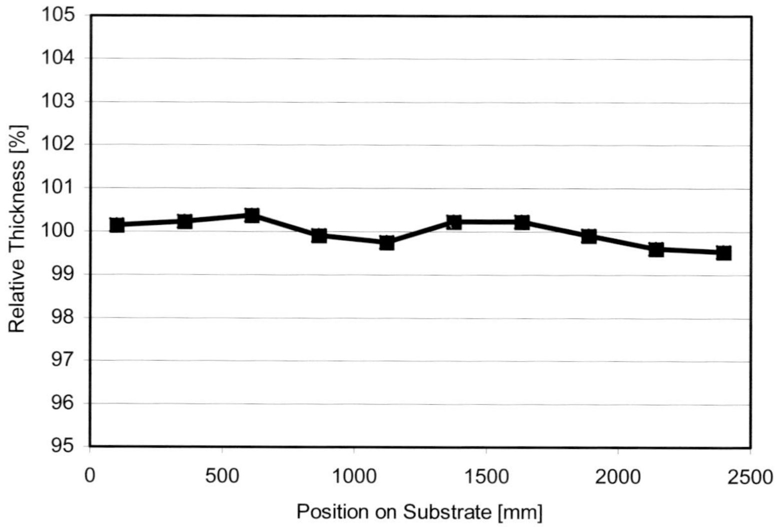

Fig. 6. Film thickness uniformity for SiO_2 deposition using a 3200 mm long dual magnetron system

In the metallic mode the target surface is metallic and the deposition rate is high. However, the deposited films are metal rich and in most cases they cannot be used for the required application.

In the reactive mode the target surface is completely covered with a thin compound layer. Stoichiometric films with very good properties are deposited. However, the deposition rates are very low compared to the metallic mode.

In order to obtain films with the correct stoichiometry in reasonable process times it is indispensable to stabilize the reactive discharge in the unstable transition region between metallic and reactive mode. Different strategies have been developed to detect the reactive gas partial pressure and keep it constant at any desired operating point in this transition region (Schiller 1984, May 2001). Residual gas analyzers (mass spectrometers) enable a direct measurement. In newer systems, the so-called "lambda probe" known from the automobile catalyst converter is employed to control the oxygen flow during reactive deposition of oxides. A further approach in wide use is optical emission spectroscopy of the plasma (plasma emission monitor or PEM). Its working principle is based on the fact that a certain amount of sputtered metal atoms is excited and emits characteristic radiation in the visible spectrum. The intensity of such lines decreases continuously with increasing reactive gas partial pressure.

In all cases, process stabilization is realized by control of reactive gas flow rate or cathode power.

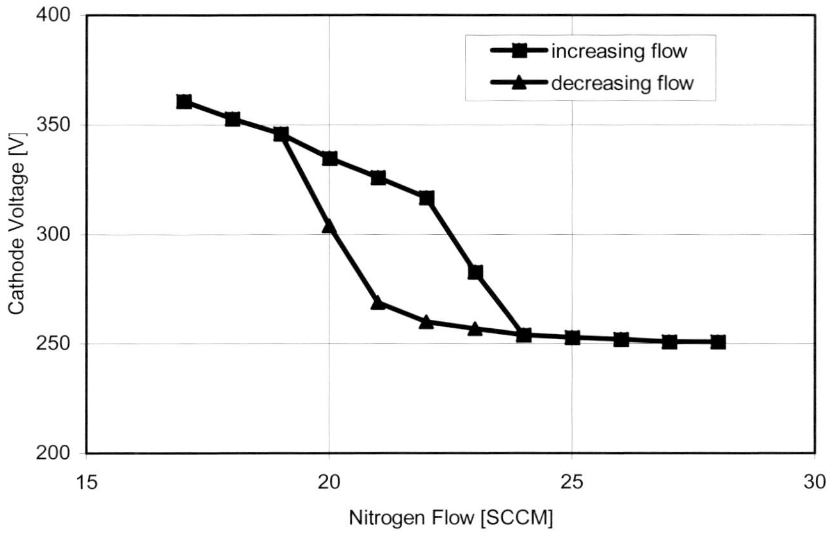

Fig. 7. Hysteresis loop for reactive DC deposition of AlN

2 Applications

2.1 The Need for Large Area Glass Coating

The fascination of glass as a construction material for various applications is founded on its outstanding properties. High volume manufacturing and processing of glass is possible in a simple and inexpensive way. Glass is extremely stable against environmental attacks and highly scratch resistant. Its high transparency in the visible spectrum permits a clear neutral colour and thus realistic view into the outside world. All these features have led to an intensive use of glass in the fields of architecture and transportation. Furthermore, glass is one of the most important materials in modern information and communication technologies. The drastically growing markets for TV screens and data displays need glass with a variety of coatings on its surface.

Glass is not perfect with regard to a lot of applications, in particular for architectural use. On one hand, its low reflection in the far infrared (room temperature radiation) causes undesired losses of thermal energy which is needed to heat buildings in colder climate regions. On the other hand, its high transmission in the near infrared (solar radiation) increases the energy necessary for cooling of buildings in hot climate zones.

In the visible part of the electromagnetic spectrum a glass pane reflects 8.4% of the impinging light (4.2% per side). In some cases this reflection is rather disturbing; in optical lens systems consisting of many individual elements it is inadmissible.

Carefully designed coatings on glass nowadays can overcome all these drawbacks. Since the end of the seventies vacuum coating technologies for the deposition of optical thin films on large area glass substrates have enjoyed a steady growth. The current worldwide installed annual production capacity of industrial vacuum coating systems for architectural and automotive glass amounts to approximately 150 million square meters. The main applications today are found in the fields of so-called low emissivity and solar control coatings.

Since the end of the eighties a new and fast growing market for glass coating has developed: the flat panel display industry. Transparent and conductive ITO layers of highest quality are the basis for liquid crystal display manufacturing.

New opportunities and challenges appear on the horizon, like electrochromic thin films or layers used for solar cells.

2.2 Low Emissivity and Solar Control Coatings

Glass does not reflect infrared radiation due to its high emissivity of $\varepsilon = 0.85$. (According to Kirchhoff's law, ε is identical with the infrared absorption of glass.) ε can be substantially reduced by any conductive coating like ITO, doped SnO_2 or a metal. Today nearly all low emissivity glazings are based on a silver film with a thickness in the 10 nm range. To maintain the high transparency of glass in the visible spectrum and to protect the silver from corrosion, additional anti-reflective and protective layers of high refractive index materials (e.g. SnO_2, ZnO, Bi_2O_3, Si_3N_4, TiO_2) have to be added. Further protection of the silver is achieved by a so-called "blocker" film. A huge variety of difference designs are available on the market. One of the simplest stacks introduced already 20 years ago is

$$\text{glass} – SnO_2\ (40\ \text{nm}) – Ag\ (10\ \text{nm}) – NiCrO_x\ (1.5\ \text{nm}) – SnO_2\ (40\ \text{nm}).$$

In a double glazing unit the coating is located on the inner glass pane at the surface opposite to the inside of the building. Figure 8 shows the optical properties of a typical low e film system compared to those of uncoated glass.

Modern silver-based low e coatings reduce the emissivity to values around $\varepsilon = 0.04$ (Schaefer 1997). The relevant number for practical use is the coefficient of heat transfer k ($Wm^{-2}K^{-1}$) describing the heat losses through an area of 1 m^2 at a temperature difference of 1 K. For a single uncoated glass pane (thickness 4 mm), k is 6 $Wm^{-2}K^{-1}$. Uncoated double glazing units reduce k to approximately 3 $Wm^{-2}K^{-1}$.

An emissivity of 0.04 corresponds to a k value of around 1.1 $Wm^{-2}K^{-1}$. Double glazing units with such a coating reduce the thermal losses through windows by 60 % compared to those of uncoated units. ε depends on the sheet resistivity of the film. This dependence is given by

$$\varepsilon = 0.0129 \cdot R - 6.7 \cdot 10^{-5} R^2 \tag{1}$$

with $R = \rho/d$ (ρ = specific resistivity, d = film thickness).

From Eq. (1) one can calculate that for $\varepsilon = 0.04$ a sheet resisitivity of approximately 3 Ω has to be achieved.

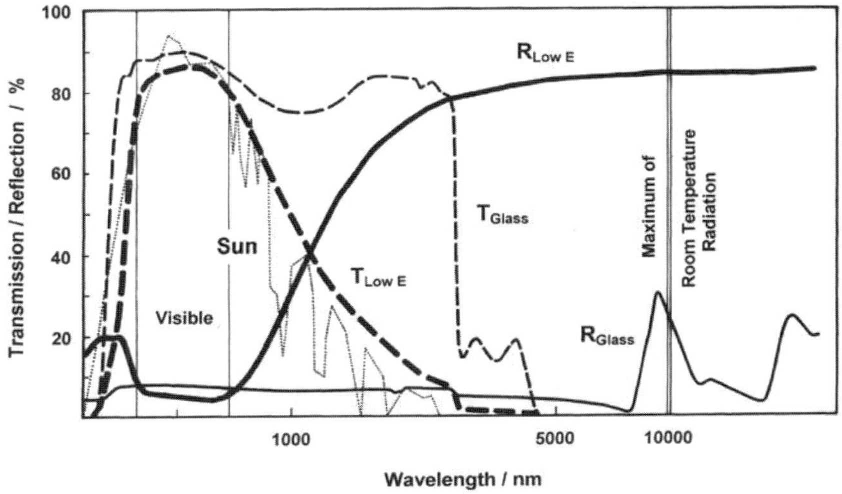

Fig. 8. The optical properties of a low emissivity coating compared to those of uncoated glass

Besides an emissivity as low as possible, further requirements on the coated glass are high light transmission (integral transmission $T_y > 80\%$ for a single pane), a neutral reflection colour, high stability against environmental attacks and high scratch resistance.

The relation between the performance of a low emissivity coating and the properties of the individual films is illustrated in Table 3. In general, thick silver layers permit reduction of ε and thus low k values. However, they reduce the transmittance in the visible and create undesired red reflection colours. The influence of the silver layer thickness on optical and electrical performance is demonstrated in Table 4.

An alternative approach to realizing low k is the use of two independent silver layers. Such a system may be

$$\text{glass} - SnO_2 - ZnO - Ag - NiCrO_x - SnO_2 - Ag - NiCrO_x - SnO_2 .$$

At a total silver thickness of 24 nm, such a stack permits a light transmission of around 79% and an emissivity of $\varepsilon = 0.04$ ($R = 3.3\ \Omega$).

Table 3: The relation between the performance of a low emissivity coating and the properties of the individual films

Property of the low e coating	**Mainly influenced by**
Light transmission	Refractive index of protective films Thickness of silver film
Colour	Refractive index of protective films Thickness of silver film
Emissivity	Conductivity of silver film
Chemical stability	Blocker film
Mechanical stability	Scratch resistance of top layer

Table 4: The influence of the silver layer thickness on the electrical and optical properties of a low emissivity coating

Thickness of silver film [nm]	8	12.5
Light transmission [%]	84	76.5
Sheet resisitivity [Ω]	10.6	5.1
Emissivity	0.13	0.07
k [W/m^2K]*)	1.41	1.23

*)k is calculated according DIN EN 673 for a double glazing unit (glass thickness 4 mm, glass distance 16 mm, Ar filled).

In Figure 9 the transmittance and reflectance curves of three different coatings are compared. There is a remarkable difference between the two single silver layer systems and the double silver layer system. The transmittance cutoff in the near infrared is much steeper for the double silver, meaning that such designs are also useful as solar control films during hot and sunny summer days. On the other hand, they gather less solar energy on a cold sunny winter day.

Pulse magnetron sputtering has stimulated new developments of low emissivity films due to the opportunity to deposit materials with enhanced properties. One of these new coatings is

$$\text{glass} - TiO_2 - ZnO - Ag - NiCrO_x - Si_3N_4\,.$$

MF sputtered TiO_2 exhibits a rather high index of refraction and thus enables a neutral reflection colour and the high density and hardness of MF sputtered Si_3N_4 results in an improved stability against environmental attacks. The emissivity of such a coating is $\varepsilon = 0.04$ at a visible integral transmission of $T_y > 80\%$.

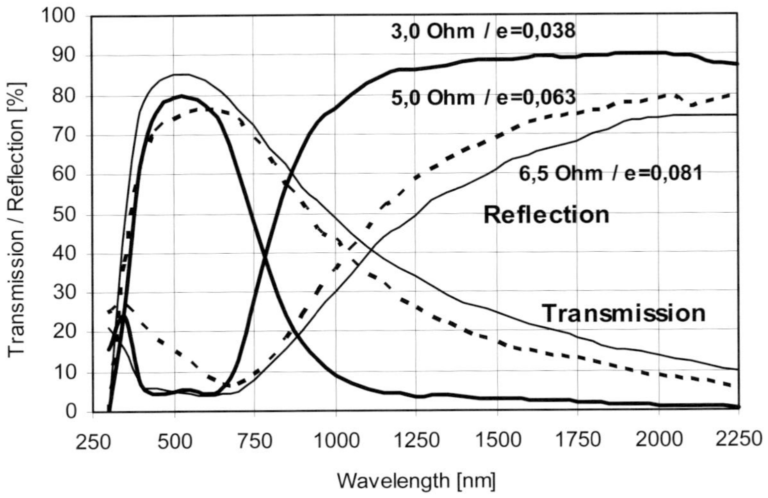

Fig. 9. Tansmission and reflection curves of different low emissivity coatings

Solar control films consist of an absorbing and reflecting material (e.g. CrN_x, TiN_x, FeN_x) embedded in layers of a highly refractive material (e.g. SnO_2). A typical system is

$$\text{glass} - SnO_2\ (10 - 100\ \text{nm}) - CrN_x\ (10 - 30\ \text{nm}) - SnO_2\ (10 - 30\ \text{nm})\ .$$

In a double glazing unit the solar control coating is located on the inner side of the outer glass pane. Depending on the thickness of the first SnO_2 layer, a lot of reflection colours like grey, silver, bronze, blue or blue-green are possible. The total solar energy transmission is determined by the thickness of the metal nitride layer. Solar control coatings are characterized by the so-called "shading coefficient" SC. SC is given by the ratio

$$SC = g/g_0\ , \tag{2}$$

where g = total solar energy transmission of coated glass and
g_0 = total solar energy transmission of clear glass (thickness: 3mm).

The value of g_0 is 0.87. Typical values for shading coefficients range from 0.2 to 0.4.

Figure 10 shows the optical properties of a solar control coating with a blue reflection colour.

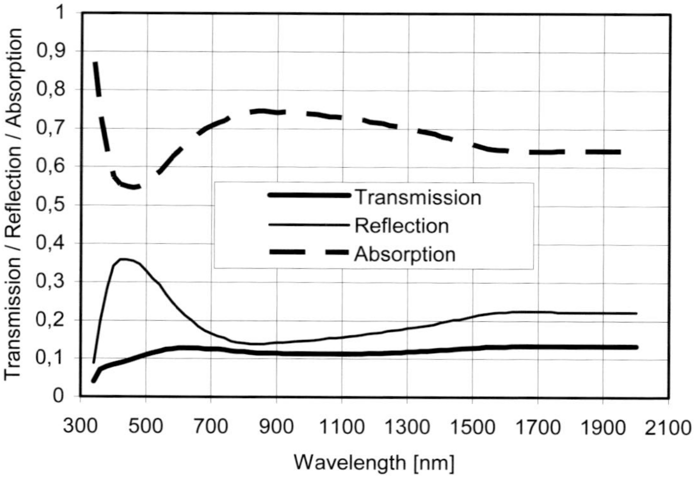

Fig. 10. Optical properties of a typical solar control coating

2.3 Anti-Reflective (AR) and Anti-Reflective Anti-Static (ARAS) Coatings

Wide-band anti-reflective (AR) coatings permit a reduction of glass reflection down to values of 0.1 – 0.3% in a wide range of the visible spectrum.

During the past decades AR coatings have continuously spread into a wide field of applications. Modern sputter technology, in particular pulse magnetron sputtering, permits deposition of these coatings also on large areas (Beißwenger 1993, Oyama 1999, Szczyrbowski 1997). Nowadays, there is a steadily increasing demand to apply them on data and TV screens (Cathode Ray Tubes) as well as flat panel displays.

Since there is also a concern to minimize the amount of electromagnetic emission through the front of the CRT, the AR coating has to be combined with a transparent conductive layer, resulting in a so-called ARAS system. The "antistatic" layer also reduces the accumulation of dust on the CRT.

In general, a wide-band AR coating uses multilayer stacks of alternating high and low refractive index dielectric thin films. As low refractive index material SiO_2 (n = 1.46) is employed, a prominent high refractive index material is TiO_2 (n = 2.20 – 2.50).

A typical specification for optical and electrical properties of an ARAS coating is given as follows:

(1) reflection R < 0.6% at any wavelength between 430 and 650 nm
(2) integral reflection R_y < 0.3% between 430 and 650 nm

(3) resistivity across the CRT < 1 kΩ

In Figure 11 the reflection curve of a four layer ARAS coating

CRT – ITO (20 nm) – SiO_2 (35 nm) – TiO_2 (110 nm) – SiO_2 (85 nm)

is shown. The given thickness values are only rough, exact numbers depend on the desired reflection colour. The integral reflection in the wavelength range from 430 to 650 nm is less than 0.2%. The resistivity measured between contact marks located at opposite edges of the screen is approximately 900 Ω. The layers are highly scratch resistant and pass all relevant chemical tests. In particular they withstand temperatures of nearly 500 °C.

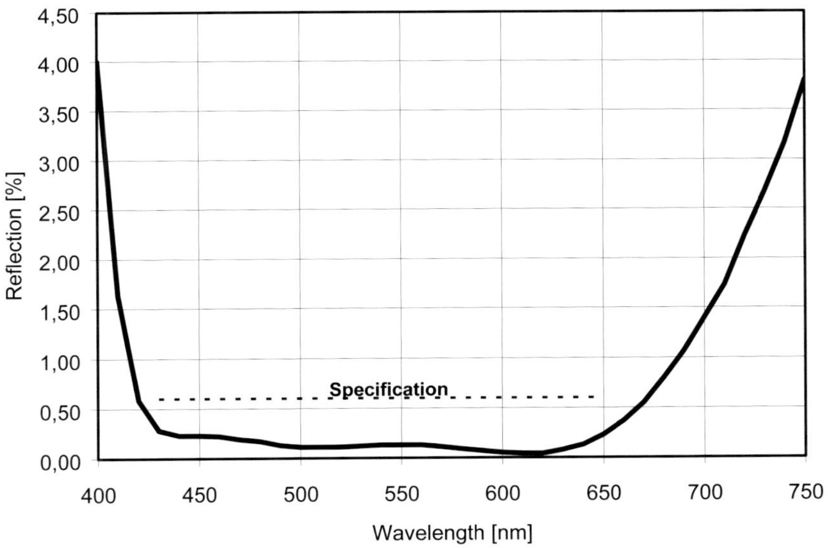

Fig. 11. Reflection curve of an ARAS coating

In the case of CRT's, the coating has to be applied only on the outer surface of the screen since back side reflection is negligible. Simple four layer AR coatings can also be used on large area float glass panes, e.g. for show cases or shop windows. In this case, the conductive ITO layer can be replaced by TiO_2.

An example is

glass – TiO_2 (15 nm) – SiO_2 (35 nm) – TiO_2 (110 nm) – SiO_2 (85 nm).

In order to obtain the highest possible reflection reduction, the coating has to be deposited on both glass surfaces.

2.4 ITO Coatings for Displays

One of the most important and most critical process steps in manufacturing liquid crystal displays (LCD's) is the deposition of ITO as a transparent electrode.

The major concerns for ITO deposition are as follows:

- low specific resistivity (< 150 μΩcm),
- high uniformity across the substrate ($\Delta d/d < \pm 5\%$, d: film thickness),
- low particle contamination,
- low manufacturing costs.

Magnetron sputtering using ceramic ITO targets of high density is the preferred deposition technology today. For achievement of lowest resistivities, the glass substrates have to be heated to temperatures up to 380 °C prior to deposition. Furthermore, film defects resulting from radiation damage by bombardment of negatively charged ions during the sputtering process can be minimized by a decrease of the discharge voltage of the cathode, thus leading to a further improvement of conductivity. By use of strong permanent magnets a discharge voltage of approximateley -240 V and a specific resistivity of 120 μΩcm have been realized.

However, for coating on colour filters substrate temperatures of more than 200 °C are not applicable due to the thermal sensitivity of the filter materials. In order to obtain the lowest sheet resistivities even for this temperature, a further reduction of the magnetron discharge voltage is necessary. Besides this, the necessity to optimize production costs calls for high utilization of the expensive ITO targets. Both requirements have led to the development of the so-called MoveMag cathode driven by a mixture of DC and RF power (Daube 1997).

The permanent magnet set of the MoveMag cathode creates a double plasma ring which is continuously shifted across the target surface, resulting in a wide erosion groove and a material utilization of more than 40%. The combination of DC and RF power results in a decrease of discharge voltage to values below 100 V. Figure 12 shows the specific resistivity, dynamic deposition rate and DC voltage for different DC/RF power mixtures. The optimum resistivity value of 150 μΩcm at a substrate temperature of 200 °C is found for an RF portion of 50%.

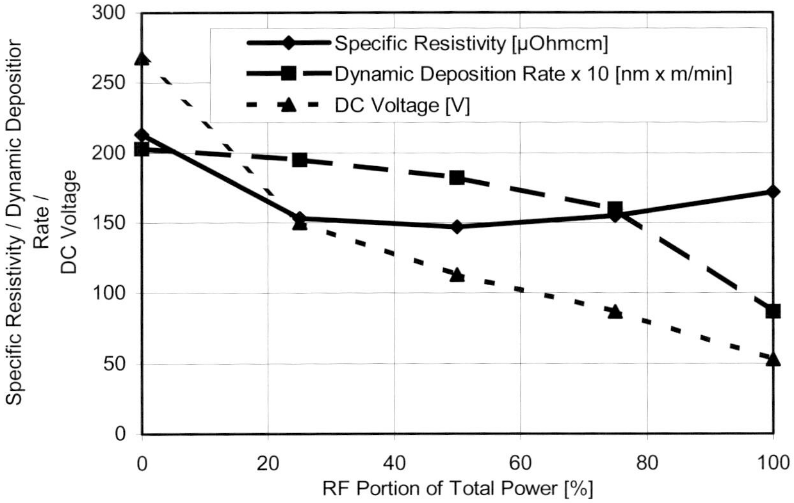

Fig. 12. Specific resistivity, dynamic deposition rate and DC voltage for a DC/RF driven MoveMag cathode as a function of RF/DC power mixture (Daube 1997)

To maintain the quality of the ITO layer for the entire lifetime of the display, a diffusion barrier consisting of SiO_2 has to be applied between the glass substrate and the ITO film. In general, the SiO_2 is deposited by means of RF or MF sputtering.

More information on ITO deposition parameters and a vertical in-line sputtering system is given in (Bräuer 1999).

2.5 Other Applications

Much effort is put into the development of glazings with variable optical and thermal properties ("smart windows"). Different technical solutions exist on the basis of electrochromic, gasochromic and thermochromic effects. By the end of the last millennium, a small amount of electrochromic architectural glazings was already introduced into the market.

Electrochromism is based on a colouration of tungsten oxide by injection of protons or other light ions. To move the ions, an electrical field has to be applied by means of transparent electrodes. A typical layer stack is

glass – ITO – WO_3 – ion conducting layer (electrolyte) – ion reservoir – ITO.

Sheet resistivities of much less than 10 Ω are necessary to realize short switching times. The applied voltage is between 1 and 3 V. The light transmission (λ = 550 nm) of an electrochromic coating may vary between 80% (bleached state) and 10% (coloured state). Transmission spectra are shown in Fig. 13.

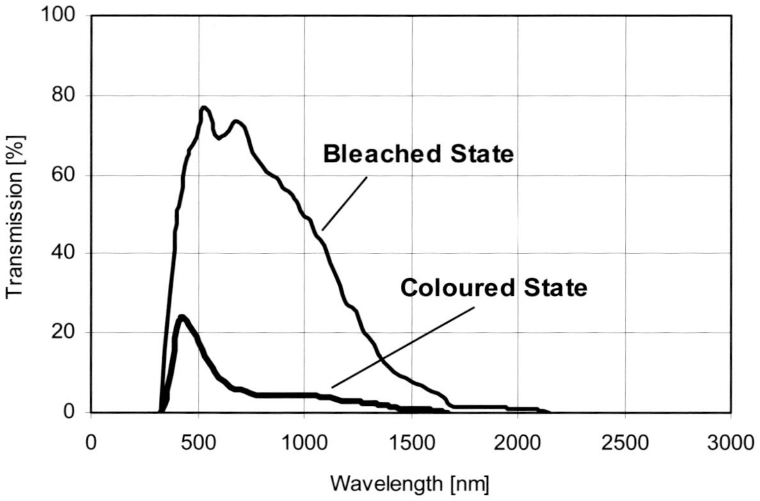

Fig. 13. Transmission spectrum of a bleached/coloured WO_3 layer on transparent conductive oxide coated glass (Wittkopf 1997)

Thin film solar cells will play a major part in the future of energy generation. In principle a solar cell needs a substrate (a glass pane, foil, metal strip or sheet), a metallic conductor (an Al or Ag film), an absorbing film and a transparent conductor exposed to the solar radiation. The current between the two conducting films is carried by electron-hole-pairs produced through the photoelectric effect in the absorbing layer. The most promising absorber materials today are amorphous silicon, copper-indium-sulfide (CuInS) or copper-indium-diselenide ($CuInSe_2$).

As an inexpensive alternative to ITO, aluminum doped ZnO films are preferably used as transparent conductors in thin film solar cells (Ellmer 1997, Szyszka 1999).

3 Equipment Technology

3.1 Coaters for Architectural Glass

Modern in-line coaters which enable high throughput are preferred for glass substrates with dimensions of up to 3.21 m x 6.00 m. Such systems are 4 m in width and up to 150 m in length. The principal layout of a "5-chamber system" is shown in Fig. 14. The machine is symmetric and consists of an entrance and exit chamber, two buffer chambers and a sputter section with two transfer chambers and the sputter modules. The entrance and exit chamber as well as the buffer chambers are separated from the sputter section by valves making it possible to keep the sputter section under high vacuum during loading and unloading of glass. The sputter chambers are equipped with magnetron cathodes (up to 30 for modern coaters) holding the sputter targets. The design of the sputter chambers is modular. They consist of identically dimensioned compartments arrayed along the direction of glass transport. Each compartment may be used as a high vacuum pumping station or sputter station. The total number of compartments and their layout depends on the type of multilayer stack to be produced as well as the minimum required cycle time of the machine. The glass panes move in a horizontal orientation supported by rolls. The design of the coater allows continuous substrate transport with a constant speed of several meters per minute throughout the sputter section.

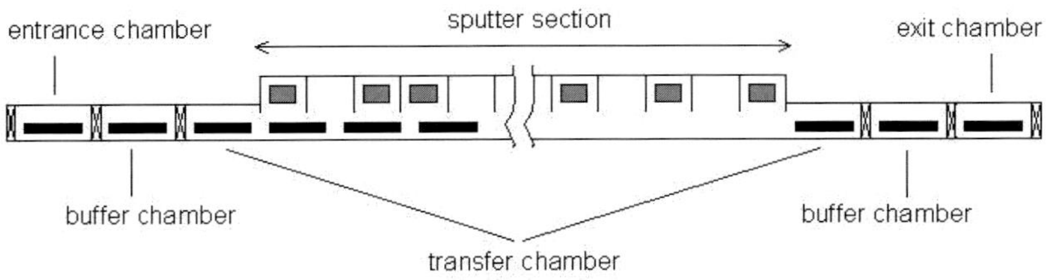

Fig. 14. Scheme of an in-line vacuum coater for architectural glass

The coating systems must meet high requirements on reliability. Uptimes (fraction of operation time during which the machine makes a good product) of 90%, maintained over uninterrupted production runs of one to two weeks, nowadays represent the minimum standard. Keeping this in mind, system modules handling evacuation, glass transport, coating and process control must be engineered and selected carefully. Sputtering chambers require cleaning between production runs in order to minimize generation of particles. Eroded sputter targets have to be replaced at the end of their lifetimes, causing venting and opening of the chamber. Today all efforts for improvement of the system have reached a state where the annual uptime is about 6000 hours, if production also takes place on weekends. The fastest machines operate at cycle times in the 30 s range, meaning that every 30 s a coated glass pane with a size of 3.21 m x 6.00 m leaves the exit chamber. The annual production capacity is then around 14 million m^2.

Fig. 15. Architectural glass coater (Von Ardenne Anlagentechnik)

3.2 Web Coating Equipment

Vacuum coating of web is a rather old technology. The first silver deposition on textiles was carried out in 1933 (Mattox 2000). Since the seventies rapidly growing markets for evaporated metal films on plastic foils have developed. Today packaging materials and thin film capacitors still dominate the market for web coating (Hartwig 1993, Fenn 1998). However, foils are of increasing interest as substrates for optical interference coatings instead of glass. To avoid the rather complex process of direct deposition of ARAS thin film stacks on CRT surfaces, these coatings are often applied to foils and glued to the screen. Other examples of optical films on flexible plastics are solar control films, greenhouse foils or materials with decorative effects.

Sputter web coating systems are much more compact and easier to handle than glass coaters. They consist of one big vacuum chamber accommodating the transport system and the sputter section. However, different parts of the chamber are separately pumped to avoid cross contamination between different processes and to reduce the influence of residual gas desorbing from the surface of the web during its unwinding.

Depending on chamber size and foil thickness, rolls with a length of several km can be processed. Typical widths range from 0.5 to 2 m. The web coater is a so-called "batch system". After loading of the roll at atmospheric pressure the chamber is pumped down and the web is coated during rewinding.

Figure 16 outlines a system with six sputter stations (Fahland 2000). In this case all stations are equipped with dual magnetron systems for pulse magnetron sputtering (of course the magnetrons can be operated also in the DC mode). Further important components of the machine are two big cooled coating drums guiding the web during deposition. Close thermal contact between the drum surface and the substrate is absolutely necessary in order to protect the foil from damage since the thermal load during sputtering caused by the energetic particles is very high. Pretreatment stations for plasma surface cleaning or modification prior to film deposition may also be part of the coater. Furthermore, equipment for in-line transmission or reflection measurement is indispensable to ensure constant quality of the product. Sputter web coaters run at typical production speeds in the range of 1 m/min.

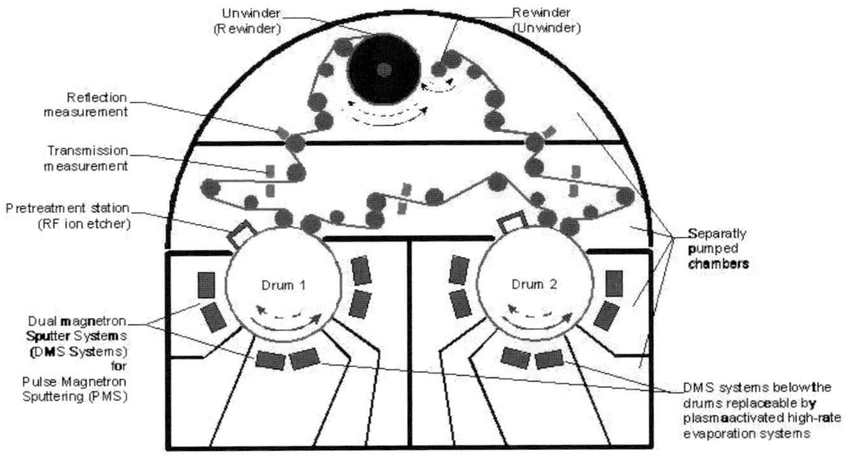

Fig. 16. Scheme of a sputter coater for production of optical interference layers on web

4 Summary and Outlook

Magnetron sputtering is the preferred production process for various coatings on large glass panes and plastic foils. During the past two decades glass and web coating has spread into a wide field of applications, namely low emissivity and solar control coatings on architectural glazings as well as functional layers for displays.

With the advent of mid-frequency powered double magnetron arrangements the most serious drawbacks in uniform deposition of high quality optical thin films on large scale substrates could be overcome, thus opening the door for new or improved products. High performance anti-reflective anti-static coatings on data display monitors nowadays are already state-of-the-art. The near future of large area thin film deposition will most probably be dominated by glazings with variable optical and thermal properties, the need for cost-effective production of thin film solar cells and a fast growing market for information displays. Pulsed plasmas have great potential to meet such demands.

References

Behrisch R, Ed. (1981) Sputtering by particle bombardment I. Topics in Applied Physics, Vol. 47, Springer-Verlag

Beißwenger S, Bräuer G, Dicken W, Szczyrbowski J (1993) Large-scale antireflective coatings on glass produced by reactive magnetron sputtering. Surf. Coat. Technol. 60, pp 624-628

Bräuer G (1999) Large area glass coating. Surf. Coat. Technol. 112, pp 358-365

Daube C (1997) private communication

Ellmer K, Wendt R (1997) D.c. and r.f. (reactive) magnetron sputtering of ZnO:Al films from metallic and ceramic targets: a comparative study. Surf. Coat. Technol. 93, pp 21-26

Fahland M, Kirchhoff V (2000) Vakuum in Forschung und Praxis 4/2000, pp 243-247

Fenn, JB (1998) Roll to Roll Sputter Web Coater – a Status Report. In: Society of Vacuum Coaters, Proc. 41th Ann. Techn. Conf.

Frey H (1995) Vakuumbeschichtung 1, VDI-Verlag

Haefer RA (1987) Oberflächen- und Dünnschicht-Technologie I, Springer-Verlag

Hartwig E (1993) The Future of Aluminum Web Coating Considering Environmental Legislation and Nonmetallic Vacuum Web Coating. Proc. 7th Intern. Conf. on Vacuum Web Coating, Miami 1993, pp 2-11

Jäger S, Szyszka B, Szczyrbowski J, Bräuer G (1998) Comparison of transparent conductive oxide thin films prepared by a.c. and d.c. reactive magnetron sputtering. Surf. Coat. Technol. 98, pp 1304-1314

Kukla R, Krug T, Ludwig R, Wilmes K (1990) A highest rate self-sputtering magnetron source. Vacuum 41, numbers 7-9, pp 1968-1970

Mattox DM (2000) A Concise History of Vacuum Coating Technology. Society of Vacuum Coaters 2000 special edition, pp 2-13

May C, Strümpfel J (2001) Balanceregelung für reaktives Magnetronsputtern zur optischen Großflächenbeschichtung. Vakuum in Forschung und Praxis 2/2001, pp 79-84

Mc Bride MW (1990) New coaters employing DC sputtering of SiO_2 for the production of optical components. In: Society of Vacuum Coaters, Proc. 33rd Ann. Techn. Conf., pp 250-256

Oechsner H (1975) Sputtering – a review of some recent experimental and theoretical aspects. Appl. Phys. 8, pp 185-198

Oyama T, Ohsaki H, Tachibana Y, Hayashi Y, Ono Y, Horie N (1999) A new layer system of anti-reflective coating for cathode ray tubes. Thin Solid Films
351, numbers 1-2, pp 235-240

Pulker HK (1999) Coatings on Glass, Elsevier

Ruske M, Bräuer G, Pistner J, Pfäfflin U, Szczyrbowski J (1999) Properties of SnO_2 films prepared by DC and MF reactive sputtering. Thin Solid Films 351, numbers 1-2, pp 146-150

Schaefer C, Bräuer G, Szczyrbowski J (1997) Low emissivity coatings on architectural glass. Surf. Coat. Technol. 93, pp 37-45

Schiller S, Heisig U, Steinfelder K, Strümpfel J (1979) Reactive DC sputtering with the magnetron-plasmatron for tantalum pentoxide and titanium dioxide films. Thin Solid Films 63, pp 369-375

Schiller S, Beister G, Sieber W (1984) Reactive high rate DC sputtering: deposition rate, stoichiometry and features of TiO_x and TiN_x films with respect to the target mode. Thin Solid Films 111, pp 259-268

Scholl RA (1993) A new method of handling arcs and reducing particulates in DC plasma processing. In: Society of Vacuum Coaters, Proc. 36th Ann. Techn. Conf., pp 405-408

Szczyrbowski J, Bräuer G, Dicken W, Scherer M, Maaß W, Teschner G, Zmelty A (1997) Reactive sputtering of dielectric layers on large scale substrates using an AC twin magnetron cathode. Surf. Coat. Technol. 93, pp 14-20

Szczyrbowski J, Bräuer G, Teschner G, Zmelty A (1997) Antireflective coatings on large scale substrates produced by reactive twin-magnetron sputtering. Journal of Non-Crystalline Solids 218, pp 25-29

Szczyrbowski J, Bräuer G, Ruske M, Teschner G, Zmelty A (1997) Some properties of TiO_2-layers prepared by mid-frequency and dc reactive magnetron sputtering. Journal of Non-Crystalline Solids 218, pp 262-266

Szyszka B (1999) Transparent and conductive aluminum doped zinc oxice films prepared by mid-frequency reactive magnetron sputtering. Thin Solid Films 351, numbers 1-2, pp 164-169

Wallace G (1999) Design considerations for the AC/DC C-MAG®deposition source and power supply system. Thin Solid Films 351, numbers 1-2, pp 21-26

Wittkopf H (1997) Electrochromics for architectural glazing applications. In: Proceedings of Glass Processing Days, Tampere, Finland, September 13-15, 1997, pp 299-303

Abbreviations

DC	-	direct current
RF	-	radio frequency
PEM	-	plasma emission monitor
ITO	-	indium tin oxide
SC	-	shading coefficient
AR	-	anti-reflective
ARAS	-	anti-reflective-anti-static
CRT	-	cathode ray tube
LCD	-	liquid crystal display
AC	-	alternating current
MF	-	mid frequency

Symbols

Y	-	sputter yield
D	-	deposition rate
ε	-	emissivity
R	-	resistivity
ρ	-	specific resistivity
d	-	film thickness
k	-	coefficient of heat transfer
T_y	-	integral transmission
g	-	total solar energy transmission of coated glass
g_0	-	total solar energy transmission of clear glass
R	-	reflection
R_y	-	integral reflection

Characterisation and Monitoring

Detlev Ristau

Laser Zentrum Hannover, Germany

1 Introduction

In the course of the rapid development of new laser systems and modern optics, optical coating technology is challenged by ever increasing quality requirements. For example, many of current laser applications in material processing, medicine, or fundamental research are dependent on optical components with low losses, high laser induced damage thresholds, and a high stability in respect to environmental influences. Besides innovative applications of fs-lasers, semiconductor lithography can be considered as one of the major pace makers for optical thin film technology which pushes the wavelength to lower limits in the UV/VUV- and even the EUV-spectral range and defines new specifications in respect to life time and contamination of optical coatings. Not least, complex spectral transfer functions have to be fulfilled by optical coatings in the emerging fields of telecommunication and applications of modern optics in precision metrology.

In order to cope with this variety of quality parameters, high expertise in deposition processes and handling of optical coatings alone is not sufficient. Companies and research institutes involved in the production of high quality coatings also have to possess an extended knowledge on the characterisation, because an optimisation and marketing of their products can be only performed on the basis of reliable and standardised characterisation techniques. Therefore, in most production environments advanced quality management routines are implemented, which assess the critical parameters of the products after each production step by adapted measurement techniques. As a response to the extreme demands of many innovative applications, approaches for advanced monitoring are under development, which provide the means to measure critical parameters and to perform corrections directly during the individual production steps.

This brief outline concentrates on some essential topics of optics characterisation and its role in the development of thin film technology. Selected International draft standards and Standards describing important quality parameters of optical coatings will be summarised and discussed in respect to their applicability. The main emphasis will be placed on the determination of the spectral transfer function and optical losses. Also, photothermal methods will be considered, which represent a wide variety of different techniques for the measurement of optical, mechanical and thermal parameters in thin films. In the second part, on-line spectrometric monitoring techniques will be outlined as an example for an advanced quality assortment in next generation thin film deposition processes. Finally, specific difficulties for the implementation of on-line monitoring devices in different deposition techniques will be discussed in respect to the achievement of predictable coating results.

1.1 Quality Parameters for Optical Laser Components

Besides the spectral transfer function, the set of important quality parameters for an optical component contains the optical losses, the surface properties, and criteria considering the various influences of the environment. Especially for laser applications, the damage threshold, which is dependent on the operation conditions of the used laser system, has to be considered. Specifications of major importance in modern optical technology, are compiled in Table 1 in connection with the corresponding standard measurement techniques. For the assessment of the basic parameters reflectance and transmittance as a function of the wavelength, a draft international standard (ISO/DIS 15368 2000) describing spectrophotometric measurement techniques is presently discussed. This standard is dedicated to the determination of the spectral transfer function using customary spectrophotometric devices, which reach a typical precision of 10^{-3}. For the quality control of coated laser components, this accuracy is often not sufficient. Therefore, also a high precision measurement technique based on a laser radiometric method has been tested as standard measurement procedure (ISO/WD 13697 2000). On the basis of a well stabilised laser source, accuracy values in the range of 10^{-4} to 10^{-5} can be accomplished for the prominent wavelengths of the Nd-Laser (1.064 μm) and the CO_2-Laser (10.6 μm). The losses of an optical component can be divided into two main channels: absorption and scattering. For both loss mechanisms, adapted standard measurement protocols have been developed and tested by extended international round-robin experiments. The laser calorimetric method (ISO 11551 1997) is recognised by most industrial companies as the standard technique for the determination of absorptance in optical laser components and routinely applied in high quality production cycles. Total scattering is covered by the standard (ISO/FDIS 13696 2001) which outlines an uncomplicated measurement with an Ulbricht sphere or a Coblentz sphere as collecting element for the radiation scattered by the specimen. Depending on the operation condition of the laser system, damage thresholds are assessed by applying the standard series (ISO 11254 2002), which defines different irradiation sequences and evaluation schemes for single shot (1 on 1-) and multiple shot (S on 1-) measurements. Also, a third part of ISO 11254 will be available, which is focussed on the certification of optical components in respect to their power handling capability. For the denotation of the surface quality, the standard series ISO 10110 (ISO 10110 1997) has been created which concentrates on the standardised preparation of drawings for optical elements and systems. Often, also the stability of the components in respect to environmental, mechanical and chemical influences has to be considered. For this purpose, a broad spectrum of different conditioning methods including climatic influences, contamination, and mechanical stress as well as chemical agents, abrasion, and adhesion tests is described in the standards ISO 9022 (ISO 9022 1996-2002) and ISO 9211 (ISO 9211 1994-2001).

Table 1. Selected quality parameters of optical laser components and the corresponding characterisation techniques.

Specification	Parameter /Unit	Standard / Measurement Principle	
Transfer Function	Reflectance	WD 13697:	precise laser method
	Transmittance	DIS 15368:	spectrophotometry
Optical Losses	Absorptance	ISO 11551:	laser calorimetry
	Total Scattering	FDIS 13696:	integration method
Laser Induced	cw-LIDT	ISO 11254-1:	cw-laser irradiation
Damage Threshold	1 on 1-LIDT	ISO 11254-1:	single pulses
(LIDT)	S on 1-LIDT	ISO 11254-2:	repetitive pulses
	Certification	DIS 11254-3:	irradiation sequence
Surface Quality	Form Tolerances	ISO 10110:	13 parts describing the
	Scratch/ Digs		preparation of drawing
	Roughness		for optical components
Stability	Abrasion	ISO 9211:	different test methods
	Environmental	ISO 9022:	21 parts describing
	Stability		conditioning methods

1.1.1 Reflectance and Transmittance

The spectral transfer function of an optical component is of fundamental importance for the quality assessment and is monitored after nearly every production step of an optical component. Transmittance and reflectance measurements are routinely performed with commercial spectrophotometers covering a spectral range from 190 nm up to 50 µm. Meanwhile the determination of transmittance in these two beam systems is uncomplicated, accurate reflectance measurements on the basis of V-N or V-W methods are dependent on reliable calibration samples and require more skilfulness. For the wavelength below 190 nm, where oxygen as a normal atmospheric gas starts to absorb the radiation, special spectrophotometers flushed with nitrogen or operating in vacuum have been developed. The principle of such a UV/VUV-spectrophotometer, which is designed for a variety of computer controlled measurement modes (Kadkhoda et al. 2000), is illustrated in Fig. 1. The system consists of four separate functional sections. The lamp compartment is equipped with two D2-discharge lamps which can be switched to the spectrometer as light sources for two wavelength intervals covering the range from 115 nm to 300 nm. For the spectral resolution of the input radiation a curved holographic grating is installed in Seya Namioka configuration imaging the entrance slit onto the exit slit of the monochromator unit. The third compartment contains a prism polariser which can be adjusted for measurements with s- and p-polarised radiation. Following the measurement beam into the sample chamber, a scraper mirror for coupling part of the measurement radiation into a reference photomultiplier is encountered. A 5-axes stage is built in the sample chamber for positioning and rotating of the specimen in respect to the beam axis. In transmission measurement mode, the signal photomultiplier (PMT) is positioned behind the specimen, and the wavelength is tuned. For an assessment of the reflectance values, the signal PMT is placed into the reflected beam directly recording the reflected radiation power without using additional reflectance insets. As a consequence of the movable signal PMT and the flexible sample stage, the spectrophotometer is equipped for the measurement of angle resolved spectrometric reflectance, transmittance and scattering with polarised light. On the basis of the control software package, also a mapping of the sample surface or an adapted positioning of beam splitter cubes are attainable with the system. For the development of wafer stepper systems in semiconductor lithography for the wavelengths of the ArF- (193 nm) and the F2- (157 nm) excimer lasers, VUV-spectrophotometers have become a common tool for testing and optimisation of UV/VUV-coatings and substrate materials.

Semiconductor technology is also the pace setter for new EUV-concepts in next generation lithography, which is expected to implement the 50 nm node in the industrial chip production during this decade. Corresponding spectrophotometers have to cover the spectral range between 1 - 100 nm, which imposes severe difficulties on the radiation sources, the beam steering and the detection of radiation. Apart form the synchrotron devices, which are available since many years and still most frequently applied, different concepts with laser-induced EUV sources (Kreuznach and Mann 2002) and discharge lamps (Lebert et al. 1999) are under development to achieve a table top spectrophotometer which can be afforded by the companies involved in EUV thin film and optics development.

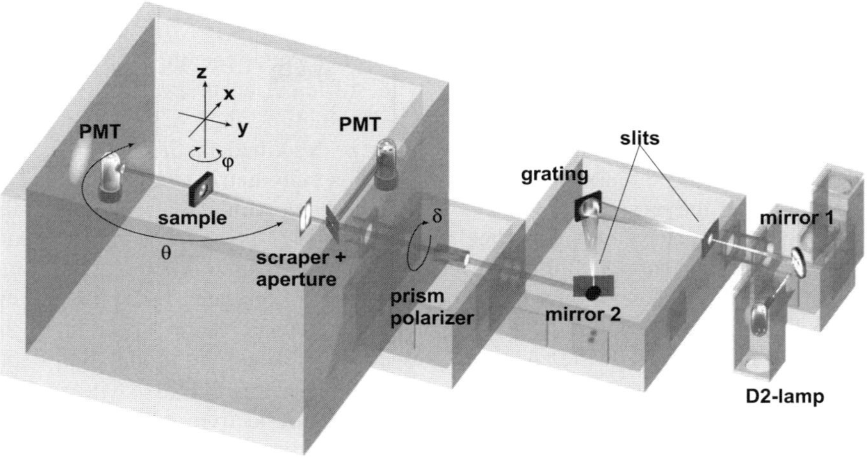

Fig. 1. Schematic view of a spectrophotometer for the wavelength range from 115 nm to 300 nm.

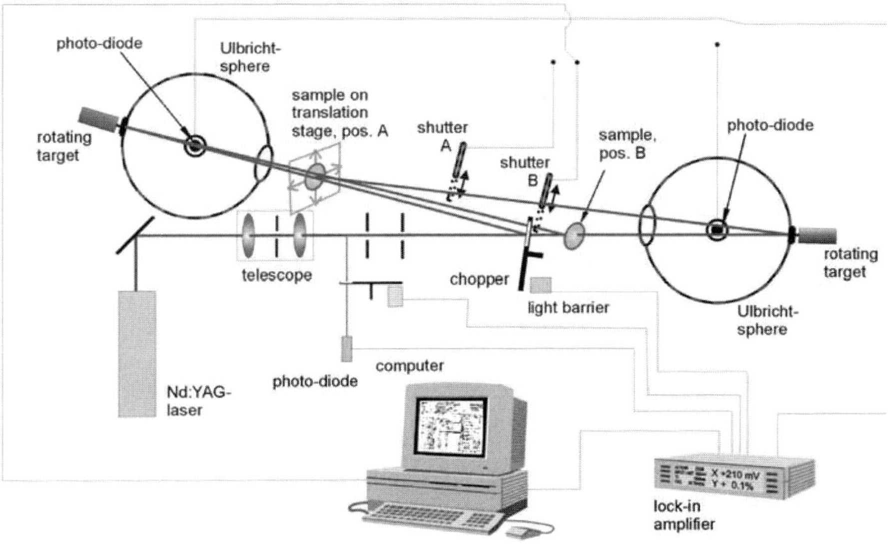

Fig. 2. Extended set-up for precision measurements of reflectance and tranmittance in accordance to ISO/WD 13697.

In laser technology repeatedly reflectors are necessary, which are certified in respect to their reflectance or transmittance values with an accuracy better than 10^{-3} beyond the capability of conventional spectrophotometers. As a high precision method well adapted for high quality laser mirrors, the cavity ring down

(CRD) technique (Anderson et al. 1984) has been cultivated and qualified by comparison to equivalent loss measurements (Henking et al. 1994). However, the applicability of the CRD-technique is restricted to mirrors with a reflectivity above 99.9%, because optics with higher optical losses generate very short ring down times which can not be resolved with sufficient accuracy. An alternative precision method, which in principle covers the entire range of reflectance and transmittance values for a single wavelength is described in the draft standard ISO/WD 13697. The fundamental principle of this radiometric method is based on a switching of the measurement beam between a direct path and a reflection at the specimen by a rotating chopper wheel (Voss et al. 1994). The measurement beam is detected by an adapted Ulbricht-sphere arrangement, which incorporates a rotating target and the detector unit. Dependent on the position state of the chopper, the detector signal corresponds to the incident measurement beam power or to the beam power reflected at the specimen, respectively. Thus, the modulation depth ΔS, which can be determined using a lock-in amplifier tuned to the switching frequency of the chopper blade, is directly related to the reflectivity of the specimen:

$$R_S = \frac{1}{R_c}\left(1 - \frac{\Delta S}{S_0}\right) \tag{1}$$

Where R_S and R_C are the reflectance values of the specimen and the chopper blade, respectively. S_0 denotes the signal of the sphere according to the incident power, which can be determined by blocking the beam path to the specimen. In view of the relatively complicated calibration procedure for the fundamental measurement arrangement, which involves three measurements using different chopper blades, alternative concepts have been thoroughly studied (Koschade and Steiger 1998; Fleig and Giesen 2000; Starke et al. 2001).

An example for an extended measurement arrangement in accordance to ISO/WD 13697, which had been developed to facilitate the calibration procedure is depicted in Fig. 2. The basic configuration described before is attained if the sample and shutters are in position A. In this case A, the collimated beam of the Nd-laser (1,064 μm) can pass the chopper directly and enters the integrating sphere (right in Fig. 2). If the chopper wheel is rotated to a reflecting position, the beam is deflected to the sample at position A and subsequently reflected into the Ulbricht sphere. In configuration B, the sample is installed behind the chopper and reflects the measurement beam into the second integrating sphere (left in Fig. 2) if the chopper wheel is at a transmitting position. For a reflecting position of the chopper blade, the beam is directly deflected into the same Ulbricht sphere. Thus, the reflectivity of the sample can be related to the signals:

$$R_S = R_c\left(1 - \frac{\Delta S}{S_0}\right) \tag{2}$$

In contrast to the relatively elaborate calibration procedure of the standard principle, this extended measurement arrangement can be calibrated by two subsequent measurements of the sample at the position A and B. Relative accuracy values below 10^{-5} could be achieved for the present set-up with a Nd:YAG-laser.

1.2 Photothermal Methods: Absorptance and Thermal Conductivity

Since more than two decades, laser based photothermal methods are applied for the characterisation of optical thin films with respect to a variety of properties (Tam 1989; Hummel and Günther 1995; Wu 1995; Welsch and Ristau 1995). The basic principle of some selected photothermal methods is illustrated in Fig. 3. A pump laser beam is impinging on the sample surface, and part of the laser energy is absorbed by the thin film system, the surface, and the bulk of substrate. Most of the absorbed energy is converted into heat, which induces a temperature profile in the sample. As a consequence of the thermal expansion of the substrate and the thin film materials, a bulge is induced on the sample surface. If the laser radiation is modulated, the height of this bulge oscillates synchronously inducing thermal wave and acoustic waves induced in the sample and its surrounding medium. The different photothermal methods can be classified corresponding to the technique applied for the detection of these laser induced effects.

One of the early detection channels employed for the measurement of absorptance is the photoacoustic gas cell microphone technique (Rosencwaig and Gersho 1976; Fernelius 1982). The acoustic waves induced by the oscillating surface can be either detected by a highly sensitive condenser microphone in the surrounding gas or by a piezoelectric transducer directly attached to the sample. For fluids, which were also tested as a transponding medium for the acoustic wave (Rosencwaig and Hindley 1981), the contribution of the thermoeleastic surface deformation to the signal has to be considered for the evaluation of the absorption measurement.

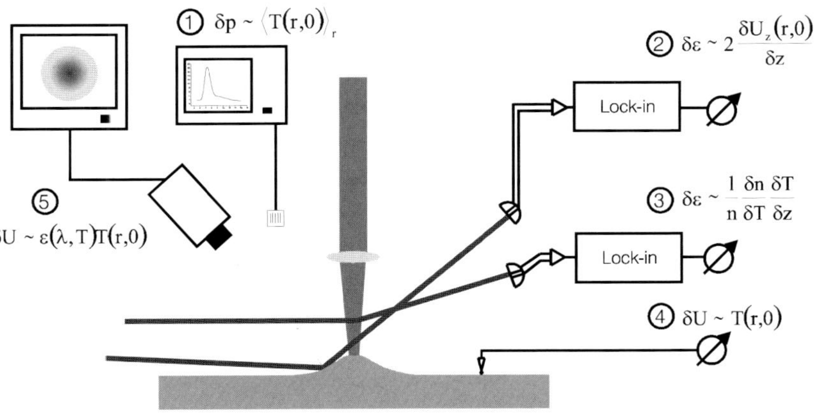

Fig. 3. Photothermal techniques: Detection schemes are illustrated for the photoacoustic technique (1), photothermal deformation or deflection technique (2), the mirage effect (3), laser calorimetry (4) and radiometry (5).

The photothermal deformation or deflection technique is one of the most prominent photothermal method in thin film technology (scheme 2 in Fig. 3) (Olmstead et al. 1983; Abate et al. 1983; Abate and Roides 1983). In this approach, the laser induced bulge is directly detected by the deflection of a probe laser beam. The deflection of the beam can be measured very sensitively down to displacement values below 1 Å by an optimised arrangement of a slit and a photo detector (Zimmermann and Welsch 1994) or different types of semiconductor position sensors. Photothermal deformation techniques are widely used for investigations in thin film properties including absorption and thermal conductivity. Also, photothermal microscopy and the measurement of pre-damage phenomena in thin films can be performed by this method.

Another detection scheme which is often used in thin film technology is the Mirage effect (Boccora et al. 1980). During irradiation of the sample, heat is conducted from the sample surface into the surrounding medium resulting in a localised gradient of the refractive index. This gradient leads to a deflection of a probe beam which is adjusted parallel to the sample surface. The Mirage effect is often used for the measurement of absorption and pre-damage phenomena in thin films, as well as for photothermal microscopy.

The theoretical background for the described measurement schemes can be developed on the basis of the thermoelastic wave equation for a solid. Under consideration of the boundary conditions appropriate for the actual experimental arrangement, two coupled differential equations have to be solved (Nowacki 1962):

$$\chi \Delta T(\vec{r},t) - \rho c \delta_t T - T_0 \alpha_{th} (3\lambda_L + 2\mu_L)\ \delta_t \vec{\nabla} \vec{U}(\vec{r},t) = \vec{\nabla} \vec{\gamma}_e \tag{3}$$

$$\mu_L \Delta \vec{U}(\vec{r},t) + (\lambda_L + \mu_L)\ \vec{\nabla}(\vec{\nabla} \vec{U}) - \rho \delta_t^2 \vec{U} - \alpha_{th} (3\lambda_L + 2\mu_L)\ \vec{\nabla} T(\vec{r},t) = 0 \tag{4}$$

Table 2: Notation for the parameters and physical quantities of Eq. 3 and 4.

Notation	Physical Quantity	Notation	Physical Quantity
χ	thermal conductivity	T_0	room temperature
c	heat capacity	$T(\vec{r},t)$	temperature field
ρ	density	δ_t	time derivative
α_{th}	thermal expansion	$\vec{U}(\vec{r},t)$	displacement field
λ_L, μ_L	Lame's modulus of elasticity	γ_e	incident radiation energy flux

The first equation (Eq. 3, for the notation see Table 2) describes the heat flux in the system and contains besides the terms of the conventional heat equation additional source terms for the generation of heat by the elastic motion of the system and by the laser beam. The second differential equation (Eq. 4) describes the strain distribution in a balance sheet of the forces acting in the system. Besides the terms describing the elastic motion, an additional term has to be inserted for the build up of thermal strain in the system by thermal expansion. Obviously, analytical solutions of these two differential equations, which are coupled by the described source terms, are of high complexity and only achievable for special boundary conditions under specific assumptions. Even assuming the availability of a special solution for a defined experimental arrangement, a variety of material parameters has to be known to derive quantitative data usable for practical applications. Therefore, the calibration of most photothermal detection schemes is difficult restricting the techniques to relative measurements. Nevertheless, for many applications the sensitivity and the spatial resolution in surface mapping of photothermal methods is unsurpassed. Also, a calibration of PTD-techniques is always feasible for individual sample types of equal thermoelastic properties allowing for absolute measurements of special sample categories.

In the special case of laser calorimetry, which is based on a direct measurement of the laser induced temperature distribution in the sample, the thermoelastic equations collapse to the fundamental heat transport equation. Thus, only the heat capacity and the thermal conductivity have to be considered as material parameters for a theoretical calibration of the measurement principle. First reports on the application of laser calorimetry for the measurement of thin film absorption date back to the early seventies (Pinnow and Rich 1973). During the last thirty years, laser calorimetry could be cultivated to a reliable and calibrated tool for the measurement of even extremely low absorption losses in optical components. As a consequence, this measurement principle had been adopted for the International Standard ISO 11551, which is dedicated the determination of absorptance in optical laser components.

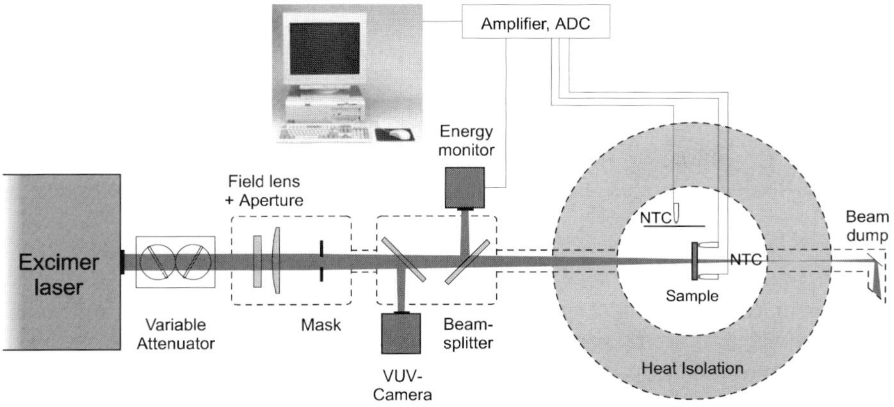

Fig. 4. Experimental set-up for laser calorimetric measurements in UV/VUV-spectral range with an excimer-laser according to ISO 11551.

For an illustration of the fundamental principle, a calorimetric set-up for the DUV/VUV-spectral region is depicted in Fig. 4 as an example. The beam of the laser source is prepared in the two chambers next to the laser output and is subsequently analysed in respect to its energy and spatial profile. To achieve a good thermal isolation from the environment, the sample is located in an isolated metal chamber of high thermal mass. The temperature sensors are directly attached to specimen by a special clamp holder which is optimised in respect to lowest possible heat losses by conduction from the sample into its mount. The measurement sequence starts with a waiting period until thermal equilibrium between the specimen and its surrounding is reached. Then, the laser irradiation of the sample is switched on and the absorption induced temperature rise is detected (see Fig. 5). After the heating period, the irradiation is switched off and the sample temperature decreases as a consequence of the heat losses by radiation, thermal conductance or convection. The absorptance can be determined by an uncomplicated evaluation of the recorded temperature curve taking into account the mass and heat capacity of the sample. For the wavelength of the CO_2-laser several international round-robin tests have been performed to ascertain the practicability and comparability of the laser calorimetric measurement principle. Good agreement within an error margin of 15% was observed for sample sets which were measured by industrial and research laboratories in many countries (Ristau et al. 1998).

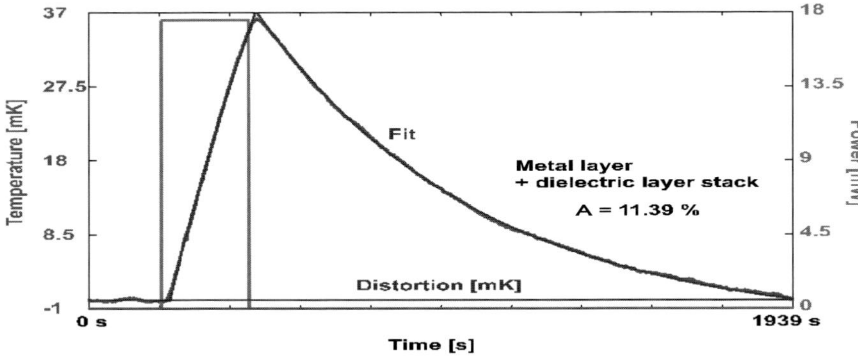

Fig. 5. Typical temperature curve for a calorimetric measurement with the arrangement in Fig. 4.

In comparison to the MIR-spectral range, laser calorimetry in the NIR/VIS-range is more demanding: Typical absorptance values of conventional components range between 10 to 100 ppm and can even reach a level around 1 ppm for layer systems deposited by advanced ion beam techniques (Rempe et al. 1992). Accordingly, very high temperature sensitivities in the μK-regime have to be achieved for calorimetric measurement devices operating in the NIR/VIS-range. In addition, materials of low thermal conductivity, as for example optical glasses or quartz, are predominantly employed for optical components in the NIR/VIS-spectral range. In practise, a low thermal conductivity induces a retardation between the start of laser irradiation and the temperature increase measured by the sensor, which imposes problems on the evaluation of the calorimetric measurement. However, most of the problems could be solved within the framework of the International Standardisation activities in laser calorimetry (Willamowski et al. 1996), and measurements in ppm- or sub-ppm-region can be performed on a routine basis today. As an example, a calorimetric record of an uncoated fused silica substrate (total absorptance: 3 ppm) is depicted in Fig. 6. Passing over to the UV/VUV-spectral range, further effects, which are related to the high energy of the photons, have to be considered. Besides surface contamination and the generation of colour centres or transient absorbing defects, non-linear effects have to be taken into account for a calorimetric measurement (Eva et al. 1996; Blaschke et al.; Apel et al. 2000).

Especially for materials exhibiting two-photon contributions, the measured absorptance values increase linearly as a function of the laser power applied during the calorimetric measurement. In addition to the effects, which are driven by partly predictable electronic transitions in the materials, ageing, contamination, and preconditioning represent another class of severe measurement problems in the UV/VUV-region. For example, the optical surface of the specimen can be cleaned by UV-irradiation under specific environmental conditions, meanwhile the same irradiation sequence may lead to contamination of the surface under typical technical vacuum conditions (Lambda Physik 2000).

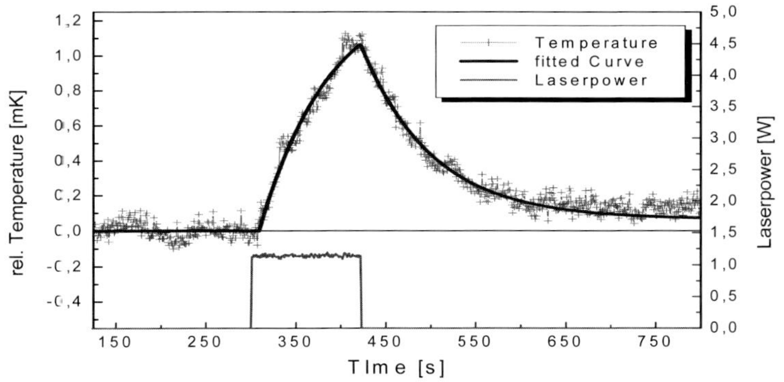

Fig. 6. Example for a laser calorimetric measurement with high sensitivity at 1.064 µm.

Photothermal techniques also open new ways to the measurement of the thermal conductivity in dielectric films, which became of increasing importance for the optimisation of thin films in high power laser applications (Guenther and McIver 1988). Different detection schemes are applied for the measurement of the thermal conductivity. Besides a direct evaluation of the stationary temperature profile, which is recorded by a pyrometric technique (Akhtar et al. 1986) and directly evaluated on the basis of the heat transport equation, probe beam deflection techniques (Welsch et al. 1992; Reichling and Grönbeck 1994) are employed in conjunction with modulated pump laser sources. For the determination of the thermal diffusivity using a modulated laser source, the behaviour of the phase as a function of the modulation frequency is evaluated. If the thermal diffusion length of the thermal wave approaches the thickness of the layer, the phase exhibits a local extreme, which is attributed to resonance effects of the thermal waves reflected at the interfaces of the layer. Another approach for the measurement of thermal conductivity is based on the thermal pulse method (Tam and Sullivan 1983; Grilli et al. 2000), which determines the travel time of the thermal wave through the thin film. Selected results of the measurement techniques are compiled in Table 3. for dielectric single layer coatings of materials often applied in optical thin film technology. Obviously, there are still severe deviations between the measured values, which may be attributed to the imperfect state of measurement technology. Furthermore, the thermal diffusivities are influenced by microstructure in the thin films, which is dependent on the different production processes and on the actual layer thickness. Also, the measurement of thermal conductivity is significantly influenced by the thermal resistance at the interface between the film and the substrate (Schwartz and Pohl 1987; Kuo et al. 1992). Currently, theoretical models for the influence of structural properties on the thermal conductivity of thin films are under investigation, but the detailed mechanisms for the reduction of thin film thermal conductivity are not completely understood until now.

Table 3: Thermal conductivity data for SiO_2, Al_2O_3, Ta_2O_5 and HfO_2 films measured by different working groups. Data measured by the reading thermal comparator (Lambropoulos et al. 1989), pyrometric (Akhtar et al. 1986), photothermal displacement method (Reichling and Grönbeck 1994) and thermal pulse method (Grilli et al. 2000; Dieckmann et al. 1994) are compared. The bulk values were obtained from tables and handbooks.

Material	Thermal Conductivity (W/mK)					
	bulk	comper.	pyrom.	displ.	th. pulse	th. pulse
SiO_2 e-beam	1.4	0.45-0.61	0.1	0.25	0.37-0.73	0.2-0.5
SiO_2 IBS		0.41-1.05				0.2-0.3
Al_2O_3	27.0	0.72	33.0	150	5.2-12.3	0.7-1.1
Ta_2O_5	-	-	0.026	0.2	2.8-15	0.3-0.6
HfO_2	17.0	0.052	$8*10^{-4}$	-	-	0.07-0.2

1.3 Total Scattering

Optical scattering is an essential loss mechanism of optical components which is mainly caused by the roughness of the optical surface, the microstructure of the coatings, and by imperfections in the bulk of the substrate. During the last decades, many efforts have been made to assess the scatter behaviour of optical components and to relate optical scattering to primary properties of the optical surface (Bennett and Mattsson 1999; Duparré 1995). Especially, in the field of scatter measurement, an enormous scientific and technical background has been built up in the 1980ies, which includes also National Standards for the measurement of total integrated (ASTM F1048 1987) and angle resolved (ASTM E1392-90 1990) scattering. As a consequence of the rapid growth of laser applications, international standardisation activities were predominantly concentrated on laser components and resulted in an ISO Draft-Standard (ISO/FDIS 13696 2001), which had been extensively tested in round-robin measurement campaigns. In contrast to the ASTM-Standards, ISO/FDIS 13696 is dedicated to measurement of Total Scattering (TS), which is defined by the total amount of radiation scattered into the 4π-fullspace by an optical component. This measurement value has been chosen in view of laser applications and does not involve additional information on the test surface.

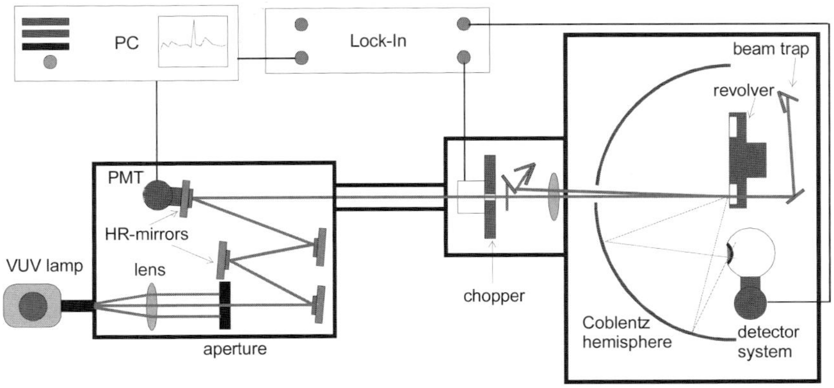

Fig. 7. TS-measurement facility for the UV/VUV-spectral range. The components are installed in chambers, which can be evacuated or flushed with pure gases for measurements at 157 nm (Kadkhoda et al. 2000).

As an example for a TS-measurement facility in accordance to ISO/FDIS 13696, an arrangement for the UV/VUV-spectral range is shown in Fig. 7. For economic reasons, a discharge lamp is employed as radiation source instead of a laser system, which is recommended in the ISO-Standard. After having run through the beam preparation and characterisation system in the first chamber of the facility, the beam is focussed onto the specimen. The sample is mounted in a computer-controlled revolver holder inside the measurement chamber, which contains also the integrating element and the scatter detector system. For the collection and integration of the radiation scattered by the specimen, an Ulbricht-sphere (Ulbricht 1900) or alternatively, a Coblentz-hemisphere (Coblentz 1913) are favoured in the ISO-Standard. In an Ulbricht-sphere the scattered radiation is integrated by multiple diffuse reflections at the inner wall of the sphere, which has to be coated with a highly diffuse reflecting material. Ulbricht-spheres can be produced in nearly every size for affordable costs and offer advantages in respect to handling, alignment, and geometry of the specimens. The application of Ulbricht-spheres at lower wavelengths is limited by the diffuse reflectance of the available coating materials for the inner wall. Therefore, below a wavelength of 200 nm, where no adequate diffuse reflecting materials can be found, Coblentz-spheres have to be employed for the collection of the scattered radiation. The TS-measurement facility in Fig. 7, which is dedicated to the wavelengths 193 m and 157 nm represents such a case and is equipped with a specially coated Coblentz-sphere. For an ideal collection of the scattered radiation, the specimen and the detector are positioned at conjugate points, forming a 1-to-1 imaging arrangement of the sample surface onto the detector. As a consequence of geometrical restrictions and imperfections of realistic Coblentz-spheres, these ideal conditions can not be fulfilled in most devices. Therefore, aberrations of the imaging have to be thoroughly considered to assess the error budget of a TS-measurement with a Coblentz-sphere (Kadkhoda et al. 1999). For sensitive measurements, Rayleigh-scattering of the environmental atmosphere may contribute significantly the scatter signal. Especially, in DUV/VUV-spectral range, evacuation or flushing with pure gases is often implemented in TS-measurement facilities to reduce the contribution of Rayleigh scattering or to suppress absorptance effects of the environment. Also, phase sensitive detection schemes and special evaluation techniques are routinely employed to improve the accuracy of TS-measurement systems. On the basis of these instrumentality, sensitivity limits below 1 ppm for the NIR/VIS-spectral range (Duparré 1995) and below 10 ppm for the UV/VUV-region (Kadkhoda et al. 2000) have been achieved, respectively.

1.4 Spectrometric On-line Monitoring

Besides ex-situ characterisation, on-line monitoring of the product quality gains of increasing importance in modern production concepts for complex optical components. Especially, in-situ characterisation of the growing coating during the deposition process opens new ways towards an integrated manufacturing of thin film optics with improved economy. On the basis of adapted process tracing algorithms, quality parameters of the growing coating system can be continuously compared to the expected behaviour, and deviations from the target can be detected in an early stage of the process, where a correction may be feasible without losing the coating batch. Also, by closing the loop between the on-line monitoring system and the regulation of the process parameters, the reliability of the manufacturing cycle would be drastically enhanced. As an ultimate development target, this concept may allow for a linear production cycle in optical coating technology. Such an integrated linear production cycle would comprise a deterministic deposition system, which, initially loaded with the target design, produces exactly the prescribed coating system.

Many approaches towards linear manufacturing of complex optical coatings are based on on-line spectrophotometric techniques. The potentiality of these concepts had been recognised already in the late 1970ies, and first spectrophotometric online monitoring systems were available at the beginning of the eighties (Vidal et al. 1979; Powell et al. 1986). However, as a consequence of the limited technical capability of the measurement equipment and especially the computers, these first approaches were restricted to data acquisition and display procedures. In order to calculate process parameters or the present state of the growing layer systems by a tracking model, the process had to be interrupted. Improved monitoring systems, which also integrate the measurement signals of other monitoring devices, were introduced during the nineties (Suillivan and Dobrowolski 1993; Bauer and Nüssler 1994; Tilsch et al. 1994). These systems were still constrained to uncomplicated on-line data reduction techniques that allowed for the determination of the switching point for the growing layers by calculating the deviation of the actual spectrum from a stored modelled data set. In the course of the recent development of computers with high calculation power and the introduction of camera chips with improved sensitivity, approaches towards sophisticated tracking algorithms have become possible during the last few years (Groß et al. 2001; Dobrowolski et al. 2002). These systems allow for a direct fitting of the measured spectral characteristic during the deposition process, resulting in a tracking of the deposited thin film design. Besides on-line spectrophotometry, research activities are also concentrated on ellipsometric measurement techniques and on the acquisition of plasma spectra in selected sputtering processes (Arens and Ristau 1998).

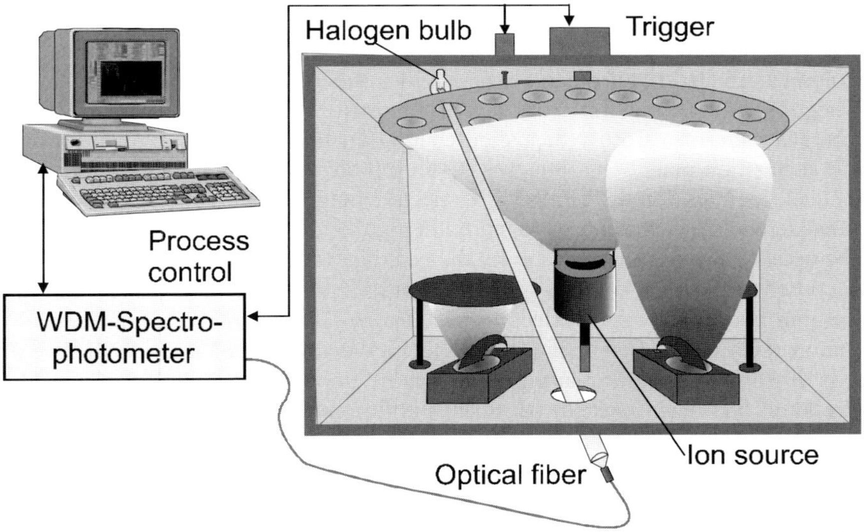

Fig. 8. Example for an in-situ spectrophotometric arrangement in conjunction with an ion assisted deposition process.

The fundamental arrangement for a spectrophotometric on-line monitoring system is illustrated in Fig. 8 for the example of an ion assisted deposition process. In the present configuration, the beam of the light source, which is installed in the deposition plant, passes through the optical component in the rotating substrate holder and reaches the input collimator of an optical fibre. The fibre is connected to a computer controlled CCD-spectrophotometer, which for example consists of several single board monochromator systems linked to the input channel by a wavelength division multiplexing (WDM) unit. By installing the WDM unit, the spectrophotometric system can be extended to more than one free spectral range of a single grating, and the measurement sensitivity can be adjusted separately for the different wavelength channels. If the measurement phases of the photometer are synchronised to the rotary motion of the substrate holder, selected components can be analysed directly. To perform a calibration of the subsequent spectrometric measurements, one position in the same ring is kept open for the back correction, and a second position is occupied by an opaque substrate for acquisition of the base line of the spectrophotometer for each revolution of the substrate holder. For the evaluation of the spectrophotometric data, a software package has been developed which calibrates the on-line spectra and traces the process by model calculations. This software package also contains algorithms enabling an automatic switching between the single layers of the coating system. In order to illustrate the benefits of on-line monitoring in thin film production, some selected results on spectrophotometric systems implemented in different deposition processes will be summarised in the following.

The production of coatings for the DUV/VUV-spectral range is still dominated by the thermal evaporation of a small class of fluoride materials including LaF_3 and MgF_2 as most prominent examples. These materials tend to form coatings with a columnar microstructure containing various defects and especially voids, which may absorb contaminants from the environment. As a consequence, the spectral characteristic of a freshly deposited coating changes during the venting cycle of the deposition plant. This imposes severe problems on the reliable production of optical coatings for the DUV/VUV-spectral range. On-line spectroscopy can be considered as a versatile tool to study these effects, which are mainly attributed to the uptake of atmospheric water by the coating systems. As an illustration, spectra recorded by a VUV-on-line spectrophotometer (Günster et al. 2001) during the venting cycle of a high reflecting system for 193 nm (angle of incidence) are summarised in Fig. 9. The presented spectra were registered after equal time periods of 3 minutes starting at a pressure of 10^{-6} mbar spectra until the atmospheric pressure had been reached. Besides the formation of the low wave-length absorption band, which is induced by the normal atmosphere, a clear shift of the spectral characteristic towards longer wavelengths can be observed. In case of the outlined thermal evaporation process, data of the on-line spectrophotometer can be employed for a prediction of the vacuum/air shift and for an exact wavelength control of the deposition cycle. As a consequence of the complex ageing behaviour of the coatings (Lambda Physik 2000), a cultivation of the on-line monitoring device to an automatic process tracing system is not very promising.

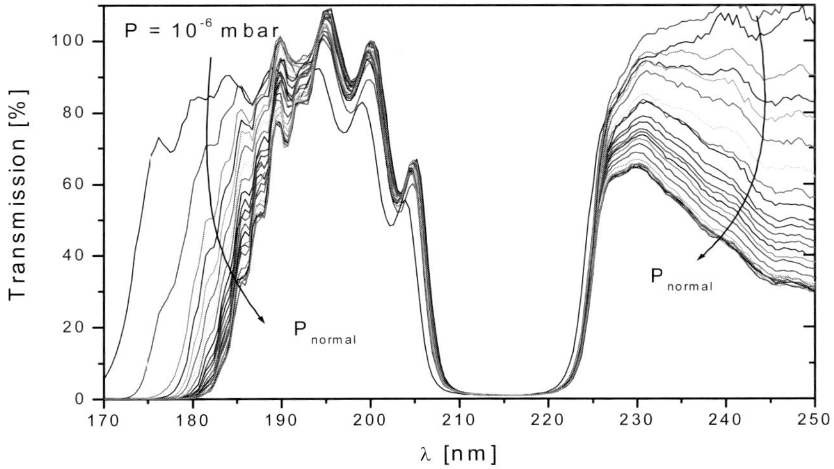

Fig. 9. On-line spectra recorded during the venting cycle of a HR-mirror (MgF_2/LaF_3) for 193 nm.

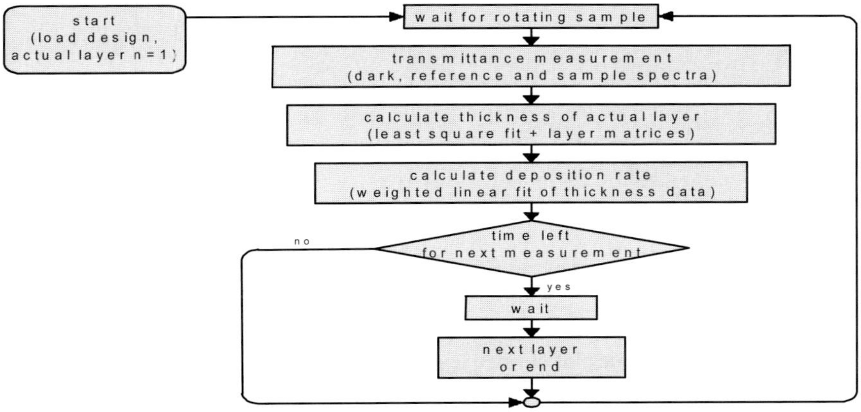

Fig. 10. Basic concept for data processing of an IBS-system.

The previous example elucidates another essential aspect in advanced thin film production: Obviously, deterministic processes, which unambiguously deliver coatings of predictable properties for a certain set of process parameters, are a obligatory prerequisite for an integrated, linear manufacturing of optical coatings.

Ion Beam Sputtering (IBS) is considered as an example for a process with high stability and is known to deliver coatings of oxide materials with unsurpassed optical quality in the NIR/VIS-spectral range. To investigate in the potentiality of IBS-Processes for an integrated production of optical coatings, an IBS-system had been equipped with a high resolution monochromator system (Starke et al. 2000) and an optimised computer system. The basic concept for data processing in the this set-up is illustrated in Fig. 10. After an initialisation step of the hardware arrangement, the target design of the layer system is loaded into the process control computer.

During the deposition, a calibrated transmittance measurement is performed within each revolution of the substrate holder. In the subsequent calculation, the actual film thickness is determined by a least-squares-fit. In this uncomplicated algorithm, the single layer temporarily grown is varied and the stack deposited before is assumed as ideal according to the target design. Typical layer thickness data fitted by the algorithm during the a deposition run for a single layer of SiO_2 within a HR-stack of TiO_2/SiO_2 is shown in Fig. 11.

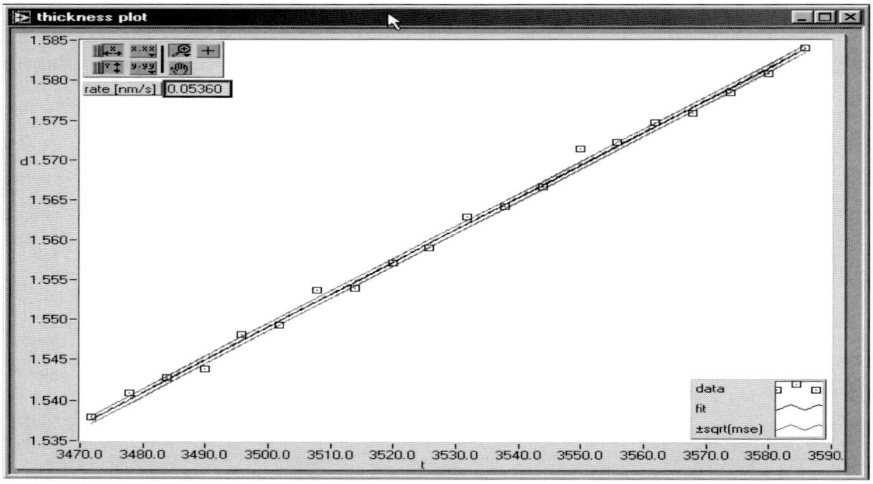

Fig. 11. Example for a time plot of fitted layer thickness.

In this case, the maximum relative uncertainty for the determination of the layer thickness is below 0.25 nm. In the final step of the process tracing algorithm the actual deposition rate is extrapolated, and the remaining time until switching between the layer materials or completion of the deposition i s determined. If this time is shorter than the time needed for the next measurement, the program waits for a defined time interval until it terminates the actual layer. Otherwise, the cycle restarts with the transmittance measurement.

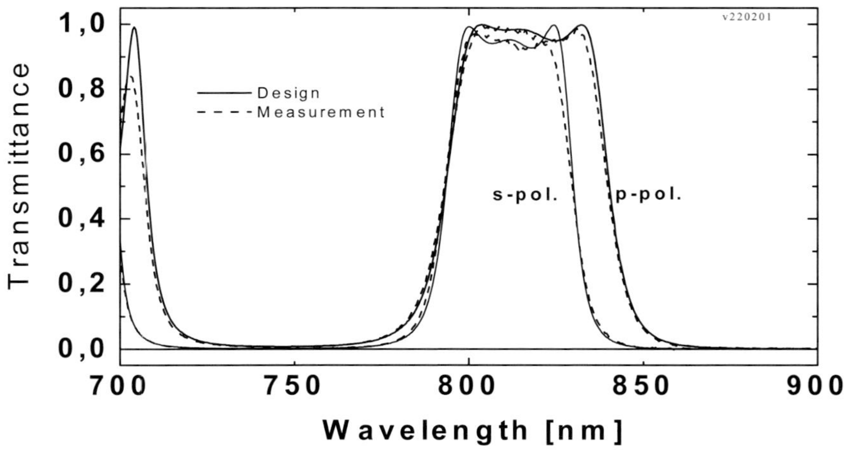

Fig. 12. Realised and target spectra of a non-polarising filter.

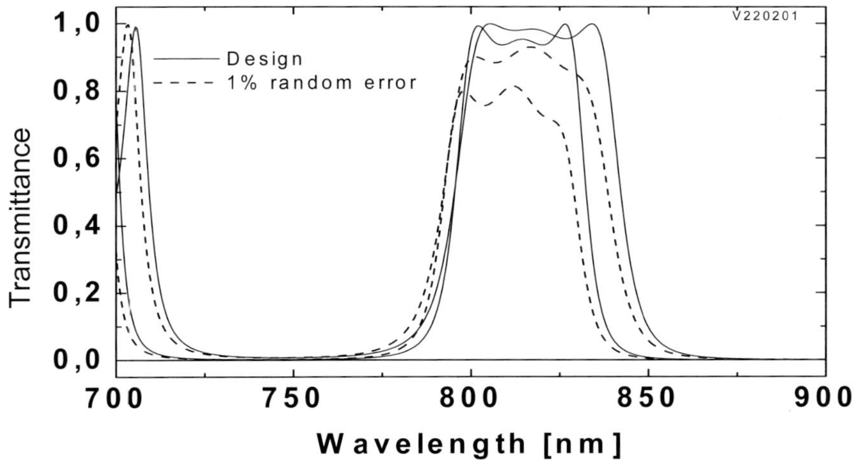

Fig. 13. Modelling with an error margin of 1%.

Once started, the deposition process is executed automatically until the coating design has been completely produced. Also, the online spectra can be stored for the purpose of quality assurement or for an evaluation and further optimisation of the design. In Fig. 12 the spectra of a non polarising edge filter produced under control of the on-line monitoring system are compared to the target design spectra. The layer system consists of 25 single layers of TiO_2/SiO_2 with three detuned cavity layers and a total thickness of 3.4 μm. In order to assess the accuracy of the on-line controlling system, the deposition run had simulated with an adapted modelling software, which takes into account specific errors of the deposition process and the monitoring system. Corresponding calculated spectra assuming a random error of 1% for the thickness of each single layer are depicted in Fig. 13 in conjunction with the target curves. A comparison clearly indicates that the random error of the on-line monitored layer system can be estimated as far lower than 1%, which represents a typical error margin achieved by conventional optical monitoring or crystal system.

2 Summary and Outlook

Characterisation and monitoring play a role of increasing importance in the production of high quality coatings. For key quality parameters of optical components, standard measurement procedures have been developed, which guarantee for readily comprehensible product specifications accepted in the global optics market. Following the future trends of optics technology and its applications towards shorter wavelength in the VUV/EUV-spectral range and towards fs-technology, new challenges are imposed on the quality management in thin film research and production. A major aspect is the qualification of ISO-standards in respect to this new horizon of innovative developments in laser technology, medicine, communication technology, life sciences, or semiconductor lithography. Besides ex-situ characterisation techniques, on-line monitoring will gain increased attention for the implementation of integrated, linear production strategies in optical coating technology. Nowadays, broad band in-situ spectrometry in conjunction with adapted deposition processes enables the reliable production of complex coatings in laboratory scale. Apart from spectrophotometry, ellipsometry and monitoring techniques for other specific process parameters are investigated. As a major prerequisite for the implementation of integrated manufacturing concept in the industrial production environment further progresses in the development of deterministic deposition processes and improved process tracing algorithms have to be achieved. This will be a further step towards an integral quality management in optics production of the future.

References

Abate JA, Roides R (1983) Spatially resolved absorption and detection of microscopic impurities in optical thin films by photothermal detection. J. de Phys. Coll. C6, No.10, pp 497-502

Abate JA, Schmid AW, Guardalben MJ, Smith DJ, Jacobs SD (1983) Characterisation of micron-sized, optical coating defects by photothermal deflection, in Laser Induced Damage in Optical Material, NBS SP 688, pp 385-392

Akhtar SMJ, Ristau D, Ebert J (1986) Thermal conductivity of dielectric films and correlation to damage threshold at 1.064 μm. NBS Spec. Publ. No.752, pp 345-351

Anderson DZ, Frisch JC, Masser CS (1984) Mirror reflectometer based on optical cavity decay time. J Appl. Opt., vol.23, No.8, pp 1238-1245

Apel O, Mann K, Heber J, Thielsch R (2000) Non-linear absorption phenomena in oxide coatings for 193 nm. Proceedings of the 31th Symposium on Laser Induced Damage in Optical Materials, SPIE vol. 3902, pp 235-241, 242-249

Arens W, Ristau D (1998) Online Monitoring of ion beam sputter processes by plasma analysis. Proceedings of „Optical Interference Coatings“, Tucson, OSA Technical Digest Series, vol.9, pp 81-83

ASTM Doc. E1392-90: Standard practice for angle resolved optical scatter measurements on specular or diffuse surfaces. November 1990

ASTM Doc. F1048-87: Standard test method for measuring the effective surface roughness of optical components by total integrated scattering. August 1987

Bauer HH, Nüssler E (1994) In-situ optical multichannel spectrometer system, SPIE 2253:423

Bennett JM, Mattsson L (1999) Introduction to Surface Roughness and Scattering. 2nd edn. Optical Society of America, Washington D.C., ISBN 1-55752-609-5

Blaschke H, Arens W, Ristau D, Martin S, Li BC, Welsch E (2000) Thickness dependence of damage threshold for 193 nm dielectric mirrors by predamage sensitive photothermal technique. Proceedings of the 31st Symposium on Laser Induced Damage in Optical Materials, SPIE vol. 3902, pp 242-249

Boccora AC, Fournier D, Badoz J (1980) Thermo-optical spectroscopy: Detection by the „mirage effect“, J Appl. Phys. Lett. Vol.36, No.2, pp 130-132

Coblentz WW (1913) Natl. Bur. Stand. (U.S.) Bull.9, p 283

Dieckmann M, Ristau D, Willamowski U, Schmidt H (1994) Measurement of thermal conductivity in dielectric films by the thermal pulse method. Proceedings of the international Symposium on Optical Interference Coatings in Grenoble, part 1, SPIE vol.2253, pp 712-719

Dobrowolski JA, Browning St, Jacobson M, Nadal M (2001) OSA Topical Meeting on Optical Coatings: Manufacturing Problem, Submitted for publication in Applied Optics, 2002. For pre-prints: http://www.osa.org/mtg_conf/2001/oic/

Duparré A (1995) Light Scattering of Thin Dielectric Films. In: Hummel RE, Guenther KH (eds) Handbook of Optical Properties, vol. 1 Thin Films for Optical Coatings, CRC Press

Eva E, Mann KR (1996) Nonlinear absorption phenomena in optical materials for the UV-spectral range. Proceedings of the International Workshop of Laser Beam and Optics Characterisation, SPIE, vol. 2870, pp 476-482

Fernelius NC (1982) Photoacoustic chopping frequency studies on uncoated zinc selenide. J Appl. Opt. vol.21, No.3, pp 481-490

Fleig C, Giesen A (2000) High precision reflectometer. Proceedings of the 5th International Workshop of Laser Beam and Optics Characterisation VDI-Verlag, pp 319-332

Grilli ML, Ristau D, Dieckmann M, Willamowski U (2000) Thermal conductivity of e-beam coatings. J Applied Physics A 71, pp 71-76

Groß T, Lappschies M, Starke K, Ristau D (2001) Systematic errors in broadband optical monitoring. Proceedings of the Conference of on Optical Interference Coatings, OSA Technical Digest pp.ME4-1 to ME4-3, July 15

Guenther AH, McIver JK (1988) The role of thermal conductivity in the pulsed laser damage sensitivity of optical thin films. Laser Optics for Intracavity and Extracavity Applications, SPIE, vol. 895, pp 246-253

Günster St, Kadkhoda P, Ristau D (2001) Online spectrophotometric characterisation of MgF2 /LaF3 - fluoride multilayer coatings production Proceedings of the Conference of on Optical Interference Coatings, OSA Technical Digest pp. ME9-1 - ME9-3

Henking R, Ristau D, Alvensleben Fv, Welling H (1994) Optical charac-teristics and damage thresholds of low loss mirrors. Symposium on Laser Induced Damage in Optical Materials, SPIE, vol. 2428, pp 281-292

Hummel RE, Günther KH (Eds) (1995) Thin Films for Optical Coatings, vol.1 Optical Properties, chap. 9, CRC Press, Boca Raton

ISO 9022 (1996-2002): Optics and optical Instruments – Environmental test methods – Parts 1-21. International Organisation for Standardisation

ISO 9211 (1994-2001): Optical Coatings, Parts 1-4, International Organisation for Standardisation

ISO 10110 (1996, 1997): Preparation of drawings for optical elements and systems. Parts 1-12. International Organisation for Standardisation

ISO 11254 (2002): Test methods for laser induced damage threshold of optical surfaces. Part 1: 1 on 1-test, 2000, Part 2: S on 1 test, 2001, Part 3 (DIS): Certification of power handling capability, International Organisation for Standardisation, ISO/TC 172/SC 9/WG 6

ISO 11551(1997): Test method for absorptance of optical laser components. International Standard. International Organisation for Standardisation

ISO/DIS 15368 (2000): Measurement of reflectance of plane surfaces and transmittance of plane parallel elements. International Organisation for Standardisation

ISO/FDIS 13696 (2001): Test Method for Radiation Scattered by Optical Components. International Organisation for Standardisation, ISO/TC 172/SC 9/WG 6

Optics and optical instruments. Lasers and laser related equipment. ISO/WD 13697 (2000): Test method for reflectance and transmittance of optical laser components. ISO/TC 172/SC 9/WG 6

Kadkhoda P, Müller A, Ristau D (1999) Total scatter losses of optical components in the DUV/VUV spectral range. Proceedings of the 31th Symposium on Laser Induced Damage in Optical Materials, SPIE vol.3902, pp 118-127

Kadkhoda P, Blaschke H, Kohlhaas J, Ristau D (2000) Investigation of transmission and reflectance in the DUV/VUV spectral range. Conference Optical Metrology Roadmap for the Semiconductor, Optical, and Data Storage Industries, SPIE, vol. 4099, pp 311-318

Kadkhoda P, Welling H, Günster S, Ristau D (2000) Investigation on total scattering at 157 nm and 193 nm. Conference Optical Metrology Roadmap for the Semiconductor, Optical, and Data Storage Industries, SPIE vol. 4099, pp 65-74

Koschade U, Steiger B (1998) Test of experimental set-up for reflectance measurement at wavelength 1.06 and 10.6 μm. Proceedings of the 4th International Workshop of Laser Beam and Optics Characterisation. VDI-Verlag, pp 634-641

Kreuznach S, Mann K (2002) Spectral characterisation of EUV radiation emitted from a laser-excited gas puff target. To be published in Elsevier Science

Kuo BSW, Li JCM, Schmid AW (1992) Thermal conductivity and interface thermal resistance of SI film on Si substrate determined by photothermal displacement interferometry. J Appl. Phys. 55:289-296

Third International UV Laser Symposium for 157 nm. Lambda Physik, Inc. Ft. Lauderdale, November 1st to 3rd, 2000

Lambropoulos JC, Jolly MR, Amsden CA, Gilman SE, Sinicropi MJ, Diakomihalis D, Jacobs SD (1989) Thermal conductivity of dielectric thin films. J Appl. Phys. 66:4230-4242

Lebert R, Bergmann K, Schriever G, Neff W (1999) A gas discharged based radiation source for EUV lithography, Micro- And Nano- Engineering 46, pp 449-452

Nowacki W (1962) Dynamic problems of thermoelasticity. Pergamon Press, Oxford

Olmstead MA, Amer NM, Kohn S, Fournier D, Boccara AC (1983) Photothermal displacement spectroscopy: An optical probe for solids and surfaces. J Appl. Phys. 32:141-154

Pinnow DA, Rich TC (1973) Development of a calorimetric method for making precision optical absorption measurements. J Appl. Opt. vol.12, No.5, pp 984-992

Powell I, Zwinkels JCM, Robertson AR (1986) Development of optical monitor for control of thin-film deposition. J Appl. Opt. 25:3645

Reichling M, Grönbeck H (1994) Harmonic heat flow in isotropic layered systems and its use for thin film thermal conductivity measurements. J Appl. Phys. vol.75, No.4, pp 1914-1922

Rempe G, Thomson RJ, Kimble HJ, Lalezari R (1992) Measurement of ultralow losses in an optical interferometer. J Opt. Lett. vol.17, No.5, pp 363-365

Ristau D, Willamowski U, Welling H (1998) Measurement of optical absorptance according to ISO 11551: Parallel round-robin test at 10.6 μm. SPIE Proceedings of the 30th Symposium on Laser Induced Damage in Optical Materials, SPIE, vol.3578, pp 657-670

Rosencwaig A, Gersho A (1976) Theory of photoacoustic effects with solids. J. Appl. Phys., vol.47, No.1, pp 64-69

Rosencwaig A, Hindley TW (1981) Photoacoustic measurement of low-level absorption in solids. J Appl. Opt. vol.20, No.4, pp 606-609

Schwartz ET, Pohl RO (1987) Thermal resistance at interfaces. J Appl. Phys. Lett. vol.51, No.26, pp 2200-2202

Starke K, Groß T, Ristau D (2000) Rapid prototyping of optical thin film filters. Conference Optical and Infrared Thin Films. SPIE, vol. 4094, pp 83-92

Starke K, Jupé M, Ehlers H, Ristau D (2001) Investigation in high precision reflectivity measurements according to ISO 13697. To be published in Proceedings of the 6th International Workshop of Laser Beam and Optics Characterisation. Munich, 18-20 June 2001

Suillivan BT, Dobrowolski JA (1993) Deposition error compensation for optical multilayer coatings. II. Experimental results-sputtering system, Appl. Opt. 32:2351

Tam AC (1989) Overview of photothermal spectroscopy. In: Sell JA (ed) Photothermal Investigations of Solids and Fluids, Academic Press, Boston, pp 1-33

Tam AC, Sullivan B (1983) Remote sensing applications of pulsed photothermal radiometry. J Appl. Phys. Lett. vol. 43, No.4, pp 333-335

Tilsch M, Scheuer V, Staub J, Tschudi T (1994) Direct optical monitoring instrument with double detection system for the control of multilayer systems from the visible to the near infrared. SPIE, vol.2253, pp 414-422

Ulbricht R (1900) Die Bestimmung der mittleren räumlichen Lichtintensität durch nur eine Messung. Elektrotech. Zeit 29, p 595

Vidal B, Fornier A, Pelletier E (1979) Wideband optical monitoring of nonquarterwave multilayer filters, J Appl. Opt. 18:3851

Voss A, Plass W, Giesen A (1994) A simple high precision method for measuring the specular reflectance of optical components. J Appl. Opt. 33:8370

Welsch E, Ristau D (1995) Photothermal measurement of optical thin films, J Appl. Opt. 34:7239-7253

Welsch E, Schäfer D, Matthias E (1992) High frequency photothermal reflectance and displacement measurements on thin film samples. In: Bicanac B (ed) Photoacoustic and Photothermal Phenomena, Springer Series in Optical Sciences vol.69, Springer Berlin, Heidelberg

Willamowski U, Groß T, Ristau D, Welling H (1996) Calorimetric measurement of optical absorption and transmissivity with sub ppm sensitivity. Proceedings of the International Symposium on Optical Systems Design and Production 2, SPIE, vol. 2775, pp 148-158

Wu ZL (1995) Overview of photothermal characterisation of optical thin film coatings. In: Laser Induced Damage in Optical Material, Proc. SPIE, 2714, pp 465-481

Zimmermann P, Welsch E (1994) Modelling of signal detection by using the photothermal probe beam deflection technique. Rev. Sci. Instrum. vol.65, No.1, pp 97-101

Abbreviations and Symbols

WD – Working Draft
DIS – Draft International Standard
ISO – International Organization for Standardisation
FDIS – Final Draft International Standard
LIDT – Laser Induced Damage Threshold
cw-LIDT – Laser Induced Damage Threshold for continuous wave operation
1 on 1-LIDT – damage threshold measurement protocol: one pulse per site
S on 1-LIDT – damage threshold measurement protocol: number of S pulses per site
UV/VUV – Ultra Violet to Vacuum Ultraviolet wavelength range (100-400 nm)
PMT – Photomultiplier
EUV – Extreme Ultraviolet wavelength range (10-100 nm)
CRD – Cavity Ring Down

ΔS – measurement signal (principle according to ISO/WD 13697)
R_S – Reflectivity of sample (principle according to ISO/WD 13697)
R_C – Reflectivity of the chopper (principle according to ISO/WD 13697)
S_0 – measurement signal according to a reflectivity of 100% (ISO/WD 13697)
Nd:YAG – Neodymium doped Ytterbium Aluminium Garnet
χ - thermal conductivity of the coating
c - heat capacity of the coating
ρ - density of the coating material
α_{th} - thermal expansion coefficient of the coating
λ_L, μ_L - Lame's modulus of elasticity of the coating
T_0 - room temperature
$T(\vec{r}, t)$ - temperature field
δ_t - time derivative
$\vec{U}(\vec{r}, t)$ - displacement field
γ_e - incident radiation energy flux
MIR – Mid Infrared wavelength range
NIR/VIS – Near Infrared to Visible wavelength range
ASTM – American Society for Testing and Materials
TS – Total Scattering (measurement principle according to ISO/FDIS 13696)
DUV/VUV – Deep Ultraviolet to Vacuum Ultraviolet wavelength range (100-200 nm)
CCD – Charge Coupled Device (semiconductor detector array)
WDM – Wavelength Division Multiplexing
IBS – Ion Beam Sputtering
HR-stack – High Reflecting stack

Mechanical Stress in Optical Coatings

Georg N. Strauss

Thin Film Technology Group, Institute of Ion Physics, University of Innsbruck, A

1 Introduction

The mechanical properties of thin films like stress, density, adhesion, hardness and abrasion resistance are, like structure and morphology, primarily dependent on the coating material and the kind of deposition process applied for their production. Such properties can therefore be influenced and modified by production technology variants and changing process parameters. Furthermore exists an interrelation between mechanical and the optical film properties.

Thin films are generally in a state of elastic mechanical stress during and after deposition. The stress behaviour of films is very important in all applications of thin films with respect to durability, stability and usability. Understanding the mechanisms that determine the formation and evolution of film stress and developing strategies to control these stresses represent some of the most important issues in the field of thin film technology. The first investigations regarding mechanical stress in thin films were carried out in 1877 (Mills 1877) on chemically deposited films, and in a more quantitative way in 1909 (Stoney 1909). Evaporated films were studied in the 1950's with the aim of understanding the origin of mechanical stress. Results of further measurements, explanations of the origin and semi quantitative calculations of tensile stresses based on a grain boundary interaction model for crystalline films have been published by different groups (Hoffman 1975) and (Pulker et al. 1979). During extended investigations it was soon found that almost all films, whatever production technology was chosen, show mechanical stresses. In most thin film applications, especially in the field of optical coatings, observed stresses should be clearly distinguished.

The intrinsic stress, a characteristic film property determined by the material and the deposition process can be tensile or compressive. Tensile stresses arise if the film tends to contract parallel to its surface and compressive stresses arise if the film expands parallel to the surface. Sign and magnitude are influenced by particle energy, deposition rate, film thickness and physical or chemical incorporation of material.

The observed total mechanical stress can consist of three components according to:

$$\sigma = \sigma_{ext} + \sigma_{therm} + \sigma_{\text{int}} \tag{1}$$

The external stress (σ_{ext}) arise from attack by external forces and the thermal stress (σ_{therm}) always arise if large differences exist in the thermal expansion coefficients between film material and substrate material. Strong effects can occur even during film production, at post deposition annealing or in warming up or cooling down procedures.

$$\sigma_{therm} = Y_f\left(\alpha_f - \alpha_S\right)\left(T_p - T_m\right) \tag{2}$$

Y_f : Youngs modulus of the film
α_f and α_s : thermal coefficient of film and substrate
T_p and T_m : temperature during evaporation process and at measurement

The last and most important part is the intrinsic stress (σ_{int}) in the film itself. Intrinsic stress is a structure and microstructure sensitive property (see also chapter "Fundamentals of Thin Film Growth" from Kaiser), which is caused by the mode of film growth and microstructural interactions, also influenced by some contaminations. Usually, the intrinsic stresses in the films are dominant, so that many investigations have dealt with them. The main mechanisms behind the origin and temporary changes of the intrinsic stresses in thin films are:

- Formation of lattice defects (microscopic voids, special arrangements of dislocations and lattice disorder)
- Variation of the interatomic spacing (occurancy of cavities or regions of higher density in amorphous films)
- Changes in the film structure and microstructure (temporarily, dependent on process)
- Incorporation of foreign materials or gas particles (e.g. atoms or molecules from the residual gas or working gas) and chemical reactions
- Recrystallization processes and phase transformations during and/or after film formation
- Differences of the lattice spacing of substrate and film during epitaxial film growth

In addition, all of these processes may occur both during condensation and growth of the film and under annealing conditions afterwards. Depending on the material and the deposition conditions obtained films can be vitreous or amorphous, but many are crystalline, particularly polycrystalline.

2 Models of the Origin of Intrinsic Film Stress

Most of the models are based on the concept that intrinsic stress is introduced into the films due to a contraction or expansion parallel to the substrate and film surface, which leads in the first case to tensile and in the latter to compressive stresses.

A satisfactory theory for the origin of intrinsic stresses has yet to be found. The wide variation in the experimental results coupled with a still improvable knowledge of structural details has led to multiple qualitative and quantitative models. The development of stress is in most cases caused by a combination of several effects, depending on the film material and the production process. First steps in this direction have been worked out by different groups (Heavens and Finegan 1957) and (Hoffman 1975), who described the appearance of film stress by means of a combined effect of surface tension and lattice defects.

2.1 Defect Model

In this model the intrinsic stress is related to point defects and defect clusters, like single vacancies and interstitial migration to a free surface. In the neighbourhood of such defects an elastic deformation of the lattice produces the mechanical stress (Story and Hoffman 1956; Hoffman and Anders 1953). The out-annealing of vacancies at temperature above 150°C leads to a stress reduction, there the total stress relief at room temperatures corresponds to a loss of about 1% vacancies. However, this point defect model is too great a simplification to account fully for the stress behaviour in films and can only be applied on lower density films showing tensile stresses.

2.2 Surface Tension Model

Surface tension can be an important contribution to the stress in thin films. Given a film with no other stress terms than the surface tension τ_1 at the film-vacuum interface and τ_2 at the film-substrate interface, the total apparent stress can be written as, where t_f is the film thickness:

$$\sigma = \frac{\tau_1 + \tau_2}{t_f} \tag{3}$$

A positive surface energy, due to single aggregates (e.g. dimers, clusters), manifests itself as a reduced lattice constant compared with bulk material and results in a tensile film stress. An increased interatomic spacing is an indication of a negative surface energy and corresponds with compressive stress. The sign of the film stress is supposed to depend on the balance of the two effects.

2.3 Grain Boundary Model

Most films deposited by evaporation onto unheated substrates develop a less dense, voids and gaps containing, columnar microstructure which is caused by surface geometric shadowing effects during deposition and maybe influenced also by crystal growth and by recrystallization during deposition of thicker films. The columns consist of fairly fine prismatic amorphous or crystalline aggregates of the film material. This growth mode is between zone 1 and 2 of the well-known MD-zone model (Movchan and Demchishin 1969). The column boundaries are in this case like grain boundaries. The observed behaviour can therefore be interpreted in terms of a stress mechanism introduced by forces at the grain boundaries of the columnar crystallites. The interactions between columns or grain boundaries are regarded for explaining the origin, and especially the occurrence, of tensile stresses (Hoffman 1975).

Frequently, evaporated films of pure elements and of many stable compounds have tensile stresses, with values depending on the kind of material, the applied deposition parameters and the film thickness. Generally the stresses reach a maximum when the last holes in the network stage of a growing film are filled. Additional thickness growth causes, along with recrystallization, a large decrease of the stresses, as can be seen in Fig. 1 (Wilcock 1967; Strauss et al. 1997; Floro et al. 2002).

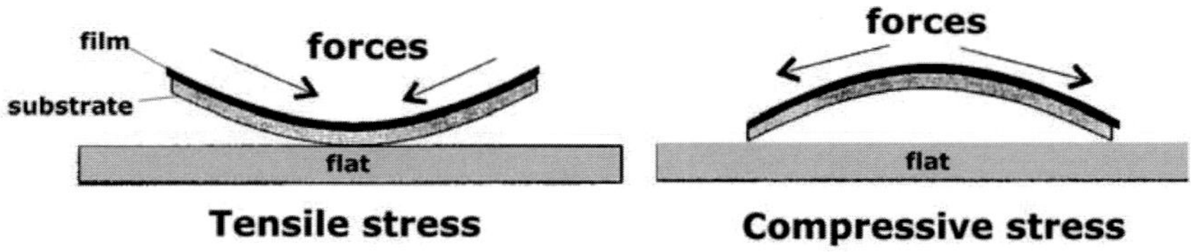

Fig. 1. Definition of compressive and tensile stress

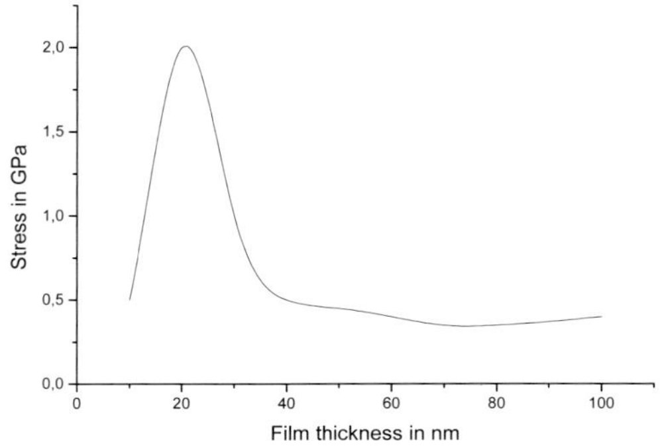

Fig. 2. General behaviour of the intrinsic mechanical stress as function of film thickness

The close approach of two free surfaces (surface energy γ_S) in the formation of a grain boundary (grain boundary energy γ_{gb}) results in a change in energy:

$$\Delta\gamma = 2\gamma_S - \gamma_{gb} \tag{4}$$

Part of the energy produces a constrained relaxation of the lattice atoms in the grain boundary by few percent of the unstrained lattice constant a. The elastic deformation is responsible for the macroscopically observed tensile stress according to Hook's law:

$$\sigma = Y * \varepsilon \tag{5}$$

where Y is Young's modulus and ε is the strain

$$\varepsilon = \frac{\Delta}{d} \tag{6}$$

Here Δ is the constrained relaxation and d is the mean crystallite diameter. The average force per unit area of boundary after relaxation must equal the average elastic stress throughout the volume of the crystallite. Considering also the Poisson ratio ν and the packing density p of the film the intrinsic stress is given by (Pulker and Mäser 1979; Pulker 1982):

$$\sigma = \frac{Y}{1-\nu}\frac{\Delta}{d}p \tag{7}$$

In situ measurements of the intrinsic stress in thin non-dissociating crystalline dielectric films also have been interpreted in terms of an interaction mechanism of forces between the grain boundaries of the columnar crystals (Seidel 1981).

Large grain sizes are associated with low stresses and small grain sizes with high stresses and it seems that the grain size is a dominant determining factor for the magnitude of the tensile stress. So grain growth is responsible for the observed decrease of the stress values as substrate temperature is raised, thickness increased and deposition rate decreased (Sun et al. 1975).

Recently a more quantitative and detailed model of the physical origin of intrinsic stress and the fundamental mechanisms that generate stress during film growth according the Volmer-Weber (V-W) mechanism has been discussed (Floro et al 2002). In the V-W growth mode, films initially grow by the nucleation of discrete islands of different crystallographic orientations, or even no crystallinity in the case of amorphous films. With ongoing deposition, the existing islands enlarge and contact each other, and new islands may nucleate, until a percolating network is achieved (see Fig.3). The film continues to grow until the substrate is completely covered, followed then by an additional growth in thickness (see also chapter "Fundamentals of Thin Film Growth" from Kaiser).

During and after deposition the film undergoes several occurrences, like defect introduction, annihilation, grain growth, recrystallization or phase transitions, which cause stress within the film. This dynamic structural and microstructural evolution processes change the density of the film while the film is rigidly attached to its substrate (Doerner and Nix 1988; Nix 1989). The film will be now in a state of mechanical stress and as a consequence of the forces and moments applied to the substrate by the film, the substrate will bend slightly, which is often used to determine the resulting film stress. The evolution of the film stresses during evaporative deposition shows a so-called CTC (compressive-tensile-compressive) behaviour, which has been directly correlated with microstructural evolution (Koch 1994; Floro et al 2001).

Fig. 3. Film growth: Nucleation – Coalescence – Continuous film (Floro et al. 2001)

The initial compressive stress occurs in the discrete-island stage of film growth, while the rapid increase in tensile stress correlates with the onset of island coalescence and grain boundary formation. The peak at the end of the tensile stress development occurs when the film becomes fully continuous, and in the final compressive stage, the continuous film is growing in thickness. CTC behaviour has been classified as a characteristic of high mobility films, whereas low mobility films exhibit only tensile stress. Surface stress is an important contribution to the stress evolution of discontinuous thin films in the discrete-island regime. Elastic deformation associated with grain boundary formation and grain growth during the coalescence process is widely regarded to be an important mechanism for tensile stress generation. The final compressive stress in continuous films can be explained by the fact that deposition occurs under highly non-equilibrium conditions in which the flux of evaporated atoms is many orders of magnitude higher than the equilibrium vapour pressure of the film at the deposition temperature. Therefore, the surface of the film is at a higher chemical potential which is the driving force for the movement of atoms into the grain boundaries and the evolution of compressive stress (Chason et al. 2001).

2.4 Peening Model

Thin films produced with PVD technologies without any ion or plasma assistance develop normally tensile film stress. Ion bombardment during deposition can reduce the tensile stress or even transfer tensile stress into compressive stress. The effects on the mechanical properties due to ion bombardment of the growing film are mainly depending on the kind, the amount and the energies of the involved ions, atoms and molecules. First descriptions concerning the development of compressive stress as consequence of an ion peening effect were carried out by d'Heurle (d'Heurle 1970) and later from different research groups starting in 1980 (Hoffman and Gaertner 1980; Cuomo et al. 1982; Huang et al. 1985; Windischmann 1987; d'Heurle and Harper 1989; Volkert 1991). Some calculations and computer simulations concerning the densification of the films due to ion bombardment and the effect on the microstructure of the films were carried out by Müller (Müller 1986, 1987, 1987-2). Based on such binary collision calculations a simplified but universal ion peening model explaining stress evolution and diminution due to ion bombardment was developed (d'Heurle and Harper 1989; Davis 1993; Carter 1994). According to this model compressive stress arises when the growing film is bombarded by ions, atoms or molecules with kinetic energies of tens or hundreds of electron volt by atomic peening processes. These knock-on processes cause atoms to be incorporated into spaces in the growing film, which are smaller than the usual atomic volume.

In order to initiate this process the impinging ions have to overcome a threshold energy E_{crit}, whose magnitude will depend on details of the film structure and composition. The compressive stress caused by implanted atoms is associated with an increase in the strain energy of the film. A reduction in the stress by movement of the implanted atoms to the film surface is energetically favoured, however, the implanted atoms are hindered from moving by repulsive forces exerted by the surrounding atoms. The implanted atoms, then, are in metastable positions within a few nanometres of the film surface. A significant fraction of the energy of a bombarding ion is transferred to violent motion, so called thermal spikes, of the atoms in the area of impact and can provide the energy required to release implanted atoms from their metastable positions within the film (Müller 1986).

With these assumptions the stress can be calculated considering that there is a balance between constrained implantation and relaxation processes so that the density n of implanted atoms is constant with time. As a result of the competition between implantation and relaxation, in general the stress rises to a maximum with increasing energy of the bombarding atoms and then decreases, as can be seen in Fig. 4.

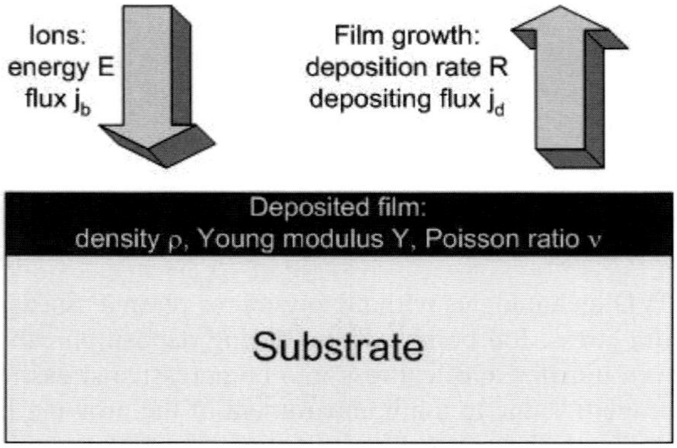

Fig. 4. Schematic diagram of a growing film with simultaneous ion bombardment of flux j_b with an energy E

As a first-order approximation, the volumetric strain ε is assumed to be proportional to the fraction n/ρ of implanted atoms in the film (Windischman 1987), where n is the density of implanted atoms and ρ is the film density. For a thin film the stress is related to the strain by

$$\sigma = \frac{Y_f}{1-\nu}\varepsilon \tag{8}$$

where Y_f is the Young's moduls of the film material and ν is the Poisson ratio. With this assumptions the stress is given by

$$\sigma \propto \frac{Y_f}{1-\nu}\frac{E^{1/2}}{j_d/j_b + k\cdot E^{5/3}} \tag{9}$$

Where j_d is the net deposition flux, j_b is the bombarding flux and $k=0.016*C_\rho*E_0^{-5/3}$. This result describes the bombardment of monoenergetic ions with kinetic energies higher than E_{crit}. A range of ions may be taken into account by integrating over all energies greater than E_{crit} to give

$$\sigma = \int_{E_{crit}}^{\infty} \sigma(E) f(E) dE \tag{10}$$

where f(E)dE is the number of ions with an energy between E and E+dE.

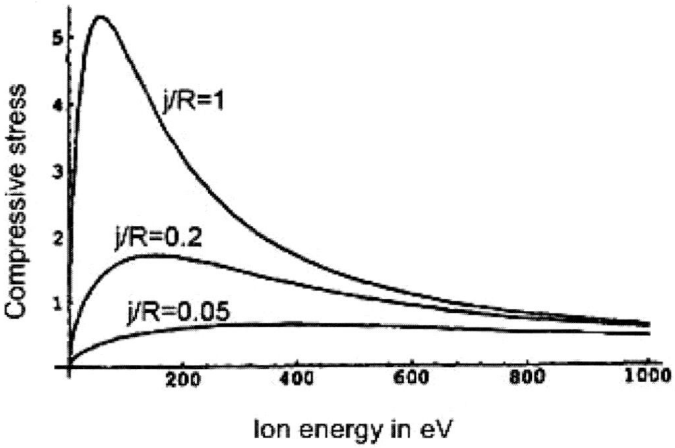

Fig. 5. Variation in compressive stress with ion energy, predicted by Eq. 9, for three values of the normalized flux j_b/j_d=1, 0.2 and 0.05, assuming an effective excitation energy of E_0=8eV and C_ρ=1, which is a material depending parameter (Davis 1993)

The model predicts that the magnitude of the compressive film stress is strongly dependent on the energy of the incident ions, with an energy dependence determined by the normalized flux j_b/j_d, where j_d is the net depositing flux and j_b is the bombarding flux. The differences in the stress-energy relationship seen in the literature can therefore be explained in terms of variations in the normalized flux. For low values of the normalized fluxes the model predicts that the film stress is proportional to the square root of the energy of the impinging ions, higher normalized fluxes, however, cause the stress to go through a maximum with increasing ion energy with a power law decrease for large energies and fluxes.

3 Methods of Measuring Stress

The determination of the intrinsic film stress is normally performed by measuring those film properties, which are changed due to the appearance of the film stress. There are several methods, which are in practical use.

The method of X-ray diffraction (XRD), and the electron diffraction method (RHEED – Reflection High Energy Electron Diffraction) determines the stress induced changes in the crystal plane distances of the lattice due to the broadening of the diffraction lines, from which the stress can be calculated. Line shifts in the diffraction diagram indicate changed lattice parameters and therefore provide information on macrostrain, whereas line broadening gives information on grain size and microstrain, i.e. lattice distortions that cause a spread around the lattice constant of the stressed material. This method is mainly used with metal films, but can also be applied for crystalline films.

Mechanical stress in solid bodies changes the inter-atomic distances and so the elastic module, which varies the velocity of sound and this can be detected by ultra sonic methods.

The piezoresistive effect consists in the change of the conductivity of a semiconductor as a consequence of an applied mechanical stress. The magnitude of the effect depends on the applied stress and its orientation with respect to the crystalline axis. The method can be applied to any pair of dielectric film-semiconductor substrate.

Several magnetic properties are connected with the atomic structure of the thin films, they can be measured to calculate the intrinsic stress.

Furthermore mechanical stress can be detected due to the change of the electrical resistance of the films or the changes in the band gab of semiconductors.

One of the well-recognized techniques to measure the stress in thin films is to determine the curvature of the bending substrate. The most general arrangements are to use a thin strip of glass clamped at one end to form a cantilever onto which the film is deposited or to use circular glass substrates or Si-wafer and determine the radius of curvature before and after deposition. The direction of the elongation of the free end indicates whether the stress is compressive or tensile.

3.1 Cantilever Methods

In these methods bending of the substrate can be measured optically (travelling light spot), electrically (capacity change), mechanically (stylus pick up) or by electro-mechanical and magnetically means (lever in a balance measures the restoring force).

In the optical deflection measurement a one side clamped substrate strip, bearing a mirror on the rear side of its free end is used. With a projection microscope the shift of a light spot caused by bending during coating is measured with a position detector in an arrangement shown schematically in Fig.6. Careful calibration of the system is required. It is possible to detect a minimum displacement of 10^{-4} cm.

It is also possible to measure the electrical capacity. In this case the cantilevered strip, silvered on its rear side, is held close and parallel to a fixed conductive plate to form part of the tuning capacity of an oscillator circuit. With the oscillator operating slightly off resonance, any change of the capacity that results from a bending of the glass strip will produce a linear change in the output of the oscillator.

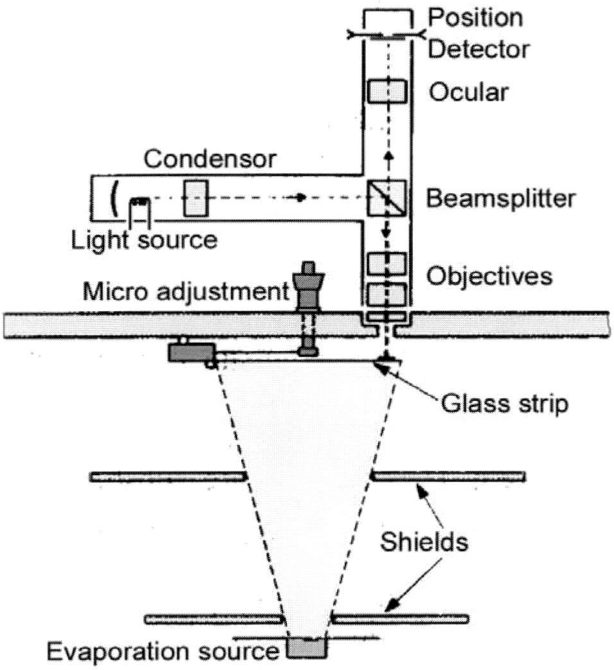

Fig. 6. Schematic diagram of an apparatus for the determination of intrinsic film stress using the cantilever method with optical detection

The electromechanical restoration method makes use of the application of a force to the free end of the strip in order to return it to the same position it was in prior to the deposition of the film. This restoring force is applied by a lever attached to the coil of an electromagnetic balance. The magnetic restoration technique is similar to the electromechanical restoration method, but in this case the restoring force is supplied by an electromagnet. This technique is limited to substrates made of, or containing some magnetic material.

3.2 Disc Methods

The interferometric method makes use of conventional optical interference techniques to observe the bending of a thin circular glass plate. In general, this technique is limited in use in that the measurements normally must be made outside the vacuum chamber. Newton's interference rings arising in monochromatic light between an optical flat (reference) and a first uncoated and than coated glass-disc can be measured from photographic prints or a CCD. The stress induced change of the radius of curvature than can be determined and together with the elastic data of the substrate the film stress can be calculated.

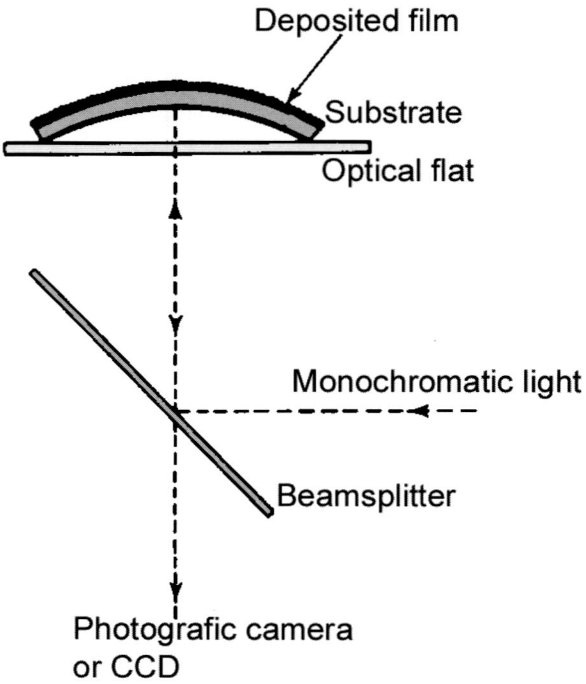

Fig. 7. Schematic representation of an apparatus for the determination of intrinsic film stress by observing Newton fringes between a stressed circular substrate and an optical flat

The change in the radius of curvature of a substrate (e.g. Si-wafer) due to progressive deposition of a stressed film is measured and the film stress value is calculated according to the following equation (Stoney 1909):

$$\sigma = \frac{Y_S t_S^{\,2}}{6(1-\nu)t_f} \frac{1}{R} \tag{11}$$

where $\frac{Y_S}{(1-\nu)}$ (in Pa) is the biaxial elastic modulus of the substrate, t_S and t_f (in m) are the substrate and film thickness, R is the radius of curvature (in m) and σ is the average film stress (in Pa).

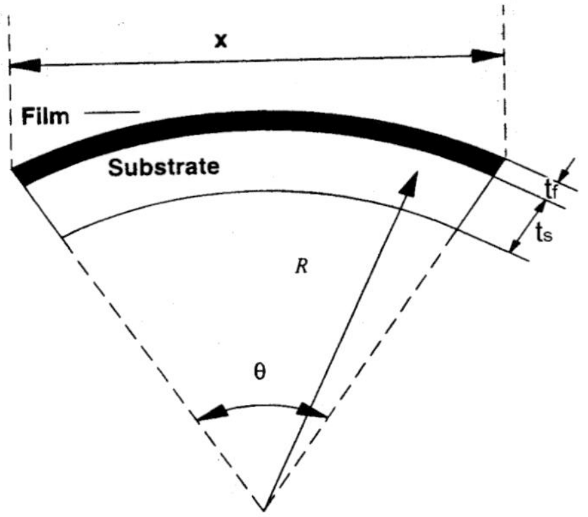

Fig. 8. Radius of curvature of a substrate due to the deposition of a film

Figure 8 shows a schematic drawing of the substrate deformed to radius R by the deposition of a film. In this case the film is under compression.

Multibeam optical stress systems are extremely sensitive and able to measure curvatures greater than 10km. Since the technique is optical and the system is mounted outside the deposition chamber, it can be used for monitoring virtually any type of process as long as there is optical access to the substrate. The principle underlying this technique is also to measure the stress-induced curvature of a substrate. A single lower power laser is used where the beam is divided into a linear array of multiple parallel output beams by transmission through a etalon, which is a piece of optical glass with parallel faces, and each face has a high reflectivity optical coating. The linear beam array then reflects off the sample and is detected on a charge-coupled device (CCD) camera. The only other optics are a spatial filter, an objective lens to focus the beam on the camera and attenuating filters to prevent saturation of the CCD (Floro et al. 1996; Taylor 1997). The primary advantage is that the optics is simple and stationary, requiring only minimal alignment during initial set-up. Simultaneous detection of the entire laser array makes the measurement inherently less sensitive to sample vibration compared with scanning mirror systems. Critical to the measurement is the use of a high resolution CCD array that enables highly accurate measurements of the spot positions. The use of image processing and data analysis routines makes it possible to detect micron-size changes in spot position.

The film stress can be calculated by a simple equation (Stoney 1909), that requires only knowledge of the film and substrate thickness, as well as the elastic modulus of the substrate:

$$\frac{1}{R} = \kappa - \kappa_0 \propto \frac{Y_f t_f \varepsilon}{Y_S t_S^{\ 2}} \qquad (12)$$

where κ is the curvature of the substrate, Y_f and Y_S are the film and substrate biaxial moduli and t_f and t_S are the film and substrate thicknesses. κ_0 is the initial curvature of the substrate prior to film growth. It can be seen that reducing the thickness of the substrate enhances the curvature resolution. The film stress is determined directly from the curvature measurement. Determining film strain requires the biaxial modulus Y_f of the film, which is rarely known with sufficient accuracy. The challenge of this technique is to accurately detect curvature in the substrate with sufficient resolution to measure the amount of stress typically found in thin films.

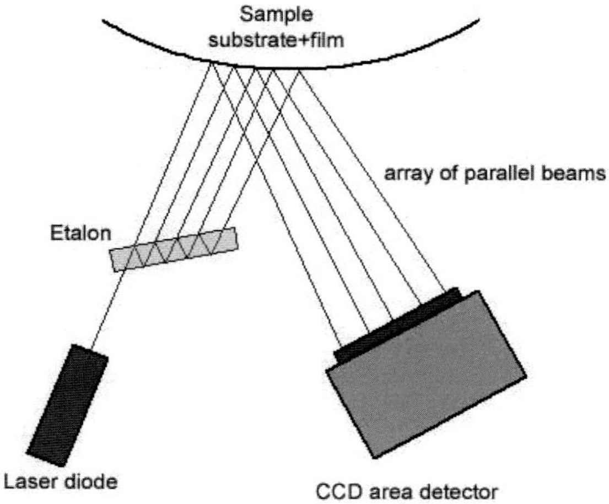

Fig. 9. Schematic diagram of a multibeam optical stress system

4 Stress Values in Dielectric Films

Metal films: Evaporated metal films in general show tensile stresses, whose values depend on the kind of material, the deposition technique and the film thickness.

Metal oxide: Metal oxide films prepared by conventional reactive evaporation (RE) have either tensile or compressive stress values, whereas those prepared by ion and plasma assisted techniques, for example ion plating (IP) and ion assisted deposition (IBAD), show nearly exclusively relatively high compressive stresses. These mostly amorphous films have a high packing density and therefore show excellent stability under changing conditions of relative humidity. The compressive intrinsic stress is a consequence of the high film density caused by higher energy of the condensing and bombarding atoms and molecules. The stress can be reduced by varying either deposition rate, gas pressure and/or the electrical parameters of the ion or plasma sources.

SiO_2: Silicon dioxide films prepared by electron beam evaporation of fused silica show a nearly constant compressive stress of 160 MPa in 500nm thick films. Ion and plasma assistance during deposition produce a relatively large compressive stress up to 700 MPa for film thicknesses larger than 150nm. Films with thicknesses below 100nm show even higher stress values up to 1,1 GPa.

SiO: The stress in silicon monoxide films is very dependent on the evaporation conditions like deposition rate, gas pressure, residual gas, substrate temperature, etc. Films produced with high deposition rates (>3nm/s) at low gas pressures ($<5*10^{-6}$ mbar) show relatively low stress values of 15 MPa tensile in films greater than 70nm thickness.

MgF_2: Magnesium Fluoride develops a relatively high tensile stress, which reaches values of about 400 MPa in films with 100nm thickness. The magnitude of the stress is influenced by chemical composition and by deposition parameters. MgF_2 films produced with reactive advanced plasma source ion plating (RAPSIP) show stress values from 90 MPa to 220 MPa tensile, where reactive evaporation (RE) can produce films with intrinsic stresses in the range from 30 MPa compressive to 120 MPa tensile (Atanassov et al. 1999). A permanent stress reduction in MgF_2 films can be achieved by the incorporation of foreign material of substances, which are insoluble in MgF_2 and have a low vapour pressure and low free surface energy. In this way a surface segregation of the doping material with low surface energy occurs, for example CaF_2 or BaF_2, and produces a pronounced decrease in grain boundary interaction which results in lower stress values (Pulker 1999).

Table 1. Stress values of SiO_2 deposited with different technologies

Process	Stress in MPa	Reference
RFMS – radio frequency magnetron sputtering	-150 to –270 -150 to –200	(Richter et al. 2001)
DCMS – direct current magnetron sputtering	-150 to –250	(Richter et al. 2001)
RAPSIP - reactive advanced plasma source ion plating	-100 to -500	(Gäbler et al. 2000)
RE - reactive evaporation	-100 to -500	(Christova and Manov 1994)
PECVD – plasma enhanced chemical vapour deposition	-70 to –80	(Ambree et al 1993) (Rutten 1989)
RIP – reactive ion plating	up to –1000	(Seddon et al 1988)
RLVIP – reactive low voltage ion plating	-500 to –1000	(Strauss et al 1997)
IBAD – ion beam assisted deposition	-10 to -470 -610 to -1900	(Robic et al 1996) (McNeil et al 1984)

Compressive stress: negative sign, tensile stress: positive sign

Table 2. Stress values of TiO_2 deposited with different technologies

Process	Stress in MPa	References
RAPSIP - reactive advanced plasma source ion plating	110 to 190	(Atanassov et al 1998)
	95	(Gäbler et al. 2000)
RE - reactive evaporation	165 to 300	(Atanassov et al 1998)
	220 to 300	(Szczyrbowski et al 1999) (Ottermann and Bange 1996)
MFMS – mid frequency magnetron sputtering	-500 to -900	(Szczyrbowski et al 1999)
SC – spin coating	100 to 250	(Szczyrbowski et al 1999) (Ottermann and Bange 1996)
PICVD – plasma impulse chemical vapour deposition	0 to 400	(Szczyrbowski et al 1999)
RIP – reactive ion plating	-150 to -400	(Szczyrbowski et al 1999) (Seddon et al 1988) (Ottermann and Bange 1996)
RLVIP – reactive low voltage ion plating	-800	(Pulker and Bühler 1985)
IBAD – ion beam assisted deposition	-540	(McNeil et al 1984)
FAD – filtered arc deposition	-700 to -2500	(Bendavid et al 1999)

Compressive stress: negative sign, tensile stress: positive sign

Table 3. Stress values of Al_2O_3 deposited with different technologies

Process	Stress in MPa	References
RFMS – radio frequency	-250 to –1200	(Roth et al 1987)
magnetron sputtering	-200 to -300	(Qing-Shan 1989)
RE - reactive evaporation	+100	(Kubovy and Janda 1977)
RIP – reactive ion plating	-210	(Seddon et al 1988)

Compressive stress: negative sign, tensile stress: positive sign

Table 4. Stress values of Ta_2O_5 deposited with different technologies

Process	Stress in MPa	References
RFMS – radio frequency magnetron sputtering	-200 to -350	(Cheng et al 1999)
RIP – reactive ion plating	-200	(Seddon et al 1988)
RLVIP – reactive low voltage ion plating	-400 to -1000	(Strauss et al 1997)

Compressive stress: negative sign, tensile stress: positive sign

Table 5. Stress values of Nb_2O_5 deposited with different technologies

Process	Stress in MPa	References
RFMS – radio frequency magnetron sputtering	+100 to -1100	
DCMS – direct current magnetron sputtering	-100 to -250	(Richter et al. 2001)
RLVIP – reactive low voltage ion plating	-200 to -500	(Edlinger and Pulker 1989)

Compressive stress: negative sign, tensile stress: positive sign

5 Correlation Between Film Stress and Optical and Mechanical Properties

Density and refractive index of isotropic materials are interrelated by the Lorentz-Lorenz equation. The mean refractive index can be calculated for the general case from

$$\overline{n_f}^2 = \frac{2\overline{\rho_f} R_{mol} + M_{mol}}{M_{mol} - \overline{\rho_f} R_{mol}} \tag{13}$$

if the mean film density ρ_f is known, R_{mol} is the molar refraction, which can be calculated from single crystal data, and M_{mol} is the molar weight. If we use the packing density p, which is defined as the ratio of the average film density ρ_f and the bulk density ρ_m, the correlation between the packing density p and the refractive index n is given by:

$$p = \frac{\rho_f}{\rho_m} = \frac{n_f^2 - 1}{n_f^2 + 2} \frac{n_m^2 + 2}{n_m^2 - 1} \tag{14}$$

A linear dependence between the refractive index n and film density ρ_f can be observed for evaporated, ion- and plasma-assisted films. A strong correlation exists between the intrinsic stress and the film density. In general, evaporated films have low density, open columnar microstructure and tensile intrinsic stress, while sputter deposited or ion plated films have higher density, close-packed microstructure and compressive stress. It is further known, that evaporated films prepared under continuous bombardment with ions of increasing kinetic energy are densified and show stress changes from tensile to compressive (Hoffman and Gaertner 1980).

One problem in highly stressed films is mechanical failures like cracking or delamination of the films. Tensile stresses cause cracks in the film perpendicular to the film plane and delamination is typical for compressive stress. Responsible for these film failures is not the film stress for itself but the elastic energy E_{elast} stored in the film. Assuming that the film stress can be considered to be biaxial and isotropic in the film plane and independent on film thickness t_f, the elastic energy E_{elast} per unit surface stored in the film is (Klockholm 1987)

$$E_{elast} = \sigma^2 \frac{(1-\nu)}{E} t_f \tag{15}$$

The mechanical stability of the film-substrate system becomes critical when the stored elastic energy approaches and exceeds a critical value E_{crit}, which is determined by a critical stress value σ_{crit}, and the Griffith criterion (Griffith 1920) with the crack length h, omitting ν and surface energy γ.

$$E_{crit} \cong \frac{\sigma_{crit}^2 h}{E} \cong 2\gamma \Rightarrow t_{crit} \cong \frac{2E\gamma}{\sigma^2} \tag{16}$$

This demonstrates that the two new surfaces formed by a crack contribute to the total surface energy with an increase of 2γ. When σ is constant and independent of film thickness t there exists a critical film thickness t_{crit} at which E_{elast} exceeds E_{crit} and film fracture will occur when $t > t_{crit}$. In the case of delamination the Barenblatt model (Barenblatt 1962) is more appropriate. Similar to Eq. 16 obtaining a critical energy for delamination leads to a critical stress value and finally to a critical film thickness

$$t_{crit} \cong \frac{\gamma_d E}{\sigma^2} \tag{17}$$

where the surface energy is given by $\gamma_d = \gamma_s + \gamma_f - \gamma_i$, (index s means substrate, f means film and i stands for the part where substrate and film were never joined).

The investigations show that, by the use of fracture theory, the criterion for film-substrate mechanical stability is dependent on γ, γ_d, σ and t and that the equations that describe the criteria for fracture and delamination have in principle the same form. To avoid catastrophic film failure, $\sigma^2 t$ must be reduced in some manner. Varying the deposition parameters or changing the process technology can reduce the film stress.

6 Summary and Outlook

Evaporation, sputtering and energetic ion and coating material deposition processes operated with different sets of parameters cause different structures and microstructures so that the films of the same material but produced by different technologies generally also have different stress values. Hence, the possibility exists of influencing stresses in a desired way.

In films deposited via vacuum evaporation, in the absence of any external agency (including contamination), tensile stresses are normally observed. Impurities in the deposition atmosphere of films deposited by evaporation cause the stresses to become compressive. Several mechanisms are probably involved in this phenomenon.

Theoretical treatments often explain the origin of compressive stress as a consequence of energetic particle bombardment, already discussed in chapter 2.4 Peening Model. Peening processes cause atoms to become incorporated into the growing film with a number quite higher than would be obtained otherwise since sufficient energetic atoms may be forced into spaces too small to accommodate them under thermal equilibrium conditions.

Generally, vacuum deposited films and sputter deposited films prepared at high gas pressures have tensile stresses which may be anisotropic with off-normal angle of incidence deposition. In low-pressure sputter deposition and ion plating, energetic particle bombardment gives rise to high compressive stresses due to the recoil implantations of surface atoms (Sun et al. 1975; Hoffmann and Thornton 1979; Thornton and Hoffman 1985). Studies of deposited films with concurrent bombardment have shown that the conversion of tensile to compressive stress is very dependent on the ratio of bombarding species to depositing species (Hoffman and Gaertner 1980). In plasma processing the residual film stress may be very sensitive to the substrate bias and gas pressure during deposition in a plasma environment (Bland et al 1974; Cuthrell and Mattox 1988). The lattice strain associated with the film stress represents stored energy and this energy along with a high concentration of lattice defects leads to a lowering of the recrystallization temperature in crystalline materials, a lowered strain point in glassy materials, a high chemical etch rate, electro migration problems, void growth in metallization lines by creep or other such mass transport effects.

Film property modifications by ion bombardment are mainly caused by momentum transfer from the ions to the film atoms. Experiments on a reactive low voltage ion plating (RLVIP) process with dielectrics showed that reactive film deposition in presence of excited and ionised atoms and molecules of higher energy leads to the formation of dense chemical compound films. The high density of such films is responsible for the relatively high refractive index and the compressive stress. Ion bombardment induced local chemical defects may increase the extinction coefficient value when high ion current densities and ion energies of several tens of eV are applied in film production. Film deposition with low ion current densities and low ion energies yield stoichiometric coatings with very low extinction coefficient, low density, low stress and low refractive index. The optical and mechanical stability of such films is, however, often insufficient in such cases. The films need to be densified by a longer lasting post-deposition heat treatment at temperature higher than 300°C. This leads to a decrease in physical thickness and an increase in the refractive index and environmental stability. Heat treatment of dense films often causes microstructural relaxation phenomena and oxidation resulting in a decrease of the film density, extinction coefficient and intrinsic stress, there the physical film thickness if increased (Lechner et al 1998; Strauss et al. 1999).

As the total gas pressure of the vacuum atmosphere rises, the mean free path length in PVD processes is reduced. The paths by which neutrals, non-decelerated particles can travel or along which ionised particles can be accelerated by electri-

cal fields are shortened. The kinetic energy of the particles impinging on the growing film decreases, which causes a decrease in growth-induced compressive stresses with rising total pressure. If the coating is deposited without applying a bias voltage, tensile stresses may even occur in the film.
If the reactive gas component of the vacuum atmosphere is increased at constant total gas pressure, the number of dissociated and ionised gas particles, which can be accelerated on to the substrate, be incorporated in the coating and contribute to the formation of the lattice, rises. As the reactive gas component increases, the propagation of reactive gas atoms in the coating will therefore also rise until certain saturation points are reached. At higher reactive gas pressures, higher compressive stresses then occur.

The coating rate of PVD processes is primarily a function of the vaporization power or sputtering power. If the particle energy is not increased, then an increase in vaporization power, and thus in the coating rate, results in a reduction in compressive stresses in PVD coatings.

As the coating thickness increases, a decrease in the growth-induced compressive stresses is generally observed. Coating failure, like delamination, may occur during or after the process where high stresses are combined with increasing coating thickness, indicating that the stresses in the interface have exceeded the limiting failure value (Pulker 1982; Hirsch and Mayr 1988; Strauss et al 1997). It will apparent that a compressive stress maximum occurs in the vicinity of the interface zone, whereas stress remains virtually constant over the coating thickness in the remainder film.

The orientation of the specimens also can have an effect on the film stress. In planar sputter processes, it was noted that compressive stresses decreased if the surface of the specimen was inclined to the direction of incidence of the particles. An inclination of 45° resulted in almost complete elimination of the stresses (Hoffman 1982).

References

Atanassov G, Turlo J, Kai Fu J, Sheng Dai Y (1999) Mechanical, optical and structural properties of TiO_2 and MgF_2 thin films deposited by plasma ion assisted deposition. Thin Solid Films 342, pp.83-92

Ambree P, Kreller F, Wolf R, Wandel K (1993) Determination of the mechanical stress in plasma enhanced chemical vapour deposited SiO_2 and SiN layers. J.Vac.Sci.Technol. B 11 (3), pp.614

Barenblatt G.I (1962) Adv.Appl.Mech. 7, pp.55-131

Bendavid A, Martin P.J, Jamting A, Takikawa H, (1999) Structural and optical properties of titanium oxide thin films deposited by filtered arc deposition. Thin Solid Films 355-356, pp.6

Bland R.D, Kominiak G.J, Mattox D.M (1974) Effect of ion bombardment during deposition on thick metal and ceramic deposits. J.Vac.Sci.Technol. 11, pp.671

Carter G (1994) Peening in ion assisted thin films – a generalized model. J.Phys.D:Appl.Phys. 27, pp.1046

Chason E, Sheldon B.W, Freund L.B, Floro J.A, Hearne S.J (2001) Phys.Rev.Lett.
Cheng W.H, Chi S.F, Chu A.K (1999) Effect of thermal stresses on temperature dependence of refractive index for Ta_2O_5 dielectric films. Thin Solid Films 347, pp.233-237
Christova K.K, Manov A.H (1994) Mechanical stress and refractive index variation in dry SiO_2. Int.J.Electronics 76, pp.913-916
Cuomo J.J, Harper J.M, et.al, (1982) Modification of niobium film stress by low-energy ion bombardment during deposition. J.Vac.Sci.Technol. 20,pp. 349
Cuthrell R.E, Mattox D.M, et.al (1988) Residual stress anisotropy, stress control, and resistivity in post cathode magnetron sputter deposited molybdenum films. J.Vac.Sci.Technol. A6(5), pp.2914
Davis C.A (1993) A simple model for the formation of compressive stress in thin films by ion bombardment. Thin Solid Films 226, pp.30
D'Heurle F.M. (1970) Metall.Trans. I, pp.725
D'Heurle F.M, Harper J..M.E (1989) Note on the origin of intrinsic stresses in films deposited via evaporation and sputtering. Thin Solid Films 171, pp.81
Doerner M.F, Nix W.D (1988) CRC Crit.Rev. Solid State Mater.Sci. 14, pp.225
Edlinger J, Ramm J, Pulker H.K, 1989 Thin Solid Films 175, pp.207-212
Gäbler D, Laux S, Kaiser N, Bernitzki H, (2000) Low stress and shift free optical coatings for ultra-high precision surfaces. Optical Society of America
Griffith A.A (1920) Phil.Trans.Roy.Soc.Lond. 221, pp.163
Floro J.A, Chason E, Lee S.R (1996) Real time measurement of epilayer strain using a simplified wafer curvature technique. Mater.Res.Soc.Symp.Proc. 406, pp.491
Floro J.A, Hearne S.J, Hunter J.A, Kotula P, Chason E, Seel S.C, Thompson C.V (2001) J.Appl.Phys. 89, pp.4886
Floro J.A, Chason E, Cammarata R.C, Srolovitz D.J (2002) Physical Origins of Intrinsic Stresses in Volmer-Weber Thin Films. MSR Bulletin, pp.19
Heavens O.S, Smith S.D (1957) J.Opt.Soc.Am, 47, pp.469
Hirsch T, Mayr P (1988) Surf.Coat.Technol. 36, pp.729-741
Hoffman R.W, Anders F.J (1953) J.Appl.Phys, 23, pp.231
Hoffman R.W (1975) The mechanical properties of non-metallic thin films. AEC Techn. Rep. 82, Atomic Energy Commision, Case Western Reserve University, Cleveland OH
Hoffman D.W, Thornton J.A (1979) Effects of substrate orientation and rotation on internal stresses in sputtered metal films. J.Vac.Sci.Technol. 16(2), pp.134
Hoffman D.W, Gaertner M.R (1980) Modification of evaporated chromium by concurrent ion bombardment. J.Vac.Sci.Technol. 17, pp.425
Hoffman D.W (1982) Film stress diagnostics in sputter deposition of metals, Proc. 7th Int.Conf. on Vacuum Metallurgy, Tokyo
Huang T.C, Lim G, Pramigani F, Key E (1985) Effect of ion bombardment during deposition on the x-ray microstructure of thin silver films. J.Vac.Sci.Technol. A6, pp.2333
Klockholm E, (1987) Delamination and fracture of thin films. IBM J. Res.Develop. 31, pp.585
Koch R, (1994) J.Phys.: Condens..Matter 6, pp.9519
Kubovy A, Janda M (1977) Thin Solid Films 42, pp.169-173
Lechner W, Strauss G.N, Pulker H.K (1998) Int. Conf. Coatings on Glass ICCG-98, Saarbrücken, Germany
McNeil J.R, Barron A.C, Wilson S.R, Herrmann W.C (1984) Ion-assisted deposition of optical thin films: low energy vs. high energy bombardment. Appl.Opt. 23, pp.552-559
Mills E.J (1877) Proc.Roy.Soc. London 26, pp.504

Müller K.H (1986) Model for ion assisted thin film densification. J.Appl.Phys. 59, pp.2803
Müller K.H (1987) Ion-beam induced epitaxial vapour-phase growth: a molecular dynamics study. Phys.Rev.B 35, pp.7906
Müller K.H (1987-2) Stress and microstructure of sputter deposited thin films: molecular dynamic investigations. J.Appl.Phys. 62, pp.1796
Movchan B.A, Demchishin A.V Study of the structure and properties of thick vacuum condensates of nickel, titanium, tungsten, aluminium oxide and zirconium dioxide. (1969) Phys.Met.Metallorg., 28, pp.83
Nix W.D, (1989) Metall. Trans. A 20A, pp.2217
Qing-Shan S (1989) Proc IPAT-89, Geneva, pp.64-69
Ottermann C.R, Bange K (1996) Correlation between the density of TiO2 films and their properties. Thin Solid Films 286, pp.32
Pulker H.K, Mäser J.M (1979) The origin of mechanical stress in vacuum deposited MgF_2 and ZnS films. Thin Solid Films, 59, pp.65
Pulker H.K (1982) Mechanical properties of optical films. Thin Solid Films, 89, pp.191
Pulker H.K (1999) Coatings on Glass 2nd revd.edn., Elsevier, pp.374
Richter F, Kupfer H, Schlott P, Gessner T, Kaufmann C (2001) Optical properties and mechanical stress in SiO_2/Nb_2O_5 multilayers. Thin Solid Films 389, pp.278-283
Robic J.Y, Leplan H, Paulean Y, Rafin B (1996) Residual stress in silicon dioxide thin films produced by ion-assisted deposition. Thin Solid Films 290-291, pp.34-39
Roth T, Kloos K.H, Broszeit E (1987) Proc IPAT-87 Brighton, UK, pp.252-257
Rutten G.M.R (1989) Proc. IPAT-89, Geneva, pp.449-454
Seddon R.I, Temple M.D, Klinger R.E, Tuttle-Hart T, LeFevre P.M (1988) Proc.Conf.Opt.Interference Coatings, Tucson, USA, pp.255-258
Seidel J (1981) Stress in thin MgF_2 films. Thesis, University of Innsbruck, Austria
Stoney G.C (1909) Proc.Roy.Soc. London A32, pp.172
Story H.S, Hoffman R.W (1956) J.Appl.Phys, 27, pp.193
Strauss G.N, Danh N.Q, Pulker H.K (1997) Mechanical stress in thin SiO_2 and Ta_2O_5 films produced by reactive low voltage ion plating (RLVIP). J.Non.Cryst.Solids, 218, pp.256
Strauss G.N, Lechner W, Pulker H.K (1999) Gas pressure influence on the optical and mechanical properties of Ta_2O_5 films produced by reactive low voltage ion plating (RLVIP). Thin Solid Films 351, pp.53
Sun R.C, Tisone T.C, Cruzan P.D (1975) The origin of internal stress in low-voltage sputtered tungsten films. J. of Applied Physics, Vol.46, No.1
Szczyrbowski J, Bräuer G, Ruske M, Bartella J, Schroeder J, Zmelty A (1999) Some properties of TiO_2 layers prepared by medium frequency reactive sputtering. Surface and Coatings Technology 112, pp.261-266
Taylor C, Barlett D, Chason E., Floro J (1997) Indust.Phys., 4 (1), pp.25
Thornton J.A, Hoffman D.W (1985) The influence of discharge current on the intrinsic stress in Mo films deposited by cylindrical and planar magnetron sputtering sources. J.Vac.Sci.Technol. A3, pp.576
Volkert C.A (1991) Stress and plastic flow in silicon during amorphization by ion bombardment. J.Appl.Phys. 70, pp.3521
Wilcock J.D (1967) Stress in thin films, Thesis, Imperial College, University of London
Windischmann H (1987) An intrinsic stress scaling law for polycrystalline films prepared by ion beam sputtering. J.Appl.Phys. 62, pp.1800

Abbrevations and Symbols

σ	mechanical stress:	σ, σ_{ext}, σ_{therm}, σ_{int}
Y	Young's modulus:	Y_f, Y_s
α	thermal coefficient :	α_f, α_s
T	temperature	
τ	surface tension	
t	thickness :	t_f, t_s
γ	surface energy:	γ_s, γ_{gb}, γ_d, γ_f, γ_i
ε	strain	
a	lattice constant	
Δ	constrained relaxation	
d	diameter	
ν	Poisson ratio	
p	packing density	
E	energy	E, E_{elast}, E_{crit}
j	flux	j_{bomb}, j_{depos}
ρ	density	ρ_f
n	particle density	
n	refractive index	
C	constants	C_ρ
R	radius	
κ	curvature	

Optical Thin Films for Micro-Components

Ludovic Escoubas and Francois Flory

Institut Fresnel (UMR CNRS 6133) – Ecole Nationale Supérieure de Physique – Domaine Universitaire de St Jérôme, Marseille, France

1 Introduction

Being able to control refractive index modulation in the three dimensions of space in order to design or fabricate new optical components is of great scientific interest. When refractive index modulation is periodic, the components generally fall into the photonic - crystal category (see chapter "Photonic Structures as Interference Devices" by C-G. Ribbing) (Krauss and De La Rue 1999). They diffract visible and infrared light as natural crystals diffract X rays.

The refractive index of an optical structure constituting a component may be controlled in the direction perpendicular to its surface by using interference optical filters made from optical thin - film stacks.

The first part of this chapter is devoted to thin materials for optical micro-components. Step index or graded index thin film fabrication techniques are presented. There exist numerous methods for controlling the optical properties in the two other dimensions of space. One paragraph discusses 2D or 3D structuration of micro-components. Processes inherited from the microelectronics industry, more particularly etching or ion implantation for local refractive index modification of materials, are particularly interesting. In the second part of this chapter ion implantation is described as an alternative technique for fabricating optical micro-components.

The properties of these components depend strongly on the scale on which the refractive index is modulated:

- If the scale is larger than the wavelength: optical sensors for chemical or biological detection (Culshaw and Dakin 1989), optical waveguides and passive or active components for telecommunications
- If the scale is of the order of the wavelength: gratings for filtering or information routing in optical télecommunications, diffractive optical elements

- If the scale is smaller than the wavelength: refractive index control and artificial anisotropy for components impacting on the polarization of light (Bloemer and Haus 1996)

Thus, by mastering the refractive index distribution in the three directions of space the behavior of optical waves, particularly the wave vector directions, can be controlled. Then one can obtain wavelength or propagation direction filtering functions.

The optical properties of structured materials in thin film form are different from those of bulk materials. These properties become crucial parameters for high-performance micro-components. Thus very accurate techniques for measuring optical properties are needed. The third part of this chapter describes techniques, mostly based on guided waves, generally suitable for optical micro-components.

An example of the usefulness of the association of planar index modulation with thin films is discussed in the last part of this chapter. Gratings embedded in thin-film stacks to improve their diffraction efficiency and grating-waveguide structures for narrow-band filtering are described.

2 Thin Materials for Optical Micro-Components

Various types of material in thin film form can be used to make planar optical micro-components: thin film oxides (example: flame hydrolysis deposited silica), ion-exchanged materials (example: Ag+ - Na+ exchange for waveguide fabrication), cristalline and active materials (example: LiNbO3 or lead lanthanum zirconate titanate PLZT), etc. Moreover, different technologies for the fabrication of photonic integrated circuits (PIC's) are available. This section will give a brief overview of the thin film fabrication processes for optical micro-components, more particularly for optical waveguides. Depending on the refractive index profile required, various fabrication methods may be employed. A general classification of the methods is obtained by classifying devices in two categories: step-index and graded-index components. Two-dimensional and three-dimensional micro-components are then obtained by structuring materials with methods such as masking and etching.

STEP-INDEX

Step-index waveguides (Koch et al. 1995) generally consist of a dielectric amorphous or crystalline thin film with a high refractive index deposited on a low refractive index substrate. This may allow light to propagate in the form of a guided mode in the high refractive index layer. Furthermore, the layer may be structured by etching.

Hence, for a few years a lot of research has been done on thin film dielectric oxides grown on silicon wafers. The availability of a mature silicon processing technology, designed for microelectronics, has led to the development of a silica based technology with doped silica thin films deposited on oxidized wafer surfaces. Waveguides and optical micro-components obtained in this way exhibit extremely low losses. One deposition technique, flame hydrolysis (FHD), was inherited from optical fibre manufacturing. Silica powder is deposited on a silicon wafer to serve as the low refractive index substrate. Then a thinner layer of silica doped with germanium oxide is deposited to form the high refractive index core of the waveguide. This material is hardened at around 1,300° C for several hours. It is then patterned using an etching process to define the shape of the micro-components. Then FHD is used again, followed by hardening, to form the cladding layer (Takahashi et al. 1991).

Another technique for fabricating silica based optical micro-components on silicon wafers has been inherited from the microelectronics industry. The silicon wafer is oxidized to a thickness of 5 to 15 micrometers using high-pressure steam. This prevents light from leaking to the high index silicon substrate. Then the high refractive index core is fabricated using low-pressure chemical vapor deposition (LPCVD) either from silane and oxygen or from tetraethylorthosilane and ammonia. To increase its refractive index, the core is doped with a few percent of phosphine. Then the wafer is annealed at approximately 1,000 °C to densify the deposited films and to release stress. Losses lower than 0.05 dB/cm have been reported using this process (Henry et al. 1989).

Waveguides and optical-micro-components may also be fabricated using materials such as Si_3N_4. The step index contrast between the core and the cladding is then much higher than with other materials, which results in more confined guided modes and consequently in smaller passive optical components. A curvature radius of the order of a few hundred micrometers may be achieved in such integrated optics components. Moreover, Si_3N_4 waveguides present confined guided modes that are matched to those commonly found in semiconductor sources such as laser diodes. This results in much lower coupling losses.

Sputtering is also very useful to fabricate step-index waveguides and optical micro-components (see chapter "Thin Film Deposition Techniques" by H K Pulker). It is obtained by firing atoms or ions (in the tens of eV to a few keV range) at a solid source target in vacuum. The bombardment induces the ejection of atoms or molecules from the target. A thin-film layer builds up slowly as the ejected atoms or molecules hit the substrate. The sputtering process is more mechanical than chemical, which allows growth of various materials on different substrates. Highly pure and uniform low-loss amorphous or polycrystalline layers are obtained using this technique (Table 1) (Lee 1986).

Table 1. Amorphous materials deposited in thin film form using sputtering: measured values of refractive indices and attenuation

Thin film material	Refractive index	Attenuation [dB/cm]
Ta_2O_5	2.20 @ 0.6328 μm	0.9
GeO_2	1.61 @ 0.55 μm	0.7
SiO_2	1.47 @ 0.6328 μm	0.7
Nb_2O_5	2.10 – 2.30 @ 0.6328 μm	1.0 – 2.0

The electron beam gun is also a widely used evaporation source. For many metal oxides, the fabrication of stoichiometric oxides requires constant O_2 partial pressure with a constant low condensation rate of the starting material. This technique is referred to as reactive electron beam evaporation (Jeschkowski and Niederwald 1997; Pulker and Guenther 1995) (see chapter "Thin Film Deposition Techniques" by H K Pulker). Feeding energy into the growing film allows elimination of low packing density, poor substrate adhesion or non-stochiometric composition. Techniques such as ion-assisted deposition (Bovard 1995) and ion plating (Pulker and Guenther 1995) using ion bombardment of the growing layer or plasma assistance respectively have emerged from research work and are used for the commercial production of thin films today.

Epitaxially grown optoelectronic III-V materials are particularly useful for the integration of optical and electronic functions into the same planar micro-component. Coupling passive optical waveguides with laser sources, photodetectors and optical modulators can be accomplished by fabricating waveguides and micro-components from semiconductors. Today a variety of highly versatile vapor and beam deposition techniques have been made available as a result of the development of the original liquid phase epitaxy: vapor phase epitaxy, molecular beam epitaxy, chemical beam epitaxy, metal organic beam epitaxy, metal organic vapor phase epitaxy (Lee 1986). Three criteria to evaluate the quality of the growth need considering: thickness uniformity, composition uniformity and doping possibilities. Indeed, layer thickness variations may lead to high wavelength shifts in the case of distributed feedback lasers (DFB) or filters. Differences in composition may induce shifts of the wavelength emitted by laser diodes. Highly concentrated doping and abrupt changes in the doping level are needed for high-speed transistors (Koch et al. 1995).

Pulse Laser Deposition (PLD) is also very useful for the deposition of crystalline films. This technique is based on fast evaporation by means of high power laser shots followed by condensation of the ablated material on the substrate. PLD is relatively fast, simple and highly versatile for depositing thin films on a wide variety of substrates. PLD permits the deposition of doped materials, single crystals and optical waveguides to serve as laser sources (Ti:sapphire, Nd:YAP, Nd YAG) (Jelinek et al. 1998). It leads to the film having the same stoichiometry as the target.

Optical polymers are clear plastics with very high transmission capability (see chapter "Coating of Plastic Optics" by L Martinu). They present many advantages over optical glass: they are light, easy to mold into various shapes and to deposit, and cheap. Besides, they are extremely versatile as molecular engineering helps modify their structure and their properties. Applications range from optical interconnects and integrated optics circuits to active layers for gas or biological sensors. The key transparent polymers for photonic applications include polymethyl methacrylate (PMMA), polycarbonate, polystyrene, allyl diglycol carbonate, cyclic olefin copolymer and polydimethylsiloxane (PDMS). Several techniques have been developed to deposit polymer films on various types of substrate. A low cost technique consists in covering the substrate with the polymer in liquid form. The liquid layer's thickness is monitored either by spinning the substrate at the rate of a few thousand revolutions per minute or dipping the substrate vertically into the liquid and removing it at a controlled rate. Depending on the polymer used, the film may need to be air dried and baked at temperatures from 60°C to 100°C for periods ranging from a few minutes to tens of hours. Using this deposition process, it is possible to attain weak optical attenuation, in the range of 0.1 to 0.3 dB/cm. However the film's thickness and its uniformity are difficult to control accurately. The light transmittance, the refractive index, the attenuation and the Abbe number give a complete description of the polymer's optical properties. Other good physical, thermal and mechanical properties, such as low water absorption, low coefficient of thermal expansion and low stress for minimizing birefringence, are also needed for photonic applications. Some properties of the key optical polymers are summarized in Table 2 (Bosc et al. 1999).

Table 2. Some optical and physical properties of the key optical polymers

	PMMA	Polystyrene	Polycarbonate	Epoxy	Photoresist
Refractive index @ 0.633 μm	1.491	1.589	1.586	1.581	1.615
Abbe number	57.4	31	30	-	-
Attenuation [dB/cm]	0.2	0.2 - 0.3	0.2 - 0.3	0.3	7
Coeff. of linear thermal expansion [10^{-5}/K]	6 - 7	6 - 7	6 - 7	-	-

GRADED-INDEX

Graded-index waveguides are generally obtained from modification of the refractive index of the material near its surface. This can be accomplished by introducing foreign species or ions into the surface. Various techniques exist: ion migration from the surface into the material, thermal in-diffusion, proton-exchange. These techniques produce a refractive-index profile which varies continuously from the surface to the non-modified substrate.

Ion exchange is the most widely used technique both in laboratories and in production (Righini and Laybourn 1995). This process has been used extensively for the fabrication of graded-index planar and channel components. Passive integrated optical devices can be fabricated in certain glass substrates using this technique. Indeed, much of the optical glass used in integrated optics applications is composite, consisting of a mixture of silica and metal oxide such as Na_2O, K_2O, Al_2O_3 and Li_2O. By diffusing ions such as Ag^+, K^+ or Tl^+ into the glass, in place of lighter ions such as Na^+ or Li^+, one can increase the refractive index in the diffused region. In this fabrication process, the glass substrate is placed in a bath of melted salts containing alkali ions such as Ag^+, K^+ or Tl^+. An aluminium contact is deposited on the top surface of the optical glass and connected to the ground potential. The melted salt is kept at a positive potential by means of an immersed electrode. Melting temperatures range between 200 and 350°C with an electric field in the 10^5 V/m range. Using this method, the increase in refractive index near the surface is about 0.1 with Ag^+ ion diffusion. Channel waveguides are made by selectively masking the glass surface. Propagation losses are of the order of a few tenths of a decibel per centimetre. Generally, multimode devices are fabricated using Tl^+ ions while single-mode waveguides are typically realized by diffusing Ag^+ or K^+. The main physical properties of ions used for exchange processes are presented in Table 3 (Koch et al. 1995; Lee 1986).

Table 3. Main physical properties of ions used for exchange processes in glass

	Ag^+	K^+	Li^+	Na^+	Tl^+
Salt	$AgNO_3$	KNO_3	$LiNO_3$	$NaNO_3$	$TlNO_3$
Melting point [°C]	212	334	264	307	206
Decomposition point [°C]	444	400	600	380	430

Lithium niobate ($LiNbO_3$) and lithium tantalate ($LiTaO_3$) exhibit acousto-optic, electro-optic, piezo-electric and 2nd order non linear properties which prove particularly useful for making optoelectronic micro-components. $LiNbO_3$ is an excellent electro-optic material (electro-optic coefficient r_{33}=30.8x10^{-10} cm/V) with high optical transmission in the visible and near infrared. Phase modulators, Mach-Zehnder interferometers or optical switches can be made from $LiNbO_3$. Thus, techniques for fabricating planar waveguides using this crystal are quite useful. Proton exchange is similar to the ion migration presented previously: the metal ions of the $LiNbO_3$ crystalline substrate are partially replaced with hydrogen ions. The substrates are placed in a solution of benzoic acid heated to a temperature between 200 and 250°C. A guiding region is formed because the protons from the acid are exchanged with lithium ions. This results in large refractive index changes (about 0.1). The fabrication of single-mode waveguides is performed in a few minutes. Propagation losses can be as low as 0.5 dB/cm (Lee 1986).

Many $LiNbO_3$ planar micro-components are also obtained using a titanium in-diffusion process. 50 to 100 nm thick titanium layers, deposited by sputtering or evaporation, are diffused into the $LiNbO_3$ at temperatures around 1,000°C with diffusion times of about 10 h. This leads to a local increase in both the ordinary and the extraordinary refractive indices of the crystal. Such diffusion produces a Gaussian distribution of titanium atoms, the refractive index being proportional to their concentration. For lithium tantalate, niobium in an argon atmosphere is generally used for in-diffusion processes (Koch et al. 1995; Lee 1986).

Ion implantation can also be used for optical micro-components both in semiconductor or dielectric materials. This technique will be described at some length in the third part of this chapter.

2D OR 3D STRUCTURED MICROCOMPONENTS

Planar optical micro-components are generally made by lithography, masking and etching processes or laser writing. These basic methods have been inherited from microelectronic VLSI technology. The techniques for structuring materials are presented schematically in Fig. 1. Lithographic tools form the mask pattern by controlling the intensity of the beam of photons or electrons impinging onto a thin layer of resist (a few nanometers in thickness). The resist is deposited on the substrate by either spinning or dipping. It can be exposed either to UV light through an amplitude mask comprising opaque and transparent regions or to a scanning electron beam which writes the desired pattern. For 3D structures, grey scale masking may be used (Stern 1997).

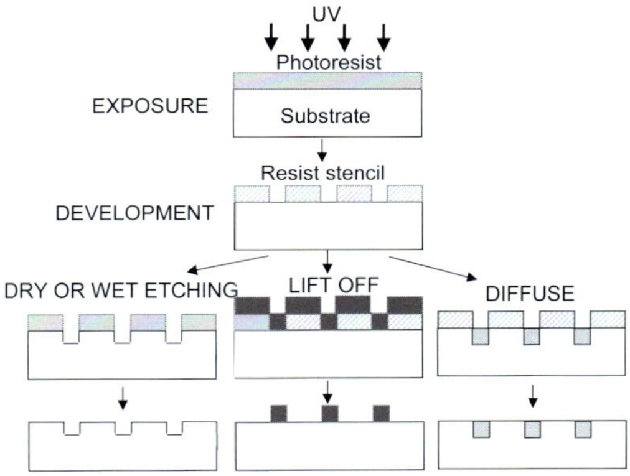

Fig. 1. Planar fabrication techniques for structuring materials

Choosing the relevant lithography method depends on the critical (smallest) dimensions of the pattern. Optical lithography has the highest throughput and can currently produce critical dimensions as small as 0.25 μm. Electron beam lithography offers the highest resolution (<10 nm) but has limited throughput and a small field size and it is costly. During irradiation the chemical composition of the resist film is altered. The excess resist material is then dissolved in a developer in order to obtain a stencil. Due to the higher spatial resolution of the electron beam, resist stencils obtained using this technique have smoother, sharper patterns than those made by optical exposure. After etching the patterns are also smoother, which is of great importance precisely for reducing losses due to scattering. Positive resists duplicate the mask's polarity whereas negative resists reverse it. Resists developed for the microelectronics industry are optimised for higher contrast. However, one has to avoid light reflection from the substrate, which may form stationary waves in photoresists. After the photolithographic step, the pattern is transferred into the substrate material by either wet or dry etching. Wet etching refers to the use of an acid bath to attack the unprotected regions of the surface. For precise geometry, another masking material such as SiO_2 is first deposited, then patterned by HF etching through a photoresist mask. Then SiO_2 thus patterned serves as a mask to etch the substrate. Classical etchants for III – V materials, with varying degrees of selectivity and crystallographic behavior, include a blend of HBr and CH_3COOH or HCl, CH_3COOH and H_2O_2. In dry etching, the material is removed from the substrate using a combination of physical sputtering and chemical reactions occurring in a plasma obtained from a glow discharge in a gas mixture. There are quite a few etching systems: plasma etchers, reactive ion etchers (RIE), magnetically-enhanced reactive ion etchers, ion beam etchers, etc. Reactive ion etching is very useful as it saves one the trouble of mask undercutting. Thus, very high lateral precision is obtained with the possibility of duplicating sub-micron features efficiently. Parameters such as gas chemistry, chamber pressure and ion energy determine the anisotropy profile, the etching rate, selectivity and uniformity. The most commonly used etching gases and materials are listed in Table 4 (Stern 1997; Koch et al. 1995; Lee 1986).

Table 4. Gases used for etching optical materials

Material	Gases
Si	HBr, Cl_2, CF_4, CF_4 - O_2, SF_6 - O_2, $SiCl_4$
Ge	CF_4, CF_4 - O_2, SF_6 - O_2
TiO_2, Ta_2O_5	CF_4 - O_2, SF_6 - O_2
SiO_2	CHF_3, CF_4 - H_2, C_2F_6
Organic Materials	O_2, O_2 - CF_4, O_2 - SF_6
GaAs	HBr, Cl_2, CF_3Cl, $SiCl_4$
InP	HBr, $SiCl_4$

3 Ion Implantation for Micro-Components

Ion implantation is a well-mastered process in microelectronics (Favennec 1993; Townsend et al. 1994). It permits localized doping of materials using a beam of energetic ions. These ions lose their kinetic energy when colliding with the material's atoms and electrons then stop inside the material. The energy used ranges from 10 keV up to one MeV. The ions are accelerated between high-voltage plates; the various species of ions are separated spatially according to their mass by passing them through a magnetic field. The position of the ion beam can be scanned across the target (Lee 1986).

Ion implantation is also an alternative technology for fabrication of optical micro-components (EerNisse 1974). Together with masking techniques, it can be used to make diffractive devices or integrated optics components. Indeed, waveguides can be obtained by ion implantation both in semiconductor and dielectric substrates by creating narrow low or high refractive index regions. Implantation of light ions (He^+, H^+, Li^+) generally causes defects on an atomic scale in the material (in the form of displaced atoms or atom ionisations leading to broken bonds) which tend to lower the refractive index (10^{-2} at most) (EerNisse 1974; Townsend et al. 1994). Thus the waveguide is bounded by a low refractive index implanted region. On the other hand, the implantation of heavy ions tends to increase the refractive index, so that the implanted region forms the waveguide. It is usually necessary to anneal the component in order to cure local defects due to the implantation process and to reduce absorption and scattering losses by oxidizing the implanted species. Doping thin film or substrate materials with the implantation of light or heavy ions has proved very useful for fabrication of telecom micro-components. Indeed, ion implantation has several advantages for optical micro-components (Arnold 1982): it allows one to modify the refractive index of bulk or thin film materials without patterning the surface; refractive index changes are determine through the dose of implanted ions; and the in-depth index profile depends on the energy of the ions.

High dose heavy ion implantation has been studied for bulk amorphous silica and $Si0_2$ layers made by Ion Assisted Deposition (IAD) (Bovard 1995). The microstructure (Fig. 2), the ion distribution in the depth of the materials and the optical properties of the implanted regions have also been investigated.

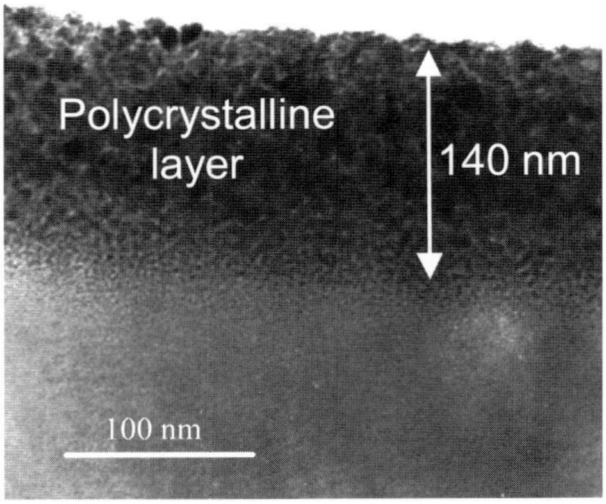

Fig. 2. Microstructure of an implanted region. Transmission Electron Microscopy – TEM - picture of the cross sections of implanted bulk silica. A polycrystalline layer is formed in the implanted region. The thickness of this layer and the average grain size increase together with the dose of ions

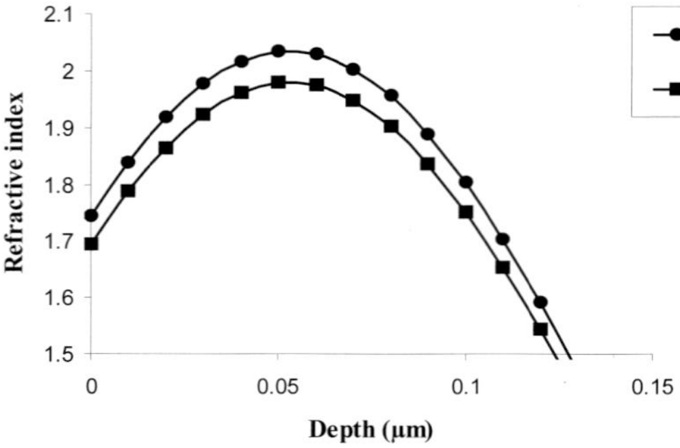

Fig. 3. Refractive index profiles of implanted fused silica at the wavelengths 0.633 and 1.55 µm

The implanted bulk silica presents a grain structure with micro-crystals whereas SiO_2 layers remain amorphous after implantation. The refractive index profile of implanted materials has been determined from guided wave measurements (Fig. 3). It has been found that it matches the distribution of ions (Fig. 4). A study of thermal annealing in air shows that the implanted materials exhibit low optical losses.

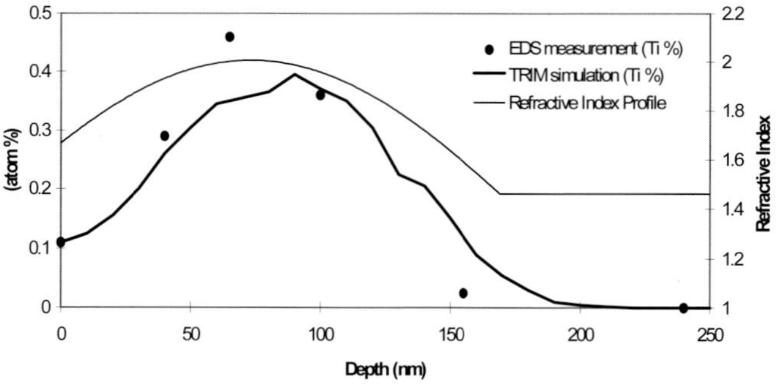

Fig. 4. Heavy ion distribution (• EDS) (▬▬ Transfer of Ions into Matter – TRIM – computation) and refractive index profile (——) at λ=632.8 nm of an implanted fused silica

4 Characterization of Thin Films for Optical Micro-Components

Accurate characterization is required in order to design optical micro-components and planar integrated optics components. Indeed, when assessing the quality of optical micro-components, it is indispensable to know propagation constants, refractive index values, the thickness of thin films, propagation losses, anisotropy, thermo-optic coefficients (dn/dT) and the thermal properties of thin films. Furthermore, feeding the characterization results back to the designers and people in fabrication units helps to improve quality and set up production. Guided wave techniques are particularly useful for accurate characterization of both the optical and thermal properties of thin films needed for fabricating optical micro-components (Flory et al. 2001; Escoubas et al. 1998). In fact, these techniques benefit from the use of guided modes which correspond to the propagation of light in resonance conditions in the layer material. Moreover, light propagates over a distance that is far greater than the thickness of the layer. As we shall see in the following paragraphs, resonance conditions and the long distance of propagation enable strong interaction between light and the material to be characterized, which results in high sensitivity to the absolute optical properties of the film and to their variations.

Different theories can help model the propagation of light in single-layer or multi-layer planar waveguides made of isotropic or anisotropic materials (Grann et al. 1994). However a very simple ray path theory gives the mode equation. Indeed, let us consider a planar waveguide, that is a thin film deposited on a substrate with a lower refractive index. Any light ray propagating in the film in such a way that the angle θ formed by this ray with the normal to the film's surface is greater than the critical angle of the interface with the smaller index difference experiences total reflection at both interfaces. Those rays which additionally fulfill the phase condition

$$2\,k\,n_f\,d\cos\theta - 2\phi_a - 2\phi_s = 2m\pi \qquad (1)$$

($k = 2\pi/\lambda$, n_f is the thin film's refractive index, d is the thin film's thickness, m is an integer which designates the mode number and ϕ_a and ϕ_s the phase changes at the film-air and film-substrate interfaces) are discrete stationary solutions which represent the guided modes supported by the thin film. Excitation of guided waves may be obtained using different methods: transverse coupling of a light beam through an exposed cross-section of the waveguide or surface coupling using devices such as prisms or gratings. The incident light on a prism base can be coupled into the waveguide by the evanescent wave underneath the prism base. The component of the wave vector which is parallel to the prism base must match the wave vector of the guided mode. The light incident on a grating can be coupled into the waveguide when the diffracted light wave vector matches that of a guided mode. In both cases, the prism or the grating provides the phase matching between the incident light beam and one of the guided modes.

Refractive Index and Thickness

The totally reflecting prism coupler technique, also referred to as the m-line technique, is commonly used to determine the optical properties of thin films (Hunsperger 1982; Pelli and Righini 1994; Ulrich and Torge 1973). It allows high coupling efficiency (up to 80% of the incoming light from a Gaussian beam). It can be used both for planar or ridge waveguides. The guided modes can be excited selectively by changing the incident angle of light. The light enters the waveguide, propagates over a short distance and is then coupled back into the prism.

The experimental set-up (Fig. 5) comprises a sample holder allowing fine tuning of the pressure of the waveguide against the prism's base (the coupling efficiency is very sensitive to the thickness of the air gap between the prism's base and the waveguide's surface). The set-up also contains a precision rotating stage to set the incident angle of light.

The m-lines (brighter spot superimposed on a dark line) appear for the different incident directions corresponding to the coupling of light into the waveguide. The incident angles called 'synchronism angles' are characteristic of the waveguide and, for a single homogeneous film, the refractive index and the thickness are easily determined from at least two of these angles (Flory 1995). This technique can also be used for multilayer waveguides. The absolute accuracy obtained using this set-up is 10^{-3} for the refractive index and 1 nm for the thickness of optical thin-film waveguides (Monneret et al. 2000).

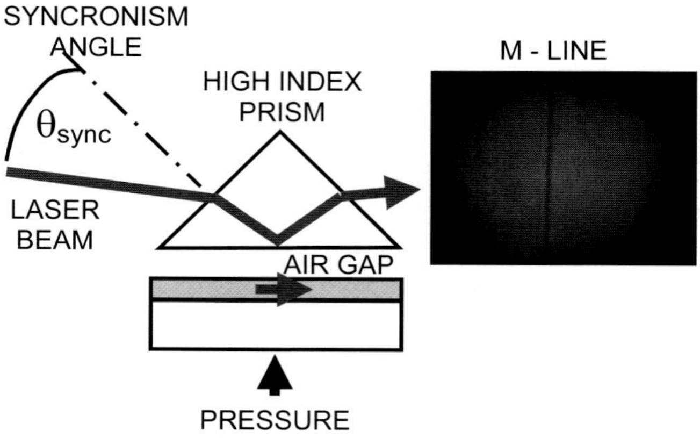

Fig. 5. Totally reflecting prism coupler technique (m-lines technique)

Propagation Losses

If we consider a modified m-line set-up with light incident on the prism's right angle (Weber et al. 1973), light can be coupled into the waveguide and not coupled back into the prism. The losses in the coating can be measured by recording the attenuation of the light coupled out of the waveguide as a result of scattering along the guided path (Mollenhauer 1994). The image of the guided light is recorded with a conventional camera and the decrease in the maxima of light intensity recorded along the path is approximated by an exponential law. Losses as low as 1 dB/cm can be measured with this type of set-up. When modelled as simple absorption, such losses correspond to an extinction coefficient of the order of 10^{-6} for a single layer.

In the case of channel waveguides, insertion losses (the ratio of the transmitted power to the incident power) can be measured by coupling the light emitted by a laser or a laser-diode through one end of the waveguide. The coupling is performed by using an optical fibre or a focusing lens. The transmitted light is then recorded at the other end of the waveguide by using a second optical fibre or an integrating sphere (Flory 1995). Positioning the input and output fibres accurately in front of the component is necessary.

Anisotropy

The propagation of light in anisotropic dielectric thin films has been described by several authors (Berreman 1972; Bloemer and Haus 1996; Teitler and Henvis 1970; Vassel 1974; Yeh 1979). A 4 x 4 transfer matrix formalism can be used to compute the reflection, the transmission and also the guided-mode conditions of anisotropic layers or multilayers (Flory et al. 1993; Hodkinson and Endelema 1990). It is well-known that the layers obtained by conventional evaporation-condensation exhibit a columnar microstructure (Fig. 6) (see chapter "Fundamentals of Thin Film Growth" by N Kaiser). The columns' tilt depends on the incident direction of the material during deposition (Guenther and Pulker 1976). Hence, being able to measure the three main refractive index values of anisotropic thin films is necessary.

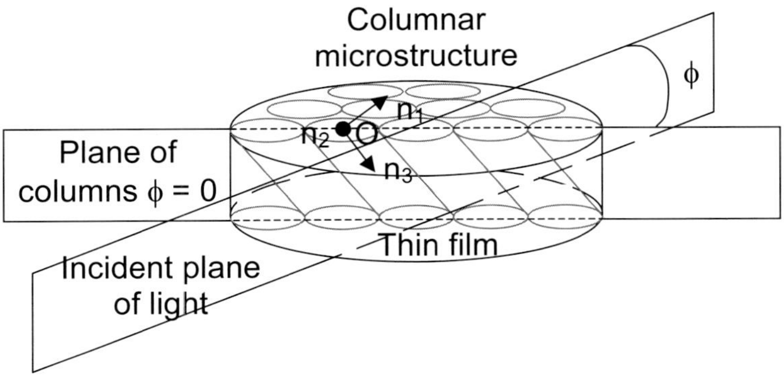

Fig. 6. Columnar microstructure of an anisotropic dielectric thin films

This measurement can be made by using the m-line technique. Indeed, call ϕ the angle between the main plane of the layer holding the axes of a row of columns on the one hand and the incident plane on the other. One can measure two-guided mode propagation constants in s polarization with the guided propagation in the plane of the columns containing the axes (On_1, On_3) ($\phi = 0$). This allows calculating the values of the refractive index n_2 (refractive index along the axis On_2 perpendicular to the plane of the columns) and the thickness of the layer. Knowing the thickness, one can measure the propagation constants of two guided modes in p polarization with the guided propagation in the plane of the columns and infer the refractive index values n_1 and n_3, respectively, along the axes On_1 and On_3. The determination of n_2 and of the thickness can be performed with an accuracy of about $2x10^{-3}$ and about 3 nm, respectively; n_1 can be obtained with an accuracy of $4x10^{-3}$ and n_3 with an accuracy of $2x10^{-3}$ (Flory et al. 1993).

Thermo-Optic Coefficient

Knowing the variations of the refractive indices of different materials in single layers or stacks relative to the temperature is crucial for designing robust narrow band pass filters needed for Dense Wavelength Division Multiplexing (DWDM). Measurements of dn/dT can be performed on various types of sample and at different wavelengths (632.8 nm and 1.55 µm) by using the m-line technique. The synchronism angles are first measured at room temperature. The system is then heated with a hot air flow until the sample's temperature stabilizes. The temperature is measured with a thermocouple placed in contact with the sample's surface. The variations of the synchronism angles of the guide at different temperatures in the range of 20-80°C are recorded and the variations of the refractive index vs temperature deduced from these measurements. The dn/dT values are obtained from linear regression on the measured refractive indices at different temperatures (Flory et al. 1993). Fig. 7 shows an example of a measurement. The changes in the refractive indices of the substrate and the prism vs the temperature are taken into account although the m-lines are mainly sensitive to the alteration of the layer's properties. A comparison between measurements of dn/dT performed on different samples at 0.6328 µm and 1.55 µm is presented in Table 5.

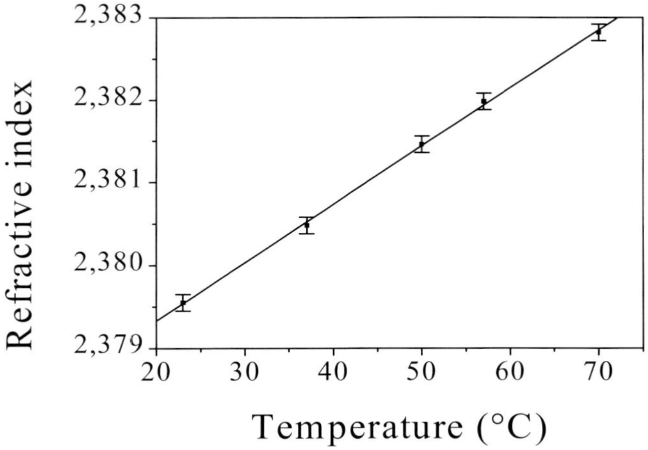

Fig. 7. Refractive index vs temperature (λ=632.8 nm) of a TiO_2 coating deposited by IAD on a BK7 substrate. dn/dT = 7.10^{-5} K^{-1}

Table 5. dn/dT values for various materials on different substrates

Material	Deposition technique	Thickness [µm]	Substrate material	λ [µm]	dn/dT [x10^{-5} K^{-1}]
Ta_2O_5	IAD	0.519	B270	0.6328	9.5
SiN_xH_y	CVD	1.470	B270	1.55	4.4
SiO_2	CVD	2.860	Silica	0.6328	7.1
TiO_2	IAD	0.533	Fused silica	0.6328	9.7
TiO_2	IAD	0.533	Zerodur	0.6328	9.2
TiO_2	IAD	0.533	BK7	0.6328	7.0

The measured values of dn/dT can be explained mainly by two phenomena: 1. The intrinsic refractive index varies because of the properties of the thin film material; 2. The thermal stress in the thin film varies as a result of the difference between the coefficients of linear expansion $1/L(\partial L/\partial T)$ (L is the length) of the substrate and the film (Drouard et al. 2002; Takashashi 1995). The measured values of dn/dT allow calculating the wavelength shifts of narrow band pass thin film Fabry-Perot filters relative to temperature. The calculated shifts are in good agreement with the measured ones (Drouard et al. 2002).

Thermal Properties

The thermal properties of bulk materials are well-known. It is generally admitted that the properties of materials in thin film form can be very different from those of bulk materials. The thermal parameters (thermal conductivity, specific heat, etc.) of thin films have yet to be measured (see chapter "Characterization and Monitoring" by D Ristau). The values that can be found in the literature for a given material can differ by several orders of magnitude, especially because these properties are very dependent on deposition technique and processes (Amsden et al. 1988; Lambropoulos et al. 1989; Ristau and Ebert 1986). Additionally, knowledge of thermal properties comes in particularly useful when dealing with the stability and the ageing of filters or with laser damage (Guenther and Mc Iver 1988). The use of a totally reflecting prism coupler (Monneret et al. 2000; Tien and Ulrich 1970; Ulrich and Torge 1973) in a set-up comprising a pump beam and a probe beam permits measurement of the thermal properties of materials in thin film form (Monneret et al. 1996). The totally reflecting prism coupler (m-line technique) is used to detect the refractive index alterations of a thin film under illumination. The main advantages of guided optics lie in the great light/material interaction length and in the use of the double resonance of the electromagnetic field in the film for both the pump and the probe beams (Fig. 8). A very high sensitivity to the film's properties is then achieved. However weak it is, the absorption of the modulated pump beam in the guide creates a periodic increase of the temperature of the guide at the modulation frequency.

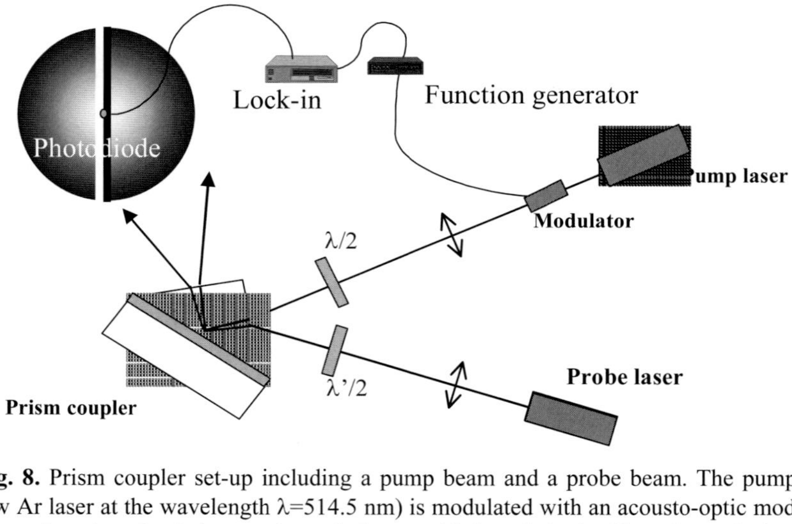

Fig. 8. Prism coupler set-up including a pump beam and a probe beam. The pump beam (cw Ar laser at the wavelength λ=514.5 nm) is modulated with an acousto-optic modulator, focused on the prism's base and coupled as a guided mode in the film. The probe beam (cw He-Ne laser at the wavelength λ'=632.8 nm) is also focused on the prism's base and coupled as a guided mode in the same place in the thin film

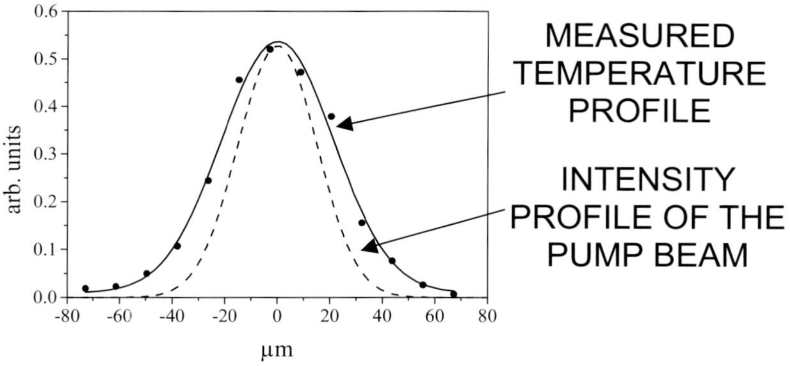

Fig. 9. Example of a measured refractive index modification profile (film of TiO_2 deposited by Ion Assisted Deposition (IAD) on a fused silica substrate). It can be noted that the Gaussian distribution of the measured profile is larger than the intensity profile of the pump beam, which signals a thermal diffusion in the thin film. The thermal diffusion length depends on the pump beam modulation frequency. Thus, for high frequencies (> 100 kHz) the measured profile is equal to the intensity profile of the pump beam

This leads to a reversible change in the refractive index of the film. This change in refractive index is assumed to be uniform in the probed region. As a consequence, a periodic shift of the m-line can be observed in the reflected spot of the probe beam. This shift is detected by a photodiode placed in the linear zone of the m-line's transverse profile. The periodic shift of the m-line is then measured with a lock-in amplifier as a periodic electrical signal. Index changes as low as 10^{-8} can be detected with this set-up. It is possible to measure the evolution of the signal as a function of the modulation frequency and it is also possible to move one beam relative to the other in order to record the distribution of the change in refractive index (that is, the transverse temperature distribution). An example of a measurement of the transverse profile of the temperature increase is shown in Fig. 9 (Huguet Chantôme et al. 2002). The temperature increase and its frequency behavior are linked to the thermal parameters (thermal conductivity, heat capacity) of the thin film.

5 Association of Gratings with Thin Films

The development of new methods to make micro-optical devices, more particularly diffractive elements to direct light or separate wavelengths, has become a very active and highly productive research topic in recent years (Astilean et al. 1998; Veldkamp and McHugh 1992). In the first part of this section we describe how thin film multilayers can be specially designed to enhance the diffraction efficiency of gratings. In the second part we set out the design and the fabrication of narrow bandpass filters obtained from resonant gratings and thin film waveguides.

Increasing the Diffraction Efficiency of Gratings

High-performance components can be obtained using techniques such as photolithography, electron beam writing, reactive ion etching, diamond turning of optical surfaces and replication (Herzig 1997). As has already been seen, ion-implantation (Favennec 1993; Flory et al. 1996) is suitable for making refractive index gratings (Flory et al. 1999). Unlike other techniques, we obtain components with quasi-plane surfaces, which allows depositing coatings on top of these while preserving the surface's planarity. Thus, ion implantation might be useful for making a wide array of diffractive optical elements (e.g. blazed gratings, Fresnel lenses, etc.). However, because of the low penetration depth of implanted heavy ions, the diffraction efficiency of index-modulated binary gratings obtained by ion implantation is low (only ~1% in s-polarization for the first transmitted order of a grating obtained in silica) and must be increased. It has been shown theoretically (Escoubas et al. 2000) that one could enhance the free-space diffraction efficiency of gratings made by ion implantation by inserting them in multilayer dielectric Fabry-Perot cavities. The grating is positioned in the spacer layer of the Fabry-Perot (Fig. 10) to use the electromagnetic field resonance. Different designs of the Fabry-Perot multilayer can be considered (different centering wavelengths of the mirrors and thicknesses of the spacer). The maximum +1 s-polarization transmitted order efficiency is found for a thickness d of the spacer for which the Fabry-Perot cavity is resonant in both the direction of diffraction and at normal incidence (which is the direction of the incident light). Indeed, it is necessary to have the strongest electromagnetic field possible on the grating, inside the Fabry-Perot cavity. This condition has to be met for both the direction of diffraction and normal incidence.

The enhancement of the free-space diffraction efficiency has been experimentally demonstrated. Three different phase gratings embedded in the spacer of seven-layer mirror Fabry-Perot structures have been fabricated with grating periods of dx = 0.982 μm, 1.000 μm and 1.013 μm respectively. +1 s-polarization transmitted order diffraction efficiencies of the gratings in the multilayer structures have been measured at normal incidence. Fig. 11 shows that the measured efficiencies of the three structures are in good agreement with the theoretical efficiency computed as a function of dx. A strong increase in the diffraction efficiency is obtained (x14).

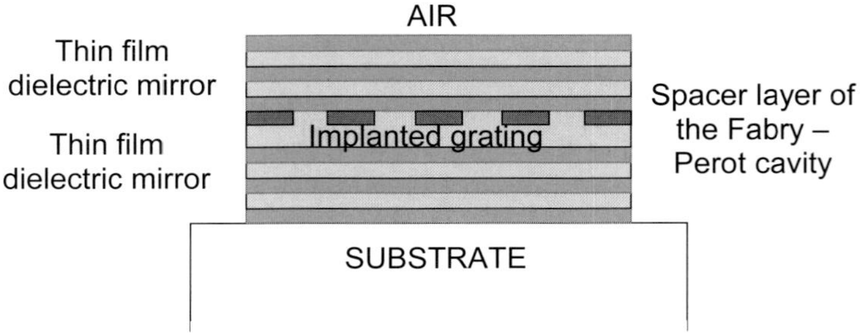

Fig. 10. Grating fabricated by ion implantation in the spacer layer of a thin-film Fabry-Perot cavity. The strong electromagnetic field in the spacer allows to improve the grating diffraction efficiency

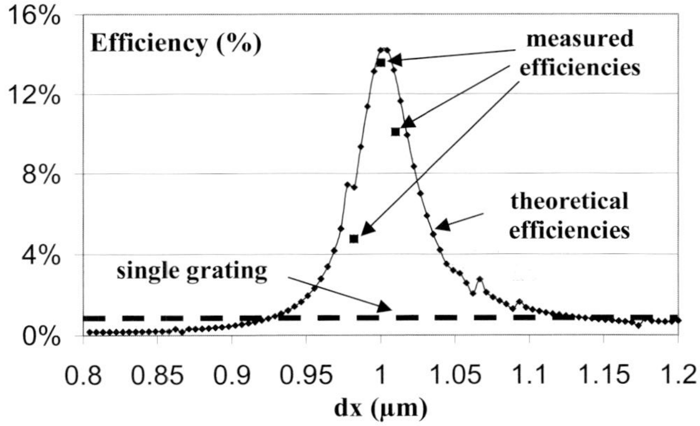

Fig. 11. Computed and measured diffraction efficiencies of the +1 s polarization transmitted order vs the grating period dx for a single grating and a grating embedded in the spacer layer of a Fabry-Perot cavity

Such components based on refractive index modulation, on the wavelength's scale, obtained using ion implantation associated with well-designed multilayer optical interference coatings may also be useful for applications to the reduction of the diffracted intensity of light, such as stealth.

Filtering Functions

Associating gratings with thin films may also be interesting for obtaining the guided-mode resonance in dielectric waveguide gratings. Indeed, a narrow resonance phenomenon may occur that causes 100 % optical energy switching between reflected and transmitted waves (Lemarchand et al. 1998; Magnusson and Wang 1992; Peng and Morris 1996; Wang and Magnusson 1993; Wang and Magnusson 1994; Wang et al. 1990). This resonance effect is observed when varying parameters such as the wavelength, the grating's thickness or the angle of incidence. Physically, this is due to coupling the propagating diffracted fields to the modes of the waveguide. For higher energy efficiency the grating is designed to admit only zero-order forward and backward-propagating waves, with all the other orders being evanescent waves. At resonance, the diffraction efficiency of the forward propagating wave approaches zero and that of the backward wave approaches one.

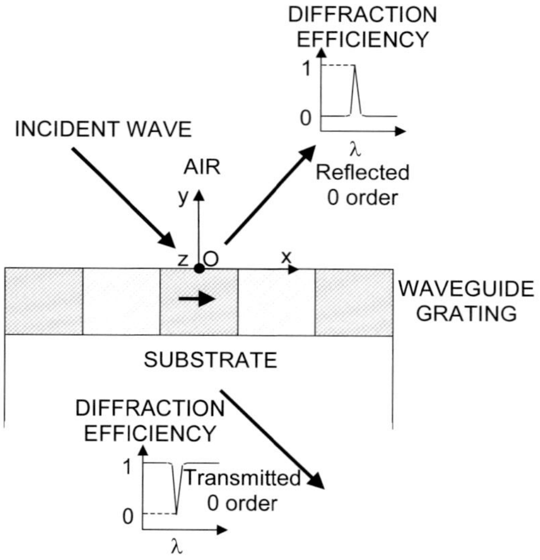

Fig. 12. Resonance filter obtained by using a single-layer waveguide grating

This guided mode resonance phenomenon can be utilized for obtaining optical micro-components such as high energy efficiency narrow linewidth small size filters suitable for applications in wavelength division multiplexing.

A basic resonance filter structure is obtained by using a single-layer waveguide grating (Fig. 12). This structure exhibits the properties of an effective thin-film layer as well as the characteristics of waveguide gratings. When the wavelength is far from resonance, the filter behaves like a simple thin film. Far from the resonance the coupling between the evanescent waves and the propagating ones is negligible because evanescent diffracted orders have a small amplitude and phase variation. At resonance, propagating waves are subject to strong coupling with evanescent waves resulting in a resonance effect stronger than the thin-film antireflection effect. For single-layer waveguide gratings, a symmetrical line-shaped filter is obtained by choosing the grating's thickness near a multiple of the half-wavelength (the resonance wavelength) in the layer. When the grating's thickness is different from a multiple of a half-wavelength of the resonance, an asymmetrical line-shaped filter is obtained. To design a multilayer guided mode resonance grating, each layer must be a quarter or a half wavelength thick and the refractive index of each region must be properly chosen so as to obtain an antireflection effect around the desired resonance wavelength.

6 Conclusion

Optical micro-components are very often key components in many systems. High bandwidth optical telecommunications, the environment, the aerospace industry, biology and health care, transportation, etc., most high-technology application fields are using or will use optical micro-components. All the research that is being done to develop new elements and to integrate them into micro-systems is paving the way for major technological breakthroughs.

Acknowledgements

The authors wish to thank Pr. Ugo Giovannini, PhD Pascal Huguet - Chantôme, Emmanuel Drouard from the Fresnel Institute Marseille and Stephane Tisserand and Laurent Roux from the Silios Technologies company for helping prepare this chapter. They also thank Pr. ag. Thierry Kakouridis for his kind assistance in helping them write this chapter in English.

References

Amsden CA, Gilman SE, Jacobs SD, Torok JS (1988) Measurement of the thermal conductivity of dielectric thin films with a thermal comparator. In: MacLeod A (ed) Optical Interference Coatings Tech Digest Series Vol 6. Optical Society of America, Washington DC, pp 78-81

Arnold GW (1982) Ion implantation effects in glasses. Radiation Effects 65: 17

Astilean S, Lalanne P, Chavel P, Cambril E, Launois H (1998) High-efficiency subwavelength diffractive element patterned in high-refractive-index material for 633 nm. Opt Lett 23 (7): 552-554

Berreman D (1972) Optics in stratified and anisotropic media : 4 x 4 matrix formulation. J Opt Soc Am 62: 502-510

Bloemer MJ, Haus JW (1996) Broadband waveguide polarizers based on the anisotropic optical constants of nanocomposite films. J of Lightwave Tech 14 (6): 1534-1540

Bosc D, Devoldère N, Bonnel M, Favennec JL, Pavy D (1999) Hybrid silica - polymer structure for integrated optical waveguides with new potentialities. Materials Science and Engineering B57: pp 155-160

Bovard BG (1995) Ion-assisted deposition. In: Flory F (ed) Thin Films for Optical Systems (Optical Engineering Series 49). Marcel Dekker, New York, pp 117 - 132

Culshaw B, Dakin J (1989) Optical Fiber Sensors: Systems and Applications II. Artech House, Norwood MA USA

Drouard E, Huguet Chantôme P, Escoubas L, Flory F (2002) dn/dT measurements performed using guided waves and application to the temperature sensitivity of WDM filters. Appl Opt 41 (16): xx-xx

EerNisse EP (1974) Compaction of ion-implanted fused silica. J Appl Phys 45: 167

Escoubas L, Flory F, Lemarchand F, During A, Roux L (2000) Enhanced diffraction efficiency of gratings in multilayers. Optics Letters 25 (4): 194-196

Escoubas L, Tisserand S, Gatto A, (1998) Le bilan des pertes dans les couches minces optiques par mesures d'absorption et d'atténuation à la propagation guidée. J Opt 29: 40-48

Favennec PN (1993) L'implantation ionique pour la microélectronique et l'optique. Collection Technique et Scientifique des Télécommunications, Masson, Paris

Flory F (1995) Guided wave techniques for the characterization of optical coatings. In: Flory F (ed) Thin Films for Optical Systems (Optical Engineering Series 49). Marcel Dekker, New York, pp 393-454

Flory F, Escoubas L, Huguet - Chantôme P, Drouard E (2001) Characterization of thin films for optical micro-components. In Optical Interference Coatings, OSA Technical Digest, pp. TuD1-1 – TuD1-3

Flory F, Endelema D, Pelletier E, Hodkinson I (1993) Anisotropy in thin films. Modelization and measurement of guided and non-guided optical properties. Application to TiO2 films. Appl Opt 32 (28): 5649-5659

Flory F, Escoubas L, Tisserand S, Nicolas E, Albrand G, Lemarchand F, Roux L (1999) Enhancement of the diffraction efficiency of titanium implanted gratings by associating them with optical interference coatings. SPIE 3738: 306-315

Flory F, Mayathaveekulchai N, Rigneault H, Zamkotsian F (1993) Characterization by guided wave of instabilities of optical coatings submitted to high power flux: thermal and third order non linear properties of dielectric thin films. Appl Opt 32: 5628-5639

Flory F, Tisserand S, Brassé L, Roux L, (October 20-25 1996) Integrated optics devices made by Ti implantation in SiO2 layers. OSA Annual Meeting, Rochester

Grann E, Moharam M, Pommet A (1994) Artificial uniaxial and biaxial dielectrics with use of two-dimensional subwavelength binary gratings. J Opt Soc Am A 11 (10): 2695-2703

Guenther AH, Mc Iver JK (1988) The role of thermal conductivity in the pulsed laser damage sensitivity of optical thin films. Thin Solid Films 163: 203-214

Guenther KH, Pulker HK (1976) Electron microscopical investigations of cross sections of optical thin films. Appl Opt 15: 2992-2997

Henry CH, Blonder GE, Kazarinov RF (1989) Glass waveguides on silicon for hybrid optical packaging. J Lightwave Tech 7: 1530-1539

Herzig HP (1997) Micro-Optics Elements, Systems and Applications. Taylor & Francis, London

Hodkinson I, Endelema D (1990) Bound modes in anisotropic multiplayer thin film waveguides. Appl Opt 29: 4424-4426

Huguet Chantôme P, Escoubas L, Flory F (2002) Guided-wave technique for the measurement of dielectric thin-film materials' thermal properties. Appl Opt 41 (16): 3127-3131

Hunsperger RG (1982) Integrated Optics : Theory and Technology. Springer-Verlag, Berlin Heidelberg New York

Jelinek M, Eason RW, Lancok J, Anderson AA, Grivas C, Fotakis C, Jastrabik L, Flory F, Rigneault H (1998) Waveguiding pulsed laser deposited Ti:sapphire layers on quartz. Thin Solid Films 322: 259-262

Jeschkowski U, Niederwald H (1997) Coating technologies. In: Bach H, Krause D (eds) Thin Films on Glass. Springer-Verlag, Berlin, pp 51-99

Koch TL, Leonberger FJ, Suchoski PG (1995) Integrated Optics. In: Bass M (ed) Handbooks of Optics, OSA (vol. I and II), Mc Graw-Hill, New York

Krauss TF, De La Rue RM (1999) Photonics crystals in the optical regime – past, present and future. Progress in Quantum Electronics 23: 51-96

Lambropoulos JC, Jolly MR, Amsden CA, Gilman SE, Sinicropi MJ, Diakomihalis D, Jacobs SD (1989) Thermal conductivity of dielectric thin films. J Appl Phys 66: 4230-4242

Lee DL (1986) Waveguide fabrication. In: Lee DL (ed) Electromagnetic Principles of Integrated Optics. Wiley J, New York

Lemarchand F, Sentenac A, Giovannini H (1998) Increasing the angular tolerance of resonant grating filters with doubly periodic structures. Opt Lett 23 (15): 1149-1151

Magnusson R, Wang SS (1992) New principle for optical filters. Appl Phys Lett 61: 1022-1024

Mollenhauer R (1994) Etude de la propagation guidée dans les multicouches optiques: mesures d'atténuation et localisation des pertes dans l'empilement. Ph.D. thesis, Aix-Marseille III University France

Monneret S, Huguet-Chantôme P, Flory F (2000) m-lines technique: prism coupling measurement and discussion of accuracy for homogeneous waveguides. J Opt A: Pure Appl Opt 2: 188-195

Monneret S, Tisserand S, Flory F, Rigneault H (1996) Light-induced refractive-index modifications in dielectric thin films: experimental determination of relaxation time and amplitude. Appl Opt 35: 5013-5020

Pelli S, Righini GC (1994) Introduction to integrated optics: characterization and modelling of optical waveguides. In: Martellucci S, Chester AN, Bertolotti M (eds) Advances in Integrated Optics. Plenum Press, New York, pp 1-20

Peng S, Morris GM (1996) Experimental demonstration of resonant anomalies in diffraction from two - dimensional gratings. Opt Lett 21 (8): 549-551

Pulker HK, Guenther KH (1995) Reactive physical vapor deposition processes. In: Flory F (ed) Thin Films for Optical Systems (Optical Engineering Series 49). Marcel Dekker, New York, pp 91–115

Righini GC, Laybourn PJR (1995) Integrated optics. In: Jha SS (ed) Perspectives in Optoelectronics. World Scientific, Singapore, pp 715-719

Ristau D, Ebert J (1986) Development of a thermographic calorimeter. Appl Opt 25: 4571-4578

Stern MB (1997) Binary optics fabrication. In: Herzig HP (ed) Micro – Optics. Taylor & Francis, London

Takahashi H, Ohnori Y, Kawachi M (1991) Design and fabrication of silica-based integrated optic 1 x 128 power splitter. Electron Lett 27: 2131-2133
Takashashi H (1995) Temperature stability of thin-film narrow-bandpass filters produced by ion assisted deposition. Appl Opt 34: 667-675
Teitler S, Henvis BW (1970) Refraction in anisotropic media. J Opt Soc Am 60: 830-834
Tien PK, Ulrich R (1970) Theory of prism-film coupler and thin-film light guides. J Opt Soc Am 60: 1325-1337
Townsend PD, Chandler PJ, Zhang L (1994) Optical Effects of Ion Implantation. Cambridge Studies in Modern Optics, Cambridge University Press, Cambridge
Ulrich R, Torge R (1973) Measurement of thin film parameters with a prism coupler. Appl Opt 12 (12): 2901-2908
Vassel MO (1974) Structure of guided modes in planar multilayers of optically anisotropic materials. J Opt Soc Am 64: 166-173
Veldkamp WB, McHugh TJ (1992) Binary Optics. Scientific American: 92-97
Wang SS, Magnusson R (1993) Theory and applications of guided - mode resonance filters. Appl Opt 32 (14): 2606-2613
Wang SS, Magnusson R (1994) Design of waveguide - grating filters with symmetrical line shapes and low sidebands. Opt Lett 19 (12): 919-921
Wang SS, Magnusson R, Bagby JS, Moharam MG (1990) Guided-mode resonances in planar dielectric-layer diffraction gratings. J Opt Soc Am A 8: 1470-1475
Weber HP, Dunn FA, Leibolt WN (1973) Loss measurements in thin-film optical waveguides. Appl Optics 12: 755-757
Yeh P (1979) Electromagnetic propagation in birefringent layered media. J Opt Soc Am 69: 742-756

Abbrevations

CW	-	continuous wave
dB	-	decibel
DWDM	-	dense wavelength division multiplexing
EDS	-	electron dispersive spectroscopy
FHD	-	flame hydrolysis deposition
IAD	-	ion assisted deposition
KEV	-	kilo electron volt
LPCVD	-	low pressure chemical vapor deposition
MEV	-	mega electron volt
M-Lines	-	modal lines
PDMS	-	polydimethylsiloxane
PIC	-	photonic integrated circuit
PLD	-	pulse laser deposition
PLZT	-	lead lanthanum zirconate titanate
PMMA	-	polymethyl methacrylate
RIE	-	reactive ion etching
TRIM	-	transfer of ions into matter
VLSI	-	very large scale of integration

Symbols

λ	-	wavelength
n_f	-	thin film's refractive index
d	-	layer thickness
ϕ_a	-	phase change at the film-air interface
ϕ_s	-	phase change at the film-substrate interface
θ_{sync}	-	synchronism angle
dn/dT	-	thermo-optic coefficient

Optical Coatings for the DUV / VUV

Roland Thielsch

Southwall Europe GmbH, Grossroehrsdorf, Germany

1 Introduction

During the last two decades, the development of powerful UV light sources, especially excimer lasers and frequency multiplied solid state lasers has lead to growing research efforts in the field of UV photon application. One of the main drivers toward the shortest possible wavelengths in the vacuum ultraviolet spectral region is semiconductor manufacturing. To produce integrated circuits with higher integration, lower power consumption and lower costs optical lithography using excimer laser based reduction wafer steppers is one of the key technologies for generation of the smallest lateral dimensions down to 0.1 µm and below (Rothschild and Ehrlich 1988, Rothschild et al. 1997, Rothschild 1998).

On going into the deep ultraviolet and vacuum ultraviolet spectral range, growing challenges result from lens and coating material properties, requirements on surface finish and handling as well as the excimer laser radiation itself like the high photon energy (5 eV at 248 nm, 6.5 eV at 193 nm, 7.9 eV at 157 nm), pulse length of 10 ns to 50 ns, maximum intensity of one pulse of some megawatts and laser pulse repetition rates up to several kHz (1000 - 4000 Hz are common for lithographic lasers). In addition, long term degradation stability of coated optical components against high energy laser radiation has to be realised. Excellent coatings have to pass 10^{10} till 10^{11} pulses with a fluence as high as 50 mJ/cm^{2}per pulse without significant alteration of their optical properties. In excimer laser beam lines of micro machining , material processing or photorefractive keratectomy devices, much higher fluences are common, which often seriously reduce the lifetime of optical components and thus limit the availability of such complex systems because of radiation induced degradation processes. Therefore, recent developments of UV-excimer laser applications have resulted in higher requirements on radiation resistant coated components at all laser wavelengths in the deep and vacuum ultraviolet spectral region. (See also Chap. "Laser Resistant Coatings" by

Stolz and Génin and Chap. "Coatings for Free Electron Lasers" by Gatto, Commandre and Couprie.)

The purpose of this chapter is to give a short and comprehensive review of some design and material aspects of multi-layer interference coatings especially for the wavelength range shorter than 250 nm.

Regarding basic research and applications in space astronomy and astrophysics, design of narrow band optical filters, general problems of optical measurements in the DUV/VUV spectral range and ultraviolet laser technology the reader is referred to the monographs of Zukic and Torr (Zukic and Torr 1995), Hunter (Hunter 1982) and Elliot (Elliot 1995).

2 Fundamentals of Optical Multi - Layer Coatings for the DUV / VUV

2.1 Special Requirements on Coatings in the DUV/VUV Spectral Range

The vacuum ultraviolet or far ultraviolet (100 nm to 200 nm) and the deep ultraviolet (200 nm to 280 nm) are only limited wavelength ranges within the ultraviolet spectral range of the electromagnetic spectrum covering roughly the wavelength range between 100 nm and 400 nm. These two narrow spectral ranges have numerous special problems for both the bulk substrate materials and the thin film coatings:

- High photon energy of the radiation (XeCl 308 nm – 4.02 eV, KrF 248 nm – 5 eV, ArF 193 nm – 6.42 eV, F2 157 nm – 7.9 eV)
- High energy density of most of the laser light sources
- Limited number of suitable substrate materials that provide sufficient transparency below 200 nm down to 100 nm
- Limited number of suitable thin film optical materials, especially high index materials at wavelengths below 250 nm
- Higher degree of absorption because the photon energy is much closer to the electronic band gap of the material and because of imperfections in the film structure, contamination, and impurities
- Higher scattering losses at interfaces, surfaces and even in the bulk materials with decreasing wavelength.

Coated optical components have to meet increasing requirements to withstand a high number of laser pulses (10^{10} to 10^{11} pulses) during their lifetime at repetition rates of several kHz without radiation induced damage and with no change or only small change in optical performance of a few percent in a harsh industrial environment. Surface shaping, grinding and polishing of the substrates as well as cleaning are inherent parts of the complex technological manufacturing chain of highly sophisticated optical components. From a practical point of view low mechanical stress, mechanical robustness during handling and ease of cleaning of exposed surfaces of the optics are also important criteria when selecting a specific material combination or technology to deposit a layer system.

2.2 Aspects of Coating Design

All thin film materials absorb in the DUV/VUV spectral range. The optical properties of a given thin film material are described by the complex refractive index $N(\lambda) = n(\lambda) + ik(\lambda)$ that consist of the refractive index $n(\lambda)$ and the extinction coefficient $k(\lambda)$. In the vicinity of the fundamental absorption edge of the materials, both quantities show pronounced dispersion, i.e. increase with decreasing wavelength. In order to precisely design a layer stack, accurate knowledge of the complex refraction indices of the used thin film materials is essential.

From a practical point of view concerning manufacture of reliable multi-layer coatings some simple rules of thumb have been established to meet the challenges of higher absorption of the materials and the higher photon energy (Kaiser 1994):

- Choice of materials (use only two or three materials in one stack)
- Optimal coating conditions for these materials in the deposition plant
- Coating design as simple as possible
- Classic quarter-wave designs preferred even at the expense of spectral performance if necessary
- Favourite design with lowest number of layers and interfaces
- Use of exact dispersion relations for materials and experimentally proven optical constants for design work
- Recognition that optical properties may vary with thickness of the layers due to structural changes.

In lithographic projection systems and other applications a wide variety of different spectral functions have to be realised by applying optical interference coatings. Antireflection coatings of various types for different wavelengths and angles of incidence, laser mirrors at normal and at 45° incidence, partial reflectors, thin film beam attenuators, polarisers, coatings with antireflection properties at one wavelength combined with partial reflection properties at another wavelength and other special coatings are the most prominent thin film devices required. Theoretically, the different spectral functions mentioned can be modelled using sophisticated design strategies and computer codes which are reviewed in chap. "Design of Optical Coatings" in this book.

Regarding the technological and economic performance of an optical system, special specifications have been established for coated optical components in the DUV/VUV. These include optical data together with projected radiation resistance figures. Table 1 reviews some of these specifications for major thin film components in 193 nm and 157 nm lithographic systems.

In micro-lithography, a rather low fluence is used in the projection optics and only a slightly higher fluence of up to $50mJ/cm^2$ per pulse in the beam lines. In other applications like photo-refractive keratotomy or precision micro-machining by excimer laser ablation the optical components are exposed to fluences of 100 mJ/cm² per pulse and higher.

Table 1. Lens material and optical coating specifications for use in 193 nm and 157 nm lithographic optics according to (Liberman et al. 1999, Otani et al. 2001).

Wavelength [nm]	Optical component	Initial target	Final target after irradiation	Irradiation life-time
193	Lens material CaF_2, SiO_2	0.997/ cm internal transmittance **		10^{11} pulses @ 1-2mJ/cm² per pulse
	AR coating	0.997 transmittance of a single AR		10^{11} pulses @ 1-2mJ/cm² per pulse
	HR coating	0.97 for HR45°		10^{9} pulses @ 5-10 mJ/cm² per pulse
157	Lens material CaF_2	0.999/cm int. transmittance	0.998/cm	10^{11} pulses @ 0.5mJ/cm² per pulse
	AR - coating	0.995 (step 1) 0.997 (step2)		
		0.99 of a both side coated CaF_2 substrate	0.98	10^{11} pulses @ 0.5mJ/cm² per pulse
	HR - coating	0.95 reflection at 0°AOI*	0.94 reflection at 0°AOI	10^{11} pulses @ 0.5mJ/cm² per pulse

* AOI – angle of incidence, ** The internal transmittance is solely determined by the transmission of light through a material of a certain thickness corrected for the reflectance losses at the surfaces.

Antireflection coatings

In a lithographic wafer scanner or stepper, the number of optical components is in the range of 60 to 70 depending on the actual optical layout. Effective reduction of the Fresnel reflection of surfaces of diverse optical elements, which is of a magnitude of about of 5% per surface, increases the optical throughput of the systems, reduces stray light and increases contrast.

For a single laser wavelength or a narrow spectral range the so called V–coating is the most efficient antireflection coating. Often, the V-coating consists of a high index and a low index layer, each with appropriately matched refractive index and matched optical thickness at the design wavelength (Macleod 1986a).

At 248 nm and longer wavelengths, V–coatings consisting of HfO_2, Al_2O_3, Sc_2O_3 layers next to the substrate and SiO_2 or MgF_2 provide appropriate antireflection properties at normal and near normal incidence. At wavelengths below 248 nm, tri-layer antireflection coatings of the design S/L1H1L2 with the L1–low refractive index layer with an optical thickness L1 close to quarter-wave optical thickness and additional low and high index layers L2, H2 with adjusted optical thickness often provide excellent antireflection properties with residual reflections R as low as 0.1 %. In Fig. 1, the measured reflectance of CaF_2 with both sides coated with tri layer AR-coatings are shown at the wavelength of the KrF (248 nm), ArF (193 nm) and F2 excimer laser (157 nm).

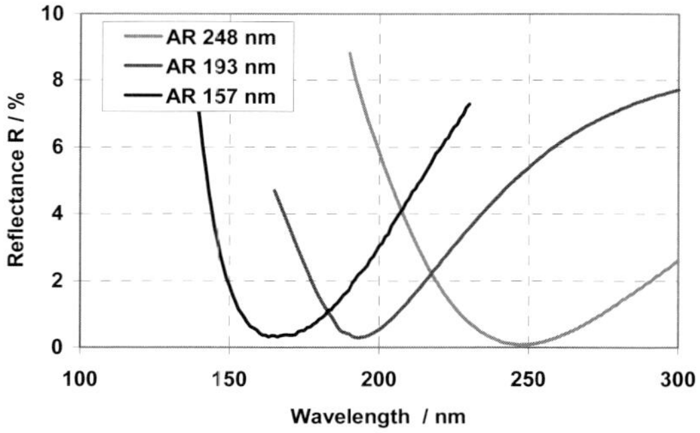

Fig. 1. Measured reflectance of CaF_2 on both sides V-coated designed for antireflection purposes at 248 nm, 193 nm and 157 nm.

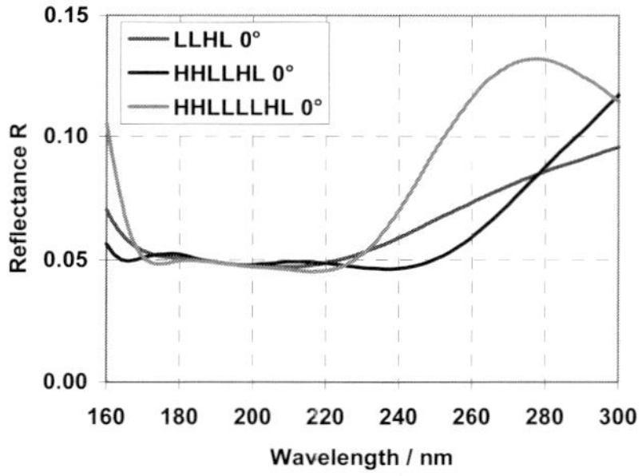

Fig. 2.

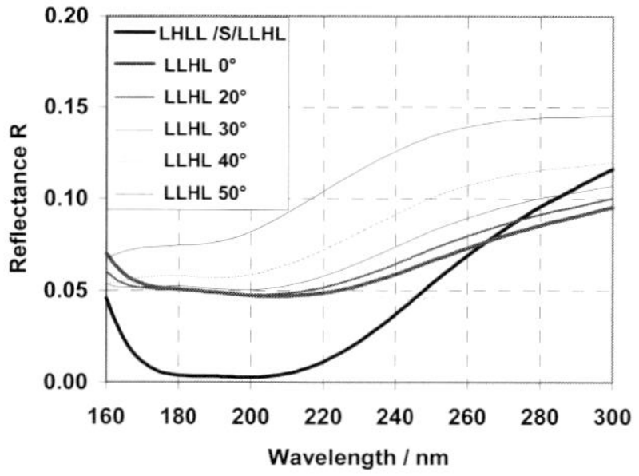

Fig. 3.

Fig. 2. Calculated reflectance R at 0° AOI for AR–coatings of the designs S/LLHL, S/HHLLHL and S/HHLLLLHL on fused silica. LaF_3 and MgF_2 were chosen as the high index layer material and the low index layer material, respectively.

Fig. 3. Calculated reflectance R of the layer system LLHL on fused silica for various angles of incidence. Calculations were performed for one side coated substrate only at AOI of 0°, 20°, 30°,40° and 50°. For comparison, the calculated reflectance of a fused silica substrate coated on both sides is also shown.

In a lithographic projection objective of high numerical aperture an essential part of the light hits the surface of a lens at not normal angles of incidence. Antireflection properties have also to be realised for high angles of incidence. Figure 2 shows the calculated reflectance of three layer systems on fused silica designed with LaF_3 as the high index material and MgF_2 as the low index coating material. These belong to the group of multiple quarter-wave coatings, with LLHL, HHLLHL and HHLLLLHL designs starting from the substrate where L is a low refractive index layer and H is a high refractive index layer of quarter-wave optical thickness, respectively. In order to limit the absorption to lower than 0.5% per layer system, the total number of quarter wave layers in the stack should be smaller than 10. Figure 3 shows the variation of the reflectance (random) of the AR–coating S/LLHL on fused silica with the angle of incidence. This layer system provides reasonable antireflective properties up to about 30° AOI.

High reflectance coatings - HR

Tuned multi-layer stacks consisting of alternative low index layers L and high index layers H of quarter-wave optical thickness (QW–stack) permit the highest reflectance at a given wavelength depending on the material properties and the number m of (HL)-pairs in the layer system.

The basic design may have either a symmetric (incident medium $(HL)^m$ H / substrate or incident medium $L(HL)^m$ / substrate) or an asymmetric form (incident medium $(HL)^m$ / substrate or incident medium $(LH)^m$ / substrate).
In the usual notation in thin film optics, $L = n_L d_L$ and $H = n_H d_H$ represent layers of quarter-wave optical thickness at the design wavelength λ_0 (Macleod 1986b, Thelen 1988, Zukic et al. 1990).

Figure 4 shows the measured reflectance of quarter-wave optical thickness all dielectric high reflection mirrors designed for 0° AOI at the wavelengths 248 nm, 193 nm, and 157 nm. The decrease in reflectance of the stacks with decreasing wavelength is attributed to the higher losses of the particular thin film materials used which where HfO_2/SiO_2 at 248 nm and LaF_3/MgF_2 at 193 nm and 157 nm.

In order to overcome the limitations of quarter-wave stacks for a given combination of thin film materials, theoretical and practical investigations have been performed to further increase the reflectance. While the extinction coefficient and therefore the losses of the high index layer are almost one order of magnitude higher than the low index material the optical thickness of the high index material can be reduced while maintaining the total optical thickness of a pair of a high index and a low index layer at $\lambda_0/2$ or in terms of phase thickness at a phase angle of π. This is the basic concept of the π-layer reflector designs which also benefit from the shift of the internal electric field distribution (Zukic et al. 1990, Apfel 1977, Newnam et al. 1983). Usually this leads to a slight increase of the reflection of the stack and to a slight narrowing of the reflection bandwidth.

In order to improve the laser radiation resistance of dielectric mirrors, addition of a half-wave layer as the first layer next to the substrate and/or on top of the stack as the last layer to the surrounding medium has been performed. These undercoat and overcoat layers do not have an optical effect on the reflectance at the design wavelength, but seem to increase the laser induced damage threshold of the stacks (Tuttle-Hart et al. 1983). (For laser induced damage thresholds see Chap.

"Characterisation and Monitoring" by Ristau and Chap. "Laser Resistant Coatings" by Stolz and Génin in this book)

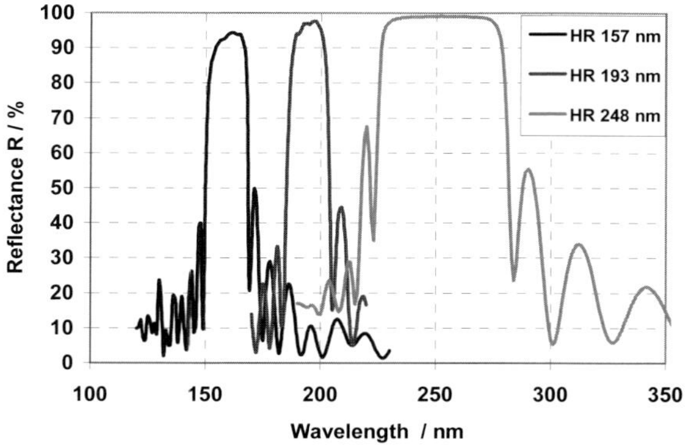

Fig. 4. Measured reflectance of quarter-wave all dielectric mirrors designed for 0°AOI at 248 nm, 193 nm and 157 nm. The HR248 nm mirror is made of HfO_2 as the high index material and SiO_2 as the low index material. The two other mirrors are all fluoride stacks consisting of LaF_3 and MgF_2 as the high and the low index materials, respectively.

Metal mirrors

In case a high reflectance in a very extended wavelength region is needed, aluminium layers offer the best solution even if the reflectance is only about 90%. Thin films of MgF_2 evaporated at low substrate temperatures immediately in the same vacuum cycle on top of a fresh deposited Al–mirror protect the Al from extensive oxidation and hence from a drop in reflectance at short wavelengths (Hass and Waylonis 1961). Typically, for a protected aluminium mirror deposited under normal high vacuum conditions (mid 10^{-6} mbar) and at a high evaporation rate of about 30 nm/s a reflectance of 88 – 90 % at 193 nm and 84 – 86 % at 157 nm is achieved 193nm (Callahan and Flint 1998). Adding dielectric layer stacks with a design 0.66L $H(LH)^m$ with m depending on the chosen materials and the desired reflectance enhances the reflectance of the metal.

3 Technological Aspects of DUV / VUV Coatings

3.1 Thin Film Materials

Compared to the visible part of the spectrum, the number of useable thin film materials in the DUV/VUV is very limited due to their intrinsic electronic properties.

A brief survey of suitable materials for use in the ultraviolet spectral range is given in Chap. “Coating Materials” in this book .

In the ultraviolet spectral range the geometrical thickness of a layer with a quarter-wave optical thickness becomes smaller with decreasing wavelength to about only a few tens of nanometers. For example, MgF_2 has a refractive index n of about 1.4 in the wavelength range below 250 nm. Therefore, the geometrical thickness of a MgF_2 quarter-wave layer at the wavelength of KrF, ArF, and F2 excimer laser is only about 44 nm at $\lambda = 248$ nm, 34 nm at $\lambda = 193$ nm, and 27 nm at $\lambda = 157$ nm, respectively. The thickness of any thin film material with a higher refractive index is even smaller. Detailed investigations of the crystalline and morphological structure of thin optical films revealed initial layer formation and transition layers in the early stages of film growth starting from the interface or the substrate surface which often have a quite different morphological and crystalline structure compared with the bulk of thick films. The thickness and the structural features in such initial and transition layers depend on the material, the interface properties and growth conditions like substrate temperature, condensation rate, pressure and gas environment and kinetic energy of the film forming constituents (Kaiser et al. 1991, Kaiser et al. 1992). in the thickness range of quarter-wave optical layers in the ultraviolet, for some thin film materials these disturbed sublayers form an essential part of the whole film influencing the physical properties of the coatings more strongly. For illustration, a sketch of the morphological structure of MgF_2 films evaporated onto amorphous substrate at three different substrate temperatures is shown in Fig. 5.

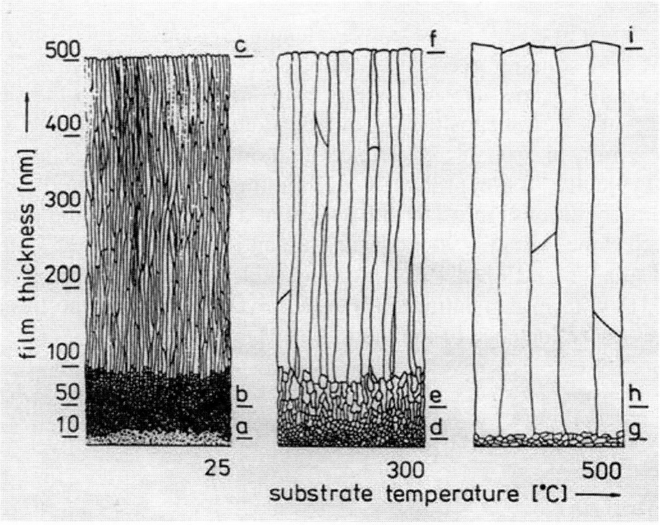

Fig. 5. Schematics of the morphological structure of MgF_2 films evaporated on amorphous substrates according to (Kaiser et al. 1992).

It is obvious, even at a substrate temperature of 300°C which is commonly applied in evaporation processes that the main part of an UV film is formed by the transition layer and thus shows some different properties compared to thicker films. Initial and transition layers at the substrate–film interface which cause optically inhomogeneous layers with thickness dependant refractive index have been reported for various fluoride films (Kolbe et al. 1992, Thielsch et al. 1998) and for metal oxide films like ZrO_2 (Cho et al. 1996) or HfO_2 (Lehan et al. 1991).

(For further information on the structure of thin optical films see Chap. "Fundamentals of Thin Film Growth" by Kaiser.)

3.2 Deposition Technologies

3.2.1 Thermal Evaporation

High vacuum electron beam evaporation and evaporation from resistively heated boats are the most common deposition technologies for the manufacture of optical interference coatings. For high temperature melting oxides, reactive electron beam evaporation with oxygen partial pressures up to several 10^{-4} mbar is applied in order to maintain the stoichiometry of the coatings and to avoid absorption losses in the UV spectral range. This technology is also used to deposit fluoride thin films. With the exception of MgF_2, electron beam evaporation is not generally best suited for the deposition of fluoride multilayer stacks in the VUV spectral range. Evaporation from resistively heated boats is frequently preferred due to the lower decomposition of the fluoride material and due to the reduced spitting of evaporation material (Graper 1971, Gloede et al. 1985).

Evaporated coatings often suffer from disadvantageous properties like columnar growth and polycrystalline structure, reduced packing density and low refractive index, adsorption of water vapour into the open voids between the columns, extensive hydrocarbon contamination during storage in air, environmental instability/shifts of the optical function of multilayer stacks, high tensile stress and high surface and interface roughness causing increased scatter losses (Pulker 1984, Macleod 1986, Kaiser et al. 1992, Kaiser and Kaiser 1994, Ritter 1976, Ennos 1966, Duppаré and Jakobs 1996).

3.2.2 Energetic Deposition Techniques

Energetic deposition techniques like ion assisted deposition (IAD) (Martin et al. 1983), plasma ion assisted deposition (PIAD) (Matl et al. 1991), ion plating (IP) (Günther et al. 1989, Waldorf et al. 1993), ion beam sputtering (IBS) (Sites et al. 1983) and magnetron sputtering (MS) (Rosenfeld et al. 1994) are used to overcome some of the structural disadvantages of evaporated layers at low or medium substrate temperatures. By proper optimisation of the deposition parameters, dense, water-free and low absorption oxide films with amorphous structure, smooth interfaces and low scatter losses can be realised even in the DUV down to 193 nm.

In the case of fluoride films difficulties in maintaining the stoichiometry of the films arise from noble gas ion bombardment induced decomposition and fluorine release of the coatings. Deficient fluorine is easily substituted by oxygen from the residual gas atmosphere (Targove et al. 1986, Kolbe et al. 1992). Despite very dense microstructure, smooth interfaces and high refractive index, fluoride coatings deposited by PIAD, IAD, IBS or MS often suffer from increased absorption at short wavelengths compared to conventional evaporated films. Efforts have even been made to compensate the fluorine release by a fluorine or oxygen reactive deposition processes. To illustrate the differences, the optical constants of AlF_3 films deposited by thermal boat evaporation, by reactive ion beam sputtering with fluorine F_2 and by reactive ion beam sputtering with oxygen added to the background gas are shown in Fig. 6. which clearly illustrate the higher extinction coefficients and thus the higher absorption at short wavelengths in the VUV of both types of reactive ion beam sputtered films. Similar results have been published on LaF_3 and MgF_2 for short wavelength while for the deposition of UV-multilayer coatings at longer wavelengths i.e. 248 nm , 308 nm or 352 nm, reactive IBS processes or IAD with fluorides have been successfully developed to grow highly reflective mirrors (Kolbe 1992, Quesnel et al. 1997, Dijon et al. 1998). For further details on evaporation and other deposition technologies, the reader is referred to Chap. " Thin Film Deposition Techniques" by Pulker.

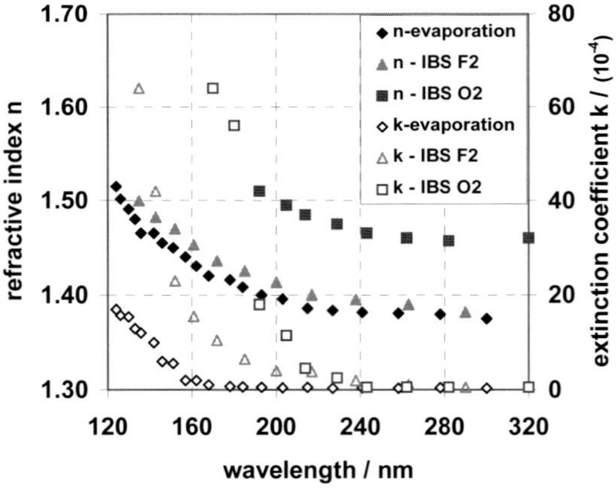

Fig. 6. Optical constants of AlF_3 films deposited by thermal evaporation, F_2 reactive IBS and oxygen reactive IBS. The reactive gases were added to the background gases. Data are taken from the literature (Schink et al.1991).

3.3 Thin Oxide Layer

Oxides like zirconia ZrO_2 (Cho and Hwangho 1996), yttria Y_2O_3 (Bezuidenhout and Pretorious 1986), scandia Sc_2O_3 (Rainer et al. 1982), lanthania La_2O_3 (Hass et al. 1959), hafnia HfO_2 (Baumeister and Arnon 1977, Lehan et al. 1991, Alvisi et al. 1999, Thielsch et al. 2002), alumina Al_2O_3 (Kaiser et al. 1995) and silica SiO_2 (Rainer et al. 1985) are the common thin film materials to deposit optical multilayer coatings down to wavelengths of 230 nm. Among them, HfO_2, Al_2O_3 and SiO_2 are the most important ones that can be deposited with high quality by different deposition technologies thus allowing the optimisation of optical properties, density, mechanical stress, hardness and other properties. Typical properties of thin films of SiO_2, Al_2O_3 and HfO_2 deposited by various techniques are summarised in Table 2.

Table 2. Properties of thin SiO_2 , Al_2O_3 and HfO_2 films in the DUV spectral range.

Film	deposition method	ion energy [eV] bias voltage [V]	substr. temp. T_s / [°C]	n, k @ λ	stress as-deposited [MPa]	Remarks
SiO_2	reactive evaporation	no	300	1,57 <5*10^{-4} @ 193	about. 1200 compr. (d=40nm)	stress depends on thickness, packing density 0.92
	PIAD	80 - 180	RT	1,56 - 1.58 12 – 20 *10^{-4} @ 193	> 200 compr.	water uptake, stress relaxation
	IBS	1000	RT	1,61 1*10^{-3} @ 193	app. 600 compr.	packing density 1.0x
Al_2O_3	reactive evaporation	no	300	1,83-1,84 0.5 - 1*10^{-2} @ 193	500 - 600 compr. (d=40 nm),	packing density 0.92
	PIAD	110 - 150	RT	1,82-1,87 8 - 15*10^{-3} @ 193	100 - 500 compr.	no water, packing density < 1
	IBS	1000	RT	1,86 4*10^{-3} @ 193	compr., several 100	packing density 1.0
HfO_2	reactive evaporation	no	300	2.20 – 2.30 @ 248 2 –10 x 10^{-3} @ 248	tensile stress	packing density 0.9
	PIAD	110-150	RT	2.20 – 2.30 @ 248 2 –10 x 10^{-3} @ 248	Compr. , several 100 MPa	Ar-ions + oxygen, packing density below 1, no water sorption
	IBS	1000	RT	2.35 – 2.4 @ 248 > 10^{-2} @ 248	Compr., several 100 MPa	packing density close to 1

* data compiled from references given in the text

3.4 Thin Fluoride Layer

Fluorides are widely used especially as low index thin film materials in the UV, the visible and the infrared spectral range. In this group of compounds MgF_2 is the most prominent fluoride material due to its low refractive index over a wide spectral range from the VUV to the mid infrared combined with low absorption and mechanical robustness when deposited at elevated substrate temperatures (Pulker 1984). Other important low index fluorides are AlF_3, cryolite Na_3AlF_6, and chiolite $Na_5Al_3F_{14}$.

High refractive index fluorides mainly belong to the family of lanthanide trifluorides. In the past, pioneering research on the properties of this group of thin film materials for use in the VUV spectral range was performed by several groups (Hass et al. 1959, Targove et al. 1986, Kolbe et al. 1992). Only a few of these materials, for example LaF_3, GdF_3 and NdF_3, were found to be useful for growing low absorption interference coatings at wavelengths of 248 nm and 193 nm (see Table 3). Additionally, LaF_3 and GdF_3 are also the most important high index materials in the wavelength range from 130 nm to 190 nm.

In Fig. 7 dispersion curves of the refractive index n and the extinction coefficient k of some fluoride thin films (MgF_2, LaF_3, GdF_3) in the VUV/DUV spectral range are shown to illustrate the small refractive index difference between the low index and the high index fluorides. The films were thermally evaporated from boats onto CaF_2 substrates at a substrate temperature above 300°C.

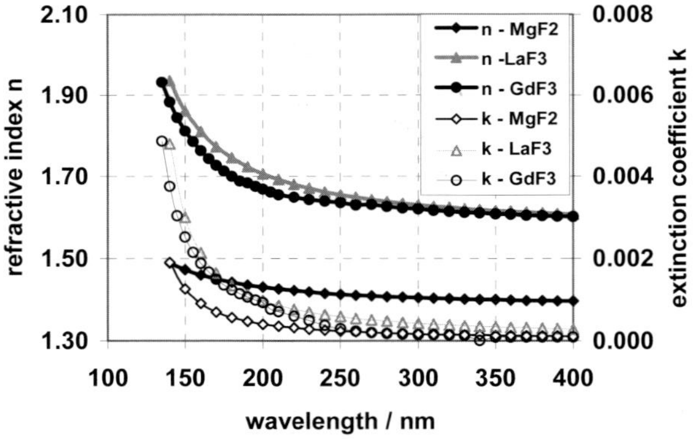

Fig. 7 Dispersion curves of the refractive index n and the extinction coefficient k of some thermally evaporated thin fluoride films in the VUV/DUV spectral range.

Table 3 Properties of thin fluoride films in the DUV / VUV spectral range.

Fluoride	energy gap [eV]	refractive index n @ λ *	ext. coefficient k @ λ *	structure *	packing density @ Ts*	Reference (for example only)
AlF_3	11	1.43–1.46 @157	0.001-0.0023 @157	amorphous	0.62 @ 75°C 0.99 @ 250°C	Niisaka et al. 2001 Barrieré et al. 1977
MgF_2	12.3	1.45–1.47 @157	0.001–0.005 @157	amorphous crystalline at higher T_s	0.7 – 0.8 @ 50°C, ~ 1 @ 300°C	Pulker 1984, Ristau et al. 2001
LaF_3	~10	1.71–1.83 @157, 1.7 @ 193	0.003–0.0095 @157, 0.0003 @ 193	crystalline, hexagonal	0.8 @ 50°C	Ristau et al. 2001, Thielsch et al. 1998 Niisaka et al. 2001
GdF_3	11	1.73–1.81 @157 1.68 @ 193, 1.63 @248	0.003–0.012 @157 ~0.001 @ 193	amorphous @ 50°C, orthorhombic @ 200°C	-	Niisaka et al. 2001, Thielsch et al. 1998

*properties depend on deposition technique and deposition conditions (rate, substrate temperature etc.)

3.5 Radiation Interaction and Environmental Stability

Radiation interactions with optical coatings include numerous effects like laser conditioning of coatings (Stolz et al. 1998, Eva et al. 1996), laser radiation cleaning of optical components in the presence of oxygen (Bloomstein et al. 1999, Heber et al. 1998), spectral shifts of coatings (Heber et al. 1998), crack formation under irradiation caused by thermal stress, colour centre formation, photo-induced post deposition oxidation and others. Usually laser radiation resistance is tested by determination of laser induced damage threshold in (1-on-1), (N-on-1) or (R-on-1) test mode (Rainer et al. 1985), photo-thermal and photo-calorimetric absorption measurements including evaluation of temporal changes and multiple photon processes (Welsch 1995) and long term or so-called "marathon" irradiation tests at low fluence levels related to the practical applications and to a gradual degradations in layer systems by the accumulated energy dose (Liberman et al. 1998).

The intrinsic electronic properties of the materials and the increased absorption in thin films compared to the bulk materials define severe limitations on the UV-laser radiation stability of the coatings. Increased linear absorption can be caused by electronic states in the electronic bandgap due to structural disorder, chemical imperfections, structural defects and impurities and radiation induced permanent and transient colour centres. Non-linear absorption by multiple photon absorption processes at 248 nm and 193 nm have been reported to occur in substrate materials like fused silica, CaF_2 and others (Eva and Mann 1996). Non-linear absorption was also measured in thin Al_2O_3 films when irradiated by ArF laser photons, whereas results on thin films of LaF_3, AlF_3 and MgF_2 did not indicate the existence of strong non–linear absorption at 193 nm (Apel and Mann 2000, Apel et al. 2000).

Marathon irradiation and degradation tests at various fluence levels have been reported in the literature at wavelengths of 308 nm (Mann et al. 1997), 248nm (Krajnovich et al. 1992), 193 nm (Heber et al. 1998,1999, Liberman et al. 1998) and recently also 157 nm for both AR and HR45° coatings (Rothschild et al. 1999, Liberman et al.2000, 2001). Test samples included oxide HR at 308 nm and 193 nm, fluoride HR at 193 nm and 157 nm, mixed oxide/fluoride AR-coatings at 248 nm and fluoride AR coatings at 248 nm, 193 nm and 157 nm. Sometimes catastrophic failure can occur after irradiation by only a few million pulses, but optimised fluoride AR coatings can survive 10^9 and more laser pulses at 15mJ/ cm^2 per pulse at 193 nm without significant degradation of their properties. Even at 157 nm, fluoride AR coatings tested at 3mJ/cm^2 per pulse can withstand irradiation of several 100 million pulses without dramatic alterations (Liberman et al. 1998, 2000, 2001; Rothschild et al. 1999). For dielectric mirrors, reported data show no significant impact on the reflectance of MgF_2/LaF_3–multilayer coatings at 193 nm after irradiation with about 240 million pulses at 70 mJ/cm^2/pulse (Heber et al. 1998, 1999) and 850 million pulses at 50mJ/cm^2/pulse (Bernitzki et al. 1998). At 157 nm, highly reflective dielectric mirrors may also pass the 10^9 pulse mark at 10 – 15 mJ/cm^2 per pulse without significant reflection degradation, while broadband metal reflectors suffered substantial reflection reductions by more than 30% (Liberman et al. 2000).

Incorporation of water and hydrocarbons into multilayer coatings during the depositions process, by air-borne contamination or by organic contamination originating from purge gas background levels or out-gassed or photo-desorbed species of resists are severe problems which can effect the optical performance of the coatings through spectral shifts and increased absorption.(Heber et al. 1998,1999, Kunz et al 2000, Bloomstein et al 2000, 2001).

Contamination becomes more important in the vacuum ultraviolet where the extinction coefficients of water and of most organic compounds are higher by orders of magnitude compared to the visible spectral range (Libermann et al. 1999). In Fig. 8 the calculated transmittance of a 2 mm thick CaF_2 sample only contaminated with thin adsorbed water layers on both sides illustrates this effect.

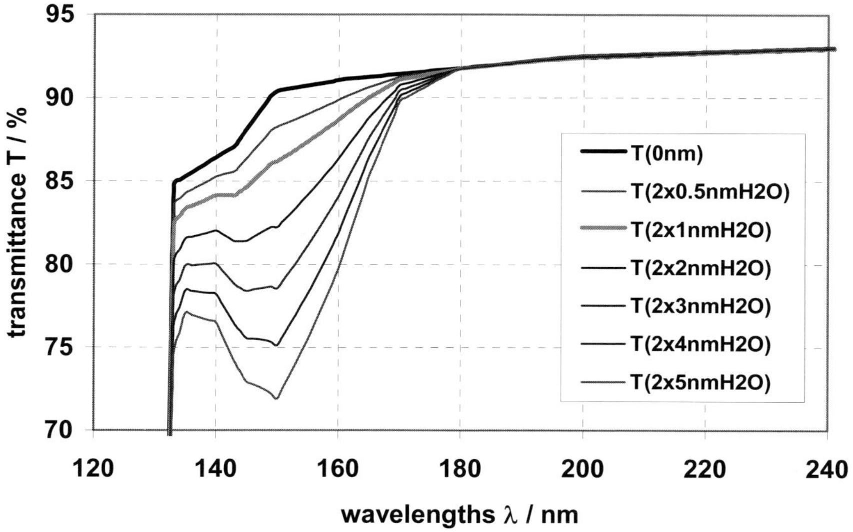

Fig. 8 Calculated transmittance of a 2 mm thick CaF_2 test sample covered with water layers of the same thickness on both side. The water is assumed to be adsorbed on both CaF_2 – surfaces.

Modelling water incorporation into optical coatings revealed that even only 1% moisture incorporation in three layer V–AR-coatings on both sides of a CaF_2 plate or in a dielectric mirror consisting of 20 HL–pairs can cause a drop of 2 % in transmittance through the antireflection plate or a 2% drop in reflectance of the model mirror at 157 nm (Liberman et al. 1999). The direct influence of water at 193 nm on the absorption of multilayer coatings is not as pronounced as it is at 157 nm. However, some materials like AlF_3 tend to react with incorporated moisture which leads to an increased absorption in the VUV and thus to a degradation of the radiation resistance with time (Barrieré and Lachter 1977, Thielsch et al. 2000). As shown in Fig. 9, an irreversible decrease of the LIDT (1-on-1) of AlF_3/LaF_3 high reflectance dielectric mirrors at 193 nm with storage time in air resulted from the reaction of AlF_3 with the adsorbed water.

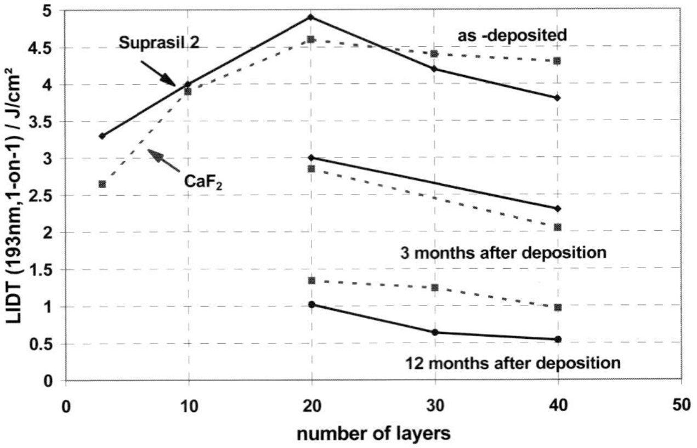

Fig. 9 LIDT (1-on-1, 20 ns, 193 nm) of AlF_3/LaF_3 dielectric quarter-wave stacks versus the number of layers after different storage times under normal laboratory conditions.

3.6 Multi–Layer Performance in the DUV – VUV Spectral Range

Despite limitation by material properties and technical difficulties, today many coated DUV/VUV optical components are on the market and in widespread use. The state-of-the-art properties of the most common optical multi-layer coatings for use at 248 nm, 193 nm and 157 nm are summarised in Tables 4 to 6.

Table 4 State-of-the-art of fluoride and oxide optical coatings at 248 nm.

Coating type	Reflectance %	Transmittance %	LIDT (1-on-1) J/cm²	Useful material combination
V-AR	0.1	99.9 max.	6 - 10	MgF_2 / LaF_3 MgF_2 / Al_2O_3
HR 0°	98	~0	12 -16	LaF_3 / Na_3AlF_6 NdF_3 / AlF_3 LaF_3 / MgF_2
	99	~0	Up to 27 on fused silica	Al_2O_3/SiO_2
	99.5%	~0	3 - 4	HfO_2/SiO_2
HR 45°	98	~0	3 – 4 up to 15 with Al_2O_3	HfO_2/SiO_2 Al_2O_3/SiO_2

Table 5 State-of-the-art of fluoride and oxide optical coatings at 193 nm.

Coating type	Reflectance %	Transmittance %	LIDT (1-on-1) J/cm²	Useful material combination
V-AR	0.1	> 99	2- 3 typical, up to 5.6 on super-polished CaF_2	MgF_2 / LaF_3 LaF_3 / AlF_3
HR 0°	99	~0	2- 4 typical, up to 5	LaF_3/Na_3AlF_6 LaF_3/AlF_3 LaF_3/MgF_2
	95	~0	~1	Al_2O_3 / SiO_2
HR 45°	97	~0	2 - 3	LaF_3 / Na_3AlF_6 LaF_3 / AlF_3 LaF_3 / MgF_2 DyF_3 / AlF_3
	92 - 93	~0	~1	Al_2O_3 / SiO_2

Table 6 State-of-the-art of fluoride optical coatings at 157 nm.

Coating type	Reflectance %	Transmittance %	LIDT (1-on-1) J/cm²	Material combination
AR	0.1	98.5	no data	MgF_2/LaF_3 MgF_2 / GdF_3
HR 0°	96	~0	no data	MgF_2 / LaF_3 MgF_2 / GdF_3
HR 45°	93	~0	no data	LaF_3 / MgF_2

4 Summary and Outlook

As a result of increasing efforts during the last two decades improved and reliable interference coatings for DUV/VUV are on the market and in widespread use in numerous applications like precision material processing, medicine and integrated circuit fabrication.

Deeper understanding of the relationships between film properties and deposition technologies as well as of the influence of environmental factors on optical performance was developed as a basis for further optimising existing deposition technologies. Evaporation is still the workhorse in the manufacture of interference coatings for the DUV/VUV spectral range, but remarkable progress has been made also by using IBS and other energetic deposition technologies especially for oxide materials.

Despite some encouraging results regarding the radiation stability of optimised, further research must be focused on evaluation of all factors affecting radiation induced failure and the radiation resistance of optical coatings at short wavelengths. Continuing demand to improve the optical performance and to increase the radiation resistance of the coatings even at the high photon energies of excimer lasers will push the state-of-the art to its physical limits.

References

Alvisi M, Scaglione S, Martelli S, Rizzo A, Vasanelli L (1999) Structural and optical modification in hafnium oxide thin films related to the momentum parameter transferred by ion beam assistance. Thin Solid Films 354: 19 - 23

Apel O, Mann K (2000) Scatter and absorption losses from DUV optics: A comparative study. In: Exarhos GJ, Guenther AH, Kozlowski MR, Lewis KL, Soileau MJ (eds) Laser-Induced damage in Optical Materials 1999. Proc. SPIE 3902, pp. 460 – 467

Apel O, Mann K, Heber J, Thielsch R (2000) Non-linear absorption phenomena in oxide coatings for 193nm. In: Exarhos GJ, Guenther AH, Kozlowski MR, Lewis KL, Soileau MJ (eds) Laser-Induced damage in Optical Materials 1999. Proc. SPIE 3902, pp. 235 – 244

Apfel HJ (1977) Optical coating design with reduced electric field intensity. Appl. Optics 16: 1880 - 1885

Barrière AS, Lachter A (1977) Optical transitions in disordered thin films of the ionic compounds MgF_2 and AlF_3 as a function of their conditions of preparation. Appl. Optics 16: 2865 – 2871

Baumeister P, Arnon O (1977) Use of hafnium dioxide in multilayer dielectric reflectors for the near UV. Appl. Optics 16: 439 - 444

Bernitzki H, Lauth H, Thielsch R, Blaschke H, Kaiser N, Mann K (1999) Current status of radiation resistance of dielectric mirrors in the DUV. In: Exarhos GJ, Guenther AH, Kozlowski MR, Lewis KL, Soileau MJ (eds) Laser-Induced Damage in Optical Materials 1998. Proc. SPIE 3578, pp. 105 – 116

Bezuidenhout DF, Pretorious R (1986) The optical properties of evaporated Y_2O_3 – films. Thin Solid Films 139 : 121 - 132

Bloomstein TM, Rothschild M, Liberman V, Hardy D, Efremow NN, Palmacci ST (2000) Laser cleaning of optical elements in 157-nm lithography. In: Progler CJ (ed) Optical Microlithography XIII. Proc. SPIE 4000, pp. 1537 – 1545

Bloomstein TM, Liberman V, Palmacci ST, Rothschild M (2001) Controlled Contamination of optics under 157-nm laser irradiation. In: Progler CJ (ed) Optical Microlithography XIV. Proc. SPIE 4346, pp. 685 – 694

Callahan GP, Flint BK (1998) Characteristics of deep UV optics at 193 nm & 157 nm. In: Exarhos GJ, Guenther AH, Kozlowski MR, Lewis KL, Soileau MJ (eds) Laser-Induced Damage in Optical Materials: 1998. Proc. SPIE 3578, pp. 45 - 53

Cho HJ, Hwangho CK (1996) Optical inhomogenity and microstructure of ZrO_2 thin films prepared by ion-assisted deposition. Appl. Optics 35: 5545 – 5552

Dijon J, Quesnel E, Rolland B, Garrec P, Pellé C, Hue J (1998) High damage threshold fluoride UV mirrors made by ion beam sputtering. In: Bennett HE, Guenther AH,

Kozlowski MR, Newnam BE, Soileau MJ (eds) Laser-Induced Damage in Optical Materials: 1997. Proc. SPIE 3244, pp. 406 - 416

Duparré A, Jakobs S (1996) Combination of surface characterization techniques for investigating optical thin-film components. Appl. Optics 35: 5052 - 5058

Elliot DJ (1995) Ultraviolet Laser Technology and Applications. Academic Press, San Diego

Ennos AE (1966) Stress developed in optical film coatings. Appl. Optics 5: 51 – 61

Eva E, Mann K (1996) Calorimetric measurement of two-photon absorption and color center formation in UV–window materials. Appl. Physics A62: 143 - 147

Eva E, Mann K, Kaiser N, Anton B, Henking R, Ristau D, Weißbrodt P, Mademann D, Raupach L, Hacker E (1996) Laser conditioning of LaF_3/MgF_2 dielectric coatings at 248 nm. Appl. Optics 35: 5613 - 5619

Gloede C, Kaiser N, Müller H, Mattheis R (1987) UV losses in MgF_2 thin films induced by the evaporation boat. Thin Solid Films 152: L139 – L143

Graper EB (1971) Evaporation characteristics of materials from an electron-beam gun. J. Vac. Sci.Technol. 8: 333 – 337

Günther KH, Loo B, Burns D, Edgell J, Windham D, Müller KH (1989) Microstructure analysis of thin films deposited by reactive evaporation and by reactive ion plating. J. Vac. Sci. Technol. A7: 1436 - 1445

Hass G, Ramsey JB, Thun R (1959) Optical properties of various evaporated rare earth oxides and fluorides. J. Opt. Soc. Am. 49: 116 – 120

Hass G, Waylonis JE (1961) Optical constants and reflectance and transmittance of evaporated aluminium in the visible and ultraviolet. J. Opt. Soc. Am. 51: 719 – 722

Heber J, Thielsch R, Blaschke H, Kaiser N, Mann KR, Eva E, Leinhos U, Görtler A (1998) Stability of optical interference coatings exposed to low fluence 193 nm ArF radiation. In: van den Hove L (ed) Optical Microlithography XI. Proc. SPIE 3334, pp. 1041 – 1047

Heber J, Thielsch R, Blaschke H, Kaiser N, Leinhos U, Görtler A (1999) Changes in optical interference coatings exposed to 193nm excimer laser radiation. In: Exarhos GJ, Guenther AH, Kozlowski MR, Lewis KL, Soileau MJ (eds) Laser-Induced Damage in Optical Materials 1998. Proc. SPIE 3578, pp. 83 – 96

Kaiser N (1994) Resistance of coated optics to UV laser irradiation. In: Abeles F (ed) Optical Interference Coatings. Proc. SPIE 2253, pp. 722 – 730

Kaiser U, Kaiser N (1994) C-adsorption behaviour of thin fluoride films. Thin Solid Films 237: 250 – 254

Kaiser U, Kaiser N, Weißbrodt P, Mademann D, Hacker E (1991) Initial layer growth in thin MgF_2-, LaF_3-, CaF_2-, and LiF-films. Le vide le coches mines, suppl. 259 : 104 – 106

Kaiser U, Kaiser N, Weißbrodt P, Mademann D, Hacker E, Mueller H (1992) Structure of thin fluoride films deposited on amorphous substrates. Thin Solid Films 217: 7 – 16

Kaiser N, Uhlig H, Schallenberg UB, Anton B, Kaiser U, Mann K, Eva E (1995) High damage threshold Al_2O_3/SiO_2 dielectric coatings for excimer lasers. Thin Solid Films 260: 86 – 92

Kolbe J, Kessler H, Hofmann T, Meyer F, Schink H, Ristau D (1992) Optical properties and damage thresholds of dielectric UV/VUV – coatings deposited by conventional evaporation, IAD and IBS. In: Bennett HE, Chase LL, Guenther AH, Newnam BE, Soileau MJ (eds) Laser-Induced Damage in Optical Materials: 1991. Proc. SPIE 1624, pp. 221 - 235

Krajnovich DJ, Kulkarni M, Leung W, Tam AC, Spool A, York B (1992) Testing of the durability of single-crystal calcium fluoride with and without antireflection coatings for use with high-power KrF excimer lasers. Appl. Optics 31: 6062 – 6075

Kunz RR, Liberman V, Downs DK (2000) Experimentation and modelling of organic photo contamination on lithographic optics. J. Vac. Sci. Technol. B18: 1306 – 1313

Lehan JP, Mao Y, Bovard BG, Macleod HA (1991) Optical and microstructural properties of HfO_2 thin films. Thin Solid Films 203: 227 – 250

Liberman V, Rothschild M, Sedlacek JHC, Uttaro RS, Bates AK, Van Peski C (1998) Marathon testing of optical materials for 193-nm lithographic applications. In: Exarhos GJ, Guenther AH, Kozlowski MR, Lewis KL, Soileau MJ (eds) Laser-Induced Damage in Optical Materials 1998. Proc. SPIE 3578, pp. 2 – 15

Liberman V, Bloomstein TM, Rothschild M, Sedlacek JHC, Uttaro RS, Bates AK, Van Peski C, Orvek K (1999) Materials issues for optical components and photomasks in 157 nm lithography. J. Vac.Sci.Technol. B17: 3273 - 3279

Liberman V, Rothschild M, Sedlacek JHC, Uttaro RS, Bates AK, Orvek K (2000) Long-term testing of optical components for 157-nm lithography. In: Progler CJ (ed) Optical Microlithography XIII. Proc. SPIE 4000, pp. 488 – 495

Liberman V, Rothschild M, Efremow NN, Palmacci ST, Sedlacek JHC, Van Peski C, Orvek K (2001) Marathon evaluation of optical materials for 157-nm lithography. In: Progler CJ (ed) Optical Microlithography XIV. Proc. SPIE 4346, pp. 45 – 51

Macleod AH (1986a) Thin-film optical filters. Adam Hilger Ltd, Bristol, sec. Ed. , pp.71 - 136

Macleod AH (1986b) Thin-film optical filters. Adam Hilger Ltd, Bristol, sec. Ed. , pp. 159 - 187

Mann K, Granitza B, Eva E (1997) Multiple-pulse damage thresholds of optical components for excimer lasers. In: Bennett HE, Guenther AH, Kozlowski MR, Newnam BE, Soileau MJ (eds) Laser-Induced Damage in Optical Materials: 1996. Proc. SPIE 2966, pp. 496-504

Martin PJ, Macleod HA, Netterfield RP, Pacey CG, Sainty WG (1983) Ion-beam-assisted deposition of thin films. Appl. Optics 22: 178 – 184

Matl K, Klug W, Zöller A (1991) Ion-assisted deposition with a new plasma source. Mater. Sci. Eng. A140, pp. 523 – 527

Newnam BE, Foltyn SR, Jolin LJ, Carniglia CK (1983) Multiple-shot ultraviolet laser damage resistance of non-quarterwave reflector designs for 248 nm. In: Bennett HE, Guenther AH, Milam D, Newnam BE (eds) Damage in Laser Materials: 1981. Nat. Bur.Stand. (U.S.) Spec. Publ. 638, pp. 363 - 379

Niisaka S, Otani M, Biro R, Ouchi C, Hasegawa M, Suzuki Y, Saito T, Saito J, Tanaka A, Sone K, Matsumoto A (2001) Development of optical coatings for 157 nm lithography (I: Coating materials). OSA Technical Digest Optical Interference Coatings 2001, pp ThB7-1 - 3

Otani M, Niisaka S, Biro R, Suzuki Y, Saito T, Saito J, Tanaka A, Ouchi C, Hasegawa M, Sone K, Matsumoto A (2001) Development of optical coatings for 157 nm lithography (II: Multilayer coatings). OSA Technical Digest Optical Interference Coatings 2001, pp ThB8-1 – 3

Pulker HK (1984) Coatings on Glass. Elsevier, Amsterdam 1984

Quesnel E, Rolland B, Muffato V, Labroche D, Robic JY (1997) The ion beam sputtering: A good way to improve the mechanical properties of fluoride coatings. Proc. 40th Annual Technical Conference Proceedings Society of Vacuum Coaters, pp. 293 - 298

Quesnel E, Dumas L, Jacob D, Peiró F (2000) Optical and microstructural properties of MgF_2 UV coatings grown by ion beam sputtering process. J. Vac. Sci. Technol. A18 : 2869 – 2876

Rainer F, Lowdermilk WH, Milam D, Tuttle Hart T, Lichtenstein TL, Carniglia CK (1982) Scandium oxide coatings for high-power UV laser applications. Appl. Optics 21: 3685 – 3688

Rainer F, Lowdermilk WH, Milam D, Carniglia CK, Tuttle Hart T, Lichtenstein TL (1985) Materials for optical coatings in the ultraviolet. Appl. Optics 24: 496 – 500

Ristau D, Günster S, Bosch S, Duparré A, Masetti E, Ferré-Borrull J, Kiriakidis G, Peiro F, Quesnel E, Tikhonravov A (2001) UV-optical and microstructural properties of MgF_2 - and LaF_3 – coatings deposited by IBS and PVD processes. OSA Technical Digest Optical Interference Coatings 2001, pp. ThA5-1 - 3

Ritter E (1976) Optical thin film materials and their applications. Appl. Optics 15: 2318 – 2327

Rosenfeld D, Sanjines R, Lévy F, Buffat PA, Demarne V, Grisel A (1994) Structural and morphological characterization of Nb_2O_5 thin films deposited by reactive sputtering. J. Vac. Sci. Technol. A12: 135 - 139

Rothschild M (1998) Photolithography below 200 nm. In: Dubowski JJ, Dyer PE (eds) Laser Applications in Microelectronic and Optoelectronic Manufacturing III. Proc. SPIE 3274 222 –225

Rothschild M, Ehrlich DJ (1988) A review of excimer laser projection lithography. J. Vac. Sci. Technol. B 6: 1 – 17

Rothschild M, Forte AR, Kunz RR, Palmateer SC, Sedlacek JHC (1997) Lithography at a wavelength of 193 nm. IBM J. Res. Develop. 41, pp. 49 - 54

Rothschild M, Bloomstein TM, Curtin JE, Downs DK, Fedynyshyn TH, Hardy DE, Kunz RR, Liberman V, Sedlacek JHC, Uttaro RS (1999) 157 nm: Deepest deep-ultraviolet yet. J. Vac. Sci Technol. B17: 3262 – 3266

Schink H, Kolbe J, Zimmermann F, Ristau D, Welling H (1991) Reactive Ion-beam-sputtering of fluoride coatings for the UV/VUV range. In: Bennett HE, Chase LL, Guenther AH, Newnam BE, Soileau MJ (eds) Laser-induced Damage in Optical Materials: 1990. Proc. SPIE 1441, pp. 327 - 338

Smith D, Baumeister P (1979) Refractive index of some oxide and fluoride coating materials. Appl. Optics 18: 111 – 115

Sites JR, Gilstrap P, Rujkorakarn R (1983) Ion beam sputter deposition of optical coatings. Opt. Eng. 22: 447 – 449

Stolz CJ, Sheehan LM, Maricle SM, Schwartz S, Kozlowski MR, Jennings RT, Hue J (1998) Laser conditioning methods of hafnia silica multilayer mirrors. In: Basu S (ed) High-Power Lasers, Proc. SPIE 3264, pp. 105 -112

Targove JD, Messerly MJ, Lehan JP, Weng CC, Potoff RH, Macleod HA (1986) Ion-assisted deposition of fluorides. In: Seddon RI (ed) Optical Thin Films II: New Developments. Proc. SPIE 678, pp. 115 – 122

Targove DJ, Lehan JP, Lingg LJ, Macleod HA, Leavitt JA, McIntyre LC (1987) Ion-assisted deposition of lanthanum fluoride thin films. Appl. Optics 26: 3733 – 3737

Thelen A (1988) Design of optical interference coatings. McGraw-Hill Book Company, New York, 1988

Thielsch R, Heber J, Jakobs S, Kaiser N, Duparré A (1998) Optical, structural, and mechanical properties of lanthanide trifluoride thin film materials for use in the DUV-spectral region. OSA Technical Digest Series 9 , pp. 116 - 118

Thielsch R, Heber J, Kaiser N, Martin S, Welsch E (2000) Critical issues on the assessment of laser induced damage thresholds of fluoride multilayer coatings at 193 nm. In: Exarhos GJ, Guenther AH, Kozlowski MR, Lewis KL, Soileau MJ (eds) Laser-Induced Damage in Optical Materials 1999. Proc. SPIE 3902,pp: 224 - 234,

Thielsch R, Gatto A, Heber J, Kaiser N (2002) A comparative study of the UV–optical and structural properties of SiO_2, Al_2O_3, and HfO_2 single layers deposited by reactive evaporation, ion assisted deposition and plasma ion assisted deposition. Thin Solid Films 410: 86 - 93

Tuttle Hart T, Lichtenstein TL, Carniglia CK (1983) Effects of Undercoats and Overcoats on Damage Thresholds of 248 nm coatings. In: Bennett HE, Guenther AH, Milam D, Newnam BE (eds) Damage in Laser Materials: 1981. Nat. Bur. Stand. (U.S.) Spec.Publ. 638, pp. 344 - 349

Waldorf AJ, Dobrowolski JA, Sullivan BT, Plante LM (1993) Optical coatings deposited by reactive ion plating. Appl. Optics 32: 5583 – 5593

Welsch E (1995) Absorption Measurements. In: Hummel RE, Guenther KH (eds) Handbook of Optical Properties Vol I Thin Films for Optical Coatings. CRC Press, Boca Raton, pp 243 - 272

Zukic M, Torr DG (1995) Extreme and far–ultraviolet narrowband and broadband filters. In: Hummel RE, Guenther KH (eds) Handbook of Optical Properties Vol I Thin Films for Optical Coatings. CRC Press, Boca Raton, pp 79 – 104

Zukic M, Torr DG, Spann JF, Torr MR (1990) Vacuum ultraviolet thin films. 2: Vacuum ultraviolet all-dielectric narrowband filters. Appl. Optics 29: 4293 – 4302

Abbreviations and Symbols

UV	ultraviolet
excimer	excited dimers
DUV	deep ultraviolet, spectral range between 200 nm and 280 nm
VUV	vacuum ultraviolet, spectral range between 100 nm and 200 nm
AOI	angle of incidence
L	low refractive index layer of quarter-wave optical thickness
H	high refractive index layer of quarter-wave optical thickness
QW	quarter-wave
IAD	ion assisted deposition
PIAD	plasma-ion assisted deposition
IBS	ion beam sputtering
IP	ion plating
MS	magnetron sputtering
substr. temp.	substrate temperature
compr.	compressive to assign compressive stress in a coating
ext. coefficient	extinction coefficient

a/o.	and other
(1-on-1) test mode	see Chap. "Laser Resistant Coatings"
(N-on-1) test mode	see Chap. "Laser Resistant Coatings"
(R-on-1) test mode	see Chap. "Laser Resistant Coatings"
LIDT(1-on-1)	laser induced damage threshold, determined in the (1-on-1) test mode
LIDT (1-on-1,20 ns, 193 nm)	laser induced damage threshold, determined in the (1-on-1) test mode, laser pulse length of 20 ns, wavelength of the laser
AR	antireflection coating
HR	high reflective coating
HR 0°	high reflective coating for 0° AOI
HR 45°	high reflective coating for 45° AOI

$N(\lambda) = n(\lambda) + ik(\lambda)$	complex refractive index
$n(\lambda)$	refractive index
$k(\lambda)$	extinction coefficient
λ	wavelength
λ_0	design wavelength of an optical multi-layer stack

Multilayer Coatings for EUV/Soft X-ray Mirrors

Sergey Yulin

Fraunhofer Institute for Applied Optics and Precision Engineering, Jena, Germany
Technical University "Kharkov Polytechnic Institute" (KhPI), Kharkov, Ukraine

1 Introduction

The demand to enhance optical resolution has pushed the search for technological innovations and improvements. Induced mainly by this aim, optics development in recent years has been focused on components for electromagnetic radiation with increasingly shorter wavelengths λ (Feigl et al. 2002). The good prospects of extreme ultraviolet radiation and soft X-rays (EUV/Soft X-rays) to be applied in next generation lithography systems (λ = 13.5 nm), microscopy in the "water window" (λ = 2.3 – 4.4 nm), astronomy (λ = 5 – 31 nm), spectroscopy, plasma diagnostics and EUV/Soft X-ray laser research have led to considerable progress in the development of different sources and optics for this spectral range.

However, the ability to use EUV/Soft X-rays (1 - 40 nm) in a constructive way with various optics used in longer wavelength ranges has been limited by the fundamental physics of matter - radiation interaction. There are two reasons for this. First, EUV/Soft X-rays are strongly absorbed by all materials and have a maximum penetration depth less than a few microns. Second, the refractive index of all materials approaches unity, resulting in very small near-normal reflectivity from any single boundary. Therefore, reflectivity enhancement is only possible by using the constructive interference of a large number of partial waves reflected from many boundaries of an artificial multilayer coating. Multilayer coatings for the EUV/Soft-X-ray spectral range can be imagined as an extension of natural crystals towards artificial structures with larger lattice spacing, or as an extension of interference coatings for the visible light towards shorter wavelengths and smaller layer thicknesses (Spiller 1994). This chapter summarizes progress and the present knowledge in preparation and characterization of multilayer coatings for the EUV/Soft X-ray range with regard to different designs, minimization of structure imperfections and thermal stability.

2 Basics of EUV/Soft X-ray Mirrors

2.1 Refractive Index in the EUV/Soft X-ray Region

The interaction between electromagnetic radiation and media is commonly described in terms of the complex refractive index $\tilde{n} = n + i\beta$, which is a function of the wavelength λ (James, 1962). The real (n) and imaginary (β) parts of $\tilde{n}$ are termed the refractive index and the extinction coefficient, respectively. The propagation of a monochromatic electromagnetic wave with a wavelength λ through isotropic and homogeneous medium in the x - direction is described by

$$E(x) = E_0 \exp(-\frac{2\pi\beta x}{\lambda}) \exp\left(\frac{2\pi i}{\lambda}(nx - ct)\right) \tag{1}$$

where E_o is the field amplitude at x = 0 and c is the speed of light in vacuum.

In the EUV/Soft-X-ray region, the refractive index n of all materials is less than unity by a small amount δ ($n = 1 - \delta$) and characterizes the refractive properties of a medium or the phase speed of the wave. The extinction coefficient β describes the absorption of the medium and is directly related to the linear absorption coefficient $\mu = 4\pi\beta/\lambda$ (the distance in which the intensity of the wave falls to 1/e times its initial value). Usually, n (δ) and β are referred to as the optical constants of the medium. In the EUV/Soft-X-ray region, the complex refractive index can be expressed by the complex atomic scattering factor $f = (f_1 - if_2)$ (Henke 1982) with

$$\tilde{n} = n + i\beta = 1 - \delta + i\beta = 1 - \frac{N_a r_e \lambda^2}{2\pi}(f_1 - if_2) \tag{2}$$

where N_a is the number of atoms per unit volume, $r_e = 2.82 \times 10^{-15}$ m is the classical radius of an electron and λ is the wavelength.

The atomic scattering factors f_1 and f_2 of all materials, for photon energies from 50 eV to 30 keV (0.04 - 25 nm), have been tabulated by Henke (Henke et al. 1993). Figure 1 shows the optical constants and atomic scattering factors as a function of wavelength for scandium. The difference in the behaviors of (f_1, f_2) and (δ, β) is due to the λ^2-factor in Eq. 2. In the EUV/Soft X-ray region all materials become more absorbing and more refracting with increasing wavelength and δ and β is in the range from 10^{-2} to 10^{-6}. The most prominent feature in Fig. 1 is the abrupt variation in the optical constants (δ and β) near the Sc-K and Sc-L absorption edges ($\lambda_{Sc\text{-}K} \cong 0.28$ nm and $\lambda_{Sc\text{-}L} \cong 3.1$ nm). Such ranges are called "anomalous dispersion regions".

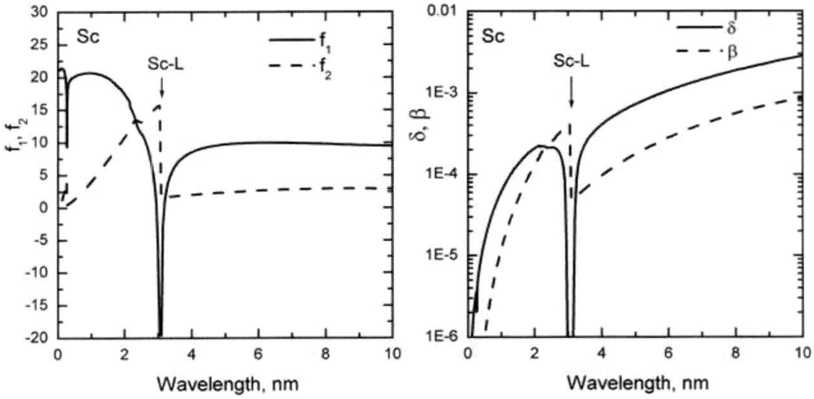

Fig. 1. Atomic scattering factors f_1 and f_2 (left), and optical constants δ and β (right) for scandium as function of wavelength.

2.2 Reflectivity of a Single Boundary

The amplitude reflection coefficients r_s and r_p for s- and p-polarization, respectively, at the boundary of two materials with complex refractive indices $\tilde{n}_1$ and $\tilde{n}_2$ are obtained from the well-known Fresnel equations and are given by

$$r_s = \frac{\tilde{n}_1 \cos\alpha_1 - \tilde{n}_2 \cos\alpha_2}{\tilde{n}_1 \cos\alpha_1 + \tilde{n}_2 \cos\alpha_2} \qquad \text{s- polarization} \qquad (3)$$

$$r_p = \frac{\tilde{n}_1 \cos\alpha_2 - \tilde{n}_2 \cos\alpha_1}{\tilde{n}_1 \cos\alpha_2 + \tilde{n}_2 \cos\alpha_1} \qquad \text{p- polarization} \qquad (4)$$

where α_1 and α_2 are the angles of incidence and refraction measured from the normal. Using Snell's law ($\tilde{n}_1 \sin\alpha_1 = \tilde{n}_2 \sin\alpha_2$) we can calculate the angle of the deflected beam inside the second medium as $\cos\alpha_2 = \sqrt{1-(\tilde{n}_1/\tilde{n}_2)^2 \sin^2\alpha_1}$. For normal incidence, the same values for the s- and p-amplitude reflection coefficients $r_\perp$ result from Eqs. 3 and 4:

$$r_\perp = \frac{\tilde{n}_1 - \tilde{n}_2}{\tilde{n}_1 + \tilde{n}_2} \qquad (5)$$

The reflectivity of the boundary between vacuum (n = 1) and a material with complex refractive index $\tilde{n} = 1 - \delta + i\beta$ (where $\delta, \beta \ll 1$) becomes

$$R_{\perp} = \left|r_{\perp}\right|^2 = \frac{\delta^2 + \beta^2}{(2-\delta)^2 + \beta^2} \approx \frac{\delta^2 + \beta^2}{4} \tag{6}$$

If the incidence medium is not a vacuum, Eq. 6 has to be rewritten as

$$R_{\perp} \approx \frac{(\Delta\delta)^2 + (\Delta\beta)^2}{4} \approx \frac{(\Delta n)^2 + (\Delta\beta)^2}{4} \tag{7}$$

where Δn and $\Delta\beta$ are the differences in the optical constants of the two materials. It is important to note that both parts of the refractive indexes in Eq. 7 contribute to the reflectivity; therefore, materials with strong absorption and refractive properties are superior reflectors in the EUV/Soft X-ray range. However the magnitudes of δ and β are very small for all available materials (Fig. 1), and the normal incidence reflectivity of a single boundary between any two materials is extremely small. The combination of the small magnitude of the normal incidence reflectivity with strong absorbsion prohibits the use of conventional transmission optics for exploitation of EUV/Soft X-rays.

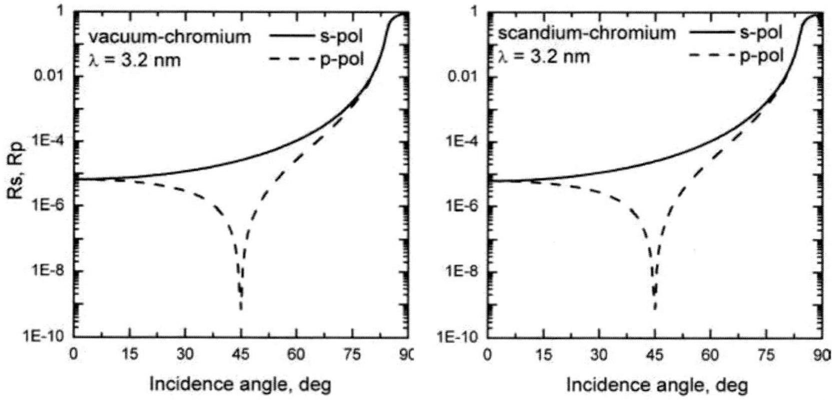

Fig. 2. Calculated reflectivities of vacuum-chromium (left) and scandium-chromium (right) boundaries at $\lambda = 3.2$ nm as functions of the incidence angle α for s- and p-polarizations.

As indicated by the Fresnel formula (Eqs. 3 and 4), a higher reflectivity can be achieved by using grazing (glancing) angles Θ ($\Theta = 90^{\circ} - \alpha$). Figure 2 shows the s - and p-reflectivities ($R_s = |r_s|^2$ and $R_s = |r_s|^2$) at λ=3.2 nm as functions of the incidence angle for the vacuum-chromium and scandium-chromium boundaries. The reflective properties of the boundaries are similar because for the wavelength λ=3.2 nm, just above the Sc-absorption edge ($\lambda_{Sc\text{-}L}$= 3.1 nm), the values δ and β are extremely small (Fig. 1) and the refractive indexes of scandium and vacuum ($n = 1$) are comparable. This behavior feature of optical constants near absorption edges is very important for the design of high-reflectance multilayer mirrors (to see 3.2). Two interesting phenomena can be observed away from normal incidence: the reflected p-polarized wave exhibits a strong minimum at about $\alpha = 45^o$ and at grazing incidence a reflectivity of almost 1 can be achieved. The first phenomenon - the R_s reflectivity increases monotonically from normal to grazing incidence, whereas R_p shows an abrupt dip – is called the polarization phenomenon. The R_p minimum occurs at Brewster's angle θ_B where the refracted and the reflected beams become perpendicular. Because the refractive index of all materials in the EUV/Soft-X-ray range is very close to unity, the minimum of R_p always appears close to 45° incidence ($\theta_B = \tan^{-1}(n_2/n_1)$). The second phenomenon, called "total reflection", occurs (according to the Snell's law) if the radiation penetrates from the high to the low refractive index material ($n_1 > n_2$). For grazing angles θ below the critical angle θ_C, the refracted beam becomes parallel to the interface and the reflectivity reaches 1. For the boundary between vacuum and any material (with $\delta/\beta >>1$), the critical angle can be calculated using Snell's law:

$$\theta_C = \sqrt{2\delta}\ . \tag{8}$$

Except for very small grazing angles ($\theta \le \theta_C$), the reflectivity from a single boundary is very low in the EUV/Soft X-ray range and the main task of multilayer coatings is to enhance the reflective properties of coatings for angles $\theta > \theta_C$ by the use of constructive interference of a large number of partial waves reflected from many boundaries. Such multilayer coatings consist of alternating layers of two materials with different refractive indexes ($\tilde{n}_1$ and $\tilde{n}_2$). Usually the two materials have a large difference in atomic number (Z) in order to maximize the difference in electron density (Eq. 2) and provide maximum reflectivity on one boundary (Eq. 7). The high-Z and low-Z (or low-density) materials are called absorber and spacer, respectively. The period of the multilayer structure H is the thickness sum of absorber and spacer layers ($H = d_A+d_S$). Figure 3 shows a schematic view and a cross-section transmission electron microscopic (XTEM) micrograph of a real Mo_2C/Si multilayer coating deposited by magnetron sputtering (Yulin et al. 2001).

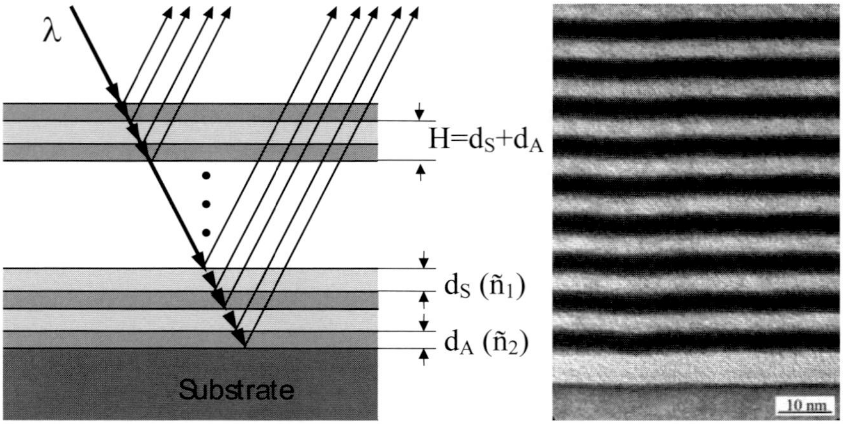

Fig. 3. Sketch of a multilayer structure (left) and XTEM micrograph of a Mo_2C/Si multilayer mirror with a period $H = d_{Si}+d_{Mo2C} = 7$ nm (right).

2.3 Reflectivity of a Multilayer Stack

The quarter-wave stack and the ideal Bragg crystal are two well-known geometrical designs using interference phenomena to obtain high reflectivity (Chakraborty 1991). For the case when absorption can be neglected for both materials (dielectric materials in the visible range), the quarter wave stack can provide maximum reflectivity with a minimum number of layers. In the EUV/Soft X-ray range, absorption is always present and some energy is dissipated inside the multilayer stack; therefore the ideal quarter-wave stack design cannot be as effective as it is in the visible range. In the ideal Bragg crystal design, the atomic planes are spaced $\lambda/2$ apart (for normal incidence) and contribute to the reflected wave. This design can give a reflectivity of almost 1 for a very large number of layers even if the atomic planes are absorbing, provided that the space between the atomic planes is absorption-free. In this case the minimization of the absorption losses is due to the fact that the atomic planes are located at the nodes of the standing wave generated by the superposition of the incoming and reflected waves. Spiller (Spiller 1972) was the first who showed that the standing wave-field within a coating could be used to minimize the absorption effect. Therefore, the basic idea in the design of multilayer coatings for EUV/Soft X-rays is to place absorbing material at the nodes of the standing-wave field and to fill the remaining space with low absorption material (Spiller 1994).

The reflectivity of any given multilayer system can be calculated by different methods (Chap. "Design of Optical Coatings" by Alexander Tikhonravov) such as the recursive method (Rouard 1937), applied successively to each following interface, or a matrix formalism (Abeles 1950), where each interface is represented by a matrix and a multilayer stack by a product of matrices. The recursive and matrix methods are both equivalent to the dynamic theory of X-ray diffraction on crystal planes (Bragg 1914). The calculation of reflectivity by the recursive method starts with a calculation of the amplitude reflection coefficients from a single film. For a single film with thickness d, the Fresnel amplitude reflection coefficients of the top rt and the bottom rb boundaries are calculated using Eqs. 3 and 4, and the amplitude reflection coefficient of the film r_f is given by

$$r_f = \frac{r_t + r_b \exp(2i\psi)}{1 + r_t r_b \exp(2i\psi)} \tag{9}$$

where the phase shift ψ represents the phase delay produced by propagation of the wave through film thickness d. For a propagation angle α_n, ψ is given by

$$\psi = \frac{2\pi d\tilde{n} \cos \alpha_n}{\lambda}. \tag{10}$$

A recursive application of Eq. 9 yields the reflectivity of any multilayer system if the results of each sublayer calculation are used for calculating the next layer (by replacing r_b in Eq. 9 by the value of r_f). The maximum reflectivity of a multilayer system is reached under the Bragg condition if wave propagation through one multilayer period H (Fig. 3) results in a phase shift equal to 2π (Eq. 10):

$$m\lambda = 2H \sin \Theta_m \left(1 - \frac{2\tilde{\delta} - \tilde{\delta}^2}{\sin^2 \Theta_m}\right)^{1/2}. \tag{11}$$

θ_m is the grazing angle of the maximum of reflectivity with refraction correction (Rosenbluth and Lee 1982) and m is the order of reflection. The value $\tilde{\delta}$ in Eq. 11 is the weighted average difference of the refractive index from unity for the two materials in the multilayer structure,

$$\tilde{\delta} = \frac{\delta_A d_A + \delta_S d_S}{d_A + d_S}, \tag{12}$$

and is directly related to the thickness ratio Γ (typical range is: $0.3 \leq \Gamma \leq 0.5$) between absorber layer thickness d_A and multilayer period H:

$$\Gamma = \frac{d_A}{d_A + d_S} = \frac{d_A}{H}. \tag{13}$$

3 Design of Multilayer Structures

3.1 Material Selection

The main aim in multilayer design is to enhance the reflectivity for EUV/Soft X - rays by use of the constructive interference effect. The design strategy and limits in the selection of suitable materials were estimated by the introduction of two parameters N_{min} and N_{max} (Spiller 1994). N_{min} is the minimum number of periods needed to achieve a reflectivity close to 1 without consideration of absorption losses, while N_{max} is the maximum number of periods that can be penetrated taking absorption into account, while still enhancing the reflectivity. By using Eq. 7, N_{min} can be estimated for normal incidence as

$$N_{min} = \frac{1}{2|r_{\perp}|} = \frac{1}{\sqrt{(\Delta n)^2 + (\Delta\beta)^2}} . \tag{14}$$

Assuming that the multilayer structure has a period $H=\lambda/2$ (Eq. 11 without refraction correction) and mainly consists of the spacer material, N_{max} can be written as

$$N_{max} = \frac{1}{2\pi\beta_S} \tag{15}$$

where β_S is the extinction coefficient of the spacer material.

The optical efficiency of chosen materials for multilayer mirror can be estimated by the ratio of N_{max}/N_{min}. For $N_{max}/N_{min}>>1$, absorption has little effect and high-reflectance multilayers can be obtained, whereas for $N_{max}/N_{min}<<1$, absorption prohibits enhancement of reflectivity by multilayer design. The spectral bandpass of the reflection peak $\Delta\lambda$ (full width at half maximum – FWHM) is limited by the required number of periods N_R contributing to the reflectivity,

$$\Delta\lambda \cong \frac{\lambda}{N_R} \tag{16}$$

and thus strongly depends on the thickness ratio Γ (Eq. 13).

3.2 High-Reflectance Mirrors

The optical performance of multilayer coatings is optimized by the selection of available materials. For high-reflectance coatings, the N_{max}/N_{min} ratio should be very high. According to this criterion, the following general strategy can be used in material selection:

- The spacer material should have the lowest extinction coefficient β_S at the desired wavelength (in this case N_{max} is maximal).

- The absorber material should provide a high reflectivity at the boundary with the spacer (Δn and Δβ have to be large for minimization of N_{min}), while still having an extinction coefficient β_A as low as possible.

These two criteria are basic and should always be used in material selection for high-reflectance multilayers. However, the layer thicknesses in EUV/Soft-X-ray multilayer structures are about two orders of magnitude thinner than those used in conventional optical coatings. Therefore the quality of the layers and their interfaces has a strong effect on the optical properties of mirrors, and very often has to be considered in material selection by an additional technological requirement:

- Ability to produce continuous layers with abrupt and smooth interfaces.

Table 1. Best spacers, most preferable absorber materials and theoretical limits for multilayer reflectivities for some EUV/Soft X-ray wavelength ranges.

Wavelengths nm	Spacer	Absorption edge, nm	Absorber materials	R_{max}, %
3.1 - 4.4	Sc	3.1	Cr, Fe, W	30 – 60
4.4 - 6.7	C	4.4	Co, Ni, Cr	30 – 55
6.7 - 8.5	B (B_4C)	6.7	La, Ru, Mo	55 – 80
8.5 – 11.3	Sr	-	Mo	35 – 70
11.3 -12.4	Be	11.3	Mo	70 – 80
12.4 – 17.5	Si	12.4	Mo, Mo_2C	50 – 77

The selection of the optimum materials for high-reflectance multilayer mirrors in the spectral range from 0.6 to 12.4 nm was first made by Rosenbluth (Rosenbluth 1988) using computer calculation with atomic scattering factors from Henke (Henke 1982). The main result of this calculation was that reflectivities of more than 40 % are achievable for multilayer mirrors in the EUV/Soft X-ray region and it is very critical for a high-reflectance design to have a suitable spacer material with low absorption at the desired wavelength. Due to the low absorption for wavelengths greater than the absorption edge (Fig. 1), the low-Z (light) elements that have absorption edges in the EUV/Soft X-ray spectral region are the most preferable candidates for the spacer materials. The selection of absorber materials is much less critical in multilayer design and similar performances can be obtained by using different high-Z materials (Kartright 1986; Stearns et al. 1991).

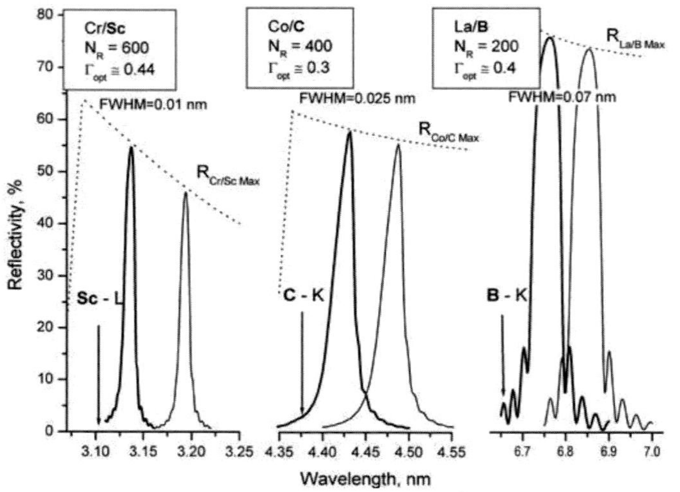

Fig. 4. Maximal normal incidence reflectivities of Sc, C and B based multilayers (dot line) and a few examples of optimized theoretical designs for ideal Cr/Sc, Co/C and La/B multilayers near the absorption edges of spacers.

The best spacers and most preferable absorber materials for some wavelength ranges and maximum achievable normal incidence reflectivities are shown in Table 1. Figure 4 shows some examples of optimized multilayer mirrors with Sc, C and B spacer materials in the 3.1 - to - 7 nm wavelength range. The maximum reflectivities for Cr/Sc, Co/C and La/B multilayer mirrors occur just above the Sc-L, C-K and B-K absorption edges, respectively. An increased distance from the absorption edges results in decreased maximal achievable reflectivities for every material pair. As shown in Fig. 4, the reduction of the required number of periods N_R (600 and 200 for $\lambda = 3.13$ nm and $\lambda = 6.75$ nm, respectively) and corresponding broadening of the reflectivity bandpass FWHM (0.01 and 0.07 nm for $\lambda = 3.13$ nm and $\lambda = 6.75$ nm, respectively) with increasing wavelength is a general feature of high-reflectance mirrors for the EUV/Soft X-ray region.

The optimum thickness ratio Γ_{opt} for maximum normal reflectivity of a periodic multilayer system with a large number of layers (infinite media) can be calculated by analytical formulas (Vinogradov and Zeldovich 1977),

$$\tan(\pi\Gamma_{opt}) = \pi\left[\Gamma_{opt} + \frac{\beta_S}{\beta_A - \beta_S}\right], \tag{17}$$

where β_S and β_A are the extinction coefficients for spacer and absorber materials, respectively.

3.3 Broadband Mirrors

New applications of EUV/Soft X-ray radiation in metrology, microscopy and astronomy make additional demands on the performance of multilayer coatings. A serious drawback of multilayer coatings concerning their application in EUV/Soft X-ray optics is their limited range of reflectivity in the spectral and angular ranges. First, the spectral FWHM of typically 0.01 - 1.0 nm (Fig. 4) covers only a small part of the output of EUV sources (Bergmann et al. 1999). Second, the fact that the reflectivity decreases significantly at incidence angles of more than a few degrees from the Bragg condition (Eq. 11) causes serious problems impeding the use of multilayer coatings on curved substrates.

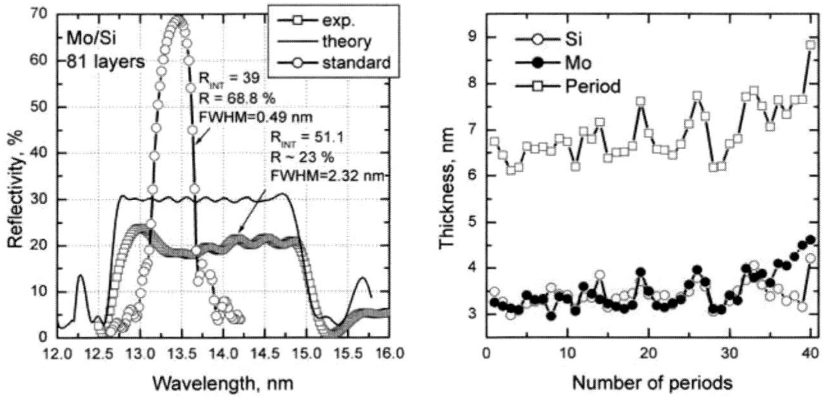

Fig. 5. Theoretical and experimental performance of a Mo/Si broadband mirror designed for the wavelength range from 13 nm to 15 nm (left) (a standard mirror with N=60 is shown for comparison) and corresponding distribution of layer thicknesses (right).

Broadband mirrors can be designed by a gradual variation of the multilayer period H through a multilayer stack (depth-graded multilayer). In this case, different wavelengths are reflected from different depths in the multilayer structure. The concept was first proposed for neutron optics under the name of supermirrors (Mezei 1988) and later developed for X-ray astrophysics (Joenson et al. 1993) and synchrotron radiation applications (Høghøj et al. 1997; Kozhevnikov et al. 2001). Current theoretical approaches to the designing of broadband EUV/Soft- X-ray mirrors are based on extensive computer calculations (Wang et al. 2000; Kuhlmann et al. 2002b) or semiempirical formulas (Joensen et al. 1993, 1997).

The stochastic design is based on a thickness optimization process for all layers inside the multilayer stack and is commonly used to obtain a broadband reflection in the EUV/Soft X-ray range. The optimized thickness distribution for 81 layers of Mo and Si and its theoretical and experimental performances are shown in Fig. 5. This Mo/Si multilayer structure was designed for the highest possible and constant reflectivity in the wavelength range from 13 nm to 15 nm. An advantage of the stochastic design is the possibility to provide constant reflectivity over wide spectral or angular ranges. The magnitude of integrated reflectivity R_{INT} is commonly used for quantitative estimation of the broadband mirror design.

A simpler way to design broadband mirrors is based on putting a few periodic multilayer stacks for different wavelengths on top of each other. The basic principle and the theoretical and experimental performances of a three-stack design (N_1=30 with H_1=7.22 nm, N_2=15 with H_2=6.90 nm and N_3=5 with H_3=6.10 nm) optimized for constant reflectivity in the angular range from 0° to 20° at the wavelength λ = 13 nm, are given in Fig. 6 (Kuhlmann et al. 2002b). The parameters of the multistack system are easily controlled by grazing incidence X-ray reflectometry, which is a genuine advantage of the multistack design.

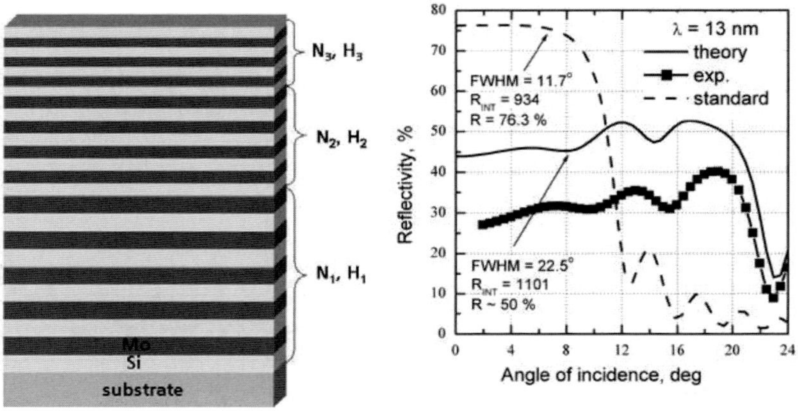

Fig. 6. Schematic design (left) and theoretical and experimental performance of a Mo/Si three-stack multilayer mirror optimized for 0° to 20° angle of incidence at λ= 13 nm (right).

3.4 Narrowband Mirrors

According to Eq. 16, the obvious way of narrowing the reflectivity bandpass of a multilayer mirror is to increase the required number of periods N_R. Therefore, the use of two low-Z materials (Slaughter et al. 1994a) or reduction of the optimal thickness ratio Γ_{opt} (Eq. 17) in a high-reflectance design (Lim et al. 2001) are two regular approaches to the design of narrowband mirrors. However, in the practical realization of both these approaches some difficulties have to be expected. In the first case, the amplitude reflection coefficient of the boundary between two low-Z materials is very small (Eq. 7) and for obtaining a high reflectivity it is necessary to deposit a greater number of layers. In the second case, the multilayer quality will be limited by the difficulty to prepare very thin (few Angstroms) continuous layers of high-Z material (especially if $\Gamma < 0.2$).

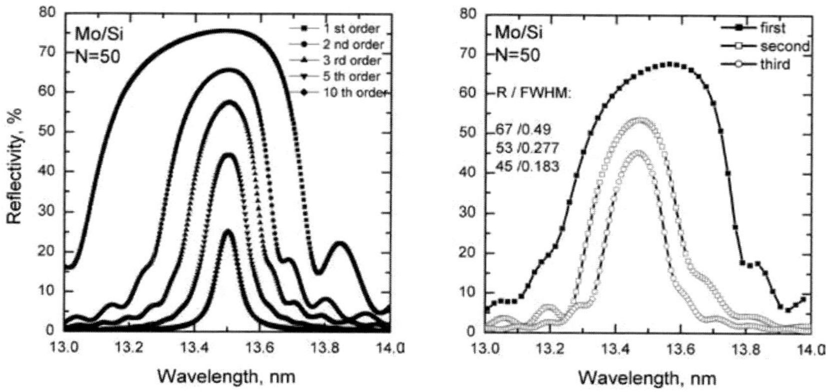

Fig. 7. Calculated (left) and measured (right) performance of Mo/Si mirrors designed for normal incidence at a wavelength of 13.5 nm as a function of reflectance order. Using the third reflectance order of a multilayer structure with period H=20.5 nm results in FWHM = 0.183 nm.

The reflectivity bandpass of any multilayer structure can be decreased by using high reflectance orders as shown in Fig. 7 for the example of a Mo/Si mirror. According to the Bragg condition (Eq. 11) for the application of high orders (m = 2, 3…), the multilayer period H has to be increased as shown in Table 2. The main advantage of this approach is the good combination of reflective and narrowband properties (R_1 = 67% with FWHM= 0.49 nm, R_2 = 53% with FWHM=0.277 nm and R_3 = 45% with FWHM= 0.183 nm for the first, second and third orders of reflectance, respectively), which can be obtained without increasing the number of layers or reducing the absorber layer thickness to a critical limit.

Table 2. Design of Mo/Si narrowband mirrors for λ=13.5 nm.

m	d_{Si}, nm	d_{Mo}, nm	Period H, nm	Γ	N	R_{max}, %	FWHM nm
1	3.92	3	6.92	0.43	50	75.6	0.631
2	10.7	3	13.7	0.22	50	65.7	0.327
3	17.5	3	20.5	0.17	50	57.4	0.220
5	31	3	34.0	0.096	50	44.4	0.137
10	64	3	67.8	0.046	50	25.2	0.080

4 Imperfections in Multilayer Mirrors

According to the theoretical predictions (Rosenbluth 1988) high normal incidence reflectivities in the EUV/Soft X-ray range can be achieved by using ideal multilayer mirrors. However, real multilayer mirrors contain structure imperfections and strongly deviate from the ideal structure (Chap. "Some Fundamentals of Optical Thin Films Growth" by Norbert Kaiser). As a result, the best experimental near-normal incidence reflectivities differ significantly from theoretical predictions (Fig. 8). The highest reflectivities (about 70 %) have been achieved near the Si-L absorption edge at λ=13.5 nm (Bajt et al. 2002a; Braun et al. 2002) and close to the Be-K absorption edge at λ=11.4 nm (Folta et al. 1999) with the material combinations Mo/Si and Mo/Be, respectively. The significant drop in reflectivity with decreasing wavelength is due to the increased effect of multilayer imperfections. Therefore, up to now the maximal experimental reflectivity of the best short-period Cr/Sc multilayer mirrors is about 15%.

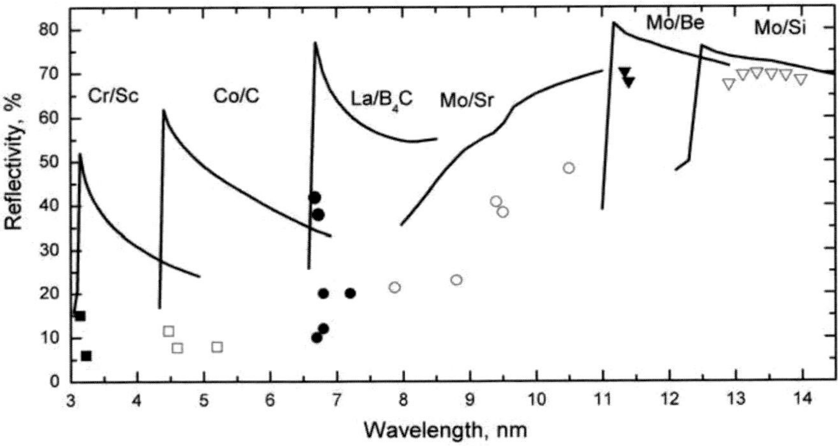

Fig. 8. Comparison of theoretical (lines) and experimental (symbols) performances for the best normal incidence multilayer mirrors in the EUV/X-ray wavelength region. Experimental results are from a database at: www-cxro.lbl.gov/multilayer/survey.html.

EUV/Soft X-ray mirrors are made of a great number of alternating layers having thicknesses thinner than conventional optical coatings by a factor of up to 100, and the required number of layers is much greater. Therefore, thickness irregularities of the multilayer stack are the first multilayer defects present in each real multilayer structure. It has been shown that such random thickness errors would result in a decrease in reflectivity and an increase in the bandwidth of the multilayer reflection (Loevezijn et al. 1996). The thickness irregularity can be minimized by in-situ thickness control during the deposition process (Spiller 1981).

It has been shown by Barbee and Spiller (Barbee 1986; Spiller 1986) that interfacial imperfections (interface roughness and diffuse intermixing) between constituent layers apparently are the primary structural defect. Their role is crucial to the optical performance of multilayer mirrors. The interface roughness of only a few Angstroms can significantly reduce the reflectivity through nonspecular scattering (Stearns and Gullikson 2000); therefore the optimization of multilayer performance requires abrupt and atomically smooth interface boundaries. Assuming that the reflectivity of one interface is very low (Fig. 2) and the interface roughness can be described by a Gaussian distribution (around the interface location), it is possible to show that the reduction of the interface reflectivity caused by roughness is an analogue to the Debye-Waller factor (DW=exp-$[2\pi(\sigma/\lambda)\sin\theta]^2$) (Debye 1913; Waller 1923). Thus as a first approximation the reduction of reflectivity R of a real multilayer can be described in terms of the ratio between the interface roughness σ and the multilayer period H:

$$R = R_0 \exp\left[-\left(2\pi m \frac{\sigma}{H}\right)^2\right]. \tag{18}$$

R_o is the reflectivity of an ideal structure ($\sigma = 0$) and m is the reflectance order. In additional, it has to be noted that although the term "roughness" originally means fluctuations of the interface, it can also be regarded as a intermixing at the interface resulting from interdiffusion, therefore the influence of the intermixing is also commonly treated with the same Debye-Waller formalism.

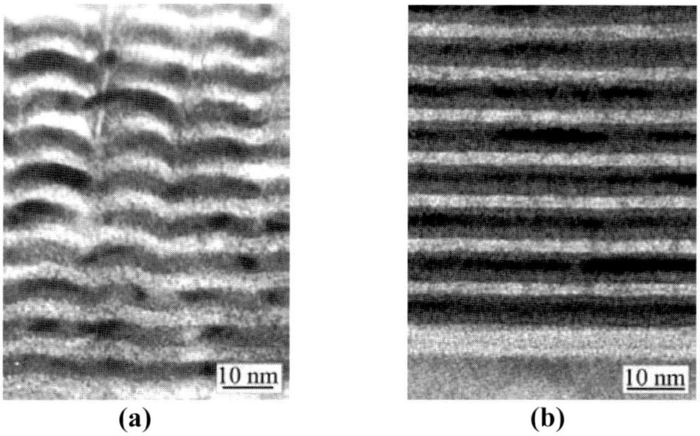

Fig. 9. XTEM micrograph of Mo/Si multilayer mirrors deposited by magnetron sputtering in Ar plasma at deposition pressures of 12 mTorr (a) and 2 mTorr (b). The bright and dark bands correspond to Si and Mo layers, respectively. Transition zones (gray) with mixed composition separate the pure Mo and Si layers (b). The film morphology exhibits a transition from columnar growth (a) at high pressure to continuous layers (b) at lower pressure.

Interface roughness and diffuse intermixing are two major interface imperfections. Both defects are shown in Fig. 9 by the XTEM micrographs of a Mo/Si multilayer structure prepared by dc magnetron sputtering at different pressures. Multilayer coatings sputtered at high pressure ($P > 6$ mTorr) or evaporated at low temperatures exhibit a column-like growth mode with the roughness increasing from the bottom to the top of the stack as shown in Fig. 9a. The degradation of the structural quality is a result of the low surface mobility of adatoms and is attributable to the low surface energy of growing layers during the deposition process

(Vernon et al. 1993). The optimization of deposition parameters provides a transition from a columnar (Fig. 9a) to a layer (Fig. 9b) interface morphology with a sufficient reduction of interface roughness.

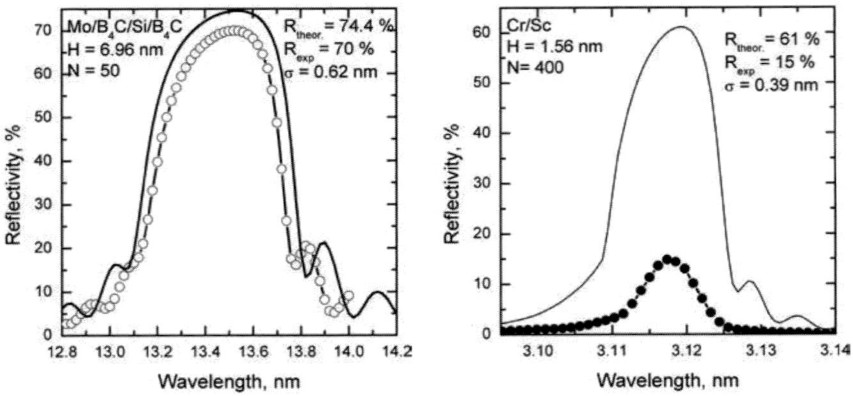

(a) (b)

Fig. 10. Theoretical and experimental performance of Mo/B_4C/Si/B_4C (a) and Cr/Sc (b) multilayer mirrors with periods H=6.96 nm and H=1.56 nm, respectively. Maximal near-normal reflectivities of 70% (by S. Bait and J. Alameda) and 15% were achieved. The drop in reflectivity from 61% to 15% for the short-period Cr/Sc mirror is a result of the increase in interface roughness σ from 0 to 3.9 nm.

However, even after optimization of the deposition process for the Mo/Si multilayer system, the XTEM micrograph indicates that, although the interfaces are relatively smooth, the pure Mo (dark) and Si (bright) layers are separated by transition regions of mixed composition (gray) (Fig. 9b). A number of studies have been performed to investigate the origin of Mo/Si intermixing and its dependence on the deposition parameters (Windt et al. 1992; Slaughter et al. 1994b; Bajt et al. 2001b; Yulin et al. 2002). As a result, the interlayer thicknesses have been minimized (down to 1.0 nm at the Mo-on-Si interface and 0.5 nm at the Si-on-Mo one) but not totally eliminated. This suggests that interlayer formation is an intrinsic property of the Mo/Si material pair and could present a fundamental barrier to achieving maximum reflectivity (Stearns and Rossen 1991). Therefore, further improvement in multilayer performance requires the complete suppression, or at least reduction, of interdiffusion processes on interfaces. This could be accomplished by the introduction of super-thin (0.3 - 0.5 nm) diffusion barriers between the absorber and spacer materials. This design strategy was recently used by several groups to improve the thermal stability of the multilayers, though always at the expense of reflectivity (Takenaka and Kawamura 1996; Feigl et al. 2001). Successful attempts to minimize the interdiffusion effect in Mo/Si multilayers for implementing a four-layer design were performed by different groups (Bajt et al. 2001a; Braun et al. 2002). A comparison of the theoretical and experimental performances for a Mo/B_4C/Si/B_4C multilayer mirror designed for near-normal incidence at the wavelength $\lambda \cong 13.5$ nm is shown in Fig. 10a.The maximum reflectivity of 70 % at wavelength λ=13.52 nm and at an incidence angle of 5° (the theoretical prediction for the ideal case is 74.4 %) was achieved through minimization of interdiffusion processes on interfaces by use of the four -layer interface-engineering multilayer design (Bajt et al. 2001a). The simulations of experimental data yielded an interface roughness of 0.62 nm.

As can be seen from Eq. 18, as the multilayer period H decreases the effect of the interface roughness σ on the multilayer reflectivity becomes dominating (Kozevnikov et al. 1994; Salashchenko and Shamov 1997; Kuhlmann et al. 2002a). For a Cr/Sc mirrors with period H=1.56 nm and number of periods N=400 a normal incidence reflectivity of 15 % was achieved experimentally (Fig. 10b). The model simulation predicts a drop in normal incidence reflectivity from R=61% to R=15 % if the interface roughness σ increases from 0 to only 0.39 nm. Due to a strong effect of interface imperfection on optical quality, the preparation of high-performance short-period mirrors remains a big challenge.

5 Deposition Methods

From the theoretical point of view, the maximum reflectivity of any periodic multilayer structure can be obtained by optimization of stack regularity, layer composition and interface structure. With these considerations multilayer deposition techniques have to meet the following basic requirements: excellent reproducibility of layer thickness inside the stack, maximum electron densitiy for high-Z and minimum density for low-Z materials and minimized interface imperfections (Chakraborty 1991). High-performance EUV/Soft X-ray mirrors have been successfully deposited with different physical deposition techniques including evaporation (Spiller et al. 1980; E. Louis et al. 1999), sputtering (Burkhart et al. 1999; Bajt et al. 2001a) and laser ablation (Mai and Pompe 1992). Of the many methods that have been applied in the past, magnetron sputtering and electron beam evaporation are the most widely used today (Chap "Thin Film Deposition Techniques" by Hans Pulker). The main differences between them are given in Table 3.

The important step in the development of electron beam evaporation techniques for preparation of high-performance EUV/Soft X-ray mirrors was thickness control improvement by in-situ monitoring of soft-X-ray reflectivity (λ = 4.44 nm) from the growing multilayer structure during the deposition (Spiller et al. 1980). It has been shown that interference maxima and minima observed during the deposition give a precise measure of the film thickness. The thickness errors can be compensated by a correction of the subsequent layers. In this monitoring system the thickness errors do not accumulate from the bottom to the top of the multilayer stack and can be minimized by proper selection of the monitoring wavelength. Additionally, the amplitude of reflectivity oscillations decreases with increasing roughness of the film surface and the development of roughness can be immediately recognized and minimized by optimization of the evaporation parameters. The biggest advantage of the magnetron sputtering method is the excellent stability of material rates over the deposition time due to good stabilization of the working pressure and power supply. In Barbee's sputtering system (Barbee 1986), the substrates are mounted on a horizontal rotating table and are moved past two separate sputter sources, a single layer being deposited on each pass by a source. Sputter sources use solid targets and can be operated in the deposition-up, deposition-down or vertical deposition mode.

Table 3. Comparison between magnetron sputtering and electron beam deposition methods for preparation of EUV/Soft X-ray multilayer mirrors.

Parameter	E-beam evaporation	Magnetron sputtering
Working pressure, Pa	10^{-4}	$(5\text{-}50)\cdot 10^{-2}$
Kinetic energy of deposited atoms, eV	~ 0.1	~ 10
Optimization of surface energy and interface performance by	Substrate temperature and ion polishing	Working pressure and ion beam assistance
Thickness control by	In-situ thickness monitoring	Time
Multilayer deposition on figured optics	Difficulty of thickness monitoring	Possible
Deposition of short-period multilayers with large number of periods (N>300)	Problematic	Possible
Deposition of compound materials	Problematic	Stoichiometries of target and coating are similar

For a large number of applications (EUV-lithography, microscopy, astrophysics) it is necessary to deposit high-reflectance coatings with good uniformity or a laterally-graded period over large-area flat or figured substrates. Therefore thickness distribution control for both uniform and graded coatings across curved surfaces up to 300 mm in diameter is the state-of-the-art for modern deposition systems. Magnetron sputtering (Folta et al. 1999; Braun et al. 2002) and e-beam (Louis et al. 2000) deposition systems have been developed in different labs to meet these requirements. Figure 11 shows a sputtering system (Soufli et al. 2001) that was recently developed in a partnership between the Lawrence Livermore National Laboratory and Veeco Instruments Inc. (Plainview, New York). The tool has a "sputter-up" geometry and five rectangular targets. During a multilayer deposition run the platter is rotated over the sources at speed ϕ while the substrate is simultaneously spun around its centre with a frequency of ω. In order to achieve the desired multilayer thickness profile, the platter speed ϕ is modulated when the substrate passes the targets. High-performance Mo/Si multilayer coatings with laterally graded period were successfully deposited on figured imaging optics for EUV lithography application by using this deposition technique (Soufli et al. 2001).

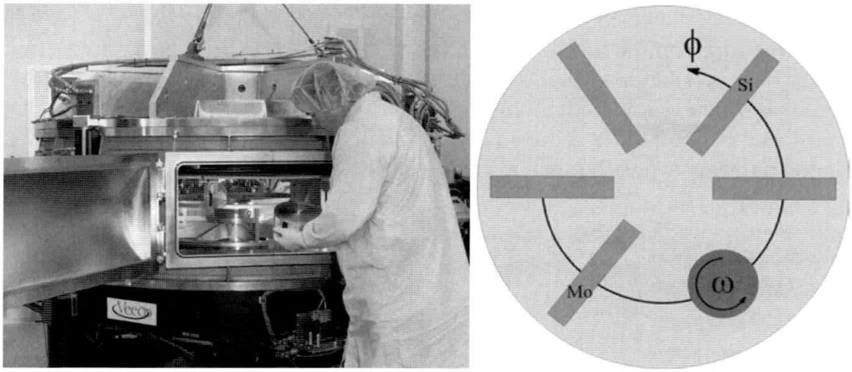

Fig. 11. General view of a sputtering tool (left) and scheme of the target-substrate arrangement inside the chamber (right). The three positions between the Mo and Si targets can be used for deposition of diffusion, buffer or capping layers (Soufli et al. 2001).

The new EUV sputter system (NESSY) of similar configuration was designed and created in the Fraunhofer Institut für Angewandte Optik und Feinmechanik (Jena). The system has a cylindrical vacuum chamber with a radius of more than 600 mm equipped with four rectangular targets (600×125 mm^2). The optimal geometry and size of NESSY allows coating three or two figured optics simultaneously with diameters of up to 300 mm or 400 mm. The thickness uniformity is better than ± 0.1% and ± 0.2% for figured optics with diameters of 150 mm and 300 mm, respectively. The extremely low base and working pressures (10^{-7} and 10^{-2} Pa) coupled with a precise load look system provide good chances for the achievement of an excellent optical performance of EUV/Soft-X-rays mirrors.

6 Mirrors with Enhanced Stability

EUV/Soft X-ray multilayer mirrors with abrupt interfaces between ultrathin layers of different materials usually represent nonequilibrium thermodynamic systems that tend to degrade with time and to undergo a gradual change of structure and optical characteristics. The instability of multilayers is especially dramatic at elevated temperatures and under the load of a power flux of photons and particles (Stearns et al. 1990; MacGowan et al. 1993). This feature of multilayers may limit the number of applications in many fields of science and technology such as plasma physics, astrophysics, X-ray lasers and EUV lithography.

The effects of high-energy photons and particles on multilayer coatings include heating to high temperatures, destruction of multilayers through diffusion intermixing of layers, restructuring of the layer structure, crystallization of amorphous layers and other structure and phase transformations (Rosen et al, 1993; Kopilets et al. 1996). The degradation of the internal structure of a 6.9 nm Mo/Si multilayer after annealing at 400°C for two hours and corresponding changes of reflective properties are shown in Fig. 12. The growth of Mo/Si interlayer regions with a corresponding contraction of the multilayer period (Fig. 12 a,b) by formation of higher density silicides Mo_XSi_Y from pure Mo and Si causes a significant reduction of resonance wavelength and reflectivity (Fig. 12 c). It has to be noted that multilayer irradiation by X-rays is accompanied by thermal heating as well as radiation damage; typically, their combined effect is stronger (Ziegler 1995).

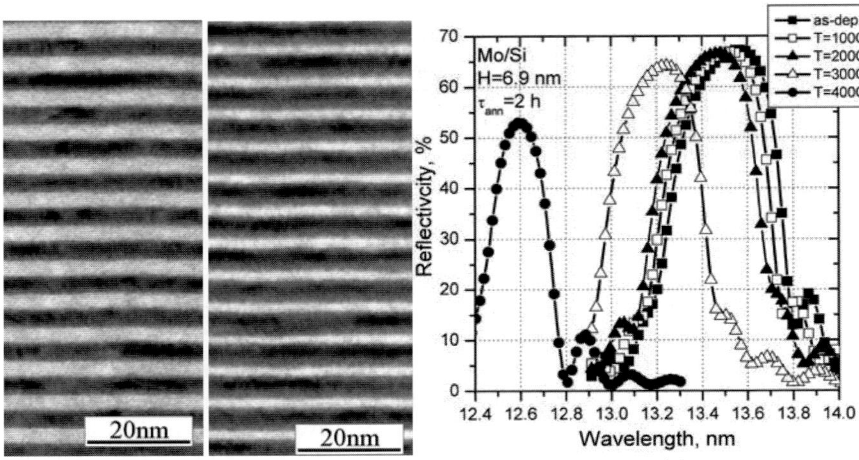

(a) **(b)** **(c)**

Fig. 12. XTEM micrographs of a 6.9 nm Mo/Si multilayer mirror in the as-deposited state (a) and after annealing at 400°C (b) for 2 hours. Annealing causes growth of the Mo/Si interlayer regions with a corresponding contraction of the multilayer period (formation of higher density silicides from pure Mo and Si). The change in the optical properties of the Mo/Si multilayer after annealing (c) agrees well with the structural degradation (a,b).

For prevention of interfacial reactions during the deposition process and under the influence of elevated temperatures two different approaches can be used. A kinetic approach to more stable multilayers implies a maximum increase in the activation energy of the processes taking place in a nonequilibrium system, so as to shift them to the highest possible temperature range. More stable multilayers can be achieved by proper selection of multilayer materials forming eutectics with a high melting temperature (Bugaev et al. 1995), by doping impurities in layers for stabilization of their amorphous structure or suppression of interdiffussion processes (Feigl et al. 2001; Yulin et al. 2001) or by insertion of diffusion barriers between layers (Takenaka and Kavamura 1996). Figure 13 shows the effect of annealing on the internal structure of a 6.9 nm Mo_2C/Si multilayer mirror (Fig. 13 a,b) and corresponding changes of reflective properties (Fig. 13 c). The replacement of molybdenum by molybdenum carbide (Mo_2C) and the transition from the Mo/Si to the Mo_2C/Si multilayer system causes a considerable suppression of interdiffussion processes between adjacent layers for annealing temperatures less than 400°C and a conservation of optical properties (Fig. 12 and 13).

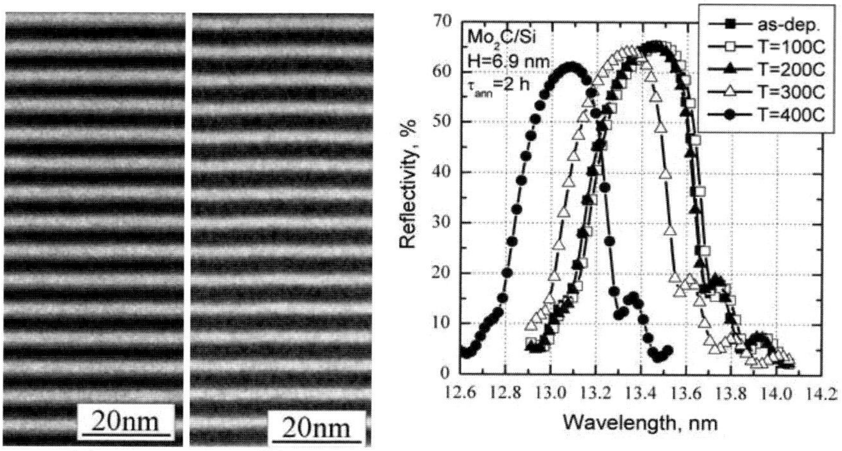

(a) (b) (c)

Fig. 13. XTEM micrographs of a 6.9 nm Mo_2C/Si multilayer mirror in the as-deposited state (a) and after annealing at 400°C (b) for 2 hours. The change in internal structure (a, b) and reflective properties (c) seems to be less than in the case of a Mo/Si multilayer mirrors.

The thermodynamic aspect of thermal and radiation stability calls for a phase and structure equilibrium of layers, minimum specific energy of interfaces, wetting of layers and similar thermal expandability of multilayer materials in all temperature ranges of interest. The difficulties in meeting all these requirements simultaneously are obvious. Therefore a satisfaction of at least the most important requirements is proposed (Kondratenko et. al 1993). Phase equilibrium at an interface has been proposed as a major requirement in multilayer material selection for enhanced stability. It is known that phase equilibrium of two phases at their interfaces may be achieved only if they are neighbors separated by a region of their

Table 4. Comparison of different multilayer systems in terms of reflective properties and maximal working temperature range.

Wavelength range, nm	Multilayer systems	Maximal achievable R_{max}, %	Maximum working temperature, °C
4.4 – 6.7	Cr (Co, Ni)/C	~ 60	≤ 500
	CrB_2/C	~ 55	~ 800
	TiC/C	~ 55	~ 800
12.4 – 17.5	Mo/Si	~ 76	≤ 250
	Mo_2C/Si	~ 75	~ 400
	$MoSi_2/Si$	~ 58	~ 450

own eutectic in the phase diagram of a binary alloy system. For example, according to the phase equilibrium diagram of Mo-Si alloys, phase equilibrium between Mo-Si compounds is possible if they are neighbors in the following sequence: Mo-

Mo_3Si-Mo_5Si_3-$MoSi_2$-Si. It has been shown that using Mo_2Si instead of pure Mo as a multilayer absorber material provides heat resistance up to a temperature of 450°C (Kondratenko et al. 1993). By a similar approach, a TiC/C multilayer system has been designed and developed for work up to a temperature of 800°C (Bugaev et al. 1995). A comparison of classical and high-resistance multilayer systems in terms of reflective properties and maximal work temperatures is presented in Table 4.

7 Concluding Remarks

Over the last couple of years remarkable progress has been made in the field of multilayer mirrors for the EUV/Soft X-ray spectral region. The progress is associated with deep understanding of the multilayer growth mechanisms, improvement of technologies for minimization of interface imperfections and intensive development of modern deposition techniques. Noteworthy are the achievement of a 70 %-level of normal incidence reflectivity at λ=13.5 nm for Mo/Si mirrors with superthin B_4C diffusion barriers as well as the possibility to deposit these high-reflectance mirrors on curved optics with excellent thickness distribution control. This advance in Si-based multilayer technology has provided a capability to meet current stringent requirements of EUV-lithographic imaging optics and further favorable development of the next generation of EUV lithographic systems. At shorter wavelengths promising results at normal incidence were achieved using new Mo/Sr, La/B_4C and Cr/Sc material combinations with peak reflectivities of 48 % at 10.5 nm, 44 % at 6.7 nm and 15 % at 3.1 nm, respectively. These advanced multilayer coatings are believed to satisfy today's demands in various application fields such as microscopy in the water window, plasma diagnostics, spectroscopy, etc. In future great effort can be expected in the following areas of research and applications:

- Search for new material combinations for multilayer mirrors with improved optical performance
- Deposition processes using low energy ion bombardment as well as intensive application of superthin diffusion barriers
- Advanced deposition systems for preparation of multilayer coatings with excellent thickness distribution control over figured large area substrates
- Reduction of residual stresses, improvement of thermal and radiation stability of multilayer coatings.

Acknowledgments

I would like to thank my colleagues T. Feigl and T. Kuhlmann for use of joint results, useful discussions and much practical advice. My thanks also go to A.I. Fedorenko and V.V. Kondratenko for introduction to the basics of multilayer theory and technology; S. Braun, S. Bajt and R. Soufli for kindly providing figures from their original publications and many others for their vivid interest and cooperation in providing me relevant information. I also gratefully acknowledge U. Kaiser and T. Gorelik from Jena University and E. Zubarev from Kharkov University for preparation of XTEM micrographs.

References

Abeles F (1950) Recherches sur la propagation des ondes electromagnetique sinusoidales dans les milieux stratifies Application aux couches minces. Ann Physique 5: 596-639

Bajt S, Alameda J, Barbee T, Clift WM, Folta JA, Kaufmann B, Spiller E (2001a) Improved reflectance and stability of Mo/Si multilayers. Proc SPIE 4506: 65-67

Bajt S, Stearns DG, Kearney PA (2001b) Investigation of the amorphous-to-crystalline transition in Mo/Si multilayers. J Appl Phys 90: 1017-1025

Barbee TW (1986) Multilayer for X-ray optics. Opt Eng 25: 893-915

Bergmann K, Schriever G, Rosier O, Müller M, Neff W, Lebert R (1999) Highly repetitive, extreme ultraviolet radiation source based on a gas-discharge plasma. Appl Opt 38: 5413-5417

Bragg WH (1914) The intensity of reflection of x-rays by crystals. Phil Mag 27: 881-96

Braun St, Mai H, Moss M, Scholz R, Leson A (2002) Mo/Si multilayers with different barrier layers for applications as EUV mirrors. J J Appl Phys 41: 4074-4081

Bugaev EA, Fedorenko AI, Kondratenko VV, Zubarev EN (1995) Thermal stability of normal incidence multilayer mirrors for the X-ray wavelength near carbon K-edge. J X-ray Science & Tech 5: 295-306

Burkhart S, Cerjan C, Mirkarimi P, Walton C, Ray-Chaudhuri A (1999) Low-defect reflective mask blanks for Extreme Ultraviolet lithography. Proc SPIE 3676: 570-577

Chakraborty P (1991) Layered synthetic microstructures as optical elements for the extreme ultraviolet and soft X-rays. Modern Physics B 13: 2133-2228

Debye P (1913) Über die Intensitätsverteilung in den mit Röntgenstrahlung erzeugten Interferenzbildern. Verh d Deutsch Phys Ges 15: 738

Feigl T, Lauth H, Yulin S, Kaiser N (2001) Heat resistance of EUV multilayer mirrors for long-time application. Microelectronic Engineering 57-58: 3-8

Feigl T, Heber J, Gatto A, Kaiser N (2002) Optics developments in the VUV-soft X-ray spectral region. Nucl. Instr. Meth. A 483: 351-356.

Folta JA, Bajt S, Barbee TW, Grabner RF, Mirkarimi PB, Nguyen T, Schmidt MA, Spiller E, Walton CC, Wedowski M, Montcalm C (1999) Advances in multilayer reflective coatings for extreme-ultraviolet lithography. Proc. SPIE, 3676: 702-709

Henke BL, Lee P, Tanaka TJ, Shimabukuro RL, Fujikawa BK (1982) The atomic scattering factor, for 94 elements and for the 100 to 2000 eV photon energy region. Atomic Data and Nuclear Tables 27, pp. 1-144

Henke BL, Gullikson EM, Davis JC (1993) X-ray interaction: photoabsorption, scattering, transmission, and reflection at E=50-30000 eV, Z=1-92. Atomic and Nuclear Data Tables 54, pp 181-342

Høghøj P, Ziegler E, Susini J, Freund AK, Joensen KD, Gorenstein P, Wood J (1997) Focusing of hard X-rays using a W/Si supermirror. Nucl Instrum Meth B132: 528-533

James RW (1962) The Optical Principles of the Diffraction of X-rays. Ox Bow Press, Woodbridge

Joensen KD, Christensen FE, Schnopper HW, Gorenstein P, Susini J, Høghøj P, Hustache R, Wood J, Parker K (1993) Medium-sited grazing incidence high-energy X-ray telescopes employing continuously graded multilayers. Proc. SPIE 1736: 239-248

Joensen KD, Gorenstein P, Høghøj P, Susini J, Ziegler E, Freund A, Christensen F, Wood J, Gutman G (1997) Broad-band hard X-ray reflectors. Nucl Instrum Meth B132: 222-227

Kartright JB (1986) Multilayer reflectors for the extreme ultraviolet spectral region. Nucl Instrum Meth A 246: 344-347

Kondratenko VV, Pershin YuP, Poltseva OV, Fedorenko AI, Zubarev EN, Yulin SA, Kozhevnikov IV, Sagitov SI, Chirkov VA, Vinogradov AV (1993) Thermal stability of soft X-ray Mo-Si and $MoSi_2$-Si multilayer mirrors. Appl Opt 32: 1811-1816

Kopilets IA, Kondratenko VV, Fedorenko AI, Zubarev EN, Poltseva OV, Ponomarenko AG, Lyakhovskaya II (1996) Evolution of structure, phase composition, and X – ray reflectivity of multilayer mirrors Mo-(B+C) after annealing at 250-1100°C. J X-ray Science & Tech 6: 141-149

Kozhevnikov IV, Fedorenko AI, Kondratenko VV, Pershin YuP, Yulin SA, Zubarev EN, Padmore HA, Cheung KC, Dorssen GE, Roper M, Balakireva LL, Serov RV, Vinogradov AV (1994) Synthesis and measurement of normal incidence X-ray multilayer mirrors oprtimized for a photon energy of 390 eV. Nucl Instrum Meth A345: 594-603

Kozhevnikov IV, Bukreeva IN, Ziegler E (2001) Design of X-ray supermirrors. Nucl Instrum Meth A460: 424-343

Kuhlmann T, Yulin S, Feigl T, Kaiser N (2002a) Cr/Sc multilayer mirrors for the nitrogen K_α-line in the water window. Appl Opt 41: 2048-2052

Kuhlmann T, Yulin S, Feigl T, Kaiser N, Bernitzki H, Lauth H (2002b) Design and fabrication of broadband EUV multilayer mirrors. Proc. SPIE 4688: 509-515

Lim YC, Westerwalbesloh T, Aschentrup A, Wehmeyer O, Haindl G, Kleineberg U, Neinzmann U (2001) Fabrication and characterization of EUV multilayer mirrors optimized for small spectral reflection bandwidth. Appl Phys A 72: 121-124

Loevezijn P, Schlatmann R, Verhoven J, Tiggelen BA, Gullikson EM (1996) Numerical and experimental study of disordered multilayers for broadband x-ray reflection. Appl Opt 35: 3614-3619

Louis E, Yakshin AE, Görts PC, Abdali S, Maas ELG, Stuik R, Bijkerk F, Schmitz D, Scholze F, Ulm G, Haidl M (1999) Reflectivity Mo/Si multilayer systems for EUVL. Proc SPIE 3676: 844 –845

Louis E, Yakshin AE, Görts PC, Oestreich S, Stuik R, Maas ELG, Kessels MJH, Bijkerk F, Haidl M, Müllender S, Mertin M, Schmitz D, Scholze F, Ulm G (2000) Progress in Mo/Si multilayer coating technology for EUVL optics. Proc SPIE 3997: 406-411

MacGowan BJ, Mrowka S, Barbee TW, Da Silva LB, Eder DC, Koch JA, Pan LS, Turner JA, Underwood JH, Young PE (1993) Investigation of damage to multilayer optics in X-ray laser cavities: W/C, WRe/C, Stainless-Steel/C, and Cr_3C_2/C mirrors. J X-ray Science & Tech 3: 231-282

Mai H, Pompe W (1992) Manufacture and characterization of soft X-ray mirrors by laser ablation. Appl Surface Science 54: 215-226

Mezei (1988) Multilayer neutron optical devices. In: Dhez P Weisbuch C (eds) Physics, Fabrication, and Application of Multilayered Structures, Plenum Press

Rouard P (1937) Etudes des proprietes optiques des lames metalliques tres minces. Annales de Physique 7: 291-384.

Rosen RS, Stearns DG, Viliardos MA, Kassner ME, Vernon SP, Cheng Y (1993) Silicide layer growth rates in Mo/Si multilayers. Appl Opt 32: 6975-6980

Rosenbluth AE, Lee P (1982) Bragg condition in absorbing X-ray multilayers. Appl Phys Lett 40: 466-468

Rosenbluth AE (1988) Computer search for layer material that maximaze the reflectivity of X-ray multilayers. Rev Phys Appl 23: 1599-1621

Salashchenko NN, Shamov EA (1997) Short-period X-ray multilayers based on Cr/Sc. Opt Commun 134: 7-10

Slaughter JM, Medower BS, Watts RN, Tarrio C, Lucatorto TB, Falco CM (1994a) Si/B_4C narrow-bandpass mirrors for the extreme ultraviolet. Opt Lett 19: 1786-1788

Slaughter JM, Schulze DW, Hills CR, Mirone A, Stalio R, Watts RN, Tarrio C, Lucatorto TB, Krumrey M, Mueller P, Falco CM (1994b) Structure and performance of Si/Mo multilayer mirrors for the extreme ultraviolet. J Appl Phys 76: 2144-2156

Soufli R, Spiller E, Schmidt MA, Davidson JC, Grabner RF, Gullikson EM, Kaufmann BB, Mrowka S, Baker SL, Chapman HN, Hudyma RM, Taylor JS, Walton CC, Montcalm C, Folta JA (2001) Multilayer optics for an extreme ultraviolet lithography tool with 70 nm resolution. Proc SPIE 4343: 51-59

Spiller E (1972) Low-loss reflection coatings using absorbing materials. Appl Phys Lett 20: 365-367

Spiller E, Segmüller A, Rife J, Haelbich R-P (1980) Controlled fabrication of multilayer soft X-ray radiation mirrors. Appl Phys Lett 37: 1048-1050

Spiller E (1981) Evaporated multilayer dispersion elements for soft X-rays. In Attwood DT, Henke BL (ed) Conf. Proc. No.75, AIP, New York

Spiller E, Rosenbluth AE (1986) Determination of thickness errors and boundary roughness from the measured performance of a multilayer coating. Opt Eng 25: 954-963

Spiller E (1994) Soft X-ray Optics. SPIE Optical Eng Press, Bellingham Washington

Stearns DG, Stearns MB, Cheng Y, Stith JH, Ceglio NM (1990) Thermally induced structural modifiction of Mo-Si multilayers. J Appl Phys 67: 2415-2426

Stearns DG, Rosen RS (1991) High-performance multilayer mirrors for soft X-ray projection lithography Proc. SPIE 1547: 2-13

Stearns DG, Rosen RS, Vernon SP (1991) Fabrication of high reflective Mo-Si multilayer mirrors by planar magnetron sputtering. J Vac Sci Tech A9: 2662-2669

Stearns DG, Gullikson EM (2000) Nonspecular scattering from extreme ultraviolet multilayer coating. Physica B 283: 84-91

Takenaka H, Kawamura T (1996) Thermal stability of Mo/C/Si/C multilayer mirrors. J Elec Spec & Rel Phenom 80: 381-384

Vernon SP, Stearns DG, Rosen RS (1993) Ion-assisted sputter deposition of molybdenum-silicon multilayers. Appl Opt 32: 6969-6974

Vinogradov AV, Zeldovich BYa (1977) X-ray and far UV multilayer mirrors. Appl Opt 16: 89-93

Waller I (1923) Zur Frage der Einwirkung der Wärmebewegung auf die Interferenz von Röntgenstrahlen. Z. Physik 17: 398

Wang Z, Cao J, Michette AG (2000) Depth-graded multilayer X-ray optics with broad angular response. Opt Commun 177: 25-32

Windt DL, Hull R, Waskiewicz K (1992) Interface imperfections in metal/Si multilayers. J Appl Phys 71: 2675-2678

Yulin S, Feigl T, Kuhlmann T, Kaiser N (2001) Damage resistant and low stress EUV multilayer mirrors. Proc. SPIE 4343: 607-614

Yulin S, Feigl T, Kuhlmann, Kaiser N, Fedorenko AI, Kondratenko AI, Poltseva, Sevryukova VA, Zolotaryov AY, Zubarev EN (2002) Interlayer transition zones in Mo/Si superlattices. J Appl Phys 92: 1216-1220

Ziegler E (1995) Multilayers for high heat load synchrotron applications. Opt. Eng 34: 445-452

Abbreviations and Symbols

μ	linear absorption coefficient
θ	grazing angle
α	incidence angle
σ	interface roughness
ψ	phase shift of wave
ϕ	rotation speed
ω	spin rotation frequency
Γ	thickness ratio
τ	time
λ	wavelength
$\Delta\lambda$	spectral bandpass (FWHM)
θ_C	critical angle
Γ_{opt}	optimal thickness ratio
c	speed of light
d	thickness
d_A	absorber thickness
d_S	spacer thickness
$f = f_1 - if_2$	atomic scattering factor
H	multilayer period
β	extinction coefficient
β_S	extinction coefficient of the spacer material
β_A	extinction coefficient of the absorber material
m	order of reflection
$\tilde{n}=n+i\beta$	complex refractive index
n	refractive index
n (δ), β	optical constants
N_a	number of atoms per unit volume
N_{max}	maximum number of periods
N_{min}	minimum number of periods
N_R	required number of periods
r	amplitude reflection coefficient
R	reflectivity
$r_\perp$	amplitude reflection coefficient for normal incidence

r_e	classical radius of an electron
r_f	amplitude reflection coefficient of film
R_{INT}	integrated reflectivity
R_{max}	maximal achievable reflectivity
r_P	amplitude reflection coefficient for p-pol.
r_S	amplitude reflection coefficient for s-pol.
T	temperature
Z	atomic number

Laser Resistant Coatings

Christopher J. Stolz and François Y. Génin

Lawrence Livermore National Laboratory, Livermore CA 95330, USA

1 Introduction

Interest in laser resistant coatings started with Maiman's invention of the ruby laser (Maiman 1960). It soon became apparent that the existing quality of optical coatings was insufficient to withstand the high-photon flux of a laser source (Glass and Guenther 1973). In 1965 the Office of Naval Research contracted with Baush & Lomb to study the damage thresholds of dielectric films and multilayer coatings to overcome the problems of absorption and coating defects (Turner 1972). In 1969, a symposium on Laser-Induced Damage in Optical Materials was formed and two years later, thin films became one of the four primary research topics. This symposium has been the premier repository of documented research in laser resistant optical thin films. Although the subject of laser damage in optical materials is a rather narrow field of research, there are some excellent books on the subject (Kozlowski 1995; Wood 1986).

The invention of the laser ring gyro changed the focus of damage resistant coating development from macroscopic thin film damage to subtle changes in film microstructure and scatter. No longer did thin films only have to survive laser exposure without catastrophic failure, but they had to simultaneously maintain low losses to maximize component lifetime.

Today, thin film laser damage research continues to focus on reducing absorption and coating defect density. The requirements of both governmental programs and industrial applications currently drive research. Inertial Confinement Fusion, Airborne Lasers and Free Electron Lasers have extremely high energy levels over meter-size optics. Long lifetime is required of much smaller components within commercial lasers used in industrial, scientific, military, and medical applications. Short pulse laser requirements are pushing the coating industry to develop coatings capable of withstanding picosecond and femtosecond pulses (Jasapara et al. 2001). The needs of the lithography industry for short wavelength coatings necessitate the development of low-absorbing materials in the deep and extreme UV region (Apel et al. 2000; Kaiser et al. 1995). Finally telecommunication optics are being limited by laser damage issues as greater numbers of photons are injected into single fibers.

2 Damage Testing

In order to understand the laser resistance of a specific coating, it is important to understand how coatings are damage tested and how damage thresholds are determined. Standardized tests and methods (see also chapter "Characterization and Monitoring" by Ristau) are critical to enable meaningful comparisons of laser damage test results between laboratories. Damage testing enables the user to investigate the cause of coating damage by examination of the damage morphology and by evaluating damage statistics. Thin film deposition processes can then be optimized for improved laser resistance through the process of damage mechanism prediction followed by experimental validation.

2.1 Damage Testing Methods

A complete description of a typical damage tester and test methods are described in ISO standard 11254. In summary, a damage tester consists of a laser focussed to a sample plane. Laser diagnostics characterize pulse length, energy, fluence, and spot diameter. Damage detection diagnostics are typically scatter systems, plasma emission detectors, acoustic detectors or optical microscopes for visual inspection of the surface after laser irradiation. High-resolution images have even been acquired by in-situ Scanning Electron Microscopy (Poulingue et al. 1999).

Typically there are four methods of damage testing illustrated in Fig. 1:

- 1:1 consists of exposing a site to a single fluence for a single shot.
- S:1 consists of exposing a site to a single fluence over multiple shots (typically 600).
- R:1 consists of exposing a site to a a fluence ramp over multiple shots (typically 600).
- N:1 consists of exposing a site to a fluence ramp over a small number of shots (typically 5-10).

For each of these testing methods, multiple sites are tested to obtain a representative damage threshold of the coating. Often, the testing procedure influences the results of the measurements. Laser conditioning by irradiating a coating at low fluence and ramping to a high fluence can improve the laser resistance of a coating as described in Sect. 4.2. Therefore, it is important to determine if unconditioned damage thresholds determined by 1:1 and S:1 tests are the appropriate measurement for the laser application or if R:1 and N:1 tests which measure the conditioned damage threshold are more suitable. Finally, the cleanliness of the testing facility can have a significant impact on the results as contaminates that drift into the test beam can create surface damage and beam modulation effects.

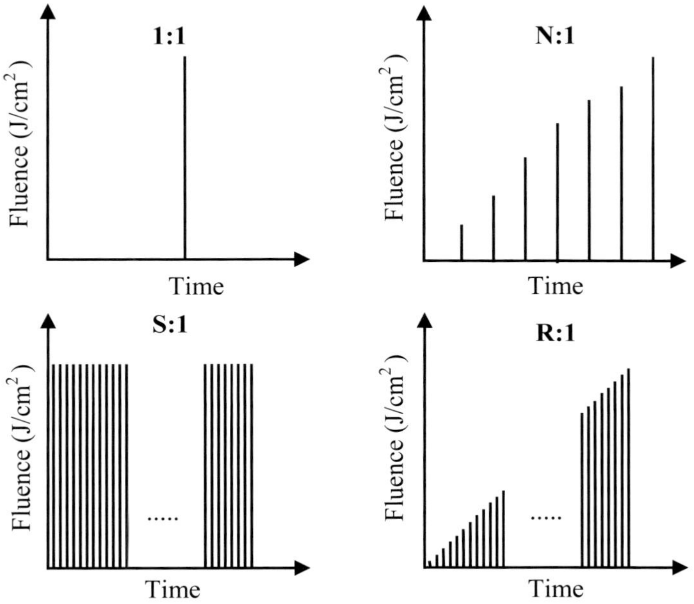

Fig. 1. Illustration of the four typical methods (1:1, N:1, S:1 and R:1) used in damage testing (Adapted from Kozlowski (1995).

2.2 Damage Testing Results

To analyse the results of a damage test, it is critical to understand the damage criteria used during the test. As the resolution of the detection system increases, the reported damage threshold can decrease. Some applications require that no detectable changes occur in the coating. Other applications may tolerate limited damage that remains below a critical size or area upon further irradiation. To alleviate this problem, an initiation and growth threshold (also referred to as functional damage threshold) can be defined for each type of damage threshold measurement.

The diameter of the test beam, or more specifically the total test area will affect the reported damage threshold. Coating defects are statistically distributed across the surface. Since different defects can have different laser resistances, damage threshold accuracy is improved by increasing the test area by using larger test beams or raster scanning a small beam over a large area (Hue et al. 1997).

Damage thresholds can be calculated over a range of pulse widths by using the scaling relation in Eq. 1. F_D is the damage threshold in the use environment, F_{Dt} is the damage threshold at the testing pulse length, τ_u is the use pulse length, τ_t is the test pulse length, and b is the scaling factor.

$$F_D = F_{Dt}\,(\tau_u/\tau_t)^b \quad (1)$$

Simple pulse length scaling exponents can only be used over a limited range of pulse lengths because damage mechanisms are pulse length dependent (Guenther et al. 1984; Walker et al. 1981). Figure 2 illustrates the change in scaling relationship as a function of pulse length and defect type (detectable versus non-detectable). A theory was developed for uncoated fused silica to explain the pulse length dependence based on electron production via multiphoton ionization, Joule heating and avalanche ionization (Feit et al. 1995). A theory for coatings is significantly less well developed due to the added complexity of a number of factors not limited to multilayer interference, scatter, microstructure, multiple materials and porosity effects. Experimental studies at nanosecond pulse lengths suggest an empirical scaling factor of b = 0.35 for coatings (Rainer et al. 1992).

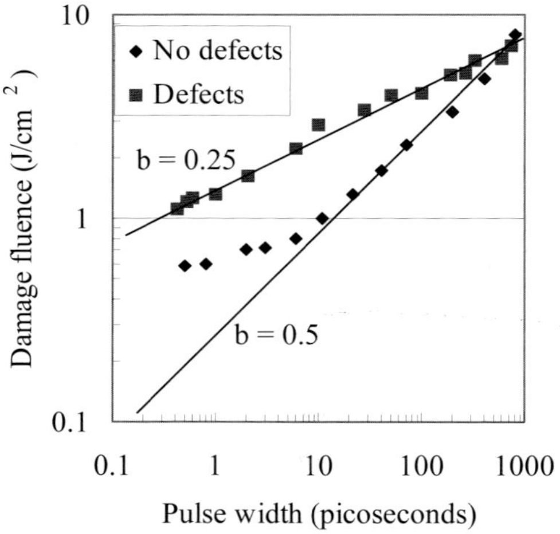

Fig. 2. Pulse length scaling factors for multilayer coatings in low defect areas (polarizer) and high defect areas (high reflector) (adapted from Stuart et al. 1995).

3 Damage Morphologies

Once laser damage occurs, the damage morphology can be studied to determine potential initiation mechanisms. For example, four different damage morphologies (nodular ejection pits, flat bottom pits, plasma scalds and delamination) are found in optical thin film high reflectors and polarizers (see Fig. 3).

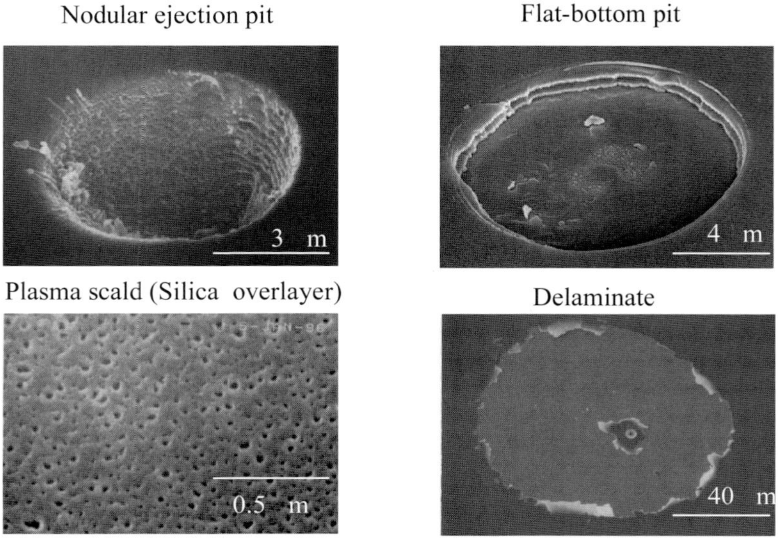

Fig. 3. SEM micrographs of typical damage morphologies of hafnia silica 1053nm high-reflector and polarizer coatings.

3.1 Characterization Techniques

Damage morphology characterization techniques vary in both cost, resolution and the type of information that is obtained. The optical microscope is the standard tool used in laser damage studies due to ease of use, data acquisition speed, absence of sample modification, trivial integration for in-situ characterization and low cost. Unfortunately the resolution of optical microscopes is diffraction limited at approximately 1 μm and no chemical, thermal or microstructural information is obtainable. When higher resolution is required, advanced microscopic techniques listed in Table 1 can be used. SEM and TEM typically require the sample to be modified by depositing a conductive layer on the dielectric thin film to prevent charging effects, although significant advances have been made in low-voltage SEM to image dielectric surfaces without sample preparation.

Photothermal microscopy (see also chapter "Characterization and Monitoring" by Ristau) is a useful tool for evaluating the thermal characteristics of coating defects described in Sect. 4 (Commandré and Roche 1995). The magnitude of observable scatter from defects does not correlate well with damage susceptibility (Fornier et al. 1997). The photothermal signal amplitude from defects, however, has a strong correlation to damage susceptibility (Reichling et al. 1998; Wu et al. 2001). This method is non-destructive, unlike damage testing, so further characterization of the defect is possible. Current photothermal microscopes are slow, however, making it challenging to probe sufficient areas of today's low-defect density laser resistant coatings.

Table 1. High-resolution instruments used for inspecting laser damage.

Instrument	Information acquired
Atomic force microscope	Dimension, charging interactions
Near field scanning optical microscope	Dimension, electric-field profile
Scanning electron microscope	Dimension, cross-section, composition
Focussed ion beam microscope	Dimension, cross-section, composition
Transmission electron microscope	Dimension, cross-section, microstructure, composition
Photothermal microscope	Absorption characteristics, composition
X-ray fluorescence microscope	Absorption characteristics, composition
Rutherford backscattering	Microstructure, composition
Secondary ion mass spectroscopy	Microstructure, composition
X-ray diffraction	Microstructure, composition

3.2 Laser-Induced Pits

Two pit morphologies typically occur in laser-damaged optical coatings. Nodular-ejection pits are formed when micron-scale defects, described in Sect. 4.1, are heated until ejection. Nodular-ejection pits are usually deep and can penetrate into the substrate. They can be irregular in shape depending on ejection severity.

Flat-bottom pits are shallow interfacial damage that occur when nanometer-scale defects are heated to a plasma state by standing-wave electric fields described in Sect. 5 (Dijon et al. 1999) The coating yields at a layer interface near the top of the coating and forms a circular defect with a flat bottom as illustrated in Fig. 4. Hafnia (HfO_2) and silica (SiO_2) coatings deposited in a reactive environment from hafnium metal (Hf) and silica starting materials have fewer interface voids than coatings deposited directly from hafnia. Coatings with poor interfaces typically have flat-bottom pits at low fluences so proper selection of starting material composition can have a favourable impact on the damage threshold (Weakley et al. 1999).

3.3 Laser-Induced Delamination

Delaminations occur when the outer layer of the coating is completely ablated from the surface. Delamination severity increases as overcoats used in coating designs become thinner. The use of silica half-wave overcoats is a standard design strategy for improving the damage threshold of high reflector coatings without degrading spectral characteristics (Bodemann et al. 1996; Carniglia 1981; Tuttle Hart et al. 1982; Lowdermilk et al. 1980; Walton et al. 1996; Wu et al. 1991). However, the thickness of Brewster's angle plate polarizer overcoats is typically determined by optimization to admittance match the coating stack and the incident medium. Polarizers with overcoats thinner than 100 nm delaminate easily while 240 nm and thicker overcoats typically do not delaminate (Stolz et al. 1997). Irradiating polarizer coatings at different angles of incidence influences the damage morphology. As the angle of incidence is decreased from 65 to 45 degrees, delamination probability and severity increases (Génin et al. 1997).

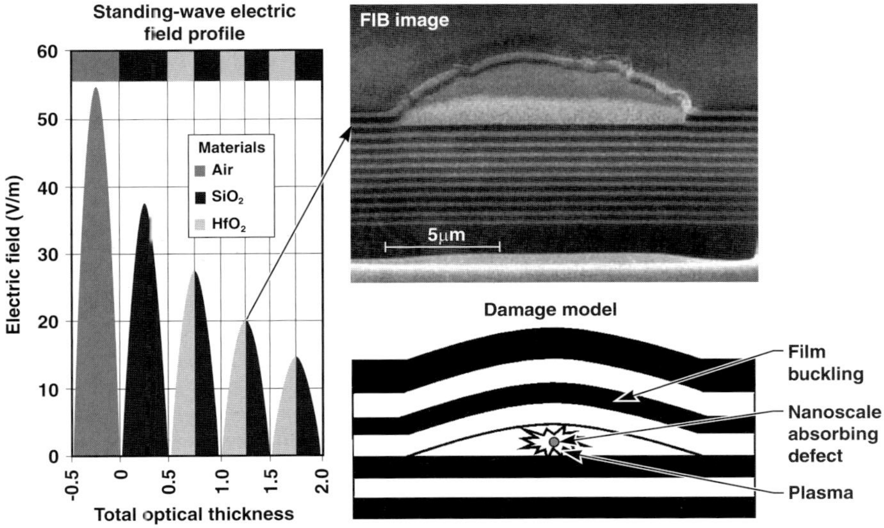

Fig. 4. Flat-bottom pits occur at interfaces with standing-wave electric-field peaks that ignite nanoscale defects to a plasma state.

3.4 Laser-Induced Scalding

Surface scalds occur when a plasma interacts with the coating surface during the damage process as illustrated in Fig. 5. Plasmas can be created by defects, contamination or air breakdown. To the naked eye, plasma scalds appear as mottled surface discoloration. When viewed with a SEM, a plasma scald is a surface with a very high density of micro-pits. Laser conditioning can minimize the severity of plasma scalding by ejecting nodules at the fluence that triggers ejection, hence creating a smaller plasma. Silica overcoats minimize the severity of plasma scalds since the depth of a plasma scald in a silica layer is typically less than 6 nm compared to plasma scalds in high index materials that have been shown to penetrate layers as thick as 130 nm (Walton et al. 1996). Although plasma scalds can be quite large in diameter, in silica layers they are stable even when exposed to high fluences and have little impact on the reflected laser beam properties (Génin and Stolz 1996).

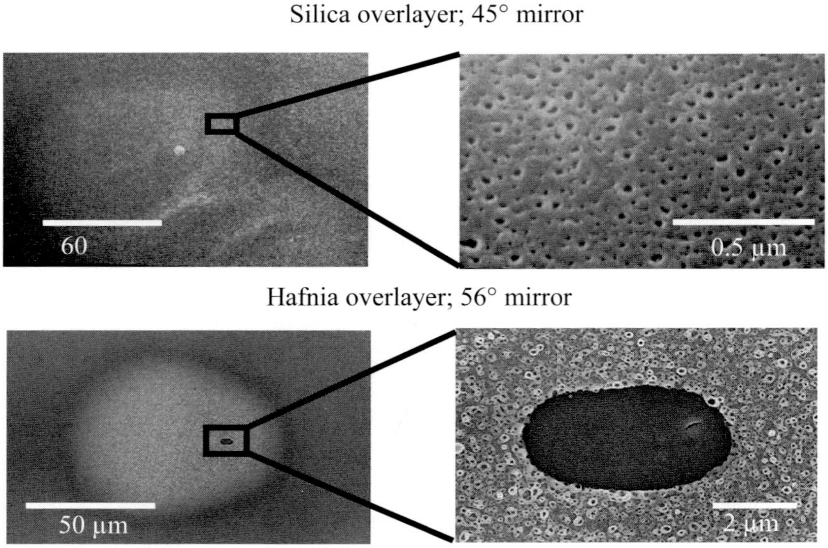

Fig. 5. SEM micrographs of plasma surface scalds of multilayer coatings with silica and hafnia top layers (adapted from Walton et al. 1996).

3.5 Laser Damage Stability

If a given laser application can tolerate some coating laser damage (Foltyn and Jolin 1984) it is important to know the damage growth rate due to repeated laser exposure to accurately predict component lifetime. Additionally, the more unstable a damage morphology, the greater the need to eliminate the mechanism that causes that particular morphology. Stability studies of the four typical damage morphologies on 1053 nm high reflector and polarizer coatings illustrate that delaminates are extremely unstable at low fluences compared to the other three morphologies (Génin et al. 1997). Even when morphologies are stable, the damage size is larger with increasing fluence. The results of these growth studies are illustrated in Fig. 6

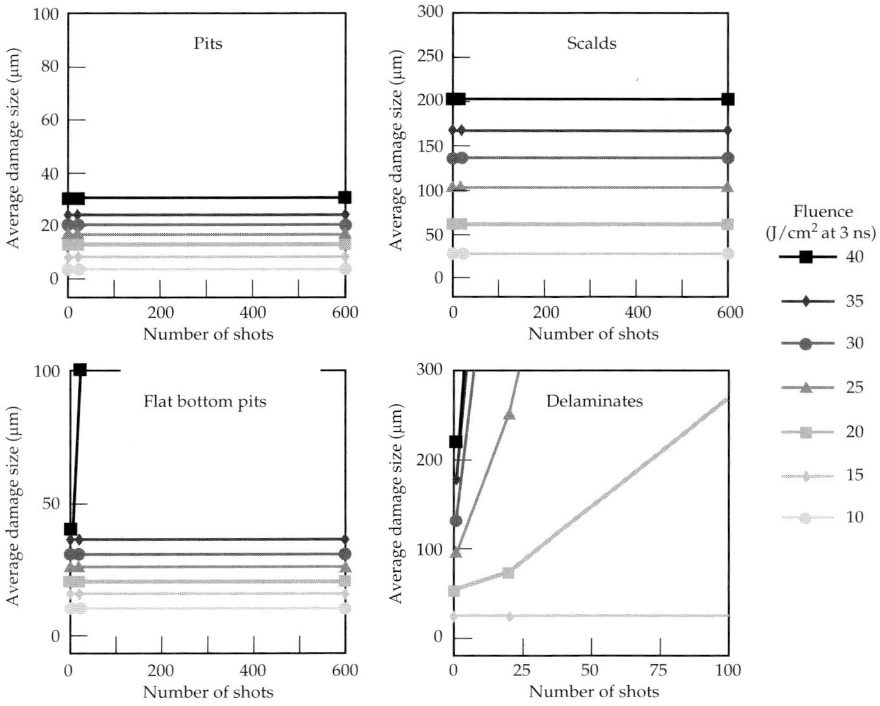

Fig. 6. Growth of various types of damage morphologies for increasing laser fluences. Delaminates tend to become unstable with continued laser exposure.

4 Defects

Coating defects can be intrinsic or extrinsic. Intrinsic defects can result from the film growth mechanism such as pores, cracks, incomplete stoichiometry, columnar microstructure or multiple crystalline phases within the material (see also chapter "Fundamentals of Thin Film Growth" by Kaiser). Extrinsic defects occur due to foreign defect seeds created during the coating process or surface contamination due to incomplete substrate cleaning.

A nodular defect is an extrinsic defect that consists of a seed embedded in the multilayer coating (Guenther 1981, Tench et al. 1994) as illustrated in the defect cross-sections in Fig. 7. The size of a nodule increases with the diameter and depth of the seed. The damage threshold of nodules decreases with increasing seed diameter, defect diameter and laser pulse length (Milam et al. 1973; Poulingue et al. 1999; Staggs et al. 1993). Therefore, defect reduction efforts should primarily focus on eliminating large (> 0.7 μm diameter) defects (Dijon et al. 1999; Kozlowski et al. 1994). Coatings such as antireflection or UV mirrors are too thin to support nodular defects of this size so damage studies of these coating types focus on nanometer-scale absorbing defect seeds (Bodemann et al. 1996; Dijon et al. 1996; Reichling et al. 1998).

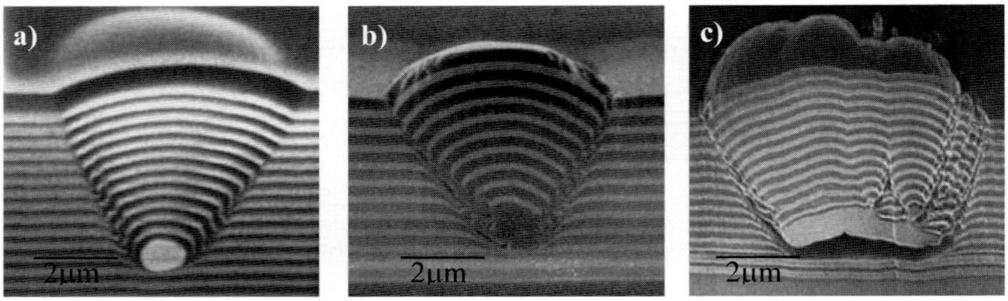

Fig. 7. Cross-sectional SEM micrographs of nodular defects embedded in hafnia silica multilayers caused by **(a)** molten HfO_2 source ejecta, **(b)** solid SiO_2 source ejecta, **(c)** chamber flakes.

4.1 Defect Sources

For e-beam coatings the most prominent defect seed source is from spatter of the starting materials (see also chapter "Coating Materials" by Friz). Particles are ejected from the source with sufficient velocity to strike and adhere to the optic thus becoming incorporated into the film during the subsequent deposition.

Oxide materials are particularly prone to spitting. Nodule seeds eject when internal voids are exposed to the electron beam. Voids are common in oxide starting materials because their low thermal conductivity results in a small melt zone. Nodule seeds are also ejected from the starting material because of volumetric stresses that occur when the starting material undergoes crystalline phase transformations. In the case of hafnia or zirconia, a monoclinic to tetragonal phase transition occurs between the water-cooled crucible and the molten pool caused by the electron beam resulting in a volumetric expansion of 3% and 7%, respectively (Kozlowski and Chow 1994).

Defect seeds can occur because of contamination due to insufficient cleaning or airborne defects attracted to the substrate prior to coating. Typically, high damage threshold coating facilities operate in clean rooms to minimize the likelihood of airborne contamination particles. Arcing of exposed high voltage surfaces in a DC magnetron deposition system has been shown to cause coating defects (Williams et al. 1992). IBS target shedding also creates particles (Knollenberg et al. 1995). Finally flakes from chamber walls and surfaces have been observed in defect cross sections (Stolz et al. 1996).

4.2 Defect Reduction and Mitigation

Numerous methods have been explored to eliminate the source of defects. A long throw-distance between the deposition source and the substrate can filter large particles by gravity (Matteson and Bowling 1988). Arc suppression circuitry has been successfully used for DC magnetron sputtering (Williams et al. 1992). Electrostatic filters have been less successful (Miller et al. 1994). Rotary filters have also been less successful. High-speed rotating systems are problematic because they are also a source of particulates.

Higher thermal conductivity starting materials, such as metal hafnium (Hf), produce coatings with three to ten times lower defect density of coatings deposited from lower thermal conductivity starting materials such as metal oxide hafnia (HfO_2) as illustrated in Fig. 8 (Chow et al. 1993, Stolz et al. 1999). Higher thermal conductivity facilitates the production of less porous plugs which minimizes spatter. The deposition plume stability of hafnium has been shown to improve by three times enabling improved layer thickness control.

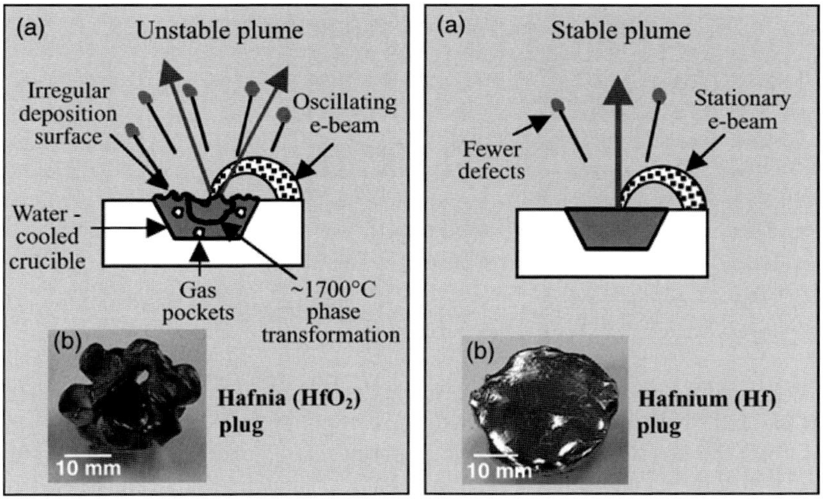

Fig. 8. (a) Schematic of the events that take place at the plug of HfO_2 versus Hf during e-beam deposition. (b) Photographs of a HfO_2 and a Hf plug illustrate porosity differences between the two starting materials.

The damage threshold of coatings can be increased by laser conditioning (Sheehan et al. 1994; Stolz et al. 1999; Wolfe et al. 1990). Laser conditioning mechanisms are not completely understood (Bercegol 1999). Nodular ejection at reduced fluence has been shown to improve coating damage resistance (Fornier et al. 1994). Laser conditioning also reduces the absorption of defects. Nodular ejection pits created during laser conditioning have a reported absorption reduction up to 156× and non-ejected defects have an absorption reduction of up to 15× (Papandrew et al. 2001).

5 Electric Field Effects

Optical thin films modify surface spectral characteristics by interference effects due to reflections at the numerous interfaces in multilayer coatings. These same interference effects are responsible for creating a standing-wave electric field (SWEF) within a coating. The magnitude of the SWEF is proportional to the square root of the laser intensity.

The maximum SWEF peak is in the incident medium. The SWEF oscillations decay exponentially through the coating stack with peaks occurring at the multilayer interfaces. The occurrence of SWEF peaks has been experimentally validated by observing a heating-induced crystalline phase transformation in TiO_2 films at the electric field peaks as illustrated in Fig. 9 (Hacker et al. 1996).

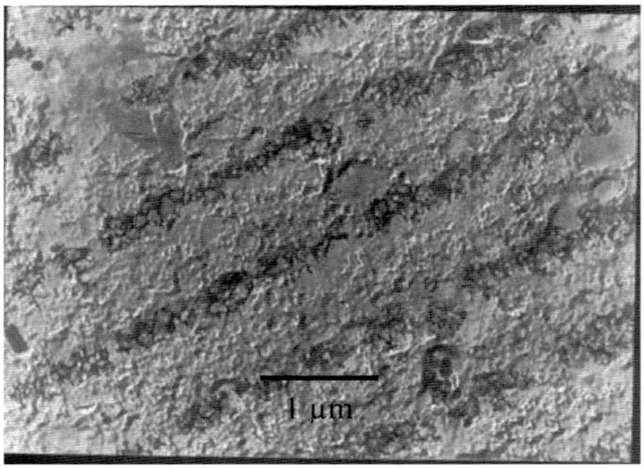

Fig. 9. TEM micrograph of a polarizer coating illustrates crystalline phase transformations in TiO_2 layers occuring at SWEF peaks (adapted from Hacker et al. 1996).

SWEF reduction techniques have been developed because coating materials have variable laser damage resistance. It is important to shift the SWEF peaks away from the layer interfaces. Coating interfaces are generally weaker due to interfacial absorption and microcrystalline phase mismatches between the two coating materials (Apel et al. 2001; Welsch et al. 1988). The laser damage resistance of higher index coating materials is typically less than that of lower index materials (Rainer et al. 1993). Therefore, methods have been determined to reduce the SWEF within the high index material at the expense of increasing the SWEF within the low index material. This technique has proven valuable in applications such as short pulse and short wavelength coatings (see also chapter "Optical Coatings for the DUV/UV" by Thielsch; chapter "Coatings for EUV/X-ray Mirrors" by Yulin; chapter "Optical Interference Coatings for Ultrafast Optics" by Tempea) where the SWEF is the dominant contributor to coating damage (Newnam et al. 1976; Carniglia 1981).

5.1 Electric Field Reduction Techniques

An example of a high reflector design employing a SWEF reduction technique in the three outer layer pairs is illustrated in Fig. 10. The SWEF in the high index layers is reduced by decreasing the thickness of the high index layers and increasing the thickness of the silica layers (Apfel 1977; Gill et al. 1978). The reflectivity of this modified design is slightly degraded since the outer six layers are no longer quarter-wave thickness. The reduction in SWEF strength of the high refractive index material is given by

$$\frac{C}{\sqrt{(m+1)C^2 - m}} \tag{2}$$

and the increase in the SWEF strength of the low refractive index material is given by

$$C\sqrt{\frac{mC^2 - (m-1)}{(m+1)C^2 - m}} \tag{3}$$

where m is the number of layer pairs that are modified and C is the contrast ratio between the high and low refractive index materials:

$$C = \frac{n_h}{n_l} \tag{4}$$

A modification to this SWEF reduction technique is a hybrid three material design. The design consists of two reflector stacks centered at the same wavelength. A high contrast ratio can be achieved in the lower mirror stack without regard for high index material durability to high electric fields. The material selection criterion for the bottom stack is a large contrast ratio such as TiO_2, Ta_2O_5, or Nb_2O_5 and SiO_2 for a minimal number of layer pairs for a specified reflectance. The top stack consists of two materials that withstand a high SWEF such as Al_2O_3, HfO_2, or Sc_2O_3 and SiO_2. The thickness of the top stack is dictated by the necessary SWEF reduction needed for the high refractive index material in the bottom stack.

The hybrid design with quarter-wave layer thicknesses has a wider specular bandwidth than the non-quarter-wave design technique which is very sensitive to layer thickness errors and wavelength range. However, nodular-ejection pits can efficiently couple high fluence light into the inherently weaker bottom stack of the hybrid design.

SWEF reduction techniques can also be applied to polarizers although the thickness constraints on the outer layers for admittance matching complicates matters (Stolz 1999). Polarizers in high-power lasers are typically used as optical isolators so the SWEF of both forward and backward propagating beams should be considered.

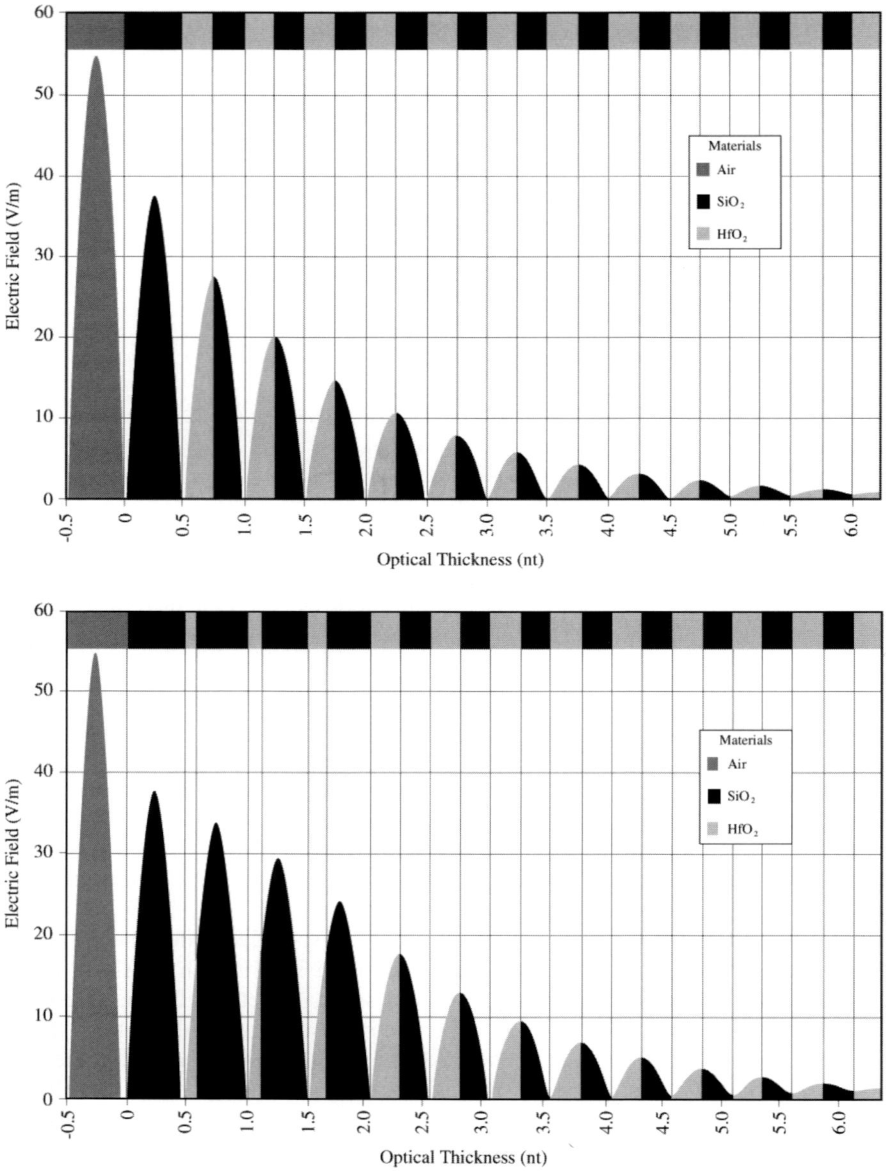

Fig. 10. SWEF profile of a quarter-wave high reflector and a non-quarter-wave high reflector incorporating SWEF reduction techniques.

The SWEF profile of a coating design can also be significantly altered by changes in the angle of incidence. For example, a Brewsters' angle BK7 thin film plate polarizer is designed to operate at an incidence angle of 56.4 degrees. When increasing the incidence angle, the SWEF peaks in the outer few layers are lowered while the SWEF is increased deeper in the coating design. The opposite occurs when the incidence angle is decreased. Consequently, a greater temperature and hence stress field difference occurs in the outer layers as the incidence angle is decreased. The damage morphology of polarizers changes as a function of incidence angle (Génin et al. 1997). At high angles, pits are formed consistent with high SWEF peaks deep within the coating. As the incidence angle is decreased, the damage morphology changes to delamination, a morphology that is typical of high stresses at the outer layers.

5.2 Electric Field Enhancement at Defects

Coating defects are a source of SWEF enhancement and hence are damage initiation sites. Bloembergen reported that micro-cracks and pores in coatings enhance the SWEF (Bloembergen 1973). For this reason, significant effort has been placed on improving the microstructure (see also chapter "Fundamentals of Thin Film Growth" by Kaiser) of coatings to increase the intrinsic damage threshold as defined as the damage threshold of coated areas free of macroscopic defects. It was later demonstrated that cracks in fused silica surfaces can lead to SWEF enhancement factors of several orders of magnitude (Génin et al. 2001).

Nodular defects can enhance the SWEF by up to four times resulting in localized heating and stresses as illustrated in Fig. 11. The SWEF enhancement is proportional to seed diameter and depth (DeFord and Kozlowski 1993; Sawicki et al. 1994). Larger diameter seeds create nodular defects with sufficient height that the incident angle at the seed edge can be greater than the angular bandwidth of the coating (Dijon et al. 1999) beyond the critical angular bandwidth of the high reflector coating. The transmission increases sharply thus enabling efficient coupling of a high fluence laser into nodular seeds.

Theoretical models have also been developed to describe light resonance effects within circular seeds, thus creating the potential for very high electric field enhancements (Libenson and Gruzdev 1998).

5.3 Electric Fields and Flat-Bottom Pits

Flat-bottom pits are interfacial damage that occur at SWEF peaks as illustrated in Fig. 4. Quarter-wave reflectors have SWEF peaks at interfaces where the high refractive index material has grown on top of the low refractive index material. By using a buried half-wave layer, the SWEF peaks can be shifted to interfaces where the low refractive index material is above the high refractive index material. Results of this study suggest preferential damage at the high over low refractive index interface suggesting a weaker mechanical interface or greater presence of interfacial nano-scale defects (Weakley et al. 1999).

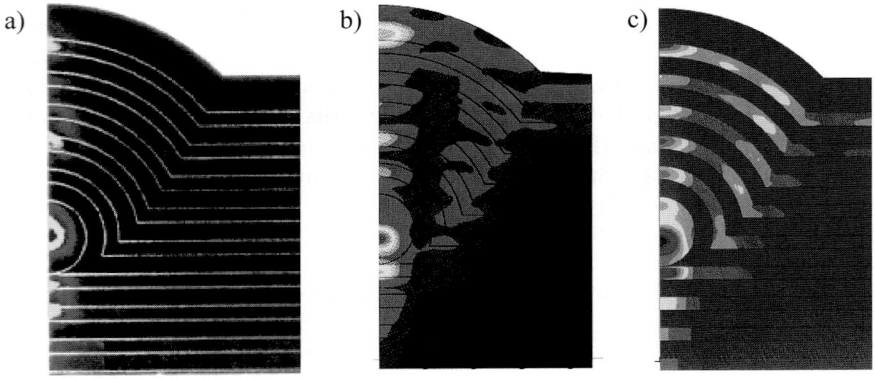

Fig. 11. Finite element calculations of a nodular defect illustrate enhancements of 4×, 1.7×, and 10,000× in the **(a)** electric field, **(b)** temperature and **(c)** stress distribution respectively.

6 Absorption

In order to fabricate laser resistant optical coatings they must be designed and manufactured from coating materials with low absorptance in the wavelength region of interest. Absorption-induced damage typically has morphologies such as blistering, delamination, melting, cracking or ionisation at higher fluences.

6.1 Bandgap

The bandgap of a material corresponds to the photon energy needed to push an electron from the valence band to the conduction band. Once this occurs, the material changes from transmissive to absorbing at the laser wavelength. Typically, high damage threshold materials have wide bandgaps for the photon energies of interest.

Metallic oxides (see also chapter "Coating Materials" by Friz) such as HfO_2, Sc_2O_3, ZrO_2, Ta_2O_5, TiO_2, Al_2O_3 and SiO_2 and fluorides such as MgF_2, LaF_2, AlF_3 and GdF_3 are typically used as laser resistant coating materials from the ultra violet to the near infrared region. Metals are typically poor candidate materials for laser resistant coatings due to their high absorption and hence ease of thermal-induced damage at relatively low fluence.

6.2 Linear Absorption

Thin films manufactured with materials that operate within the bandgap can still absorb. Improper stoichiometry such as the presence of SiO_x within the SiO_2 film due to insufficient oxidation during thin film growth can occur within a coated layer and may lead to higher absorption. Contamination at the substrate film interface can be absorptive. Cracks, pores, and voids between columnar structures can harbour impurities that absorb at the laser wavelength of interest. Enough heat can be generated during irradiation of absorbing films that thermal breakdown of the coating material can occur leading to thin film damage.

6.3 Nonlinear Absorption

Two-photon absorption occurs when the combined energy of two photons is sufficient to push an electron into the conduction band thus creating an absorptive free carrier (Van Stryland et al. 1993). The equation describing two-photon absorption from the intensity dependent version of Lambert Beer's law is

$$A = \beta \, dz + \beta_2 \, dz \, I \tag{5}$$

where β and β_2 are the linear and two-photon absorption coefficients, I is the incident intensity and dz is the sample thickness. The two-photon absorption coefficient is orders of magnitude lower than the linear absorption coefficient. Multi-photon absorption (Göppert-Mayer 1931) is the general case of more than one photon striking an electron simultaneously and exciting it to the conduction band. Although the probability of multi-photon absorption is low, the probability increases with fluence. The farther the test wavelength is from the edge of the bandgap of the material, the greater the number of photons that must simultaneously arrive and hence the lower the probability of multi-photon absorption. Therefore, multi-photon absorption usually becomes significant with short pulses at extremely high intensities and for UV coatings due to the proximity of the edge of the bandgap (Apel et al. 2000; Jasapara et al. 2001) (see also chapter "Optical coatings for the DUV/VUV" by Thielsch).

7 Thin Film Deposition Technology

The deposition method can have a significant impact on the damage resistance of a coating. Energetic processes such as IAD, IBS, IP and CVD produce extremely dense films that are environmentally stable and can have significantly fewer intrinsic defects (see chapter "Thin film deposition techniques" by Pulker). However, the extrinsic defects are more tightly bound in coatings deposited with energetic processes. Nodular ejections of tightly bound defects are significantly more catastrophic leading to collateral damage and high damage growth rates. This is consistent with the absence of the laser conditioning effect observed for these types of film. However, in defect-free regions of the coating the damage threshold of coatings deposited with energetic processes can be very high. Therefore, these technologies are well suited to small-aperture optics. In low-fluence applications such as laser ring gyros, where stability is critical, ion beam sputtered coatings are a superior choice over e-beam coatings.

Sol-gel coatings are used predominately in antireflection applications due to high damage thresholds, ease of scaling up the process to large optics and low cost (Thomas 1994). Sol-gel coatings are formed when a colloidal suspension is spun onto an optical surface or an optic is dipped into the solution. Once dried, a porous film remains that when properly prepared can be very damage resistant. Unfortunately these films have low abrasion resistance and are susceptible to environmental changes such as humidity and low vapour pressure contamination. A refractive index of 1.22 can be achieved so a single quarter-wave thickness layer can be used to yield low reflectivity given by :

$$R = \left(\frac{n_f^2 - n_i n_s}{n_f^2 + n_i n_s} \right)^2 \tag{6}$$

where n_f is the refractive index of the film, n_i is the refractive index of the incident medium and n_s is the refractive index of the substrate. Laser resistant high reflector coatings have also been produced. However, coating stress and thickness non-uniformity are still significant issues limiting the usefulness of this technology to large areas (Andrew et al. 2002).

8 Conclusions

Laser resistant coatings can be fabricated when the mechanisms and sources of each of the damage morphologies such as nodular ejection pits, flat bottom pits, plasma scalds and delaminations are understood. Standardized tests are needed to determine the laser resistance of a coating for its particular application. A multitude of characterization techniques exist to examine damage morphologies to determine likely damage mechanisms. Defects and absorption must be minimized to produce laser resistant coatings. The use of metallic starting materials deposited in a reactive environment to fabricate wide bandgap oxides has been shown to improve process stability, reduce nodular seeds and improve interface quality to minimize flat-bottom pits. Other sources of seeds such as chamber flakes and arcing must be addressed to improve coating laser resistance. Some defects can be ejected benignly to reduce collateral damage during laser conditioning. Although many different coating technologies can be utilized, optic size and coating type play significant roles in the method ultimately selected. Design strategies such as SWEF reduction techniques and sufficiently thick silica overcoats can also improve laser resistance.

References

Andrew JE, Bazin NJ, McInnes HA, Morris AJ, Porter K (2002) Aperture scaling of 351 nanometre high reflectivity sol gel based mirror coatings. In: Exarhos GJ, Guenther AH, Kozlowski MR, Lewis KL, Soileau MJ and Stolz CJ (eds) Laser-Induced Damage to Optical Materials: 2001 SPIE 4679: 271-281

Apfel JH (1977) Optical coating design with reduced electric field intensity. Appl. Opt. 16: 1880-1885

Apel O, Mann K, Zöller A, Götzelmann R, Eva E (2000) Nonlinear absorption of thin Al_2O_3 films at 193 nm. Appl. Opt. 39: 3165-3169

Apel O, Schulz-Grosser M, Leinhos U, Kennedy M, Mann K, Schuhmann R (2001) Interfacial absorption of DUV coatings. In: Exarhos GJ, Guenther AH, Kozlowski MR, Lewis KL, and Soileau MJ (eds) Laser-Induced Damage to Optical Materials: 2000 SPIE 4347: 17-23

Bercegol H (1999) What is laser conditioning? A review focused on dielectric multilayers. In: Exarhos GJ, Guenther AH, Kozlowski MR, Lewis KL, and Soileau MJ (eds) Laser-Induced Damage to Optical Materials: 1998, SPIE 3578: 421-425

Bloembergen N (1973) Role of cracks, pores, and absorbing inclusions on laser induced damage thresholds at surfaces of transparent dielectrics. Appl. Opt. 12: 661-664

Bodemann A, Kaiser N, Kozlowski MR, Pierce E, Stolz CJ (1996) Comparison between 355 nm and 1064 nm damage of high grade dielectric mirror coating. In: Bennett HE, Guenther AH, Kozlowski MR, Newnam BE, Soileau MJ (eds) Laser-Induced Damage to Optical Materials: 1995. SPIE 2714: 395-404

Carniglia CK (1981) Oxide coatings for one micrometer laser fusion systems. Thin Solid Films 77, 225-238

Chow R, Falabella S, Loomis GE, Rainer F, Stolz CJ (1993) Reactive evaporation of low-defect density hafnia. Appl. Opt. 32, 5567-5574

Commandré M, Roche P (1995) Characterization of absorption by photothermal deflection. In: Flory FR (ed) Thin films for Optical Systems. Marcel Dekker, New York, pp 329-365

DeFord JF, Kozlowski MR (1993) Modeling of electric-field enhancement at nodular defects in dielectric mirror coatings. In: Bennett HE, Guenther AH, Kozlowski MR, Newnam BE, Soileau MJ (eds) Laser-Induced Damage to Optical Materials: 1992. SPIE 1848: 455-470

Dijon J, Hue J, Disgecmez A, Quesnel E, Rolland B (1996) Thin films laser damage mechanisms at the YAG third harmonic. In: Bennett HE, Guenther AH, Kozlowski MR, Newnam BE, Soileau MJ (eds) Laser-Induced Damage to Optical Materials: 1995. SPIE 2714: 416-424

Dijon J, Poulingue M, Hue J (1999) Thermomechanical model of mirror laser damage at 1.06 μm. Part 1: nodule ejection. In: Exarhos GJ, Guenther AH, Kozlowski MR, Lewis KL, Soileau MJ (eds) Laser-Induced Damage to Optical Materials: 1998. SPIE 3578: 387-396

Dijon J, Ravel G, André B (1999) Thermomechanical model of mirror laser damage at 1.06 μm. Part 2: Flat bottom pits formation. In: Exarhos GJ, Guenther AH, Kozlowski MR, Lewis KL, Soileau MJ (eds) Laser-Induced Damage to Optical Materials: 1998. SPIE 3578: 398-407

Feit MD, Rubenchik AM, Shore BW, Stuart BC, Perry MD (1995) Laser-induced damage in dielectrics with nanosecond to subpicosecond pulses: II. Theory. In: Bennett HE, Guenther AH, Kozlowski MR, Newnam BE, Soileau MJ (eds) Laser-Induced Damage to Optical Materials: 1994. SPIE 2428: 469-477

Foltyn SR, Jolin LJ (1984) Catastrophic versus microscopic damage: applicability of laboratory measurements to real systems In: Bennett HE, Guenther AH, Milam D, Newnam BE (eds) Laser-Induced Damage to Optical Materials: 1983. NBS SP 688: 493-500

Fornier A, Cordillot C, Ausserre D, Paris F (1994) Laser conditioning of optical coatings: some issues in the characterization by atomic force microscopy. In: Bennett HE, Chase LL, Guenther AH, Newnam BE, Soileau MJ (eds) Laser-Induced Damage to Optical Materials: 1993. SPIE 2114: 355-365

Fornier A, Cordillot C, Bernardino D, Lam O, Roussel A, Amra C, Escoubas L, Albrand G, Commandré M, Roche P, Cathelinaud M, Gatto A (1997) Characterization of optical coatings: Damage threshold/local absorption correlation. In: Bennett HE, Guenther AH, Kozlowski MR, Newnam BE, Soileau MJ (eds) Laser-Induced Damage to Optical Materials: 1996. SPIE 2966: 292-305

Génin FY, Stolz CJ (1996) Morphologies of laser-induced damage in hafnia-silica multilayer mirror and polarizer coatings. In: Morin M, Giesen A (eds) Third International Workshop on Laser Beam and Optics Characterization. SPIE 2870: 439-448

Génin FY, Stolz CJ, Reitter T, Kozlowski MR, Bevis RP, Von Gunten MK (1997) Effect of electric-field distribution on the morphologies of laser-induced damage in hafnia-silica multilayer polarizers. In: Bennett HE, Guenther AH, Kozlowski MR, Newnam BE, Soileau MJ (eds) Laser-Induced Damage to Optical Materials: 1996. SPIE 2966: 342-352

Génin FY, Stolz CJ, Kozlowski MR (1997) Growth of laser-induced damage during repetitive illumination of HfO_2-SiO_2 multilayer mirror and polarizer coatings. In: Bennett HE, Guenther AH, Kozlowski MR, Newnam BE, Soileau MJ (eds) Laser-Induced Damage to Optical Materials: 1996. SPIE 2966: 273-282

Génin FY, Salleo A, Pistor TV, Chase LL (2001) Role of light intensification by cracks in optical breakdown on surfaces. J Opt Soc Am A. 32: 5567-5574

Gill DH, Newnam BE, McLeod J (1978) Use of non-quarter-wave designs to increase the damage resistance of reflectors at 532 and 1064 nanometers. In: Glass AJ, and Guenther AH (eds) Laser-Induced Damage in Optical Materials: 1977. NBS SP 509: 260-270

Glass AJ, Guenther AH (1973) Laser induced damage of optical elements - a status report. Appl. Optics 12: 637

Göppert-Mayer M (1931) Über Elementarakte mit zwei Quantensprüngen. Ann Phys 9: 273-294

Guenther KH (1981) Nodular defects in dielectric multilayers and thick single layers. Appl. Optics 20: 1034-1038

Guenther KH, Humpherys TW, Balmer J, Bettis JR, Casparis E, Ebert J, Eichner M, Guenther AH, Kiesel E, Kuehnel R, Milam D, Ryseck W, Seitel SC, Steward AF, Weber H, Weber HP, Wirtenson GR, Wood RM (1984) 1.06 μm laser damage of thin film optical coatings: a round robin experiment involving various pulse lengths and beam diameters. Appl. Optics 23: 3743-3752

Hacker E, Lauth H, Weiβbrodt P (1996) Review of structural influences on the laser damage thresholds of oxide coatings. In: Bennett HE, Guenther AH, Kozlowski MR, Newnam BE, Soileau MJ (eds) Laser-Induced Damage to Optical Materials: 1995. SPIE 2714: 316-330

Hue J, Génin FY, Maricle SM, Kozlowski MR (1997) Towards predicting the laser damage threshold of large-area optics. In: Bennett HE, Guenther AH, Kozlowski MR, Newnam BE, Soileau MJ (eds) Laser-Induced Damage to Optical Materials: 1996. SPIE 2966: 451-462

Jasapara J, Nampoothiri AVV, Rudolph W, Ristau D, Starke K (2001) Femtosecond laser pulse induced breakdown in dielectric thin films. Phys Rev B 63: 0451171-0451175

Kaiser N, Uhlig H, Schallenberg UB, Anton B, Kaiser U, Mann K, Eva E (1995) High damage threshold Al_2O_3 dielectric coatings for eximer lasers. Thin Solid Films 260: 86-92

Knollenberg RG, Long D, Lopez S (1995) An in-situ fiber optics sensor for monitoring particle microcontamination during an IBS optical coating process, in Optical Interference Coatings, Vol. 17, 1995 OSA Technical Digest Series (Optical Society of America, Washington DC), pp 124-126

Kozlowski MR (1995) Damage-resistant laser coatings. In: Flory FR (ed) Thin Films for Optical Systems. Marcel Dekker, New York, pp 521-549

Kozlowski MR, Chow R (1994) The role of defects in laser damage of multilayer coatings. In: Bennett HE, Chase LL, Guenther AH, Newnam BE, Soileau MJ (eds) Laser-Induced Damage to Optical Materials: 1993. SPIE 2114: 640-648

Kozlowski MR, Tench RJ, Chow R, Sheehan L (1994) Influence of defect shape on laser-induced damage in multilayer coatings. In: F. Abelés (ed) Optical Interference Coatings. SPIE 2253: 743-750

Libenson MN, Gruzdev VE (1998) Electrodynamic resonances and instabilities in transparent medium and their role in laser-induced damage. In: Exarhos GJ, Guenther AH, Kozlowski MR, Soileau MJ (eds) Laser-Induced Damage to Optical Materials: 1997. SPIE 3244: 11-19

Lowdermilk WH, Milam D, Rainer F (1980) Damage to coatings and surfaces by 1.06 μm pulses. In: Bennett HE, Glass AJ, Guenther AH, Newnam BE (eds) Laser-Induced Damage in Optical Materials: 1979. NIST SP 568: 391-403

Maiman, TH (1960) Stimulated optical radiation in ruby. Nature 187: 493-494

Matteson S, Bowling RA (1988) Particulate contamination in atomic and molecular-beam deposition systems. J Vac Sci Technol A 6: 2504-2507.

Milam D, Bradbury RA, Bass M (1973) Laser damage threshold for dielectric coatings as determined by inclusions. Appl. Physics Letters 23: 654-657

Miller MD, Chow R, Loomis GE (1994) Electrostatic reduction of particulates for laser resistant hafnia coatings. In: Bennett HE, Chase LL, Guenther AH, Newnam BE, Soileau MJ (eds) Laser-Induced Damage in Optical Materials: 1993. SPIE 2114: 426-431

Newnam BE, Gill DH, Faulkner G (1976) Influence of standing-wave fields on the laser damage resistance of dielectric films. In: Glass AJ, and Guenther AH (eds) Laser-Induced Damage in Optical Materials: 1975. NIST SP 435: 254-271

Papandrew AB, Stolz CJ, Wu ZL, Loomis GE, Falabella S (2001) Laser conditioning characterization and damage threshold prediction of hafnia/silica multilayer mirrors by photothermal microscopy. In: Exarhos GJ, Guenther AH, Kozlowski MR, Lewis KL, Soileau MJ (eds) Laser-Induced Damage to Optical Materials: 2000: SPIE 4347: 53-61

Poulingue M, Leplan H, Dijon J, Rafin B, Ignat M (1999) Generation of defects with diamond and silica particles inside high reflection coatings: influence on the laser damage threshold. In: Amra C, Macleod HA (eds) Advances in Optical Interference Coatings. SPIE 3738: 325-336

Poulingue M, Dijon J, Ignat M, Leplan H, Pinot B (1999) New approach for the critical size of the nodular defects: the mechanical connection. In: Exarhos GJ, Guenther AH, Kozlowski MR, Lewis KL, Soileau MJ (eds) Laser-Induced Damage to Optical Materials: 1998. SPIE 3578: 370-380

Poulingue M, Dijon J, Garrec P, Lyan P (1999) 1.06 μm laser irradiation of high reflection coatings inside a scanning electron microscope. In: Exarhos GJ, Guenther AH, Kozlowski MR, Lewis KL, Soileau MJ (eds) Laser-Induced Damage to Optical Materials: 1998. SPIE 3578: 188-195

Rainer F, DeMarco FP, Staggs MC, Kozlowski MR, Atherton LJ, Sheehan LM (1994) Historic perspective of fifteen years of laser damage thresholds at LLNL. In: Bennett HE, Chase LL, Guenther AH, Newnam BE, Soileau MJ (eds) Laser-Induced Damage to Optical Materials: 1993, SPIE 2114: 9-22

Rainer F, Atherton LJ, Campbell JH, DeMarco FP, Kozlowski MR, Morgan AJ, Staggs MC (1992) Four-harmonic database of laser-damage tesing. In: Bennett HE, Chase LL, Guenther AH, Newnam BE, Soileau MJ (eds) Laser-Induced Damage to Optical Materials: 1991. SPIE 1624: 116-127

Reichling M, Bodemann A, Kaiser N, (1998) Defect-induced laser damage in oxide multilayer coatings for 248 nm. Thin Solid Films 320: 264-279

Sawicki RH, Shang CC, Swatloski TL (1995) Failure characterization of nodular defects in multi-layer dielectric coatings. In: Bennett HE, Guenther AH, Kozlowski MR, Newnam BE, Soileau MJ (eds) Laser-Induced Damage to Optical Materials: 1994. SPIE 2428: 333-342

Sheehan LM, Kozlowski MR, Rainer F, Staggs MC (1994) Large-area conditioning of optics for high-power laser systems. In: Bennett HE, Chase LL, Guenther AH, Newnam BE, Soileau MJ (eds) Laser-Induced Damage to Optical Materials: 1993. SPIE 2114: 559-568

Staggs MC, Kozlowski MR, Siekhaus WJ, Balooch M (1993) Correlation of damage threshold and surface geometry of nodular defects in HR coatings as determined by insitu atomic force microscopy. In: Bennett HE, Chase LL, Guenther AH, Newnam BE, Soileau MJ (eds) Laser-Induced Damage to Optical Materials: 1992. SPIE 1884: 234-242

Stolz CJ (1999) Brewster's angle thin film plate polarizer design study from an electric field perspective. In: Amra C, Macleod HA (eds) Advances in Optical Interference Coatings. SPIE 3738: 347-353

Stolz CJ, Tench RJ, Kozlowski MR, Fornier A (1996) A comparison of nodular defect seed geometeries from different deposition techniques. In: Bennett HE, Guenther AH, Kozlowski MR, Newnam BE, Soileau MJ (eds) Laser-Induced Damage to Optical Materials: 1995. SPIE 2714: 374-382

Stolz CJ, Génin FY, Reitter TA, Molau N, Bevis RP, Von Gunten MK, Smith DJ, Anzellotti JF (1997) Effect of SiO_2 overcoat thickness on laser damage morphology of HfO_2/SiO_2 Brewster's angle polarizers at 1064 nm. In: Bennett HE, Guenther AH, Kozlowski MR, Newnam BE, Soileau MJ (eds) Laser-Induced Damage to Optical Materials: 1996. SPIE 2966: 265-272

Stolz CJ, Sheehan LM, Maricle SM, Schwartz S, Hue J (1999) A study of laser conditioning methods of hafnia silica multilayer mirrors. In: Exarhos GJ, Guenther AH,

Kozlowski MR, Lewis KL, Soileau MJ (eds) Laser-Induced Damage to Optical Materials: 1998. SPIE 3578: 144-152

Stolz CJ, Sheehan LM, Von Gunten MK, Bevis RP, Smith D (1999) The advantages of evaporation of hafnium in a reactive environment to manufacture high damage threshold multilayer coatings by electron-beam deposition. In: Amra C, Macleod HA (eds) Advances in Optical Interference Coatings. SPIE 3738: 318-324

Stuart BC, Herman S, Perry MD (1995) Laser-induced damage in dielectrics with nanosecond to subpicosecond pulses I. Experimental. In: Bennett HE, Guenther AH, Kozlowski MR, Newnam BE, Soileau MJ (eds) Laser-Induced Damage to Optical Materials: 1994. SPIE 2428: 568-578

Tench RJ, Chow R, Kozlowski MR (1994) Characterization of defect geometries in multilayer optical coatings. J. Vac. Sci. Technol. A 12: 2808-2813

Thomas IM (1994) Sol-gel coatings for high power laser optics-past, present and future. In: Bennett HE, Chase LL, Guenther AH, Newnam BE, Soileau MJ (eds) Laser-Induced Damage to Optical Materials: 1993. SPIE 2114: 232-242

Turner F (1972) Ruby laser damage threshold in evaporated thin films and multilayer coatings. In: Glass AJ, and Guenther AH (eds) Laser-Induced Damage in Optical Materials: 1971. NIST SP 356: 119-123

Tuttle Hart T, Lichtenstein TL, Carniglia CK, Rainer F (1982) Effects of undercoats and overcoats on damage thresholds of 248 nm coatings. In: Bennett HE, Guenther AH, Milam D, Newnam BE (eds) Laser-Induced Damage in Optical Materials: 1981. NBS SP 638: 344-349

Van Stryland EW, Sheik-bahae M, Said AA, Hagan DJ (1993) Characterization of nonlinear absorption and refraction in advanced materials. In: Etemad S (ed) Nonlinear Optical Properties of Advanced Materials: SPIE 1852: 135-150

Walker TW, Guenther AH, Nielsen P (1981) Pulsed laser-induced damage to thin-film optical coatings-Part I: Experimental & Part II: Theory, IEEE J. Q. E. QE-17: 2041-2065.

Walton CC, Génin FY, Chow R, Kozlowski MR, Loomis GE, Pierce E (1996) Effect of silica overlayers on laser damage of HfO_2-SiO_2 56° incidence high reflectors. In: Bennett HE, Guenther AH, Kozlowski MR, Newnam BE, Soileau MJ (eds) Laser-Induced Damage to Optical Materials: 1995. SPIE 2714: 550-558

Weakley SC, Stolz CJ, Wu ZL, Bevis RP, Von Gunten MK (1999) Role of starting material composition in interfacial damage morphology of hafnia silica multilayer coatings. In: Exarhos GJ, Guenther AH, Kozlowski MR, Lewis KL, Soileau MJ (eds) Laser-Induced Damage to Optical Materials: 1998. SPIE 3578: 137-143

Welsch E, Walther HG, Schäfer D, Wolf R, Müller H (1988) Correlation between morphology, optical loss and laser damage of MgF_2-SiO_2 multilayers. Thin Solid Films 156: 1-10

Williams FL, Petersen Jr. GA, Schmell RA, Carniglia CK (1992) Observation and control of thin-film defects using in-situ total internal reflection microscopy. In: Bennett HE, Chase LL, Guenther AH, Newnam BE, Soileau MJ (eds) Laser-Induced Damage to Optical Materials: 1991. SPIE 1624: 256-270

Wolfe R, Kozlowski MR, Campbell JH, Rainer F (1990) Laser conditioning of optical thin films. In: Bennett HE, Chase LL, Guenther AH, Newnam BE, Soileau MJ (eds) Laser-Induced Damage in Optical Materials: 1989. NIST Spec. Publ. 801, 360-375

Wood, RM (1986) Laser Damage in Optical Materials, Adam Hilger Series on Optics and Optoelectronics

Wu ZL, Reichling M, Fan ZX, Wang ZJ (1991) An understanding of the abnormal wavelength effect of overcoats. In: Bennett HE, Chase LL, Guenther AH, Newnam BE, Soileau MJ (eds) Laser-Induced Damage to Optical Materials: 1990. SPIE 1441, 200-213

Wu ZL, Stolz CJ, Weakley SC, Hughes JD, Zhao Q (2001) Damage threshold prediction of hafnia-silica multilayer coatings by nondestructive evaluation of fluence-limiting defects. Appl. Opt. 40: 1897-1906

Abbreviations

AFM	Atomic Force Microscope
CVD	Chemical Vapor Deposition
DC	Direct Current
e-beam	Electron Beam
FIB	Focused Ion Beam
IAD	Ion Assisted Deposition
IBS	Ion Beam Sputtering
IP	Ion Plating
NSOM	Nearfield Scanning Optical Microscope
PTM	Photothermal Microscope
SEM	Scanning Electron Microscope
SWEF	Standing-Wave Electric Field
TEM	Transmission Electron Microscope
UV	Ultraviolet

Symbols

β	Linear absorption coefficient
1:1	Single shot unconditioned laser-induced damage threshold
A	Total absorption
C	Contrast ratio between high and low refractive index
b	Damage threshold scaling factor
dz	Sample thickness
I	Incident intensity
m	Number of layer pairs
N:1	Conditioned ramped laser-induced damage threshold (5-10 shots)
n_f	Refractive index of the film
n_h	Refractive index of high index material
n_i	Refractive index of the incident medium
n_l	Refractive index of low index material
n_s	Refractive index of the substrate
τ_t	Test pulse length

τ_u	Use pulse length
R	Reflectance
R:1	Conditioned (ramped) laser-induced damage threshold (many shots)
S:1	Multiple shot unconditioned laser-induced damage threshold
F_D	Damage threshold at use pulse length
F_{Dt}	Damage threshold at test pulse length
β_2	Two-photon absorption coefficient

Coatings for UV- Free Electron Lasers

Alexandre Gatto

Fraunhofer Institute for Applied Optics and Precision Engineering, Jena, Germany

1 UV-Free Electron Lasers

Free Electron Lasers (FELs) generate tuneable, coherent, high power radiation currently spanning wavelength emissions from the millimetre to the UV and potentially to the X-ray spectral range. The FEL principle is based on the amplification of an electromagnetic wave by a high energy relativistic electron beam, hence the term free electron, as opposed to inversion population with bound atomic or molecular states, as for usual so called "quantum lasers" (Svelto 1998). FELs can have the optical characteristic properties of quantum lasers such as high spatial coherence and near diffraction limited radiation beams.

Historically, during the seventies and eighties FEL operations began and grew up strongly in the far infrared, infrared (IR) and visible spectral region (Decaon et al. 1977, Billardon et al. 1983), where technical requirements concerning optics and accelerator technology are less stringent compared to operations at shorter wavelength. Nowadays, of the various types of FELs, those based on storage rings (SRFELs) have a combination of characteristics (Couprie 2000), such as GW power ps pulses, single transverse mode and tuneable wavelength at a MHz repetition rate (Walker 1999, Walker et al. 2001), that make them attractive for performance of original experiments in the UV and deep ultraviolet (DUV) spectral range. UV-SRFELs are specific and sometimes unique tools for performance of advanced applications in spectroscopy, photochemistry and photobiology or semiconductor surface studies for example (Marsi et al. 1997).

Commercial optics are available for the IR region. However, operating FELs in the DUV or Vacuum-UV (VUV) requires overcoming severe and crucial technological problems related to optics and accelerator technology. At such short wavelengths, the problem of designing robust mirrors that sustain broad synchrotron radiation becomes a hard limitation on the use of oscillators. Therefore, a transition to high gain, single pass amplifiers using Self-Amplified Spontaneous Emission (SASE) is nowadays presented as an interesting and competitive alternative (Roßbach 1996, Andruszkow et al. 2000).

Nevertheless, high performance DUV optics have been developed to create low loss, high reflectivity dielectric mirrors with long lifetimes under harsh and broad band synchrotron radiation environments. That is the topic of this chapter. A brief introduction to FEL principles and to the characteristics of synchrotron radiation which constitute the specific radiation received by a FEL's cavity mirror are discussed in Sect. 1. Sect. 2 treats typical damage observed on SRFEL optics. As concrete examples, in Sect. 3 a review is given on recent optics developed for the FEL project at ELETTRA (Trieste, Italy), Europe's first "third generation" high brightness synchrotron radiation source for the VUV/Soft X-ray region. In February 2001, lasing at 189.7 nm was achieved, the shortest wavelength obtained so far with FEL oscillators. Finally, future trends and perspectives are presented in Sect. 4.

1.1 Storage Ring Free-Electron Lasers

The operating principle of a storage ring FEL can briefly be described as follows. As depicted in Fig. 1, an electron beam in a storage ring circles through the alternating static magnetic field produced by undulator magnets.

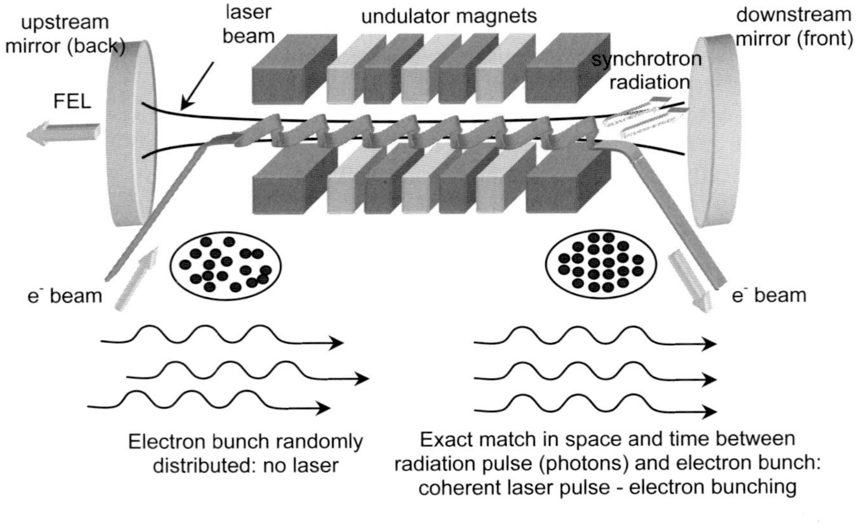

Fig. 1. Basic layout of a storage ring FEL. Displayed are the three constituents: electron beam, undulator magnets and optical cavity. Schematic of the match between radiation pulse (photons) and electron bunch to obtain a coherent emission (electron bunching).

Each time the electrons move trough the magnetic field they are forced to follow a wiggled path and therefore emit synchrotron radiation. This radiation contains a broad set of spectral lines that covers a rather wide range of wavelengths starting from the far IR and going to the hard X-ray region. By storing the undulator radiation between two mirrors (the optical cavity) located outside the undulator (no electron impinges the mirrors), it is possible to "favour" a particular wavelength out of the spectrum. The electron phases are initially distributed randomly with respect to the electromagnetic wave phase in the cavity. Some of the electrons gain and some loose energy because of the electron-photon interaction. Electromagnetic radiation at this wavelength can then be amplified if there is an exact match in space and time between the radiation pulse (photons) travelling back and forth in the optical cavity and the electron bunch circling in the storage ring. If the resonance condition, which contains the electron energy, the undulator parameter and the radiation wavelength, is fulfilled, energy can be exchanged between electron bunch and radiation pulse (photons) so that the radiation amplification results in a coherent laser pulse. This amplification process is called electron bunching (Freund and Antonsen 1996). When casting this resonance condition for radiation amplification into mathematical form, equation (1) is obtained, which links the emitted synchrotron radiation wavelength λ_r to the electron and undulator parameters:

$$\lambda_r = \frac{\lambda_u}{2m\gamma^2}\left(1 + \frac{K^2}{2} + \gamma^2\theta^2\right) \tag{1}$$

Here the Lorentz factor γ determines the emission energy range and spatial period λ_u of the undulator and the deflection parameter K proportional to λ_u B_u allows the radiation to be tuned continuously. Parameter m accounts for the harmonics emitted by the undulator and angle θ between the observer and the propagation axis accounts for the circular symmetry of the emitted radiation.

Thus, the electromagnetic waves radiated by the electrons are coherent and in phase with the electromagnetic wave of the cavity. The emitted photons of the fundamental wavelength stored in the optical cavity, interacting constructively with the electrons circulating in the ring, generate a laser pulse at the fundamental wavelength.

1.2 Principal FEL Properties

Many of the properties of FELs can be inferred from the above short presentation. The main properties of these sources are listed below (Levy et al 1994):

- *Tunability:* The amplifying medium (electron beam) allows broad tunability. In contrast to the electrons in quantum lasers, which are bound to certain discrete energy levels (Svelto 1998), the electrons in the accelerator move freely through the undulator and their energy is only determined by the machine parameters. This difference makes the FEL principle theoretically applicable to a wide range of radiation wavelengths and in particular to those spectral regions in which no other lasers exist.

- *High peak and average power:* GW watt peak powers have been demonstrated in the IR spectral region. Average available powers extend to some several kW in the IR to several hundreds of W in the UV.
- *High brilliance:* Typical brilliances up to 10^{26} photons/s/mm²/mrad²/0.1band width can be achieved with storage ring FEL sources. For the machine implemented on the so-called "third generation synchrotron source", the number of coherent photons per second in the UV can reach 10^{22} photons/s (brilliance of FELs several order of magnitude above that of quantum laser).
- *Flexible pulse structure:* The laser radiation follows the pulse structure of the electron beam. Picosecond pulses with sub-picosecond jitters can be produced.
- *Laser characteristics:* Single transverse mode, high spatial and temporal coherence and flexible polarisation properties (linear polarised for planar undulator, circularly polarised for helical undulator) are achievable.
- *Broad wavelength coverage:* Because the gain medium is transparent, FELs in principle can produce laser radiation up to the X-ray region (1 to 10 nm). In practice, electron beam energy, current, brilliance and energy spread requirements become more stringent as the wavelength decreases.
- *Size and cost:* FELs require an electron accelerator with its associated shielding. Devices are expensive and space consuming.

1.3 Synchrotron Radiation Impinging the Front Cavity Mirror

Synchrotron radiation take place because accelerated charge radiates. This fundamental principle applies in particular to the relativistic electrons (amplifying medium of FEL) that are generally used in the storage rings of light sources. Intense radiation can be produced by using special undulators (Attwood 1999).

In Fig. 1 the optical cavity of a FEL composed of a downstream mirror (called the front mirror) and an upstream mirror (called the back mirror) is shown. The downstream mirror (front) is exposed to both the spontaneous synchrotron radiation and the FEL beam and therefore to much more damage problems than the upstream one. Radiation impinging the front mirror has a broadband spectrum extending from the IR to hard X-ray spectral region either continuously (bending magnets) or containing a series of harmonic peaks (cf. Eq. (1)) with a high mean spectral flux (up to 10^{16} photons/s/mm/mrad^2 for the third generation synchrotron). Fig. 2 shows a computation of the relative photon flux versus photon energy received by the front mirror of a FEL cavity for three different kinds of undulators (amplifying media).

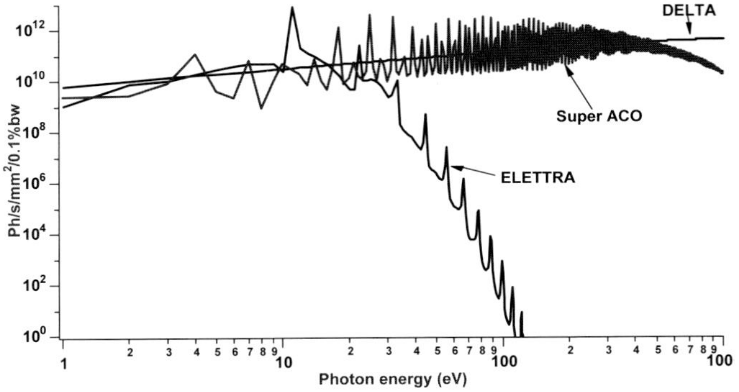

Fig. 2. Computation of the number of photons per second and mm² in 0.1 % of bandwidth versus photon energy received by the front mirror of an FEL cavity. DELTA: bending magnet, Super ACO: linear undulator, ELETTRA: helical undulator (by D. Garzella, LEL Super ACO group, Orsay).

A large part of this spectrum is not filtered in the FEL optical cavity and it is partially absorbed by the mirrors (especially the front one). This feature, combined with the hostile environment surrounding the mirrors in the vacuum chamber (hydrocarbon molecules and metal outgassing products), contributes to the enhancement of cavity total losses and thus to a decrease of laser lifetime.

Table 1 summarises the principal orders of magnitude of the radiation received by a FEL mirror compared to an excimer laser mirror.

Table 1. Principal characteristic orders of the radiation received by a UV FEL mirror compared to an excimer laser mirror.

	UV FEL	Excimer laser
Active medium	undulator + e^- beam **free electrons**	gas + discharge **bound electrons**
Amplification	electron bunching	population inversion
Saturation	overbunching	population inversion
Pulse length	10^{-12} s - 10^{-11} s	10^{-9} ns - 10^{-8} ns
Repetition rate	few MHz	few kHz
Pulse energy	several 10^{-5} J	few 10^{-3} J
Radiation received by the mirror	**synchrotron radiation + FEL radiation photon energy: [1 eV, 10 keV]**	gas discharge + laser radiation photon energy: < few tens of eV

Photon energy impinging the mirror cavity can reach several keV. To quantify the "dose of synchrotron irradiation" received by the mirrors, the integral of the current circulating in the ring over the irradiation time is conventionally calculated and given in amperes per hour.

2 The UV-FEL Optic Challenges

There is an increasing interest in the generation of intense, tuneable and coherent radiation in the UV-VUV wavelength region to be used both as source for basic research and for technological applications. Storage ring FELs represent a practical technical approach to producing photons with these characteristics. At various laboratories around the world the operating wavelength is being extended to shorter values. In particular, because of continual progress in the performance of particle accelerators and in multilayer mirror technology, marked improvement has been obtained during the past three years. After a wavelength of 212 nm was obtained at NIJI-IV ("Niji" means "rainbow" in Japanese) [Tsukuba, Japan] (Yamada et al. 1999, 2000), the Duke FEL [Duke University, Durham, USA] was able to lase below 200 nm (Litvinenko 1999). More recently, the European FEL at ELETTRA [Trieste, Italy] produced laser radiation below 190 nm (Trovò et al. 2002). SRFELs can also reach a level of reliability sufficient for original experiments to be performed in various domains, as demonstrated with Super ACO [Super Anneau de Collision d'Orsay, Orsay, France] (Couprie et al. 1994, Marsi et al. 2000) and UVSOR [Ultra Violet Source of Radiation, Ozaki, Japan].

The mirrors that compose the optical cavity (mainly the front one), which are in ultrahigh vacuum, and their performances in terms of robustness and resistivity to ageing are key factors for operation at shorter wavelengths in the DUV spectral region (Gatto et al. 2001). Processes such as increasing absorption losses and mirror heating (Garzella et al. 1995, 1996), by which the mirrors of the optical cavity are degraded, were investigated previously. Both of these phenomena are due to the extreme environment that comprises the synchrotron radiation and the residual gases present in the optical cavity. The electromagnetic radiation impinging upon the mirrors of the FEL optical cavity is quite different from the quasi monochromatic radiation that issues from sources like pulsed excimer laser (Gatto et al. 2001).

2.1 Synchrotron Radiation Induced Damage

Optics for FEL cavities have to be tested in terms of robustness and resistance to synchrotron radiation, especially in the DUV-VUV region where optics are becoming a limiting factor. Therefore, the development of new light sources during the last years like the SRFELs creates new requirements on and challenges to the UV absorption, to the packing density as well as to the radiation stability of the optical coatings used as mirror systems in the optical cavity of the laser.

SiO_2 and HfO_2 are among the most important oxide thin film materials for the manufacture of interference coatings in the DUV spectral region down to 248 nm or even down to 200 nm in the case of SiO_2 and Al_2O_3 (see the Chap. "Optical coatings for DUV/VUV" by Roland Thielsch). These materials have already been the subject of many studies even for their UV laser radiation resistivities but mainly in the context of the development of radiation resistant defect-free dielectric mirrors for high energy Nd-YAG applications (355 nm) or other laser sources in the UV spectral range (Bodemann et al. 1995).

In damage or ageing studies, mirrors of the optical cavity are usually exposed to electromagnetic energy at a defined wavelength, which is usually that of the laser cavity. Several national projects (National Ignition Facility, USA; Mega Joule, France) have led to active studies on damage phenomena at 1064 nm or at 355 nm in different materials or on optical components (see the Chap. "Laser Resistant Coatings" by C.J. Stolz and F.Y. Génin). In these cases, optics are generally exposed to a monochromatic pulse or a series of monochromatic pulses with a duration of a few ns. The photon energies range from 1.1 eV for a YAG laser at 1064 nm to 7.9 eV for a F_2 laser at 157 nm.

In the present experiments concerning SRFEL optics, the radiation source responsible for damage is not monochromatic because in SRFEL conditions the front mirror of the FEL cavity receives all synchrotron radiation, typically noncoherent and polychromatic. As detailed in Sect. 1, photon energy impinging on the mirror can extend from 1 eV to several keV. The temporal structure of synchrotron radiation is also different. Optics are exposed to a series of micro-pulses of a few tens of μJ in energy with a repetition rate up to several MHz. Therefore, these sections are devoted to typical damage observed on SRFEL optics and inband and out-band irradiation suffered by mirrors in FEL operation.

2.2 Typical Damages Observed on FEL Optics

In this section is presented a short list of damages observed on optical components irradiated with synchrotron radiation. Typically, such irradiation can lead to increased values of surface optical roughness as exposed in 2.2.1., which can be correlated to a change of crystalline structure as showed in 2.2.2. A carbon contamination is a measurable factor limiting the optic's lifetime and can be quantified by X-ray photoelectron spectroscopy (XPS), as resumed in 2.2.3.. Experiments have been performed on metal oxide samples developed by Plasma and Ion Assisted Deposition (PIAD) on UV fused silica substrates.

2.2.1 Atomic Force Microscopy Inside-Outside a Damaged Area

The studied samples are High Reflective (HR) coating at 248 nm (HfO_2/SiO_2). These mirrors have been irradiated with a dose of 979.5 mAh at the DELTA dipole (Dortmund, Germany). The damage area appears as a "burned" zone of 1 mm × 2 mm. Fig. 3 shows a montage with two atomic force microscopy (AFM) mappings taken inside and outside a damaged area (Torchio et al. 2001).

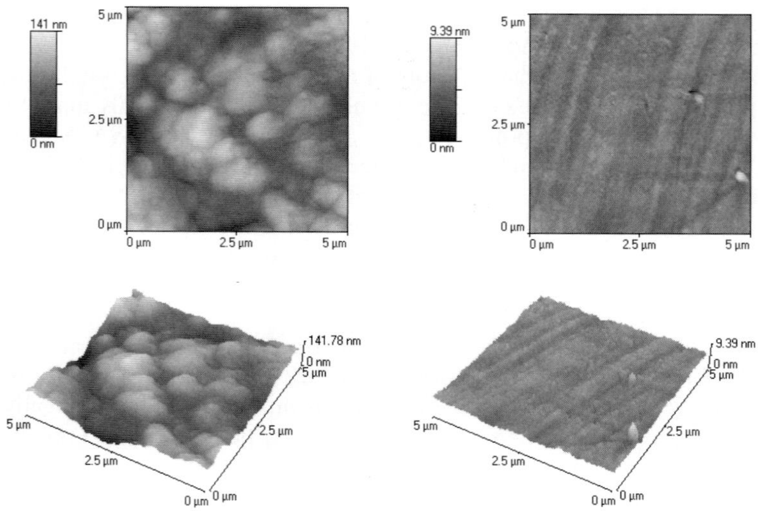

Rms. roughness: 26.30 nm	Rms. roughness: 0.40 nm
AFM image inside the damaged area	AFM image outside the damaged area
Crystalline HfO_2 structure	Amorphous HfO_2 structure

Fig. 3. AFM comparison inside/outside a synchrotron radiation damaged area. HR coating (HfO_2/SiO_2) irradiated at the Delta dipole (dose: 979.5 mAh) (by P.Torchio, Institut Fresnel, Marseille).

On the right part of Fig. 3 the AFM image taken outside the damaged area reveals a smooth surface with a RMS roughness of 0.4 nm (identical roughness obtained on a non irradiated sample). The left part of Fig. 3 shows an image obtained in the damaged area. The roughness is drastically increased by a factor of 50. The coating is still present (no delaminating of the stack) but a structural modification can be expected that will be confirmed in the next section.

2.2.2 Localised X-ray Diffraction Investigation

Large-angle X-ray scattering (LAXS) of Cu Kα radiation (λ = 0.154 nm) in symmetric θ–2θ geometry was systematically applied to obtain spatially localised information about the change in crystallisation degree of superficial HfO_2 layers in respect to the applied dose. Measurements were performed on a series of identical mirrors irradiated with increasing doses by the Delta dipole in Dortmund. Results concern measurements performed inside irradiated areas with respect to the dose. Fig. 4 shows X-ray diffraction curves in LAXS configuration (diffracted intensity in counts versus the angle) measured on damaged areas of sample irradiated with 101, 156 and 979 mAh, respectively. Diffraction curves have been recorded in the same experimental conditions. In each case the parameters for the acquisition and investigated area are identical.

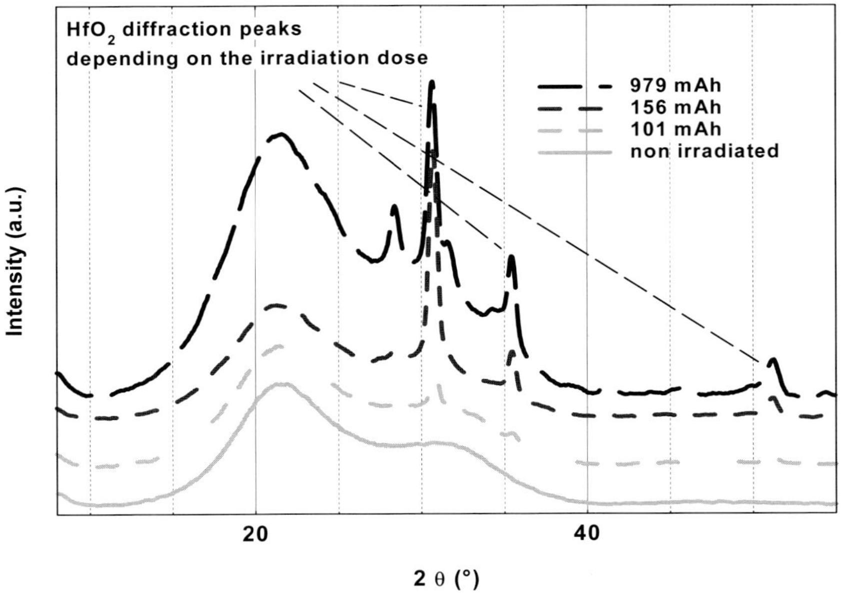

Fig. 4. Large-angle X-ray scattering performed on identical high reflectivity mirrors at 250 nm (HfO_2/SiO_2) irradiated with increasing doses at the DELTA dipole (101 mAh, 156 mAh, 976 mAh).

Before irradiation the HfO_2 superficial layer has an amorphous structure. Then a modification of the HfO_2 structure (from amorphous to crystalline) is measured, increasing with the dose of synchrotron radiation.

2.2.3 Carbon Contamination

Metal oxide layers of HfO_2, SiO_2 and Al_2O_3 have been irradiated at the Super ACO Storage Ring (Orsay, France). The respective doses were 95 mAh for the HfO_2 layer, 65 mAh for the SiO_2 layer and 95 mAh for the Al_2O_3 layer. XPS measurements were performed on irradiated samples (Al Kα radiation, 150-1487 eV, photon energy = 150 eV). Fig. 5 shows the carbon cover measured for the three single layer materials in the irradiated area compared to the carbon cover outside the irradiated area. Values obtained inside the irradiated area are between 3 and 8 times higher than those measured outside the irradiated area.

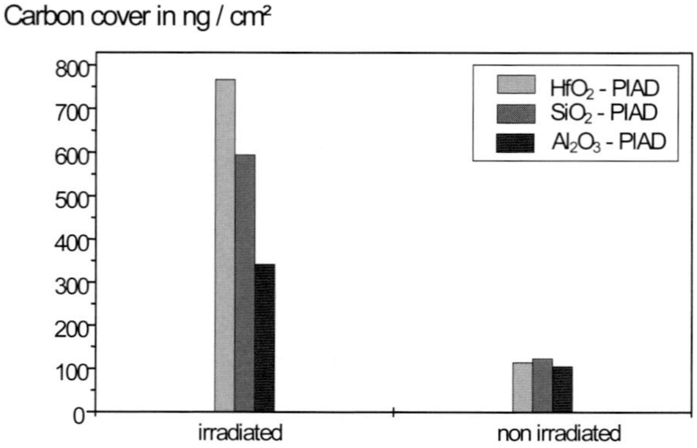

Fig. 5. Carbon cover in irradiated and non irradiated areas for 50 nm thick HfO_2, SiO_2 and Al_2O_3 single layers deposited on UV fused silica (by Peter Jörchel, Jenoptik, Laser, Optik, Systeme GmbH).

Carbon contamination (out gasing products, residual gases) is clearly measurable on coating surfaces exposed to synchrotron radiation and appears as an active ageing factor for the optics.

2.3 In-Band and Out-Band Irradiation

With respect to coating design, studies have been realised on the correlation between laser damage thresholds and standing wave electric field distributions in the multilayer coating. The standing wave electric field is a function of the refractive index of the coating layers, the irradiation wavelength and the thickness of the coating layers. In a quarter-wave design at normal incidence (QWOT is the quarter wavelength optical thickness at λ) the extrema in the standing wave electric field pattern at wavelength λ occur at layer interfaces. Because laser damage could be associated with discontinuities (high absorption centres, low mechanical stability) in film properties at layers interfaces, design coatings with electric fields reduced at the interfaces have been developed (see Chap. "Laser Resistant Coatings" by C.J. Stolz and F.Y. Génin). In the FEL case, independent of the chosen design, the radiation source in the FEL cavity is not monochromatic because the front mirror of the cavity receives all the synchrotron radiation. Wavelengths of light impinging on this mirror are numerous. Therefore the electric field distribution is also multichromatic in the stack. Such a distribution can lead to serious problems of localised losses generated by in-band and out-band radiation coming from the synchrotron radiation source.

The front mirror of a FEL resonator is exposed to the entire synchrotron radiation during the operation of the laser. Therefore, the electric field distribution which takes place in the multilayer differs quiet greatly from a mirror installed in a quasi monochromatic cavity. It varies with the multiple irradiation wavelengths λ_r of the synchrotron radiation. Usually, for a mirror irradiated by a monochromatic wavelength inside its band of maximal reflectivity, the electric field distribution versus thickness is maximal in the first layer interface and decreases rapidly until the substrate where the intensity (the square of the electric field amplitude) is practically equal to zero. Here, to depict the electric field distribution obtained in the volume of a mirror exposed to synchrotron radiation, the normalised electric field distribution is computed for a series of wavelengths λ_r including the band of maximal reflectivity versus the thickness d. It is assumed that the each different wave of wavelength λ_r constituting the synchrotron radiation impinging on the mirror has equal energy and results in a linear electric excitation for each independent wavelength. It leads to a 3D diagram (Fig. 6) where $|E|^2/|E_0|^2$ (E : electric field amplitude in the stack, E_0 : amplitude of the incoming wave) can be computed versus λ_r and d (expressed in number of QWOT layers). Real dispersion data of materials including absorption are used.

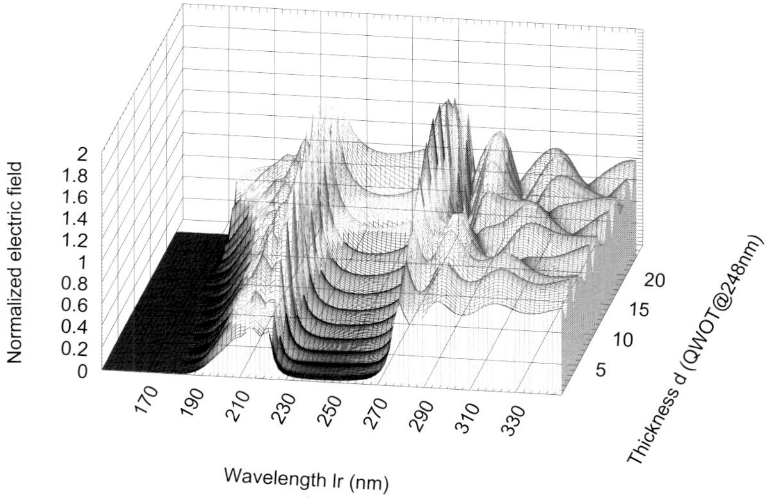

Fig. 6. Simulation of the normalised electric field distribution in a HR at 248 nm / 0° mirror (HfO_2/SiO_2) irradiated by synchrotron radiation in-band and out-band of the maximal reflectivity spectral range.

Here three different regions can be considered:

- λ_r < *170 nm*: Absorbing region. In this spectral domain HfO_2 is already strongly absorbing. The electric field distribution in the volume of the stack is zero. Therefore, the optical intensity is also zero in the volume and all the energy is assumed to be deposited at the surface of the stack and can activate carbon adsorption, for example.
- λ_r > *290 nm*: Transmitting region. In this spectral domain the residual reflection is relatively small (lower than 20 %) and thus the transmission is greater than 80 %. The electric field distribution is out of any resonance

and oscillates around a normalised average value of 0.7 until the IR spectral region.

- λ_r = *225 nm and 280 nm*: Maximum of the electric field distribution. In this spectral region where R is minimal the intensity of the electric field is maximal in the volume of the component. Thus the medium layers appear as preferential damage regions.

In conclusion, operation of dielectric mirrors under synchrotron radiation results in both the contamination of the coatings, even in ultrahigh vacuum, by carbon containing gas residuals and from the VUV-X-ray part of the undulator radiation. These conditions cause absorption to increase during operation and can lead to achromatic damage. Due to the multi-wavelength electric field excitation of the mirror caused by synchrotron radiation, not only can resonance inside the band of maximal reflection occur but also out-band. Depending on the wavelength, optical energy is localised at the surface or in the multilayer. This property, combined with the hostile environment surrounding the mirrors (hydrocarbon molecules and metal out-gasing products), contributes to enhancement of the cavity losses and, as a consequence, to a decrease in the laser's lifetime.

3 Coating Development

3.1 High Performance DUV Optics for SRFELs

Mirror technology is a crucial issue for SRFELs, which have stringent requirements in terms of mirror degradation because of the VUV-X-ray content of the undulator radiation, as described above. To ensure a wide tunability range and high output power of SRFELs, UV optics are important for their performance. Low total losses may permit laser oscillation at shorter wavelengths and more FEL out put power. The SRFEL power, of major importance for laser utilisation, increases with beam energy. The drawback is that higher beam energy induces a higher mirror degradation rate with a significant increase in the absorption of the mirror. Even if new facilities operate with much higher gains, allowing the loss values for the laser threshold to be relaxed, it is still important to reduce the optical losses as much as possible. Several previously reported tests (Yamazaki et al. 1993, Garzella et al. 1996) gave interesting indications. A higher degradation rate can occur for mirrors manufactured by thermal evaporation, whereas energy assisted techniques (ion beam sputtering (IBS), advanced plasma sources (APS) or PIAD, etc.) have lower degradation rates (Gatto et al. 2000) by as much as a factor of 50 in some cases. Results were correlated and attributed to higher degrees of amorphism and homogeneity of the films.

3.2 Substrates

Figure 7 shows the transmission in the VUV-DUV region of usual substrates employed in FEL optic development. A global overview of substrate materials for use in the ultraviolet spectral range is given in Chap. "Optical Coatings for the DUV/VUV" of this book.

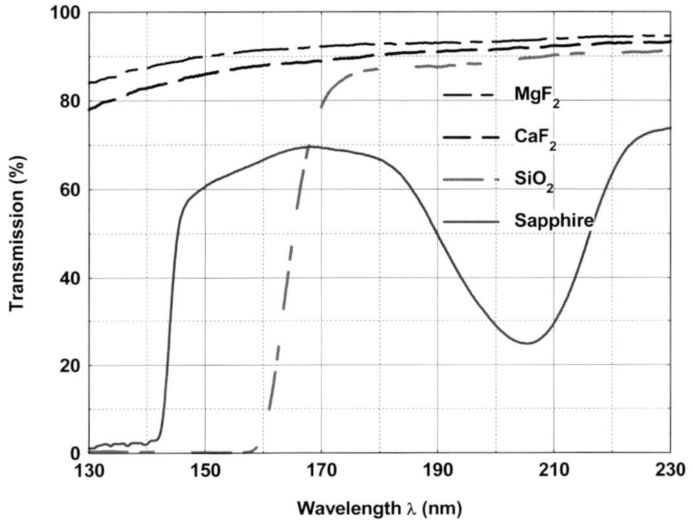

Fig. 7. Transmission in the DUV-VUV region of substrates employed in FEL optic development.

Due to its high thermal conductivity, sapphire is also a good candidate for high reflective coating for FEL. For our applications devoted to UV-FEL optics, the transmission of the substrate is an important factor to obtain more FEL output power. Low total losses including those of the substrates allow laser oscillation at shorter wavelength and greater efficiency. The stability of the substrate against broad band radiation is also of importance. It implies limitation of photo induced absorption due to colour centre or defect formation.

3.3 Coating Materials

Figure 8 and 9 show the refractive index n and the extinction coefficient k of SiO_2, HfO_2 and Al_2O_3 deposited by PIAD and IBS (IBS layers are developed at Laser Zentrum Hannover) in standard conditions for FEL coating development (a review is given by Martin Friz in Chap. "Coating Materials" of this book).

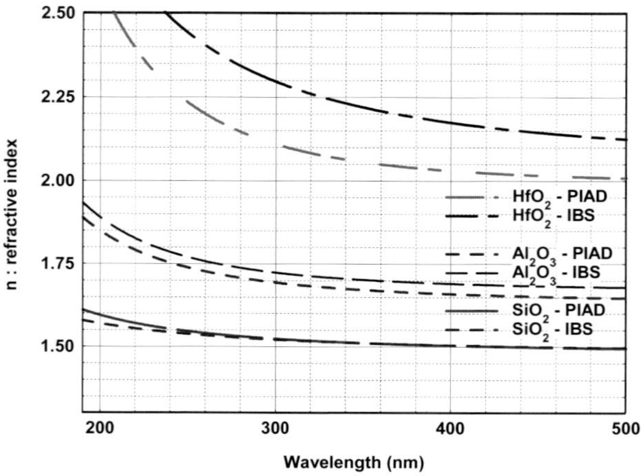

Fig. 8. Refractive index of SiO_2, HfO_2 and Al_2O_3 (mechanical thickness ~ 50 nm) deposited by PIAD and IBS at standard conditions for FEL optics development.

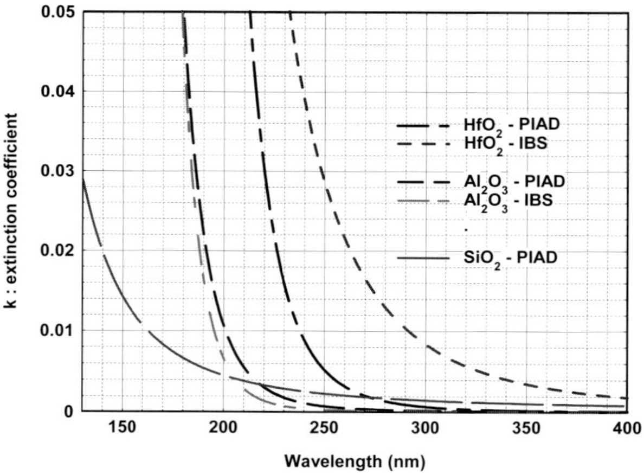

Fig. 9. Extinction coefficient k of SiO_2, HfO_2 and Al_2O_3 (mechanical thickness ~ 50 nm) deposited by PIAD and IBS at standard conditions for FEL optic development.

Optical interference coatings for UV applications usually have to meet a variety of specific requirements such as low absorption and scattering losses, high resistance to damage by high-fluence excimer laser radiation and long-term optical stability. Fulfilling these requirements represents a big challenge in mirror technology for the UV-VUV range (see Chap. "Optical Coatings for DUV/VUV" by Roland Thielsch). Thermal evaporation with controlled substrate temperature and reactive gas partial pressure give good performance, in terms of high stability and packing density, down to 180 nm and is an ultralow loss evaporation process (Kaiser et al. 1995, Heber et al. 1999).

3.4 Multilayer Prototypes

As concrete and illustrative examples, a review is given of several multilayer prototypes developed for operational tests on the ELETTRA SRFEL. The main characteristics are summarised in Table 2. The wavelength ranges over which lasing has been obtained extend from 350 nm to 190 nm.

Table 2. Overview of the prototype mirrors designed for the SRFEL cavity at ELETTRA.

Mirror (technique)	λo [nm]	Layer materials	Substrate	Lasing range [nm]	Max extracted power [mW]
Dual band HR (PIAD)	220 380	Al_2O_3/SiO_2 HfO_2/SiO_2	Sapphire Silica	217.9-224.1 355.7-362.7	10 at 27 mA
Dual band HR (IBS)	220 380	Al_2O_3/SiO_2 HfO_2/SiO_2	Si	355.7	2 at 23 mA
Single band HR (PIAD)	250	HfO_2/SiO_2	Sapphire Silica	240.4-264.7	Low power
Single band HR T ~ 1.1 %. (PIAD)	250	HfO_2/SiO_2	Sapphire	236.5-268.7	200 at 38 mA 100 at 20 mA
Single band HR T ~ 5 %. (PIAD)	250	HfO_2/SiO_2	Sapphire	242.2-266.5	330 at 36 mA
Single band HR (IBS)	190	Al_2O_3/SiO_2	Sapphire Silica	189.7-200.3	16.5 at 31 mA

3.4.1 Lasing at 220 nm

Dual band mirrors have been tested (Walker 1999, Walker et al. 2001). The aim was to achieve two lasing operations around 380 nm and 220 nm with the same optics. The component consisted of a superstructure that had two bands of maximal reflection centred at 220 nm and 380 nm at normal incidence. The design was $S/(H_1L)^{11}\ H_1\ (LH_2)^{24}$ (S is the substrate, H_1 = QWOT HfO_2 at λ_1, $H_2 = Al_2O_3$ at λ_2, and L = QWOT SiO_2, where QWOT is the quarter wavelength optical thickness at $\lambda_1 = 200$ nm and $\lambda_2 = 380$ nm; 71 layers in total) deposited on a silica substrate calculated for both 220 and 380 nm at 0° of incidence. After alignment, lasing was first achieved at 350 nm by use of conventional undulator settings (Walker et al. 2001). After changing the undulator gap, lasing at 220.5 nm was immediately obtained without any transverse or longitudinal mirror adjustment. Changing the undulator gap allowed lasing to occur over the range 217.9-224.1 nm. Figure 10 shows that the reflectivity of the 220 nm band is stable even after 3 FEL operations.

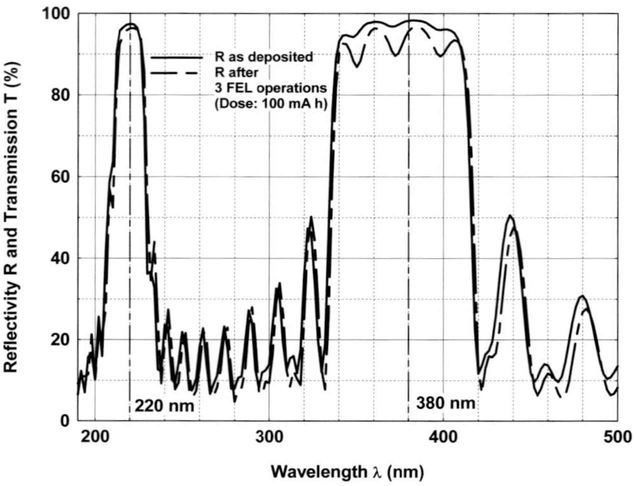

Fig. 10. Spectral characteristics R reflection, T transmission of a HR 220/0° and 380/0°, Al_2O_3/SiO_2 and HfO_2/SiO_2 deposited by PIAD measured after deposition and after 3 FEL operations (dose: 100 mAh).

Nevertheless, the band at 350 nm presents a significant reduction in reflectivity (~ 5 %) at 3 different points distributed between 350 nm and 400 nm. They correspond to 3 maxima of the electric field distribution which occur in the second band of reflectivity (350 nm - 400 nm) of the stack when the component is irradiated by the broad spectrum synchrotron radiation. Figure 11 gives the 3D repartition of the electric field distribution versus wavelength and optical thickness in the case of the dual band. Three maxima at 355, 375 and 395 nm of the electric field distribution are present, where a more intensive degradation is observed. These maxima are correlated to the mimima of the reflectivity of the component measured after FEL operation and therefore to the maxima in total losses (*total losses = 100% - R - T*) measured after FEL operation.

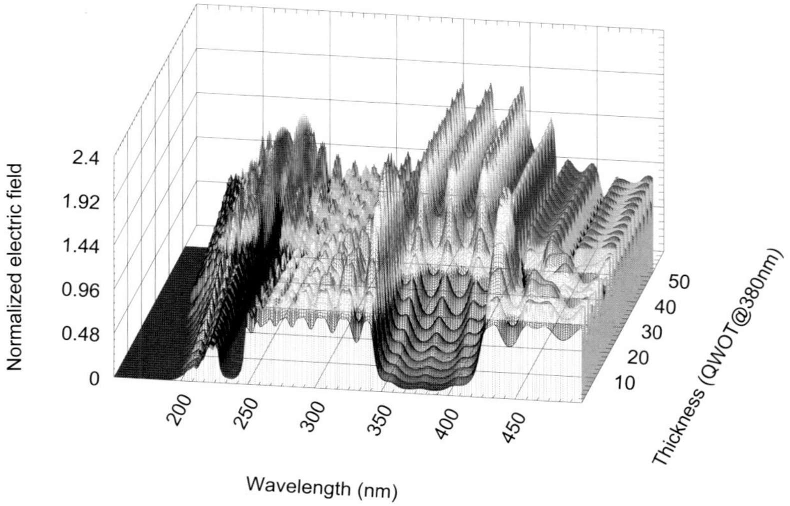

Fig. 11. Simulation of the normalised electric field distribution in a dual band mirror (HR 220/0° and 380/0°, Al_2O_3/SiO_2 and HfO_2/SiO_2 deposited by PIAD) irradiated by synchrotron radiation. Three maxima in the second reflection band can be correlated to the three minima of the reflectivity of the component measured after FEL operation.

3.4.2 Output Power and Tunability Around 250 nm

Conventional HR designs give theoretically and actually a maximal reflectivity (R) but a minimal transmission (T) equal to *100% - R – (total losses)* in a first approximation. For a thin film on UV fused silica, as in the dual band mirror described above, the transmission is ~ 0.2 % when the reflectivity is maximal. Lower values can be obtained when the lower limit of the substrate transmission range or an absorption band is approached, as is the case for fused silica around 165 nm (Fig. 7) and sapphire around 140 nm (absorption edge) or 205 nm (absorption band). To explore the possibility of increasing the extracted power, reduced quarter wave mirrors have been designed with tailored transmission. A HR normal incidence mirror at 240 nm was first designed with $S/(HL)^{11}H$ on silica and sapphire, where S is the substrate, H = QWOT HfO_2 and L = QWOT SiO_2. R and T achieved in the reflectance band were ~ 98.5 % and ~ 0.1 %, respectively. Then two reduced designs $S/(HL)^7H$ for design 1 "medium transmission" and $S/(HL)^5H$ for design 2 "higher transmission" were calculated and used to tailor the desired transmission. Figure 12 shows the reflectivity and transmission curves with 1.1 % and 5.0 % of transmission measured on the test samples respectively for design 1 and 2 (respective reflectivity, 96.5 % and 93.0% at 245 nm). Together with a low transmission (T_1 = 0.1%, R_1 = 98.5% at 245 nm) back mirror studies were carried out with a medium transmission (T_2 = 1.1%, R_2 = 96.5 % at 245 nm) and a higher transmission (T_2 = 5.0%, R_2 = 93.0% at 245 nm) front mirror.

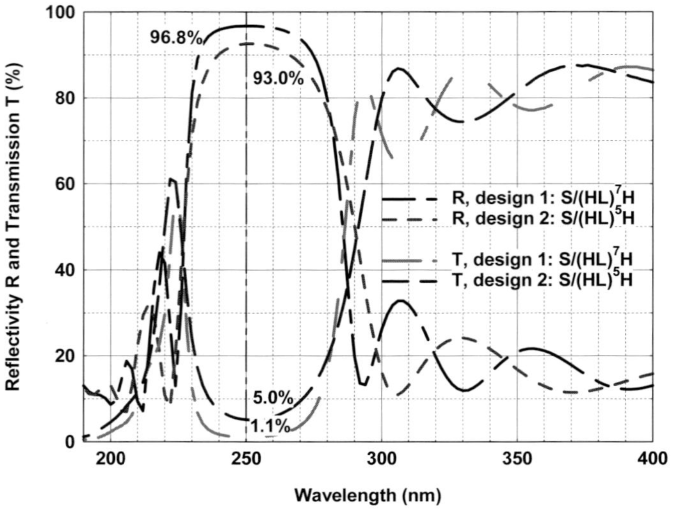

Fig. 12. Design 1 "medium transmission" at 240 nm, HfO_2 /SiO_2, $(HL)^7$ on sapphire, lasing: 251 nm, extracted power: 100 mW for 20 mA. Design 2 "higher transmission" at 240 nm, HfO_2 /SiO_2, $(HL)^5$ on sapphire, lasing: 252 nm, extracted power: 330 mW for 36 mA.

A maximum of 330 mW of extracted power at 250 nm was recorded with 36 mA beam current with the higher transmission mirror (Table 1) and the HR mirror as back mirror. Figure 13 shows the power as a function of wavelength for both front mirror types. With medium transmission, a tunability of 32.4 nm was obtained in the range 236.5-268.7 nm.

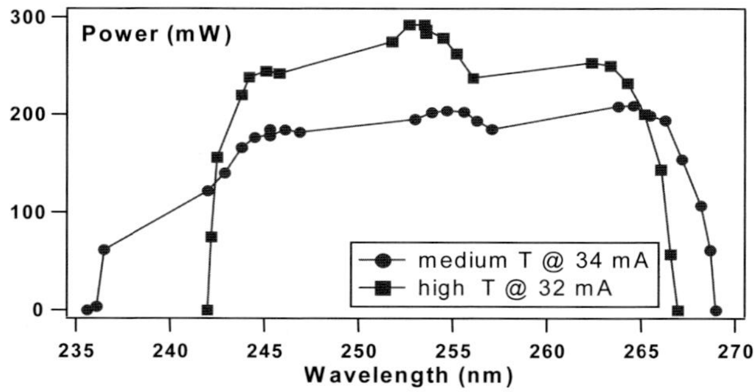

Fig. 13. Measured output power as a function of wavelength with medium and higher transmission mirrors (Trovò et al. 2002).

The useful width is of the order of 32 nm (24 nm) with the medium (higher) transmission mirrors, respectively. The gaps in the data are due to the fact that the wavelength was changed by making only a change of undulator gap, leaving the dispersion parameters fixed. This causes the laser to occasionally jump from one fringe of the optical spectrum to another.

Figure 14 shows the measured extracted power from both front mirror types as a function of beam current circulating in the electron ring.

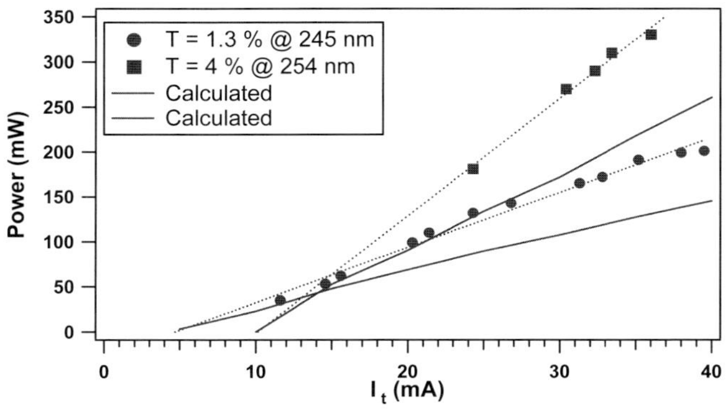

Fig. 14. Measured and calculated output power as a function of total beam current with medium and higher transmission mirrors (Trovò et al. 2002).

With both designs, increased output power up to 40 % and increased tunability up to 25 % have been achieved around 250 nm.

3.4.3 Progress Towards Short Wavelength: Lasing at 190 nm

To reach wavelengths shorter than 200 nm, a design derived from those used for ArF radiation excimer lasers (193 nm) has been employed. PIAD mirrors with $S/(HL)^{24}H$ (H = QWOT Al_2O_3; L = QWOT SiO_2 on silica and sapphire) design and Al_2O_3/SiO_2 mirrors produced by IBS at the Laser Zentrum Hannover were developed for a central wavelength of 195 nm. Lasing was obtained without difficulty using standard alignment procedures. The wavelength was tuneable (by undulator gap variation) over the range 189.7-200.3 nm. Figure 15 shows the measured spectrum at close to the shortest wavelength obtained, currently the shortest wavelength reached by an FEL oscillator. Lasing below 200 nm with the same mirrors has been obtained on three successive FEL runs. A stream camera measurement of the laser pulse duration provided a linewidth $\Delta\lambda/\lambda$ of $3.1.10^{-4}$ FWHM and the micropulse length measured 7.7 ps. The wavelength bandpass is thus sufficiently narrow to permit the use of the laser for most applications without any need for further monochromatisation. A maximum output power of 16 mW was measured for 31 mA beam current at 195 nm.

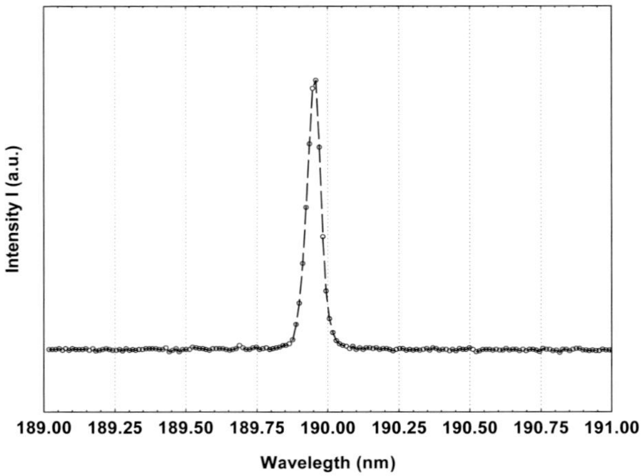

Fig. 15. Wavelength spectrum of the SRFEL at 190 nm, λ = 189.95 nm (Trovò et al. 2002).

4 Conclusion and Perspectives: Towards SRFEL below 150 nm

The development of optics for SRFELs tackles the problems of components under very broadband irradiation. To develop optics with performance at high average power and doses, a key element for FEL operation, it is necessary to identify and remove causes of damage at the surface and in the volume of the coating. This should lead to the development of original designs taking into account not only the fundamental wavelength for which the laser cavity is tuned but all the synchrotron radiation harmonics which vary also spatially at the surface of the component.

As a direct example, energy assisted prototype mirrors have been successfully used on the SRFEL at ELETTRA. Lasing at 220 nm was obtained with dual band oxide mirrors. Dedicated high transmission (HfO_2/SiO_2) mirrors have been designed that handle 330 mW of maximal output power at 254 nm and have a tunability of 32 nm. Recently 189.9 nm was obtained with Al_2O_3/SiO_2 IBS mirrors, the shortest wavelength obtained so far on SRFEL.

Modern third generation synchrotron sources cover the VUV and extended spectral regions and provide opportunities to explore scientific research with important needs for radiation in this region, such as pump probe photoemission, single photon ionisation detection and photoelectron spectroscopy. However, there is currently no SRFEL below 190 nm. As an alternative, it is possible to sustain the FEL lasing process in a scheme of so called Self-Amplified Spontaneous Emission (SASE) in a single pass and without optic cavity (Roßbach 1996). Requirements on the electron beam properties are more stringent but recent pilot experiments have demonstrated the practical feasibility of constructing such a light source in the VUV and X-ray spectral range (Andruszkow et al. 2000).

Research and development are currently going on to run SRFEL below 150 nm with robust optics. Figure 16 shows spectral reflectivity measurements for 193, 160, and 136 nm HR mirrors. New manufacturing technologies for DUV-VUV have been successfully developed and have excellent performance for excimer wavelengths at 248, 193, and 157 nm and shorter (Heber et al.1999, Thielsch et al. 2000). Components derived from such designs, developed on substrates like CaF_2, MgF_2, or sapphire, are interesting candidates for use as SRFEL mirrors in the VUV. An FEL oscillator with normal incidence mirrors may also be possible down to 100 nm using protected Al mirrors. All of these pathways are under investigation. From 100 nm to about 60 nm, SiC mirrors would permit lasing using a FEL with a gain of 10. Moreover, high performance Mo-Si multilayers are being developed by magnetron sputtering (see Chap. "Coatings for EUV/X-ray Mirrors" by Sergey Yulin). Realisation of multilayer design and deposition processes can result in a reflectivity greater than 68% at wavelengths near 13 nm (Kaiser et al. 2000, Yulin et al. 2001), although such mirrors might restrict the tunability. This wavelength region is still a far prospect but potentially very attractive for high gain SRFEL cavities (Feigl et al. 2001, 2002). Even if the reflectivity of the component is sufficient for laser operation, thermal distortion and synchrotron radiation damage may prevent or restrict the use of mirrors. This is an open area that would benefit from further research and would be important for lasing operations below 100 nm.

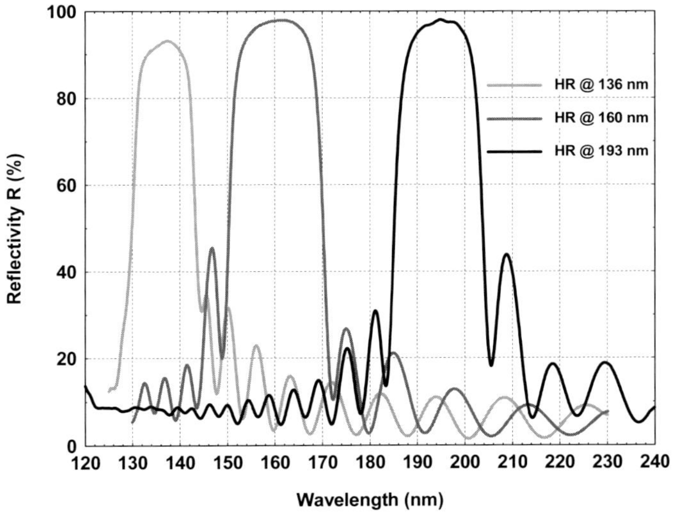

Fig. 16. Reflectivity measurements for 193 nm, 160 nm and 136 nm HR coatings.

References

Andruszkow J et al. (2000) First observation of self-amplified spontaneous emission in a free-electron laser at 109 nm wavelegth. Phys. Rev. Lett. 85:3825-3829

Attwood D (1999) Soft X-rays and extreme ultraviolet radiation: principles and applications. Cambridge University Press.

Billardon M, Elleaume P, Ortega JM, Bazin C, Bergher M, Velghe M, Petroff Y, Deacon D.A.G, Robinson K.E, Madey JMJ (1983) First operation of a storage-ring free-electron laser. Phys. Rev. Lett. 51:1652-5

Bodemann A, Kaiser N, Kozlowski MR, Pierce E, Stolz CJ (1995) Comparison between 355 nm and 1064 nm damage of high grade dielectric mirror coatings. Proc. SPIE 2714:395-404

Couprie ME, Merola F, Tauc P, Garzella D, Delboulbe A, Hara T, Billardon M (1994) First use of the UV super-ACO free-electron laser: fluorescence decays and rotational dynamics of the NADH coenzyme. Rev. Sci. Inst. 65 (5):1485-95

Couprie (2000) New ultrafast X-ray sources and their applications. Compte Rendu de l'Académie des Sciences de Paris, t.1, Série IV:329-345

Deacon DAG, Elias LR, Madey JMJ, Ramian GL, Schwettman HA, Smith TI (1977) First operation of a free electron laser. Phys. Rev. Lett. 38:892-894

Feigl T, Lauth H, Yulin S, Kaiser N (2001) Heat resistance of EUV multilayer mirrors for long-time applications. Microelectronic Engineering 57-58:3-8

Feigl T, Heber J, Gatto A, Kaiser N (2002) Optics developments in the VUV-soft X-ray spectral region. Nucl. Instrum. Methods Phys. Res. A. 483:351-356

Freund H.P and Antonsen T.M. Jr (1996) Principles of free electron lasers, Chapman and Hall, London, UK.

Garzella D, Couprie ME, Hara T, Nahon L, Brazuna M, Delboulbé A, Billardon M (1995) Mirror degradation and heating in storage ring FELs. Nucl. Instrum. Methods Phys. Res. A. 358: 387-391

Garzella D, Couprie ME, Billardon M (1996) Lasing at 300 nm and below: optical challenges and perspectives. Nucl. Instrum. Methods Phys. Res. A. 375:39-45

Gatto A, Thielsch R, Kaiser N, Hirsch M, Garzella D, Nutarelli D, De Ninno G, Renault E, Couprie ME, Torchio P, Alvisi M, Albrand G, Amra C, Marsi M, Trovò M, Walker R, Grewe M, Robert S, Roger JP, Boccara C (2000) Toward resistant UV mirrors at 200 nm for free electron lasers: manufacture, characterizations, and degradation tests. Proc. SPIE 4102:261-75

Gatto A, Kaiser N, Thielsch R, Garzella D, Hirsch M, Nutarelli D, De Ninno G, Renault E, Couprie ME, Torchio P, Alvisi M, Albrand G, Amra C, Marsi M, Trovò' M, Walker RP, Grewe M, Roger JP, Boccara C (2001) Achromatic damage investigation on mirrors for UV free electron lasers. Proc. SPIE 4347:535-546

Heber J, Thielsch R, Blaschke H, Kaiser N, Leinhos U, Görtler A (1999) Changes in optical interference coatings exposed to 193 nm excimer laser radiation. Proc. SPIE 3578:83-96

Kaiser N, Uhlig H, Schallenberg UB, Anton B, Kaiser U, Mann K, Eva E (1995) High damage threshold Al2O3/SiO2 dielectric coatings for excimer lasers. Thin Solid Films 260:86-92

Kaiser N, Yulin S, Feigl T (2000) Si-based multilayers with high thermal stability. Proc. SPIE 4146:91-100

Levy D.H. et al. (1994) Free electron lasers and other advanced sources of light: scientific research opportunities. National Research Council, Washington, US.

Litvinenko V (1999) The OK-4/Duke storage ring FEL lasing in the deep-UV. Nucl. Instrum. Methods Phys. Res. A. 429:151-158

Marsi M, Couprie ME, Nahon L, Garzella D, Bakker R, Delboubé A, Nutarelli D, Roux R, Visentin B, Grupp C, Indlekofer G, Panaccione G, Taleb-Ibrahimi A, Billardon M (1997) Two color experiments combining free electron laser and synchrotron radiation. Nucl. Instrum. Methods Phys. Res. A. 393 (1-3):548-551

Marsi M, Belkhou R, Grupp C, Panaccione G, Taleb Ibrahimi A, Nahon L, Garzella D, Nutarelli D, Renault E, Roux R, Couprie ME, Billardon M (2000) Transient charge carrier distribution at UV-photoexcited SiO/sub 2//Si interfaces. Physical Review B. 61 (8):R5070-3

Roßbach J (1996) A VUV free electron laser at the TESLA Test Facility at DESY. Nucl. Instrum. Methods Phys. Res. A. 375:269

Svelto O, Principles of lasers (4th edition). Plenum Press, New York.

Thielsch R, Gatto A, Heber J, Kaiser N (2000) A comparative study of the UV - optical and structural properties of SiO2, Al2O3, and HfO2 single layers deposited by reactive evaporation, ion assisted deposition, and plasma ion assisted deposition. Proc. SPIE 4102:276-288.

Torchio P, Gatto A, Alvisi M, Albrand G, Kaiser N, Amra C (2001) High-reflectivity dense UV mirrors. In Optical Interference Coatings, OSA Technical Digest, pp. ThA7-1 – ThA7-3.

Trovò M, Clarke JA, Couprie ME, Dattoli G, Garzella D, Gatto A, Giannessi L, Gunster S, Kaiser N, Marsi M, Poole MW, Ristau D, Walker RP (2002) Operation of the European Storage Ring FEL at ELETTRA down to 190 nm. Nucl. Instrum. Methods Phys. Res. A. (in press)

Walker RP (1999) European Project to Develop an UV/VUV FEL Facility on the ELETTRA Storage Ring. Nucl. Instrum. Methods Phys. Res. A. 429: 179-184.

Walker RP, Clarke JA, Couprie ME, Dattoli G, Eriksson M, Garzella D, Giannessi L, Marsi M, Poole MW, Renault E, Roux R, Trovò M, Werin S, Wille K (2001) The European UV/VUV storage ring FEL at ELETTRA: First operation and future prospects. Nucl. Instrum. Methods Phys. Res. A. 467-468:34-37

Yamada K, Sei N, Yamazaki T, Ohgaki H, Litvinenko VN, Mikado T, Sugiyama S, Kawai M, Yokoyama M (1999) Lasing down to the deep UV in the NIJI-IV FEL. Nucl. Instrum. Methods Phys. Res. A. 429:159-64

Yamada K, Sei N, Ohgaki H, Mikado T, Sugiyama S, Yamazaki T (2000) Lasing towards the VUV in the NIJI-IV FEL. Nucl. Instrum. Methods Phys. Res. A. 445:173-177

Yamazaki T, Yamada K, Sugiyama S, Ohgaki H, Sei N, Mikado T, Noguchi T, Chiwaki M, Suzuki R, Kawai M, Yokoyama M, Owaki K, Hamada S, Aizawa K, Oku Y, Iwata A, Yoshiwa M (1993) First lasing of the NIJI-IV storage-ring free-electron laser. Nucl. Instrum. Methods Phys. Res. A. 331 (1-3):27-33

Yulin S, Kuhlmann T, Feigl T, Kaiser N (2001) Damage resistant and low stress EUV multilayer mirrors. Proc. SPIE 4343:607-614

Abbreviations and Symbols

AFM	-	atomic force microscopy
DUV	-	deep ultraviolet
FEL	-	free electron laser
FWHM	-	full width half maximum
HR	-	high reflectivity
IBS	-	ion beam sputtering
IR	-	infra red
PIAD	-	plasma and ion assisted deposition
QWOT	-	quarter wavelength optical thickness
SASE	-	self amplified spontaneous emission
SRFEL	-	storage ring free electron laser
VUV	-	vacuum ultraviolet
XPS	-	X-ray photoelectron spectroscopy
B_u	-	magnetic field of the undulator
d	-	layer thickness
E	-	amplitude of the electric field in the stack
E_0	-	amplitude of the electric field in the incident medium
K	-	deflection parameter
k	-	extinction coefficient
m	-	harmonic order
n	-	refractive index
R	-	reflectivity
T	-	transmission
γ	-	Lorentz factor
λ	-	wavelength
λ_r	-	synchrotron radiation's wavelength
λ_0	-	central wavelength
λ_u	-	spatial period of the undulator
θ	-	angle of incidence

Optical Coatings on Plastics

Ludvik Martinu and Jolanta E. Klemberg-Sapieha

Ecole Polytechnique, Montreal, Quebec, Canada

1 Introduction

Synthetic organic polymers (plastics) are increasingly used in optics for replacement of the traditionally used glass. This trend is stimulated by the beneficial properties of plastics, namely their high strength-to-weight ratio, ease of manufacture of desired shapes, suitability for mass production, mechanical flexibility when in the form of films, webs or fibers and relatively low cost. Optical plastics have found widespread use in both high-technology and consumer products; the most frequent applications are screens, windows, windshields, ophthalmic lenses, lenses for optical instruments, displays, scanner systems, decorative coatings, flexible mirrors, security devices and optical waveguides

Beside their incontestable advantages, the use of optical plastics also creates important technical challenges when they are provided with optical coatings and when employed in environments in which mechanical, chemical, radiation and other effects are important. This stems from the difference between the physical properties of the coatings and of the substrates. Of particular importance is the large difference between the respective coefficients of thermal expansion (*CTE*), low thermal stability, hardness, H, and elastic modulus, E, of plastics, their high water absorption (*WA*) and the low energy of their surfaces, which frequently leads to adhesion problems.

When coating optical plastics, the following criteria must be considered: (i) choice of the appropriate deposition technique; (ii) surface pre-treatment for control of adhesion; (iii) choice of coating materials and suitability of their optical, mechanical and other properties; and (iv) selection of appropriate tests for performance evaluation. Many applications stimulate the use of coatings with additional functions (functional coatings) such as surface hydrophobicity or hydrophilicity, impermeability to gases or vapors (barrier coatings) and electrical conductivity. In fact, two types of coating/substrate systems are frequently used:

(a) inorganic optical coatings deposited directly on the plastic (often flexible webs); (b) rigid plastic with a hard (scratch resistant) layer followed by an optical filter (e.g., antireflective, AR, coating and a hydrophobic (or hydrophilic) top layer. Good examples of (b) are ophthalmic lenses.

The area of optical coatings on plastics has been very active over the past decade (see, for example, Proc. SVC 1990-2002), but only a small number of review articles are available on this subject (Guenther 1988). Other relevant reviews deal with the surface treatment of plastics for improved adhesion (Liston et al. 1993; Wertheimer et al. 1996, 1999a; Mittal and Pizzi 1999), with film deposition technologies (Rossnagel et al. 1990; Cuomo et al. 1990; Vossen and Kern 1991; Bunshah 1994; Simpson and Lewis 1997) and with plasma enhanced chemical vapor deposition (PECVD) for optical films and coatings (Martinu and Poitras 2000).

The aim of the present work is, on the one hand, to underline the complexity of the fabrication of optical coatings on plastics, and on the other hand, to review recent advances in the deposition techniques, mostly using low pressure plasma in the engineering of surfaces and interfaces and in the optimization of the optical, mechanical and other functional properties of different optical systems. Sect. 2 gives a brief summary of the basic characteristics of both the plastics and the coating materials. Sect. 3 describes various fabrication techniques with a particular emphasis on the control of energetic interactions during the film growth and polymer surface modification. Sect. 4 focuses on the optical and mechanical properties and their relationship to the film/substrate interface (or "interphase"). Numerous examples of different coating/plastic systems for various applications are described in Sect. 5. This chapter mainly deals with transparent coatings for interference filters (including AR coatings), while omitting the metal films most frequently used for mirrors and decorative and information storage (recording) applications. The purpose is not to give a complete bibliography on the subject, but rather to describe the general trends and illustrate them with appropriate examples. Complementary subjects are dealt with in other parts of this book.

2 Challenges in Optical Coatings on Plastics

2.1 Optical Plastics

Different plastic materials have been developed for optical applications based on their optical transparency, mechanical performance, ease of manufacture of different shapes, etc. (Harper 1992). For each generic type of plastic, however, the properties may vary over a wide range depending on the additives and the manufacturer, and even varying from one lot to another (Guenther 1988). In addition, the optical performance can be affected by birefringence. Table 1 gives a summary of the most often applied optical plastics, their properties and applications, while Table 2 reviews their most frequently reported refractive index, n, density, ρ, and other basic characteristics. Comparison is made with different forms of SiO_2 (glass and quartz) and with another form of carbon, natural diamond.

The principal characteristics that distinguish plastics from inorganic optical coatings are their high values of *CTE* (about two orders of magnitude) and their very low heat distortion (or service) temperature, *ST*. When depositing films at an elevated substrate temperature, T_s, the plastic will shrink more, imposing significant compressive thermal stress, σ_t, a phenomenon that can deteriorate adhesion.

Table 1. Summary of selected optical plastics

Polymer	Abbr.	Properties	Application
Polycarbonate	PC	- high transparency - high impact strength - poor resistance to moisture and oils	- optical components - windows - screens - spark discharge control
Poly(methyl-methacrylate)	PMMA	- high UV resistance - high transparency - resistance to chemicals - high brittleness	- windows - optical components
Polyethylene-terephthalate	PET	- dimension stability - high strength - wear resistance - wide temperature range	- windows - glazings - automotive - architecture
Polytetrafluoro-ethylene	PTFE	- high chemical resistance - low friction, no adhesion - resistance to weathering - wide temperature range	- aerospace
Fluorinated ethylene propylene	FEP	- low friction, no adhesion - high chemical resistance - wide temperature range - high (UV-VIS) transparency	- Hubble Space Telescope - glazings - medical instruments
Polyvinylidine fluoride	PVDF	- good thermal stability	- aerospace - array transducers

Abbr. Abbreviation.

Plastics are soft and hence prone to mechanical damage by scratching during cleaning, handling and in different applications. They also develop higher electrostatic surface charges than optical glass, thus attracting dust particles. Plastics are susceptible to radiation damage, such as ultraviolet (UV, $\lambda < 400$ nm), vacuum ultraviolet (VUV, $\lambda < 200$ nm), and, particularly, ionizing particle radiation (Wertheimer et al. 1999b). Energetic photons can be harmful in some deposition and treatment processes. However, when appropriately controlled they can be applied for engineering the film/substrate interface (Sect. 4.2 and 4.3).

Plastics can absorb considerable amounts of water (see Table 2). The sorption and desorption processes are diffusion controlled and hence rather slow. Desorption of water may contaminate the deposition process and its diffusion to the film/substrate interface can deteriorate adhesion. This stimulates further work on development of low *WA* plastics (Guenther 1988) and on surface treatment processes (Wertheimer et al. 1999a).

Table 2. Properties of selected optical plastics

Polymer	Trade name	*n*	ρ g/cm^3	*WA* %	*CTE* µm/m^{0}C	*ST* ^0C	*H (DSI)* GPa	*E* GPa
PC	Lexan	1.58	1.20	0.15	68	125-135	0.2; 118*	2.1
	Makrolon	1.58	1.20	0.15	70	115-125	122*	2.3
PMMA	Plexiglas	1.49	1.19	0.30	60	71- 85	0.2; 93*	3.0
	Lucite	1.49	1.17	0.45	52	82- 91	63*	1.7
PET	Mylar	1.66	1.40	0.05	59	80-100	130*	3.8
PTFE	Teflon	1.35	2.17	0.03	86	260-290	58*	0.5
FEP		1.34	2.10	0.01	140	60-204	25*	0.4
PVDF		1.42	1.78	0.03	140	68-170	75*	1.8
Quartz		1.46	2.20		0.40		9.5	110
Glass		1.52	2.18		0.75		7.2	68
Diamond		2.42	3.52		1.80		98	980

n refractive index @550nm, ρ density, *WA* water absorption, *CTE* coefficient of termal expansion @200C, *ST* service temperature, *H* depth sensing indentation (DSI) hardness; *Rockwell hardness, *E* elastic modulus.

2.2 Optical Coating Materials

Large numbers of materials are available to coat plastic substrates. Basic characteristics of transparent films obtained by the more traditional physical vapor deposition (PVD) techniques (in particular, sputtering, electron beam evaporation, EBE and ion beam assisted deposition, IBAD) are summarized in Table 3. The *n* values cover the entire range necessary for successful applications in multilayer interference filters. In most cases, *H* and *E* are comparable to or higher than those for glass. When deposited directly on the plastic substrate, their mechanical match appears difficult due to σ_t and due to frequently high values of the intrinsic compressive stress, σ_i.

During the past decade there has been much progress in the fabrication of optical coatings by PECVD (for a review, see Martinu and Poitras 2000). The main advantages of the PECVD processes and films are as follows:

(i) Wide range control of plasma-chemical reactions and plasma-surface interactions allows one to optimize the film composition and microstructure: the films generally possess high packing density (> 98%) and are therefore hard and environmentally stable. The desired optical effect can easily be complemented in the same deposition system by another function, for example, enhanced scratch resistance (Rostaing et al. 1993; Rats et al. 1999; Beckmann et al. 2001), gas or vapor impermeability (Da Silva 1998; Affinito et al. 1999) and surface hydrophobicity (or hydrophilicity) (Klemberg-Sapieha et al. 1998a).

(ii) PECVD is suitable for fabrication of films with a desired intermediate *n* value and of inhomogeneous (graded index optical films in which *n*(*z*) varies continuously as a function of depth *z* (e.g., rugate filter designs, see Sect. 5). In addition, the absence of sharp interfaces leads to a compensation of internal stresses and hence better adhesion.

(iii) PECVD films generally show better mechanical performance on polymeric substrates than their PVD counterparts; beside their different microstructure, this is believed to be primarily due to the existence of a physically thick, graded interfacial region ("interphase", see Sect. 4), the presence of which substantially enhances adhesion and scratch, abrasion, and wear resistance.

Table 3. Characteristics of selected PVD optical coatings obtained by sputtering or ion assisted techniques

Materials	***n* @ 550nm**	***H* DSI [GPa]**	***E* [GPa]**	**σ_i [GPa]**
TiO_2	2.20-2.60	7-12	110	+0.3 to –0.4
Nb_2O_5	2.30-2.40			
Ta_2O_5	2.10-2.25	9-12	120-150	-0.2 to –0.5
ZrO_2	1.95-2.15	10-15	140-180	<-1.5
ITO	1.95-2.10			
Si_3N_4	1.90-2.05	15-21	140-160	-0.2 to –0.7
Y_2O_3	1.85-1.95			
Al_2O_3	1.65-1.75	20-22	180-220	
SiO_2	1.43-1.47	7-9	80 -120	-0.2 to –0.6
MgF_2	1.38-1.42			+0.3 to +0.8

n refractive index @550nm, *H* depth sensing indentation (DSI) hardness, *E* elastic modulus, σ_i intrinsic stress.

The characteristics of the most frequently studied PECVD materials are summarized in Table 4. PECVD materials can be either inorganic or organic in nature, depending on the precursor gas or vapor and on the fabrication conditions. Many films tend to contain a certain concentration of hydrogen (for example, amorphous hydrogenated silicon dioxide, SiO_2, [H] ~ 10-15 at. %, nitride, $SiN_{1.3}$, [H] ~ 15-20 at. %, and oxynitride, SiO_xN_y) and some of them may contain residual carbon (typically 0.1-5.0 at. %) when deposited from organosilicone or organometallic precursors (for example, SiO_2, Ta_2O_5, TiO_2). A particular category of low-index films from organic precursors are the so-called plasma-polymerized fluorocarbons (PPFC), hydrocarbons (PPHC) and organosilicones (PPOS). Hard hydrogenated amorphous carbon (a-C:H) films, frequently referred to as diamond-like carbon (DLC), are included for completeness. The latter material absorbs in the visible due to a high concentration of carbon in its sp^2 hybridization and it cannot be readily applied in filter designs in the visible region. However, it is often used for applications in the NIR or for mechanical protection (e.g., optical windows for laser barcode readers in supermarkets or information storage recording discs). In the literature, the boundary between DLC and PPHC is not clearly defined; however, typically H_{DLC}> 10-15 GPa (Martinu 1997).

Table 4. Characteristics of selected PECVD optical coatings

Materials	***n* @550nm**	***H* DSI [GPa]**	***E* [GPa]**	**σ [GPa]**
TiO_2	2.14-2.50	9-11	100	+0.5 to –0.2
Ta_2O_5	2.10-2.20	9-11	120-150	+0.2 to -0.3
$SiN_{1.3}$	1.80-2.00	15-20	150-160	+0.2 to –0.6
SiO_xN_y	1.50-1.90	10-16	100-130	+0.1 to –0.4
SiO_2	1.44-1.48	7-10	90-120	+0.1 to –0.2
PPOS	1.50-1.75	2-7		
PPHC	1.50-1.70	2-4		
PPFC	1.38-1.41			
a-C:H	1.60-2.10	15-40	160-240	-0.2 to <–2.0

n refractive index @550nm, *H* depth sensing indentation (DSI) hardness, *E* elastic modulus, σ_i intrinsic stress.

Another important category is coatings applied to plastics by a wet process using dipping, spinning or spraying (Samson 1996). They frequently constitute the hard coat. Such films are often obtained from organosilicones, which are cured by heat or UV radiation and which typically contain about 10-30 % of carbon. In many cases, *n* of the hard coat is adjusted to the same *n* value of the plastic substrate in order to avoid interference effects (Beckmann et al. 2001). These layers represent a transition from the organic plastic to the inorganic coating and they may greatly reduce the mismatch between the materials characteristics. In this sense their properties are close to those of PPOS.

2.3 Characterization of Optical Coatings on Plastics

Very often the characteristics of optical coatings on plastics are incomplete due to inadequate use or description of the methods used for analysis. In this section, we give a brief overview of the approaches that should be taken into account.

Optical Properties

Difficulties associated with the deposition of optical coatings on plastics may give rise to significant inhomogeneities and hence problems with the determination of optical characteristics (Martinu and Poitras 2000; Poitras and Martinu 2000). On the one hand, "accidental" inhomogeneities represent a non-deliberate consequence of the fabrication process; these include: (a) fluctuation of the deposition parameters during film fabrication, (b) surface and interface roughness and (c) formation of an interphase. On the other hand, deliberate inhomogeneities are obtained by either (i) continuous variation of the deposition parameters in order to obtain specific $n(z)$ profiles (for example, in rugate filters or graded index devices, see Sect. 5), or (ii) doping the matrix material with dopants or clusters of another medium in order to obtain nanocomposite films (Biederman and Martinu 1990; Dalacu and Martinu 2000).

Usually, optical properties are determined either from spectrophotometric (reflectance, $R(\lambda)$, and/or transmittance, $T(\lambda)$) or ellipsometric measurements, among which variable angle spectroscopic ellipsometry (SE or VASE), combined with R and T measurements, appears to be most powerful. Ellipsometry deals with the determination of the relative phase change of a reflected polarized light beam as opposed to absolute intensity measurements in spectrophotometry, making it more sensitive to very small changes in the optical properties at the surface of the sampled material and to possible birefringence of the plastic substrate. Spectrophotometry is more appropriate when evaluating the performance of a coating system such as an optical filter. For post-process characterization of complex optical coatings, both methods share the same difficulties related to the models used for reproducing the measured data and to the optimization algorithms (for more details, see Martinu and Poitras 2000).

Mechanical Properties

Mechanical behavior is probably the most studied characteristic of films on plastics (Martinu 1997). In order to assure its successful performance and its stability, each film/plastic system must possess a suitable combination of mechanical properties. This includes good adhesion, low stress, high hardness and toughness and good tribological properties such as scratch and wear resistance and surface friction. The coating design is often a compromise between the mechanical and optical properties.

There exist numerous tests to evaluate the mechanical properties, but they are frequently qualitative or semi-quantitative, they pertain to a specific application, and it is often difficult to compare the results from one laboratory to another. In fact, apparent lack of standardisation calls for the development of appropriate testing methods and approaches. Basic mechanical tests related to coated plastics are briefly described below.

- **Adhesion**

Adhesion can be expressed quantitatively by the adhesion force, F_A, necessary to separate two materials joined together by a common interface, or by the work of adhesion, W_A, which corresponds to the energy required to separate two surfaces. Three categories of methods are available (Ohring 1992): (i) tensile-type pull-off test, which measures F_A necessary to detach a cylindrical rod glued to the coating surface; and (ii) shear-type adhesive tape peel test which usually provides only qualitative information. For semi-quantitative measurements one can apply tapes with different adhesive strengths and peel it at a controlled angle or combine it with the cross-hatch test (ASTM-D3359-97); (iii) scratch test, which has received widespread scientific and industrial acceptance since it provides quantitative, semi-quantitative and qualitative measures of adhesion.

The scratch test is very attractive since it avoids possible contamination of the interface by the glue present in the previous tests and it requires only a small area for relatively fast measurements while allowing for better experimental statistics. During scratching the indenter of known tip radius, R_s, is pulled across the coating surface under progressively increasing load. The adhesion is related to the "critical" load, L_c, corresponding to the first failure. In addition, other mechanical characteristics can be obtained at the same time, namely, scratch resistance, friction coefficient, scratch hardness and the failure mode (for a review, see Martinu 1997 or Blees et al. 2000). L_c can be related to adhesion by (Ohring 1992)

$$F_A = K H L_c / \pi R_s^2 \tag{1}$$

where H is the hardness of the film and K is a coefficient depending on the film/substrate combination ($0.2<K<1.0$).

Different models have been proposed to correlate L_c with W_A; they consider (i) fully plastic deformation, (ii) fully elastic deformation and (iii) the more realistic elasto-plastic behavior. The latter model accounts for three contributions to the critical stress leading to the coating damage: (i) elasto-plastic indentation stress and effective ploughing stress in the surface region, (ii) adhesion friction stress due to the interaction at the indenter-sample interface and (iii) the film internal stress. Consequently, W_A can be expressed in terms of the coating thickness, d, elastic modulus, E, and the scratch width, w (Burnett and Rickerby 1987, Bull and Rickerby 1990):

$$W_A = (d/2E)(8L_c/\pi w^2)^2. \tag{2}$$

- **Stress**

The total stress in the film/substrate systems is given by the sum of σ_t and σ_i. While σ_t depends on the difference of the *CTE* values and on T_s, the σ_i values are related to the film microstructure (Ohring 1992). The latter aspect and determination of stress from the Stoney equation and other methods are described in Chap. "Mechanical stress in optical coatings" by Strauss.

- **Hardness**

Hardness implies the ability of a material to resist plastic deformation under applied load. Hard materials usually possess a high cohesive energy and short bond length (Ohring 1992; Bhushan 1995). In practice the measured H values depend on the coating microstructure and the presence of defects and imperfections. Hardness is usually determined by indentation techniques and H is calculated as a ratio between the applied maximum normal load and the surface area of the permanent imprint.

Inorganic optical coatings are hard (arbitrarily, harder than glass; see Tables 2 to 4), they exhibit significant elasto-plastic deformation behavior and they are relatively thin ($d \sim \lambda$). In order to avoid the effect of substrate the penetration depth must be less than $d/10$. Therefore, depth-sensing indentation tests are increasingly used, in which the H and E values are assessed from the load-displacement curve and from knowledge of the tip geometry (Oliver and Pharr 1992). H values measured for multilayer or inhomogeneous optical coatings represent the composite hardness, which usually obeys the mixture law (Bull 2001).

- **Tribological Properties**

Tribological processes involve two contacting surfaces in relative motion. In the case of hard optical coatings on soft plastic substrates, the tribological properties include scratch, abrasion and wear resistance, friction and lubrication. These characteristics are important for handling and manipulating the specimens and for their respective functions (ophthalmic lenses, for example, see Sect. 5). In real contact conditions it is common that several mechanisms act at the same time. This includes mainly adhesive, abrasive, fatigue and chemical wear (Holmberg and Matthews 1994).

Wear is defined as the volume of the removed material divided by the normal load and the moved distance. The main material characteristics important in the control of the tribological processes are H, E, d, coating adhesion and surface roughness. The latter parameter plays a vital role in micro and nanotribology: the contact between two surfaces occurs at discrete spots, at the asperities level.

Different tests can be used to evaluate the tribological properties, often with the aim of simulating the working environment; they can be divided into two groups: (i) Sliding tests such as pin-on-disc, pin-on-flat, triboindenter and atomic force microscopy (AFM) (Raymond et al. 2001). (ii) Other, frequently semi-quantitative, wear tests such as taber (dry) test, particle air-blast, rubber wheel, steel wool or cheesecloth tests, often followed by surface haze evaluation (ASTM-D1044-99).

3 Deposition Processes

3.1 Ion Bombardment in Thin Film Growth

During the deposition of optical films on plastics T_s has to remain low. Therefore, additional steps must be taken in order to increase the film density and stability, as dictated by structure zone models (SZM). This is generally achieved by bombarding the growing films with energetic ions while controlling their energy, E_i; typically, E_i<1 keV (Harper et al. 1984). Ion bombardment related phenomena give rise to the disruption of growth nuclei, to the suppression of columnar structure and hence, to material densification. The general applicability of SZM has since been clearly demonstrated for sputtered as well as PECVD optical films (Martinu at al. 1994) while the role of the flux of ions has been accentuated (Musil 1992; Martinu et al. 1994; Kelly and Arnell 1998).

It appears that the key parameter to quantitatively describe the ion bombardment is the energy, E_p, delivered to the growing film per deposited particle, expressed in a simplified form as (Musil 1992):

$$E_p = E_i\ (\Phi_i/\Phi_m). \tag{3}$$

E denotes energy and Φ particle flux and the indices i and m refer to ions and condensing precursor species.

The need to tailor the film microstructure at low T_s frequently requires controling, possibly independently, the E_i and (Φ_i/Φ_m) values. Therefore, a majority of the new or novel film fabrication processes increasingly apply additional plasma excitation or other forms of plasma-based techniques. Good examples of these trends are microwave (MW) plasma-enhanced sputtering, plasma ion assisted evaporation, PECVD (Martinu and Poitras 2000) and plasma surface modification (Wertheimer et al. 1999a). E_i values are proportional to the difference between the surface bias potential, V_B, and the plasma potential, V_p.

In the pressure range generally used for sputtering and plasma processing the ions lose part of their energy due to elastic, inelastic and charge transfer collisions in the sheath and they usually exhibit a relatively broad ion energy distribution function (IEDF). IEDFs can be evaluated in the process chamber by implementing a multigrid electrostatic ion energy analyser (IEA) or a quadrupole mass spectrometer integrated with an IEA (Zabeida et al. 2000), as schematically illustrated in Fig. 1.

3.2 PVD Techniques

Traditionally, PVD techniques, such as EBE and magnetron sputtering, have been extensively applied for coating plastic substrates (Guenther 1988). Enhanced ion bombardment is achieved by ion beams (IBAD), by the control of V_B or by adjusting the magnetic field configuration, such as in unbalanced magnetrons (for more details, see Chap. "Thin film deposition techniques" by Pulker).

Examples of IEDFs in different configurations are shown in Fig. 1. A typical mean E_i in a IBAD system using a DC plasma ion source is around 100 eV, while the values vary slightly for different gases (Fig. 1b). In a parallel plate radiofrequency (RF, 13.56 MHz) system, using a discharge in Ar at 60 mTorr as an example, the E_i value on the grounded substrate holder is around 20 eV (Fig. 1c) while E_i can reach many hundreds of eV on the capacitively coupled RF-powered electrode (Fig. 1d). In the latter case the IEDF is structured due to sheath modulation. In fact, the IEDFs, the plasma density, n_e, V_B and V_p values (Fig. 1c and 1d) are very similar to those encountered in reactive ion etching, RF sputtering and PECVD processes.

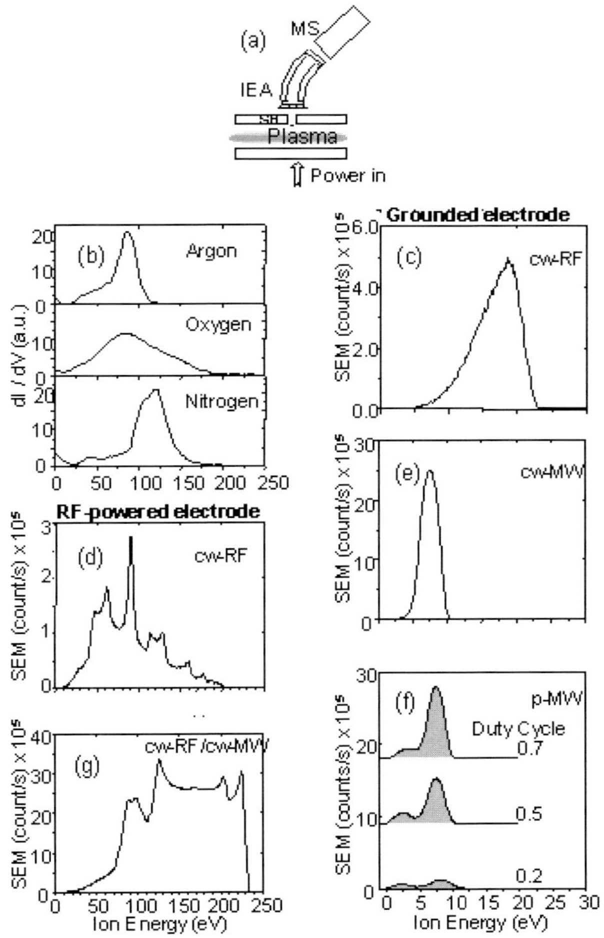

Fig. 1. Examples of IEDF in different configurations of reactors used for PVD and PECVD: **(a)** schematic illustration of a parallel-plate system with an integrated mass spec-

trometer (MS)/ion energy analyser (IEA) and a substrate holder (SH); **(b)** IEDF in an IBAD system with a cold plasma ion source (Zabeida et al. 2000). The following IEDFs pertain to Ar^+ ions in high frequency plasma in Ar at 60 mTorr: **(c)** grounded SH in cw-RF discharge; **(d)** RF-powered SH in cw-RF discharge; **(e)** grounded SH in cw-MW discharge; **(f)** grounded SH in pulsed MW discharge with different duty cycles; **(g)** RF-powered SH in dual-mode cw MW/RF discharge (modified after Martinu and Poitras 2000).

3.3 PECVD Processes and Techniques

There has been significant interest in "novel" deposition techniques for optical coatings using plasma excitation, in particular PECVD (for a review, see Martinu and Poitras 2000). In addition to the advantages already mentioned in Sect. 2, PECVD processes can also address other important aspects of optical film deposition and materials optimization. This includes: (i) high deposition rates ($r_D \sim$ 1-10 nm/s or more), which are substantially higher than for other, more traditional PVD techniques, in particular for high index materials; (ii) possibility to coat different substrate shapes (flat, hemispherical and cylindrical, interior of tubes, etc.); and (iii) to enhance film performance on plastic substrates by forming an “interphase” due to energetic surface interactions, in particular VUV crosslinking and interface mixing.

In the simplest configuration, optical films can be deposited in an RF parallel plate reactor. The r_D values on the grounded electrode are much lower than on the RF-powered electrode (usually 5 to 10 times). Relatively high E_i values on the powered electrode at $T_s \sim 25^\circ$C lead to high ρ, H and $|\sigma|$. In RF discharges alone the dissociation rate in the gas phase may be insufficient, sometimes resulting in increased optical absorption due to a substoichiometric composition. Typically, $V_p \sim$ 20 V, leading to $E_i \sim$ 25 eV on the grounded electrode (Fig. 1c). On the RF-powered electrode, Φ_i is low and E_i may reach several hundred volts (Fig. 1d) often leading to a high stress (Martinu et al. 1994).

Most successful systems for optical coatings use a microwave (MW, usually 2.45 GHz) plasma capable of yielding high r_D, Φ_i, n_e and dissociation rates. In a simple MW reactor the substrate is placed on a grounded or floating substrate holder, facing a MW window (usually, fused silica with a low water content) through which the MW power is supplied from different devices such as linear or horn antennas or standing or travelling wave (slow wave) applicators (Wertheimer et al. 1996). Typically, $V_p \sim$ 10 V, generally yielding $E_i \sim$ 10 eV such as in the continuous wave (cw) mode (see Fig. 1e).

Returning to eq. (3), it appears desirable to maintain E_i relatively low (sufficient for film densification) while increasing Φ_i/Φ_m. This implies that an independent or a selective control of E_i, Φ_i/Φ_m and of the bulk plasma processes is needed. This is impossible in conventional cw plasma reactors, which use a single power source to excite the discharge. Therefore, two approaches to obtain selectivity have been explored: (i) pulsed-mode discharges and (ii) RF-induced surface biasing in the so called dual-mode or dual frequency plasma deposition.

In the frequently used pulsed MW plasma, two plasma regimes can be distinguished during each pulse cycle: high density plasma during the "T_{on}" period and decaying plasma during the "T_{off}" period. As a consequence, the IEDF adopts a bi-modal shape (see Fig. 1f) with the high-energy peak corresponding to ions generated during the T_{on} period and the low-energy peak being due to ions arising from the T_{off} period. The ratio of the peak intensities depends on the duty cycle, $D = T_{on}/(T_{on} + T_{off})$, which, in turn, enables one to tune the plasma-surface interactions in deposition, etching or surface modification processes (Zabeida et al. 1999).

Pulsed MW plasma with low pulsing frequency, $f_p = 1/(T_{on}+T_{off}) < 100$ Hz, and low D ($D<0.2$) have become prominent for optical and functional coatings (Klemberg-Sapieha et al. 1990; Segner 1995; Zabeida et al. 2000). On the one hand, high power during the T_{on} period leads to high n_e, Φ_i and dissociation rate values and hence to a high process rate. On the other hand, longer T_{off} periods allow more time for the precursor gas to fill the volume in front of the substrate, thus leading to enhanced film uniformity. This process has achieved the highest level of sophistication at Schott Glaswerke GmbH (Mainz, Germany) in their so-called plasma impulse CVD (PICVD) process (Segner 1995).

The dual-mode microwave/radiofrequency (MW/RF) plasma approach has been developed at Ecole Polytechnique in Montreal (Martinu et al. 1989; 1994). The substrates are placed on the RF-powered substrate holder facing the MW window, through which the MW power is applied using a linear applicator. A possibility to selectively control E_i and Φ_i values is illustrated by the IEDFs in Fig. 1d and 1g. Other concepts include a remote MW/RF design (Rostaing et al. 1993) and electron cyclotron resonance (ECR) configurations, in which a magnetic field is applied in conjunction with the MW discharge in order to further increase the dissociation rate and n_e, but at lower pressures (10^{-3}torr) (Bulkin et al. 1997).

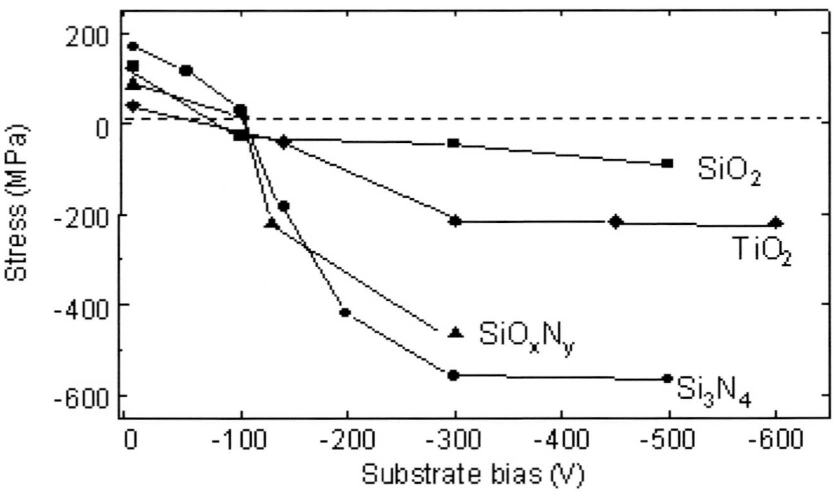

Fig. 2. Evolution of stress in different PECVD films with increasing V_B (and hence E_i) in MW/RF plasma (after Klemberg-Sapieha et al. 2000).

Clearly, E_p in Eq. (3) can be adjusted to the same level by combining low and high E_i and Φ_i / Φ_m values. However, experience suggests that good quality (dense, hard, chemically stable, low stress) films are preferably deposited under conditions of low (10-30 eV) or intermediate (~100 eV) E_i, sufficient for densification, but using high Φ_i. This reduces the microstructural damage and gas entrapment, generally yielding low $|\sigma|$ values. The effect of ion bombardment on stress in different PECVD films obtained in MW/RF discharge is documented in Fig. 2. Transitions from a tensile stress at low E_i through a zero stress region and development of $-\sigma_t$ at high E_i values can be clearly observed.

4 Plasma Interaction with Polymer Surfaces

4.1 Radiation Effects of Plasma on Plastics

Plasma is a rich source of energetic species such as ions, radicals and photons, the latter covering the entire IR-visible-UV-VUV-soft X-ray region (for a review, see Wertheimer et al. 1999b). VUV photons play an important role in the interaction with polymer surfaces (Liston et al. 1993; Wertheimer et al. 1996). Strong VUV radiation has been systematically studied for λ values above 100 nm (Holländer et al. 1995) in which the discharges containing H_2 (strong Lyman α line at 121 nm and molecular bands), O_2 (strong resonant line at 130 nm) and He (intense line at 57 nm) (see Fig. 3) appear to be the most effective. In this context, intense VUV features due to the excitation of impurities (fragments of H_2, H_2O, hydrocarbons) desorbed from the chamber walls and from the polymer itself are also very important.

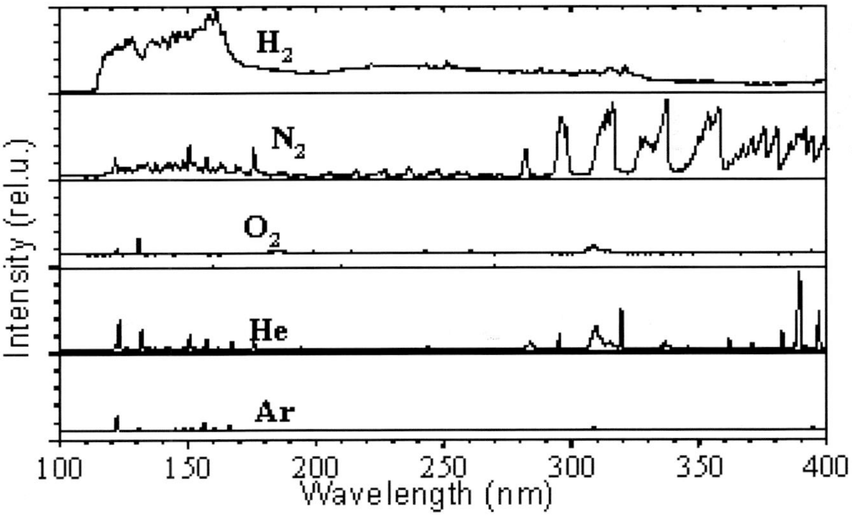

Fig. 3. Optical emission spectra in the VUV-VIS region from MW plasma excited in different gases at 2 Torr (after Holländer et al. 1995).

Energetic radiation effects in the substrate materials are inevitable, especially when surface pretreatment for surface cleaning, activation and modification becomes an integral part of the film fabrication process. Photon energies from several eV to more than 20 eV can exceed those of covalent bonds (for example, 3.2 eV for C-C, 3.6 eV for C-H, 4.2 eV for C-O). In fact, their energy often coincides with the absorption bands in the near and far UV regions. The energetic photons can penetrate to depths of several hundreds of nanometers (20 nm at $\lambda = 80$ nm, 1 µm at $\lambda = 200$ nm) compared to bombarding ions which affect the first one or two monolayers.

For polymers, degradation mechanisms include chain scission, photo-Fries rearrangements, recombination of photolytic radicals and photo-induced oxidation (when O_2 is present) (Rivaton and Gardelle 1998). They contribute to cross-linking or branching, hence creating a denser top layer (about 50-100 nm thick) characterized by a slight increase of the near-surface n value (for example, $\Delta n \sim 0.02$ in PC and $\Delta n \sim 0.08$ in PET) (Bergeron et al. 1998). In many cases, VUV radiation can lead to photon-induced ablation (etching) (Fozza et al. 1999) and volatilization of molecular fragments (Hong et al. 2002). The absorption also depends on the type and concentration of additives (UV blocking agents, antioxidants, etc.) (Gächter et al. 1993), which are designed to protect the polymer against UV, but which may enhance interaction with VUV in vacuum.

4.2 Surface Treatment of Plastics for Improved Adhesion

Control of adhesion is particularly important in coating plastic substrates. A number of theories have been proposed to describe the mechanisms of adhesion; they include (i) mechanical interlock theory, (ii) theory of formation of electrostatic forces or an electrical double layer, (iii) theory of weak boundary layers, (iv) absorption theory (thermodynamics, wettability or acid-base interactions), (v) diffusion theory related the interface rheology and (vi) chemical bonding theory. Given the complexity of adhesion, until now there is no universal theoretical model that can adequately explain all experimental observations. In practice, several mechanisms may act simultaneously.

There are numerous methods to modify polymer surfaces to improve adhesion (Mittal and Pizzi 2000). They include wet-chemical or mechanical treatments, exposure to flames, UV radiation, corona discharges, ion beams and intermediate adhesive layers. Among these, low pressure plasma modification appears very powerful since it can address all of the adhesion mechanisms mentioned above due to a possibility to control both the chemical and the physical interactions (Liston et al. 1993; Wertheimer et al. 1996, 1999a). There are four major effects of plasma on polymeric surfaces: (i) surface cleaning; (ii) ablation (or etching); (iii) crosslinking and (iv) surface chemical functionalization. All of them are always present to a certain degree, but one may dominate depending on the polymer, gas and operating conditions.

Adhesion enhancement by plasma pretreatment is illustrated by an example in Fig. 4 where PECVD $SiN_{1.3}$ was applied as a first layer in multilayer AR coatings or in multilayer filters on PC and PMMA (Klemberg-Sapieha et al. 1997, 1998b). After pre-treatment of PC in MW plasma, the highest adhesion was found when using N_2 (see the results for $d = 0.5$ μm in Fig. 4a). Based on complementary surface analyses, adhesion improvement could be related to the formation of new N and O containing groups, which react with Si to form Si-N-C and Si-O-C bonds at the interface. Interestingly, adhesion was further improved by exposing PC to a He plasma (see Fig. 4a), confirming a favorable effect of the VUV radiation.

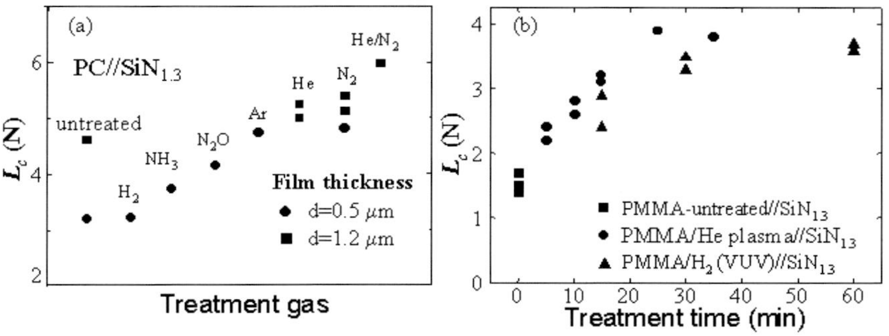

Fig. 4. Adhesion in terms of the critical load, L_c, for $SiN_{1.3}$ films on polymers: **(a)** effect of plasma pre-treatment of PC in different gases using MW plasma and VUV radiation from H_2 plasma (after Klemberg-Sapieha et al. 1997); **(b)** effect of treatment time of PMMA in He plasma and VUV radiation from H_2 plasma (after Klemberg-Sapieha et al. 1998b).

The role of VUV radiation in the optimization of surface treatment proved indispensable in the case of a polymer that exhibits cohesive failure at the interface, namely PMMA (Klemberg-Sapieha et al. 1998b) (see Fig. 4b). Indeed, optimum adhesion could be reached when VUV-rich direct MW plasma in He was used or the surface was exposed to the radiation emanating from the H_2 plasma (strong emission between 120 and 170 nm, see Fig. 3) through a MgF_2 window with a cut-off wavelength of 112 nm. The main contribution to adhesion in this case is VUV-induced etching of the surface (Holländer et al. 1995), leading to its cross-linking and mechanical stabilization (Martinu 1997).

It is interesting to mention that a similar result of adhesion improvement due to N_2 plasma treatment has been observed for other systems, such as SiO_2 on PC, PMMA, or PP, or metals on different polymers. This effect has been related to the formation of strong covalent C-N-coating bonds at the interface (Gerenser 1988; Klemberg-Sapieha et al. 1997).

4.3 Characteristics of the "Interphase"

Energetic interactions of plasma with the exposed polymer surface lead to substantial modification of the physical and chemical properties of the near-surface region and they affect the initial stage of the film growth. Due to the phenomena involving energetic photons, ions and reactive species, the interfacial region (interphase) is structured, as illustrated in Fig. 5 for the particular case of a $SiN_{1.3}$ film on PC. Similar structure has been observed for other combinations of materials, including SiO_2, PET and PMMA. Using both non-invasive optical (*in situ* and *ex situ* SE, photometry, infrared spectroscopy) and invasive methods, the interphase has been found to be several tens of nm thick (Klemberg-Sapieha et al. 1997; Da Silva et al. 1998a; Bergeron et al. 1998, 2000). It consists of a cross-linked layer (region 2 in Fig. 5 as discussed in Sect. 4.2) followed by a transition layer (region 1) formed by intermixing the growing film with the substrate materials (Da Silva et al. 1998b) and possibly by voids (Bergeron et al. 1998). In the case of $SiN_{1.3}$, n increases from 1.59 for PC to 1.80 while H increases by two orders of magnitude from 0.2 GPa for bulk PC to ~ 2 GPa for the cross-linked surface layer and up to 18 GPa for $SiN_{1.3}$ (Klemberg-Sapieha et al. 2000; Raymond et al. 2001). This inhomogeneity generally leads to better adhesion and better tribological properties.

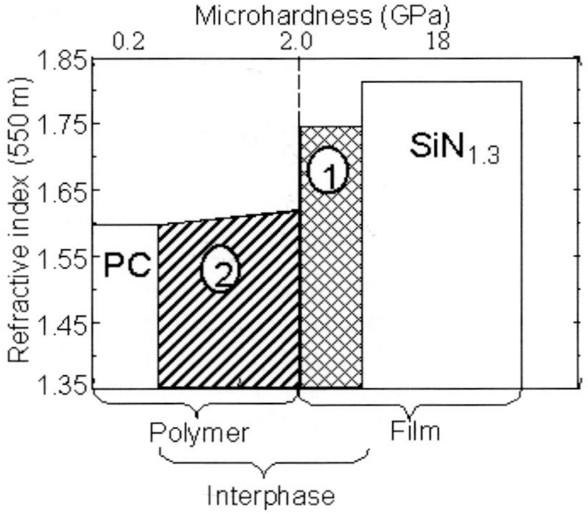

Fig. 5. Illustration of a structured interfacial region (interphase) between a plasma deposited film (here $SiN_{1.3}$) and a polymer substrate (PC). Schematic illustration of the $n(z)$ profile in the interfacial region shows: (i) a crosslinked layer due to VUV emitted light (2), and (ii) a transition layer (1) (modified after Klemberg-Sapieha et al. 2000).

Formation of the interphase has been observed also for situations when no specific plasma pre-treatment has been performed. Indeed, the energetic conditions are still present when the film starts growing (nucleation), while the plasma is always a rich source of energetic photons, particularly due to the presence of hydrogen originating from the precursor, from the polymer material itself or from residual water vapor. This situation is quite different from PVD processes in which hydrogen concentration in the gas phase and energetic VUV radiation are less intense. In fact, no interphase (below detection) has been observed for inorganic films, such as SiO_2 on PET, fabricated by PVD (Da Silva et al. 1998b). This may explain the generally observed better mechanical performance of PECVD coatings on plastics compared to their PVD counterparts.

5 Applications of Optical Coatings on Plastic Substrates

As shown in Sect. 2, PVD and PECVD can provide high, low, and medium index materials for any simple or complex optical coating system. The designs can be based on three types of approaches (for an overview, see Martinu and Poitras 2000): (i) multilayer (step index) design using two or more materials, (ii) inhomogeneous (graded index) design, and (iii) quasi-inhomogeneous design, when thin ($d<<\lambda/4$) layers with varied compositions are consecutively applied.

In most cases the optical coatings are deposited on either flexible or rigid substrates, while each of them has its specificities and area of use. Flexible substrates are most frequently applied in glazings for energy control, for transparent conductive oxides (TCO), AR and antistatic films, for displays, etc. The main application of rigid substrates is in the ophthalmic industry, screens, windows and optical instrumentation. In addition, numerous new applications are emerging.

5.1 Antireflective Coatings and Optical Filters

AR coatings are traditionally the most frequently applied systems, which typically consist of 3 to 5 layers in the visible region (Dobrowolski 1995). When plastic substrates are used, such as in ophthalmic applications, the AR coatings are usually deposited onto a hard coat. However, if grown directly on the polymeric substrate, special considerations should be taken, especially those related to adhesion and scratch resistance (Rostaing et al. 1993; Klemberg-Sapieha et al. 1997; Lee et al. 1997; Martinu 1997; Schulz et al. 2001, 2002) and to the presence of the interphase, as already mentioned above (Bergeron et al. 2000; Poitras and Martinu 2000). An example of a four layer AR system on PC is illustrated in Fig. 6. An optimized W design between 450 and 650 nm using $SiN_{1.3}$ and SiO_2, brings the original single-side reflection of 5.0% down to about 0.8% (Fig. 6a). After PECVD, a green shift of the reflectance spectrum has been observed. This discrepancy has been interpreted and confirmed by introducing a $n(z)$ gradient, namely the presence of an interphase, in the first layer (Fig. 6b), a procedure which led to a good match with the experimental data (Fig. 6c). Benefiting from the knowledge about the existence of the interphase, a re-optimized design was then capable of further reducing the overall reflectance to ~0.7% (Fig. 6d).

Recently, medium and large scale roll-to-roll deposition of AR coatings onto webs has been described using microwave PECVD (Ellison et al. 1997), plasma assisted evaporation (Uyama et al. 2001) and sputtering processes (Krautz et al. 2001).

As mentioned earlier, PECVD is well-suited for the fabrication of inhomogeneous (graded-index) coatings, such as broadband AR quintic layers or rugate filters. In this case the n value is continuously varied between the high and low indices, mostly by changing the gas composition (Greenham et al. 1993; Poitras et al. 1996), although variation of plasma density or of the bias potential is also possible (Poitras et al. 1996). In such approaches, the $n(\lambda)$ dispersion and r_D value must be known at each instant of the deposition process. Materials such as $SiN_{1.3}/SiO_2$ and TiO_2/SiO_2 (Poitras et al. 2002) systems have been tested with success.

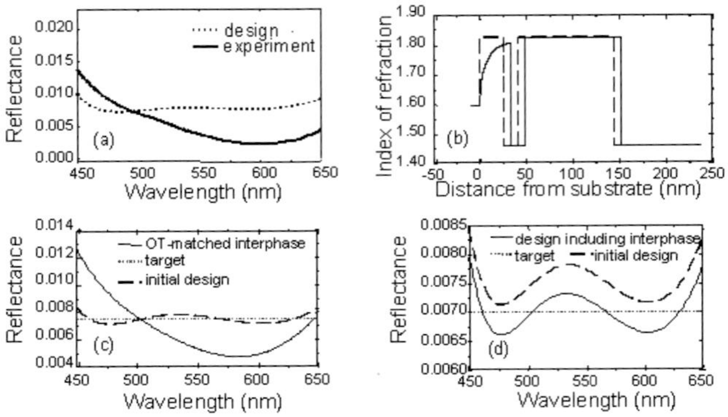

Fig. 6. Reflectance spectra of plasma-deposited AR coatings on PC: **a**) design and experiment; **b**) modified n(z) profile; **c**) reflectance according to a); **d**) reoptimized design (modified after Poitras and Martinu 2000).

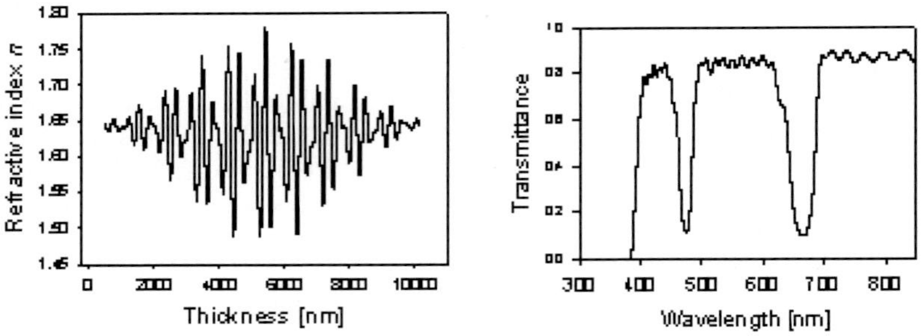

Fig. 7. Double-band rugate filter on PC: **a**) design n(z) profile; **b**) resulting transmittance spectrum. The total filter thickness is 11 μm (after Poitras and Martinu 2000).

Rugate filter design with apodization leads to elimination of harmonics and suppression of side-lobes. The design of rugate filters is often carried out by inverse Fourier-transform methods, which do not account for $n(\lambda)$ dispersion. Therefore, special approaches have to be applied when considering multi-band rugate designs. Figure 7 shows an example of a designed $n(z)$ profile and the corresponding measured $T(\lambda)$ spectrum of a PECVD double-band rugate filter deposited on PC using SiO_xN_y (Poitras et al. 2002). The total thickness of this filter was about 11 μm. Complex rugate filter designs, the effect of apodization and deposition of quasi-inhomogeneous filters on glass have been studied both theoretically and experimentally (Greenham et al. 1993; Segner 1995). In the case of plastics the inhomogeneous design has a beneficial effect on the mechanical behavior; the absence of abrupt interfaces and uniform distribution of stress lead to superior tribological properties and adhesion (Rats et al. 1999; Raymond et al. 2001).

5.2 Rigid Substrates - Ophthalmic Lenses

Plastic ophthalmic lenses and light assemblies (indoor reflectors, car light bezels) are probably the most well-known applications of optical coatings on plastics; they have a very long history, but remain a very active area due to numerous challenges. It is estimated that only in the US, 160 million people wear glasses, 80 million glasses are purchased every year and more than 90% of them are made out of plastics. Low index PMMA (n = 1.49) has been considered for such applications since the 1930s, but it was dialyl diglycol dicarbonate, known as CR39 (trademark of the PPG company) that became popular after World War II, while the high index PC (n = 1.59) has been increasing in its importance since the 1980s due to its toughness, impact resistance and thermal stability (Bohling et al. 2000).

Plastics are soft and hard coats have to be provided to protect them. This is most frequently achieved by dipping, spinning or spraying 3-5 μm thick silica-type materials: the original brittle films have been replaced by thermally cured polysiloxanes, which contain organosilane modifiers leading to less brittleness due to the crosslinkling of the organic component, or by UV cured urethane/acrylate hard coats, which provide better adhesion but lower hardness (Bohling et al. 2000). Promising hard coats have also been fabricated by PECVD using organosilicone precursors (Hora and Wohlrab 1993; Wohlrab and Hofer 1995; Beckmann et al. 2001). In this case, control of the film growth allows one to enhance adhesion to the soft substrate by adjusting the gradient from the organic to the inorganic character by, for example, changing the oxygen content (Beckmann et al. 2001). In addition, optical interference between the substrate and the hard coat can be diminished by depositing an optical matching layer ($n_{PC} = 1.59 > 1.52 > 1.46 = n_{SiO2}$). Mechanical properties of hard coat materials have been tested by the scratch method, pointing out the importance of controlling the interface properties (Martinu 1997, Blees et al. 2000).

In many cases ophthalmic lenses are provided with AR coatings whose major role is to reduce double (ghost) images and backside reflections. The AR films are usually deposited in box coaters using IBAD, but plasma-IAD, sputtering as well as PECVD have also been tested (Pongratz and Zöller 1992; Schulz et al. 2001, 2002). AR coatings can also be integrated with a thin conductive layer, such as indium tin oxide, ITO, which substantially reduces electrical charge retention; surface resistances, $R < 10^{10}\ \Omega/$ have been reported (Blacker et al. 2000).

In many instances the lenses are provided with top coats which are highly repellent to water and dust and which are either hydrophobic or hydrophilic but very thin (5-10 nm). Hyrophobic top coatings are frequently obtained from wet solutions (by dipping) or by evaporation, but PEVCD films of PPFC have also shown promising results (Klemberg-Sapieha et al. 1998a). The surface properties of such films are compared in Table 5. Surface water contact angles for PECVD and non-

PECVD coatings appear comparable, but the former exhibit somewhat better long-term stability, possibly due to their higher cross-linking, as confirmed by surface analysis. Good smudge resistance has also been reported by using organosilicones with well controlled O_2 concentrations and discharge power levels (Hora and Wohlrab 1993).

Table 5. Surface characteristics of hydrophobic fluorocarbon layers prepared by different techniques (after Klemberg-Sapieha et al. 1998a)

Fabrication method	F/C	R_a [nm]	γ_C [mJ/m]	Θ_a [deg]	Θ_s [deg]	Θ_r [deg]	n
Ion source	1.43	1.6	~20	118	95	64	1.44
Plasma polymerization	1.56	3.1	~20	123	109	67	1.38
Evaporation	1.71	3.4	~20	122	109	68	1.50
Dip coating ("Nano Film")	-	-	~25	138	118	90	-

F/C fluorine to carbon ratio, R_a surface roughness, γ_C surface energy, Θ_a advancing contact angle, Θ_s static contact angle, Θ_r receding contact angle, *n* refractive index @550 nm.

Quality control of ophthalmic lenses involves a complex characterization procedure while considering the following properties and tests (Bohling et al. 2000): (i) abrasion resistance – oscillating sand abrasion or Bayer testing, steel wool and/or tumble abrasion. These tests are frequently followed by the haze test (ASTM-D1044); (ii) durability – boiling water (often saline water) followed by rapid cooling, humidity cycling, solvent exposure, soft (340 nm) or hard (314 nm) UV radiation exposure, etc; (iii) impact resistance – dropping a steel ball of a known diameter from a height corresponding to a minimum energy of 0.2 J; and (iv) adhesion – cross hatch/adhesive tape peel test (ASTM-D3359-97) using a cutter providing a 5x5 grid of 1 mm^2 squares. Generally, the adhesion test is performed also at the end of the test (ii). Slightly different qualification tests are required for other systems, such as "wipe-resistant" coatings for reflectors consisting of a reflecting metal layer sandwiched between two hard coats (2-4 μm thick) on a PC substrate (Beckmann et al. 2001).

5.3 Displays

In recent years booming interest in flexible substrates has been stimulated by their application in the flat panel display technology, including touch screens. In this respect, the main advantages of plastics are conformality (flexibility), impact resistance, light weight and thinness and the possibility of high volume roll-to-roll process manufacturing. Flexible displays for portable devices include personal digital assistants, digital cameras, camcorders, mobile telephones and remote control units, many of which also incorporate a touch sensor as an input device. Future applications consider wearable PCs, card PCs, E-books and smart phones. A large number of prototype displays have been reported based on different technologies, such as organic light emitting diodes (OLEDs), polymer dispersed liquid crystals (PDLC), cholesteric LCs, electroluminescence, electrophoretics and polarizer-based displays (Cairnes et al. 2001; see also Chap. "Optical Coatings for Displays" by Li Li).

PET, PC and PES (polyether sulphone) are frequently used as substrates for displays. However, novel polymeric materials are presently considered, this includes polyethylene tetranaphthalate (PEN), triacetylcellulose (CTA or TAC), cyclopolyolefines (COP) and cyclopolyolefine copolymers (COC). The latter ones are preferably cast in order to eliminate birefringence and they possess many advantages regarding their optical properties, heat distortion temperature and water uptake.

OLEDs are an essential part of future all-organic optoelectronic devices (Martin 2001; Weaver et al. 2001). They comprise a multilayer structure consisting of ITO ($R < 40\ \Omega/$ is typically required), followed by organic layers containing an electro-phosphorescent dopant (Tang and Slike 1987). This includes, for example, tertiary amine (50 nm thick hole transport material) and aluminium trisquinalate (electron transport layer) backed by a low work function cathode (e.g., Ag:Mg or Cu). The total thickness is typically 100 nm.

Important components of OLEDs are water vapor permeation barriers in which a permeation rate of ~ 10^{-6}g/cc/day at 25°C is required in order to assure a lifetime of 10,000 hours of continuous operation at video brightness. Recently, oxygen and water permeabilities below < 0.005 cc/m^2/day and 0.005 g/cc/day, respectively, have been reported (Affinito et al. 1999). Trends in this area have been reviewed (Chatham 1996) and the suitability of PECVD has been underlined (Da Silva et al. 1998a).

ITO is commonly used as the conducting electrode material in display and other applications. ITO is transparent ($T > 85\%$), while this high bandgap (> 3 eV) oxide is made conductive by creating electron degeneracy due to dopants or by making them non-stoichiometric; the latter effect gives rise to oxygen vacancies thus creating n-type carriers.

The commercially available ITO on PET is usually deposited by DC sputtering in a roll-to-roll process (Cormia et al. 1998; Lewis and Paine 2000) using an In_2O_3/SnO_2 (90/10) target. Typical R values required in the display industry range from 80 to 500 $\Omega/$ (thickness 5-150 nm). Recently, its mechanical properties have been measured in detail, in particular wear and crack propagation (Cairnes et al. 2001).

5.4 Coatings for Security Devices

Development of security devices is becoming increasingly important given the fact that world trade includes perhaps 5-7% of counterfeit or pirated goods. It has been reported that the worldwide impact is about $250-$350 billion a year, but the recent estimate made in Europe by the Center for Economics and Business Research puts this number at \$1 trillion, a fact further stimulating the development of anti-counterfeiting or brand authentication devices.

Besides electronic article surveillance (EAS), optical variable devices (OVD) are becoming popular. They are of two types: diffractive optical variable image devices (DOVID) and interference security image structures (ISIS) (Bishop 2001). Their fabrication steps include development of the exposed photopolymer plate holding the artwork followed by metallization to make the master, which is used to produce the shim. The shims are mounted on a roll and the pattern is embossed into the polymer surface, which is then metallized. DOVIDs can be of four kinds, namely reflective, semireflective, patterned and transmission. The originally used aluminium is being replaced by the more durable chromium or by gold to yield an appropriate color effect.

ISIS devices are applied in numerous areas. The best examples are probably the anti-counterfeiting devices on currency. The original design developed at the National Research Council in Ottawa (Dobrowolski et al. 1989) includes multilayer dielectric structure that is hot stamped on the banknote. The present design consists of five layers on a black hot melt adhesive (Black | ZrO_2 | SiO_2 | ZrO_2 | SiO_2 | ZrO_2). The advantage of this type of structure is its brightness due to the absence of absorption. However, the cost is relatively high. Another OVD concept is the color shifting pigments used in inks. In this case, a five-layer stack is deposited on a polymer web, then stripped of and milled to size. Each flake consists of a central reflecting layer followed by a transparent film and an absorber, which assures the desired color shifting effect (for example, Cr | MgF_2 | Al | MgF_2 | Cr) (Phillips and Nofi 1999). In the USA the new series of banknotes features a green to black color change and different inks and optical effects are used on the currency in over 60 countries.

Other devices include retroreflection, conducting printing inks and window security systems. In the latter a change in conductivity (broken ITO conductor on PET) indicates an abuse.

OVDs are frequently deposited on a release layer in order to be transferred to another substrate or liberated for pigment milling. Different release materials can be used that provide appropriate adhesion (or delamination) after the film deposition. They include synthetic wax (Glick et al. 2001), acrylic latex-based release agents and aqueous solutions of low molecular weight styrene copolymer. Fabrication of photochromic pigments using VO_2 has been reported, where the photochromic effect can be achieved in VO_2 only when annealing it to 400°C (Pitt et al. 2001). In such cases only thermally-resistant substrate materials can be used. This includes polyimide and a high temperature vermiculate-based release agent formed by mica-like AlMg silicate minerals (often used as fire retardant agents with an *ST* up to 1100°C) in chemically defoliated aqueous dispersions.

5.5 Energy Control Glazings

Conventional glazings have found widespread application in the automobile industry and in architecture for energy control. Recently, solar infrared reflecting (SIRR) glazings replace the older "tinted" (absorbing) glass (Finley 2001). SIRR glazing, which can be applied in windshields, sidelights and backlights, offers high heat load reduction, rejecting up to 60% of the sun's energy, while saving fuel and increasing the customer's comfort. One can distinguish two categories of glazings: (i) reflective solar control, and (ii) "high-T/low-E" coatings. In the case of glass substrates, SIRR glazings consist of pyrolytic Co-Cr-Fe-O (spinel) or electroless Ag/Cu films. Privacy glazings in some cars, such as minivans, are typically absorbing, made of TiN, inconel or stainless steel. The "high-T/low-E" coating stacks form a spectrally selective filter, allowing in visible light while reflecting solar and thermal infrared. The dielectric layers that antireflect the silver layer (decreased reflectance, increased transmission) are typically 20 to 40 nm thick and are made of high index materials (e.g., TiO_2, SnO_2, ZnO_2, Si_3N_4).

The SIRR system consists of a coating laminated between glass lights with polyvinyl butaryl (PVB), where it is protected against deterioration by exposure to the atmosphere. Alternatively, SIRR is deposited on PET, which is then sandwiched between two PVB films and two glass substrates. Roll-to-roll deposition of such glazings using magnetron sputtering has been described (Krautz et al. 2001). As an example, the requirements in Europe are $T \geq 75\%$ for the windshield and $T \geq 70\%$ for the sidelights. The conductive glazings with $R = 2\text{-}10\ \Omega/$ can add an additional function in applications such as antennas and heated windshields, heater/deicer, etc. Increasingly important are dynamic glazings with adjustable optical transmission (Mathew et al. 1997; Lampert 2000). The electrochromic effect is obtained in WO_3, which may be laminated with polyaniline.

6 Conclusions and Perspectives

The area of optical coatings on plastics is presently in full expansion, driven by multiple technological opportunities in the fields of displays, glazings for energy control, security devices, ophthalmic lenses and numerous others, including future all-organic devices for molecular optoelectronics. It has been pointed out throughout this paper that the successful application of optical films on polymeric substrates resides in in-depth understanding of the compatibility of the coating materials with plastics, mostly considering the large differences in the thermal expansion coefficients, durability, and mechanical and structural stability.

Traditionally, PVD and wet chemical techniques have been used for film fabrication, but novel methods applying different types of plasma excitation are becoming increasingly attractive. This includes, in particular, control of the energy and flux of bombarding ions, and energetic UV and VUV photons. This allows one to selectively control polymer surface crosslinking, chemical effects and interfacial phenomena (interphase), leading to enhanced adhesion and good materials performance. This further stimulates research and development in the areas of the diagnostics of plasma-surface interactions and process control and monitoring.

References

Affinito JD, Gross ME, Maurnier PA, Shi MK, and Graff GL (1999) Ultrahigh rate, wide area, plasma polymerized films from high molecular weight/low vapor pressure liquid or solid monomer precursors. J Vac Sci Technol A 17: 1974

ASTM-D1044-99 Standard Test Method for Resistance of Transparent Plastics to Surface Abrasion

ASTM-D3359-97 Standard Test Methods for Measuring Adhesion by Tape Test

Beckmann R, Nauenberg KD, Naumann T, Patz U, Ieked G, and Hagedorn H (2001) A new high-rate deposition process for scratch- and wipe-resistance coatings for optical and decorative plastic parts. Proc 44th Ann Tech Conf of the Society of Vacuum Coaters (SVC). Philadelphia, pp 288-294

Bergeron A, Klemberg-Sapieha JE, and Martinu L (1998) Structure of the interfacial region between polycarbonate and plasma-deposited $SiN_{1.3}$ and SiO_2 optical coatings studied by ellipsometry. J Vac Sci Technol A 16: 3227-3234

Bergeron A, Poitras D, and Martinu L (2000) Interphase in plasma-deposited silicon nitride optical films on polycarbonate: In-situ ellipsometric characterization. Opt Eng 39: 825

Bhushan B (ed) (1995) Micro/Nano Tribology. CRC Press, New York

Biederman H, and Martinu L (1990) Plasma polymer-metal composite films. In: d'Agostino R (ed) Plasma Deposition, Treatment and Etching of Polymers. New York, Chap 4: 269-320

Bishop CA (2001) The use of vacuum deposited coatings for security applications. Proc 44th Ann Tech Conf of SVC, Philadelphia, pp 425-430

Blacker R, Bohling D, Coda M, and Kolasey M (2000) Development of intrinsically conductive antireflection coatings for the ophthalmic industry. Proc 43rd Ann Tech Conf of SVC, Denver, pp 212-216

Blees MH, Winkelman GB, Balkenende AR, and den Toonder JMJ (2000) The effect of friction on scratch adhesion testing: application to a sol-gel coating on polypropylene. Thin Solid Films 359:1

Bohling D, Coda M, Blacker R, Burton C, Gove R, Murphy P, Threlfall I, and Samson F (2000) Abrasion resistant and optical thin film coatings for ophthalmic lenses. Proc 43rd Ann Tech Conf of SVC, Denver, pp 222-229

Bulkin P, Bertrand N, Drévillon B, Rostaing JC, Delmotte F, Hugon MC, and Agius B (1997) Plasma enhanced chemical vapour deposition of silica thin films in an integrated distributed electron cyclotron resonance reactor. Thin Solid Films 308-309: 69-74

Bull SJ (2001) Interface engineering and graded films: structure and characterization. J Vac Sci Technol A 19: 1404-1414

Bull SJ, and Rickerby DS (1990) New developments in the modeling of the hardness and scratch adhesion of thin films. Surf Coat Technol 42: 161

Bunshah RF (ed) (1994) Handbook of Deposition Technologies for Films and Coatings. 2nd edn. Noyes Publications, Park Ridge, NJ

Burnett PJ, and Rickerby DS (1987) The relationship between hardness and scratch adhesion. Thin Solid Films 154: 403-416
Cairnes DR, Paine DC, and Crawford GP (2001) The mechanical reliability of sputter-coated indium tin oxide polyester substrates for flexible display and touchscreen applications. Proc 44th Ann Tech Conf of SVC, Philadelphia, pp 160-165
Chatham H (1996) Oxygen diffusion barrier properties of transparent oxide coating on polymeric substrates. Surf. and Coatings Technol 78: 1
Cormia RL, Fenn Jr JB, Memarian H, and Ringer G (1998) Roll-to-roll coating of indium tin oxide – a status report. Proc 41st Ann Tech Conf of SVC, Chicago, pp 452-457
Cuomo JJ, Rossnagel SM, and Kaufman GR (eds) (1990) Handbook of Ion Beam Processing Technology. Noyes Publications, Park Ridge, NJ
Dalacu D, and Martinu L (2000) Spectroellipsometric characterization of plasma-deposited Au/SiO_2 nanocomposite films. J Appl Phys 87: 228
Da Silva-Sobrinho AS, Latrèche M, Czeremuszkin G, Klemberg-Sapieha JE, and Wertheimer MR (1998a) Transparent barrier coatings on polyethylene terephthalate by single- and dual-frequency plasma-enhanced chemical vapor deposition. J Vac Sci. Technol A 16: 3190-3198
Da Silva-Sobrinho AS, Schüler N, Klemberg-Sapieha JE, Wertheimer MR, Andrews M, and Gujrathi SC (1998b) Plasma-deposited silicon oxide and silicon nitride films on polyethylene-terephthalate: A multitechnique study of the interphase regions. J Vac Sci Technol A 16 : 3190-3198
Dobrowolski JA (1995) Usual and unusual applications of optical thin films - an introduction. In: Hummel RE and Guenther KH (eds) Handbook of Optical Properties, vol I. CRC Press, Boca Raton, pp 5-35
Dobrowolski JA, Ho FC, and Waldorf A (1989) Research on thin film anti-counterfeiting coatings at the NRC of Canada. Applied Optics 28: 2702
Ellison T, Dotter B, Izu M, and Ovshinsky S (1997) New high speed, low cost roll-to-roll antireflectivity coating technology. Proc 40th Ann Tech Conf of SVC, pp 309-312
Finley JJ (2001) The evolution of solar infrared reflective glazing in automobiles. Proc 44th Ann Tech Conf of SVC, Philadelphia, pp 193-203
Fozza AC, Klemberg-Sapieha JE, and Wertheimer MR (1999) Vacuum ultraviolet irradiation of polymers. Plasmas and Polymers 4: 183-205
Gächter R, Müller H, and Klemchuk PP (eds) (1993) Plastics Additives Handbook. 4th edn. Hanser, Munich
Gerenser L (1988) An X-ray photoemission spectroscopy study of chemical interaction at silver/plasma modified polyethylene interface: correlation with adhesion. J Vac Sci Technol A 6: 2897-2903
Glick I, Dobrowolski JA, Fahland M, Karlsson P, and Seltmann W (2001) Development of three-layer OVD coating based on dual-magnetron reactively sputtered dielectric films. Proc 44th Ann Tech Conf of SVC, Philadelphia, pp 431-437
Greenham AC, Nichols BA, Wood RM, Nourshargh N, and Lewis KL (1993) Optical interference filters with continuous refractive index modulations by microwave plasma-assisted chemical vapor deposition. Opt Eng 32: 1018-1024
Guenther KH (1988) Coating of plastics – coatings on plastic. SPIE Vol. 896: 134-139
Harper CA (1992) Handbook of Plastics, Elastomers and Composites. McGraw-Hill, New York

Harper JME, Cuomo JJ, Gambino RG, and Kaufman HR (1984) Modification of thin film properties by ion bombardment during deposition. In: Auciello O, and Kelly R. (eds) Ion Bombardment Modification of Surfaces. Elsevier, New York, p 127

Holländer A, Klemberg-Sapieha JE, and Wertheimer MR (1995) Vacuum-ultraviolet induced oxidation of polymers: polyethylene and polypropylene. J Polym Sci-A; Polym Chem 33: 2013-2025

Holmberg K, and Matthews A (1998) Coatings Tribology. In: Dawson D (ed) Tribology Series 28. Elsevier, Amsterdam

Hong J, Truica-Marasescu F, Martinu L, and Wertheimer MR (2002) An investigation of plasma-polymer interactions by mass spectrometry. Plasmas and Polymers 7:245-260

Hora R, and Wohlrab (1993) Plasma polymerization: A new technology for functional coatings on plastics. Proc 36th Ann Tech Conf of SVC, pp 51-57

Kelly PJ, and Arnell RD (1998) Development of a novel structure zone model relating to the closed-field unbalanced magnetron sputtering system. J Vac Sci Technol A 16 (5): 2858

Klemberg-Sapieha JE, Küttel OM, Martinu L, and Wertheimer MR (1990) Dual microwave/radio frequency plasma deposition of functional coatings. Thin Solid Films 193/194: 965-972

Klemberg-Sapieha JE, Poitras D, Martinu L, Yamasaki NLS, and Lantman CW (1997) Effect of interface on the characteristics of functional films deposited on polycarbonate in dual-frequency plasma. J Vac Sci Technol A 15 : 985-991

Klemberg-Sapieha JE, Martinu L, Fridman V, and Morton DE (1998a) Characterization of hydrophobic thin films prepared using a cold cathode ion source. Proc of the 41st Ann Tech Conf of SVC, Chicago, pp 138-143

Klemberg-Sapieha JE, Martinu L, Yamasaki NLS, Lantman CW (1998b) Plasma-induced stabilization of PMMA surfaces for enhanced adhesion of plasma-deposited coatings. Mater Res Soc Symp Proc 544: 277

Klemberg-Sapieha JE, Dahl S, and Martinu L (2000) Nano- and micromechanical characterization of optical coatings fabricated by PECVD. Proc 43rd Ann Tech Conf of SVC, Denver, pp 234-238

Krautz H, Weissling H, Böhme T, Thielsch R, Hecht C, Kammer M, Langer A, Gawer O, and Milde F (2001) Grossflächen-beschichtung von Kunststofffolie für Automobilverglasung und Architekturanwendungen. Neues Dresdner Vacuumtechnisches Kolloquium (NDVaK), VDI, Dresden, pp 1-15

Lampert CM (2000) Smart switchable materials for the new millennium - windows and displays. Proc 43rd Ann Tech Conf of SVC, Denver, pp 165-170

Lee C-C, Hsu J-C, and Jaing C-C (1997) Optical coatings on polymethyl methacrylate and polycarbonate. Thin Solid Films 295: 122-124

Lewis BG, and Paine DC (2000) Applications and processing of transparent conducting oxides. MRS Bulletin 25: 22

Liston EM, Martinu L, and Wertheimer MR (1993) Plasma surface modification of polymers for improved adhesion: a critical review. J Adhesion Sci Technol 7: 1091

Mathew JGH, Sapers SP, Cumbo MJ, O'Brien NA, Sargent RB, Raksha VP, Lahaderne RB, and Hichwa BP (1997) Large area electrochromics for architectural applications. J of Non-Crystalline Solids 218: 342-346

Martin PJ (2001) The coming of molecular optoelectronics. Proc 44th Ann Tech Conf of SVC, Philadelphia, pp 1-3

Martinu L (1997) Plasma deposition and testing of hard coatings on plastics. In: d'Agostino R, Fracassi F, Favia P (eds) Plasma Treatments and Deposition of Polymers. Kluwer, Dordrecht NATO ASI Vol. 346: 247-272

Martinu L, and Poitras D (2000) Plasma deposition of optical films and coatings: a review. J Vac Sci Technol A 18: 2619-2645

Martinu L, Klemberg-Sapieha JE, and Wertheimer MR (1989) Dual-mode microwave/radio frequency plasma deposition of dielectric thin films. Appl. Phys. Lett. 54: 2645-2647

Martinu L, Klemberg-Sapieha JE, Küttel OM, Raveh A, and Wertheimer MR (1994) Critical ion energy and ion flux in the growth of films by PECVD. J Vac Sci Technol A 12: 1360-1364

Mittal KL, and Pizzi L (eds) (1999) Adhesion Promotion Techniques. Marcel Dekker, New York

Musil J (1992) Inter-relationships process parameters-microstructure-film properties, a key to new materials and applications. Proc of the Int Symp on Elementary Processes and Chemical Reactions in Low Temperature Plasma. Casta, Slovakia, p 177

Ohring M (1992) The Materials Science of Thin Films. Academic Press, New York

Oliver WC, and Pharr GM (1992) An improved technique for determining hardness and elastic modulus using load and displacement sensing indentation experiments. J Mater Res 7: 1564-1583

Phillips RW, and Nofi M (1999) Colors by chemistry or by physics?. Proc 42nd Ann Tech Conf of SVC, Chicago, pp 494-459

Pitt AM, Bennett RM, and Higgins DE (2001) Novel route for the production of high performance pigments. Proc 43rd Ann Tech Conf of SVC, Denver, pp 438-443

Poitras D, and Martinu L (2000) Interphase in plasma-deposited films on plastics: effect on the spectral properties of optical filters. Appl. Opt. 39 : 1168

Poitras D, Leroux P, Klemberg-Sapieha JE, Gujrathi SC, and Martinu L (1996) Characterization of inhomogeneous Si-based optical coatings deposited in dual-frequency plasma. Opt Eng 35: 2693-2699

Poitras D, Larouche S, and Martinu L (2002) Design and plasma deposition of dispersion-corrected multiband rugate filters. Appl Opt 41: 5249-5255

Pongratz S, and Zöller A (1992) Plasma ion assisted deposition: a promising technique for coatings. J Vac Sci Technol A 10: 1897-1904

Proceedings of the Annual Technical Conferences of the Society of Vacuum Coaters (SVC), (1990-2002) Volumes 34-46. Useful reference for technological aspects

Rats D, Hajek V, and Martinu L (1999) Micro-scratch analysis and mechanical properties of plasma-deposited silicon-based coatings on polymer substrates. Thin Solid Films 340: 33-39

Raymond M-A, Larouche S, Zabeida O, Martinu L, Klemberg-Sapieha JE (2001) Tribological properties of PECVD optical coatings. Proc 44th Ann Tech Conf of SVC, Philadelphia, pp 301-305

Rivaton A, Gardelle J (1998) Photo-oxidation of aromatic polymers. Angew Macromol Chem 261/262: 173

Rossnagel SM, Cuomo JJ, and Westwood WD (eds) (1990) Handbook of Plasma Processing Technology. Noyes Publications, Park Ridge, NJ

Rostaing C, Coeuret F, Drevillon B, Etemadi R, Godet C, Huc J, Parey JY, and Yakovlev VA (1993) Silicon-based protective transparent multilayer coatings deposited at high rate on optical polymers by dual-mode MW/rf PECVD. Thin Solid Films 236: 58-63

Samson F (1996) Optical lens coatings. Surf Coat Technol 81: 79-86

Schulz U, Munzert P, Kaiser N (2001) PVD coating of plastics for optical applications. Surf Coat Technol 142-144: 507-511

Schulz U, Schallenberg UB, Kaiser N (2002) Antireflection coating design for plastic optics. Applied Optics 41: 3107-3110

Segner J (1995) Plasma impulse chemical vapor deposition. In: Flory FR (ed), Thin Films for Optical Systems. Marcel Dekker, New York, Chap 7: 209

Simpson J, and Lewis KL (1997) Trends in optical coating. Proc 40th Ann Tech Conf of SVC, Denver, pp 248-254

Taga Y (1997) Recent progress in coating technology for surface modification of automotive glass. J of Non-Crystalline Solids 218: 335-341

Tang CW, and Van Slike SA (1987) Organic electroluminescent diodes. Appl Phys Lett 51: 913

Uyama H, Harada T, Takahaski T, and Tomikawa N (2001) Web coating formation of antireflection films on flexible substrates by evaporation. AIMCAL, Proc 15th Bakish Int Vacuum Web Coating Conf

Vossen JL, and Kern W (1991) Thin Film Processes II. Academic Press, New York

Weaver MS et al.(17 authors) (2001) Flexible organic LED displays. Proc 44th Ann Tech Conf of SVC, Philadelphia, pp 155-159

Wertheimer MR, Martinu L, and Liston EM (1996) Plasma sources for polymer surface treatment. In: Glocker D, and Shah SI (eds) Handbook of Thin Film Process Technology. IOP, Bristol. Chap E3.0

Wertheimer MR, Martinu L, Klemberg-Sapieha JE, and Czeremuszkin G (1999a) Plasma treatment of polymers to improve adhesion. In: Mittal KL and Pizzi A (eds) Adhesion Promotion Techniques. Marcel Dekker, New York, pp 139-153

Wertheimer MR, Fozza AC, and Holländer A (1999b) Industrial processing of polymers by low pressure plasmas: the role of VUV radiation. Nucl Instrum Methods Phys Res B 151: 65

Wohlrab C, and Hofer M (1995) Plasma polymerization of optical coatings on organic substrates: equipment and processes. Proc 38th Ann Tech Conf of SVC, Chicago, pp 222-231

Zabeida O, Klemberg-Sapieha JE, Martinu L, and Morton D (1999) Study of ion bombardment effects on optical films and polymer surfaces using a cold cathode ion source. Proc 42nd Ann Tech Conf of SVC, Chicago, pp 267-272

Zabeida O, Hallil A, Wertheimer MR, Martinu L (2000) Time resolved measurements of ion energy distributions in dual-mode pulsed microwave/radiofrequency plasma. J Appl Phys 88: 635

Abbreviations and Symbols

Abbreviations

a-C:H	hydrogenated amorphous carbon
AFM	atomic force microscopy
AR	antireflective
cw	continuous wave
CTA or TAC	triacetylcellulose
COC	cyclopolyolefine copolymers
COP	cyclopolyolefine
DC	direct current
DLC	diamond-like carbon
DOVID	diffractive optical variable image device
EAS	electronic article surveillance
EBE	electron beam evaporation
ECR	electron cyclotron resonance
FEP	fluorinated ethylene propylene
IBAD	ion beam assisted deposition
IEA	ion energy analyser
IEDF	ion energy distribution function
ISIS	interference security image structures
ITO	indium tin oxide
LC	liquid crystal
MS	mass spectrometer
MW	microwave
NIR	near infrared
OLED	organic light emitting diode
OVD	optical variable device
PC	polycarbonate
PDLC	polymer dispersed liquid crystals
PECVD	plasma enhanced chemical vapor deposition
PEN	polyethylene tetranaphthalate
PES	polyethersulphone
PET	polyethylene – terephthalate
PMMA	polymethylmethacrylate
PPFC	plasma polymerized fluorocarbons
PPHC	plasma polymerized hydrocarbons
PPOS	plasma polymerized organosilicones
PTFE	polytetrafluoroethylene
PVB	polyvinyl butaryl
PVD	physical vapor deposition
PVDF	polyvinilidine fluoride

RF	radiofrequency
SE	spectroscopic ellipsometry
SH	substrate holder
SIRR	solar infrared reflection
SZM	structure zone model
TCO	transparent conducting oxides
UV	ultraviolet
VASE	variable angle spectroscopic ellipsometry
VUV	vacuum ultraviolet
WA	water absorption

Symbols

CTE	coefficient of thermal expansion
D	duty cycle
d	film thickness
E	elastic (Young's) modulus
E_{i}	ion energy
E_{p}	energy delivered per deposited particle
F_{A}	adhesion force
f_{p}	pulse frequency
H	hardness
k	extinction coefficent
L_{c}	critical load for failure
n	refractive index (real part)
$n(\mathrm{z})$	refractive index depth profile
n_{e}	plasma (electron) density
R_a	surface roughness
$R\,(\lambda)$	optical reflectance
R	surface (sheet) resistance
r_{D}	deposition rate
R_{S}	tip radius
ST	service temperature
$T\,(\lambda)$	optical transmittance
T_{s}	substrate temperature
V_{B}	surface bias potential
V_{p}	plasma potential
w	scratch width
W_{A}	work of adhesion

Greek Symbols

Φ_i	flux of ions
Φ_m	flux of condensing particles
γ_C	surface energy
λ	wavelength
ρ	density
Θ_a	advancing contact angle
Θ_s	static contact angle
Θ_r	receding contact angle
σ_i	intrinsic stress
σ_t	thermal stress

Interference Coatings for Ultrafast Optics

Gabriel Tempea, Vladislav Yakovlev and Ferenc Krausz

Institut für Photonik, Technische Universität Wien, Vienna, Austria

1 Optical Components for Dispersion Control: an Overview

Shortly after the invention of the laser, phase locking of the axial eigenmodes of the laser cavity, briefly referred to as mode locking, opened the way to the generation of ultrashort laser pulses. Numerous applications benefited from the availability of short pulses: time-resolved (pump-probe) spectroscopy, precision material processing and large-bandwidth telecommunication are only a few of the many examples. Last but not least, ultrafast laser sources proved to be an invaluable laboratory tool for investigating new regimes of nonlinear optics.

Triggered by the prospects of these exciting applications, the pursuit of efficient sources capable of generating ever shorter pulses has lasted for decades. Until the mid eighties, state-of-the-art ultrafast lasers employed organic dyes as active media, culminating in the generation of 27-fs pulses directly from a laser oscillator (Valdmanis et al. 1985). Negligible for ps-pulses, the effect of dispersion (group delay dispersion, GDD, the second derivative of the phase picked by a broadband field upon linear propagation with respect to the angular frequency) becomes strong in the sub-100-fs regime.

Along with the large spectral bandwidth, accurate dispersion control is a fundamental prerequisite for sub-100-fs pulse generation. Although scientists had been trying to exploit the large gain bandwidth of organic dyes for a decade, it was the advent of the first low-loss dispersive delay lines made up of pairs of prisms (Fork et al 1984, Martinez et al. 1984) that allowed reproducible generation of optical pulses well below 100 fs.

The invention of Ti:sapphire as an ultra-broadband laser medium (Moulton 1986), that of Kerr-lens mode-locking (Spence et al 1991) and that of spectral broadening schemes relying on guided nonlinear propagation (Tomlinson et al. 1984, Nisoli et al. 1996) has brought the generation of pulses approaching the theoretical minimum pulse duration (one oscillation of the optical field, i.e. 2.7 fs at 800 nm, the centre wavelength of the Ti:sapphire gain curve) within reach. From that moment on, the history of ultrashort pulse generation became, to a great extent, the history of the development of broadband dispersive delay lines. A landmark in this process was set by the finding that dielectric multilayer coatings (used as highly efficient broadband reflectors ever since the discovery of lasers) could also be employed for control of group velocity dispersion. Used initially as a degree of freedom in the Fourier synthesis of complex multilayer filters (Dobrowolski and Lowe 1978, Verly and Dobrowolski 1990, Druessel et al. 1993), the frequency dependence of the phase change upon reflection can be itself engineered to meet the needs of ultrashort pulse generation and manipulation. As early as 1964 Gires and Tournois recognized that the group delay dispersion experienced by broadband radiation in a thin-film multilayer resonant cavity could be controlled by means of the different storage times of different frequency components of the incident radiation in the cavity. The bandwidth of the dispersive multilayer reflectors could be significantly enhanced by controlling GDD upon reflection by means of a frequency dependent penetration depth in a non-resonant multilayer reflector (Szipöcs et al. 1994), rather than by means of a frequency dependent storage time in a resonant cavity (Gires and Tournois 1964). This is the underlying concept of the widely used dispersive dielectric multilayer reflectors dubbed “chirped mirrors” (CMs).

As compared to traditional delay lines (consisting of prisms and/or grating pairs), the bandwidth and accuracy of dispersion management can be substantially improved by CMs, at the expense of the amount of GDD that can be introduced. Hence CMs are suitable in applications where low amounts of GDD need to be compensated. Such applications include laser oscillators, extra-cavity fibre compressors or optical parametric amplifiers. Replacement of traditional delay lines with CMs is not feasible in chirped pulse amplification schemes in which pulses need to be stretched and compressed by many orders of magnitude using large amounts of positive and negative GDD.

In Sect. 1.1 to 1.3 we describe the underlying physics as well as the performance and limitations of different dispersive multilayer mirrors employed in delay lines for ultrafast optics. We start with a brief description of Gires-Tournois interferometers and close the section with the latest generation of dispersive multilayer optics that provide accurate GDD control over more than a full optical octave. The most efficient methods for the pre-design of dispersive mirrors are summarized in Sect. 2. Accounting not only for reflectance but also for the phase properties of the multilayers in the optimisation process called for the development of dedicated implementations of numerical refinement algorithms. Some of them we outline in Sect. 3. The potential for improving the design and manufacturing of chirped mirrors as well as the opportunities they open in ultrafast optics (of which not all have been exploited so far) are addressed in Sect. 4.

1.1 Gires-Tournois Interferometers. Resonant Dispersive Mirrors

The first proposal to use multilayer structures for dispersion control dates back to as early as 1964. The structure described by Gires and Tournois consists of a partial reflector, a spacer and a second mirror with ideally 100% reflectance; the component is thus very similar to a Fabri-Perot interferometer in which one of the cavity mirrors is replaced by a 100% reflector. Consequently, Gires-Tournois interferometers (GTIs) exhibit constant high reflectivity (within the bandwidth of the reflector) while the group delay (GD) introduced upon reflectance is given by

$$\tau(\omega) = \frac{(1-R)t_0}{1+R-2\sqrt{R}\cos(\varphi-\omega t_0)} \qquad (1)$$

where R is the reflectivity of the first mirror, φ is the phase change on reflection on the second (100%-reflecting) mirror and t_0 is the round-trip time of the cavity. Obviously $t_0 = 2d/c$, d being the optical thickness of the spacer and c the velocity of light in vacuum. The maximum GD is given by:

$$\tau_{max} = t_0 \frac{1+\sqrt{R}}{1-\sqrt{R}} \qquad (2)$$

for each resonant frequency of the cavity and decreases for frequencies deviating from resonance. Thus, negative GDD will be introduced by GTIs at wavelengths slightly shorter than the resonance wavelength. As most optical materials have weak high order dispersion away from the absorption edge, usually constant negative GDD is required. The range in which the GDD of GTIs remains fairly constant is rather limited.

Figure 1 shows the GD and GDD introduced by a typical GTI having the stack formula $(HL)^{10}$ H 2L H_1 (with H=TiO_2, L=SiO_2 and H_1= TiO_2 for the full line, Ta_2O_5 for the dashed curve and Si_3N_4 for the dots) and a resonant wavelength of 800 nm. The bandwidth (i.e. the wavelength range within which the GDD of the GTI remains fairly constant) and the average GDD value can be varied, but not independently. By reducing the reflectance of the upper mirror, the range within which the GD varies linearly (i.e. GDD remains approximately constant) is enlarged. However the slope of the GD variation and thus the absolute value of the GDD decreases. This behaviour is illustrated in Fig. 1. The index and thus the reflectance of the top layer was increased from 1.88 (corresponding to Si_3N_4) to 2.1 (Ta_2O_5) and 2.3 (TiO_2).

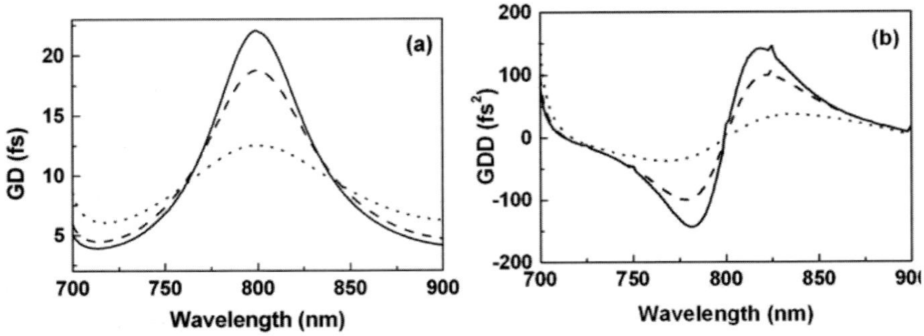

Fig. 1 GD **(a)** and GDD **(b)** introduced upon reflection by a GTI as described in the text

Recently multilayer structures consisting of multiple-cavity GTIs have been proposed (Goblutovic et al. 2000, Szipöcs et al. 2000). These components which could generically be called resonant dispersive mirrors (RDMs) employ a Bragg multilayer[1] as back reflector while the front section of the mirror consists of two or three resonant cavities. The reflectance of RDMs can be comparable to that of a Bragg reflector while the bandwidth over which constant GDD can be introduced is much larger as compared to single-cavity GTIs, being only limited by the wavelength range in which the GDD of the Bragg reflector remains small (~ 150 nm @ 800 nm). RDMs offer an alternative to CMs in applications in which a sub-200-nm bandwidth is sufficient and losses are critical.

1.2 Chirped Mirrors

Standard dielectric Bragg reflectors (BRs) are widely used in femtosecond laser applications owing to their large reflectance bandwidth and to the weak wavelength dependence of the phase shift they impose on the reflected beam in their high reflectance (HR) range. Figure 2 depicts the reflectance (a) and the GDD (b) of a Bragg mirror described by the stack formula $(HL)^{20}H$ with $H=TiO_2$ and $L=SiO_2$; the reference wavelength was chosen as 800 nm, the centre wavelength of the gain curve of Ti:sapphire. The absolute value of GDD increases towards the extremities of the HR range. For pulse bandwidths approaching that of BRs it was found that GDD errors rapidly increasing towards the boundaries of the HR range impose limitations on the shortest pulse duration (Laporta and Magni 1985).

[1] Bragg multilayers or reflectors are quarter-wavelength stacks of alternating high and low-index materials. The Bragg wavelength is equal to twice the optical thickness of the Bragg period consisting of a low and a high-index layer.

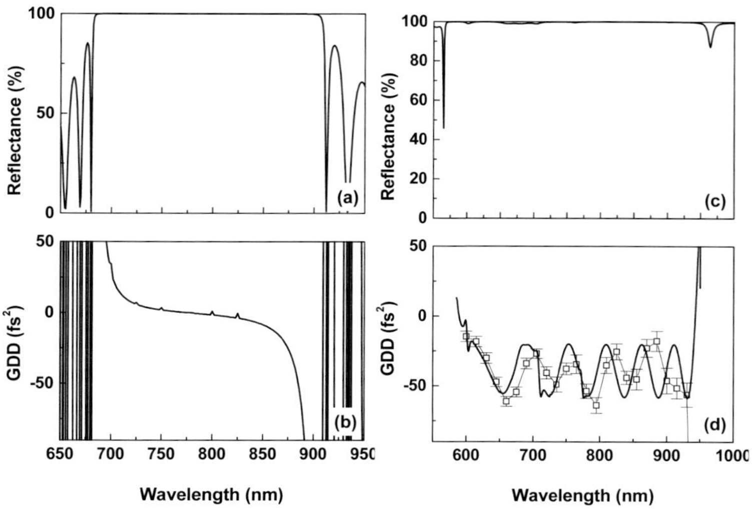

Fig. 2 Reflectance **(a)** and GDD **(b)** of a Bragg mirror as described in the text and reflectance **(c)** and GDD (**d**, line: calculation, squares: measurement) of a broadband chirped mirror consisting of 53 TiO_2 and SiO_2 layers

The physical origin of the increased GDD introduced by BRs at the boundaries of the HR spectral range can be easily understood by analysing the electric field intensity distribution in the multilayer as a function of wavelength (Fig. 3a). In the spectral regions where transmittance tends to be higher, the penetration depth of the electric field increases and consequently the delay experienced by the wavepackets upon reflection becomes larger. This observation led to the finding that the GDD introduced by multilayer mirrors can be controlled by means of the wavelength dependence of the penetration depth of the electric field. A mirror introduces negative GDD (as required by most applications) if short wavelength components are substantially reflected already by the top layers whereas long-wavelength wavepackets have to penetrate deeper into the multilayer structure before being reflected, thus experiencing a longer delay. This kind of behaviour is illustrated in Fig. 3b for a typical chirped mirror design, the reflectance and GDD of which are shown in Fig. 2c and Fig. 2d, respectively.

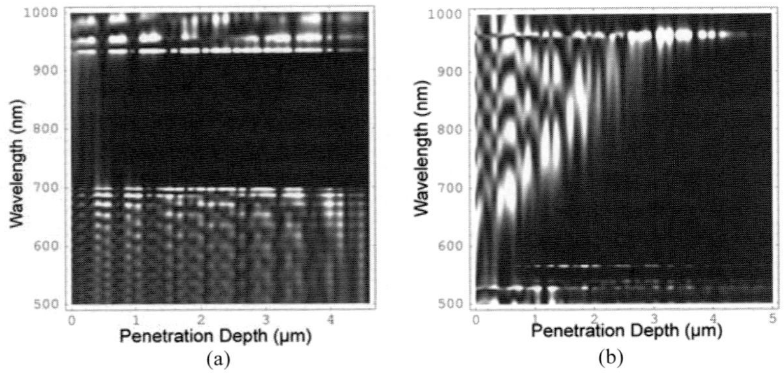

Fig. 3 Electric field intensity (arbitrary units) plots for a standard Bragg reflector **(a)** and a chirped mirror **(b)**. The reflectance and GDD curves for the same designs are shown in Fig. 2

One would expect that the desired variation of the penetration depth (i.e. depth increasing with increasing wavelengths) could be obtained by reducing the period of the Bragg reflector from the substrate towards the top layer. Whereas this technique had been successfully used for enhancing the reflection bandwidth (Penselin and Steudel 1955, Turner and Baumeister 1966), this consideration is not sufficient for the design of negative GDD mirrors because resonant trapping of the wavepackets in the multilayer plagues the GDD of designs relying merely on a linear variation of the Bragg period (this issue is addressed in detail in Sect. 2). Dispersive mirror designs obtained with various techniques do show a trend of linear decrease of the Bragg frequency towards the incident medium, but the thickness of individual layers can deviate significantly from this tendency.

Invented in 1994 (Szipöcs et al. 1994), CMs were initially designed by means of computer optimisation, starting from multilayers having linearly chirped multilayer period across the coating structure. This approach was time consuming (hours of computing time were required for obtaining the first CM designs) and led to designs with limited bandwidth (up to 200 nm at 800 nm). With the development of quasi-analytical pre-design methods (described in Sect. 2) mirrors having bandwidths in excess of 300 nm at 800 nm could be designed, manufactured and employed to set new records in ultrashort pulse generation. Using the mirror depicted in Fig. 2c and 2d to compress the continuum produced by focusing 0.1-TW, 5-fs pulses in air, 4-fs pulses were generated (Cheng et al. 1998). Similar results, both in terms of the mirror specifications (Szipöcs et al. 1997, Mayer et al. 1997, Matuschek et al. 1999, Tempea et al. 1998) and duration of the generated pulses (Baltuska et al. 1997, Sutter et al. 1999, Morgner et al. 1999) were obtained by several groups, indicating that the limits of CM design had been reached and a fundamentally new approach was required in order to further increase the bandwidth of all-reflective negative delay lines.

1.3 Dispersion Control over a Full Optical Octave: Mirror Pairs, TFI and BASIC Mirrors

The replacement of design techniques relying exclusively on computer optimisation with algorithms providing a functional starting structure (see Sect. 2) has allowed increasing the bandwidth over which high reflectance and dispersion control are provided to about 160 THz at 800 nm. Extending the bandwidth beyond this limit has only been possible at the expense of rapidly increasing unwanted oscillations in the frequency dependent group delay dispersion. These fluctuations can be partially suppressed by using pairs of mirrors with complementary dispersion characteristics (Szipöcs and Köházi-Kis 1997, Laude and Tournois 1999, Kärtner et al. 2001). Several techniques have been proposed for obtaining mirrors with mutually compensating GDD fluctuations. For bandwidths of about 160-180 THz the period of the oscillations is fairly constant. Thus, one can use identical mirrors at different angles of incidence, such that the spectral blue-shift of the GDD-curve of the mirror used at the larger angle of incidence equals the half-period of the GDD oscillation (Sutter et al. 1999). Alternatively, the same mirror design can be manufactured twice, at shifted reference wavelengths.

Much larger bandwidth (approaching one optical octave at 800 nm) can be achieved by optimising the two mirrors simultaneously and defining the target for the overall GDD of the pair (Laude and Tournois 1999). An analytical method for determining suitable starting designs for the mirror pair optimisation was proposed by Kärtner et al. in 2001. The disadvantage of mirrors optimised in pairs is that although the designed overall GDD curve is sufficiently smooth, the GDD curve of each mirror taken separately shows very large oscillations, with amplitudes of few hundreds of fs^2. These oscillations are the result of Gires-Tournois resonances in the stack that make the design very sensitive to manufacturing errors. A small deviation in the thickness of one of the resonator spacers results in a very large error in the phase shift of the reflected beam because phase errors accumulate during multiple roundtrips. For this reason, the resultant GDD of manufactured mirror pairs showed large ($\approx$+/- 100 fs^2) oscillations around the target value (Kärtner et al 2001, Ell et al. 2001).

Both the bandwidth limitation and the undesired dispersion oscillations are consequences of imperfect impedance matching between the multilayer and the incident medium, air. This problem can be overcome by replacing air with glass as the incident medium. This solution has been independently recognized and simultaneously published by researchers of the Vienna group (Tempea and Krausz 2000) and by the group from ETH Zürich (Gallman et al. 2000). The new approach overcomes the drawbacks described above and allows the design of *single* mirrors introducing smooth GDD over a comparable bandwidth.

In negative GDD CM designs, the penetration depth of wavepackets is increased with increasing wavelength by increasing (on the average) the Bragg period of the stack towards the substrate. Additional modulation of the layer thickness is essential (see Sect. 2) in order to suppress multiple internal reflections within the mirror's interfaces. Because of the large refractive index difference between the multilayer and air, a near wavelength-independent reflectance of about 3

% occurs at the front interface. This gives rise to unwanted interference between rays reflected at the air interface and those penetrating deeper into the coating (Fig. 4a), leading to strong distortions of the reflectance and particularly of the GDD of the mirror. In Sect. 1.1 it was shown how this effect is exploited in Gires-Tournois mirrors (Gires and Tournois 1964) for producing large negative GDD over a narrow bandwidth. In the case of broadband mirrors, such resonant interference effects lead to a fast oscillation of the GDD around the target curve, impairing the quality of the compressed pulses. This adverse effect can be alleviated by coating an anti-reflection (AR) filter (Kärtner et al 1997) or a narrow-band-stop filter (Tempea et al 1998) on top of the CM. Efficient suppression of GDD distortions calls for a reflectance significantly less than 0.001 % (!) at the air interface. This requirement can be met by AR coatings or bandpass filters only over a limited bandwidth (Dobrowolski et al 1996), restricting the bandwidth usable for GDD control from a single CM to 150-160 THz at 800 nm (Matuschek et al. 1997, Mayer et al. 1997, Tempea et al. 1998). Furthermore, residual interference effects, often enhanced by manufacturing errors, can lead to large fluctuations of the GDD versus frequency curve.

The above-mentioned drawbacks can be overcome by preventing the beam reflected at the front surface and that reflected by the multilayer from interfering with each other. This can be achieved either by tilting the front interface (the interface to the incident medium, which is most commonly air) with respect to other interfaces of the multilayer structure, as sketched in Fig. 4b or, alternatively, with two surfaces having different radii or centres of curvature (Matuschek et al. 2000). Two embodiments of this concept were proposed. Back-side-coated (BASIC) mirrors are deposited on a thin wedge and illuminated from the substrate side (Matuschek et al. 2000). Tilted-front-interface (TFI) mirrors are coated on standard (thick) optical substrates that ensure good surface quality. Subsequently a thin wedged plate is contacted to the multilayer's surface. Details of the technology of fabrication and post-processing of both kinds of mirrors are given in (Matuschek et al. 2000) and (Tempea et al. 2001).

In this manner spurious interference between the beam reflected at the air-dielectric interface (dotted line) and those reflected by internal interfaces (full line) can be completely eliminated. Demands on the AR coating on the tilted-front surface can be substantially relaxed because residual reflectance at this interface merely introduces some loss, without influencing the phase shift experienced by the reflected main beam. To keep the overall reflectance high, the front surface reflectance can be reduced below 0.2 % over 500-1000 nm without introducing notable phase distortions by using an AR coating consisting of merely 13 layers.

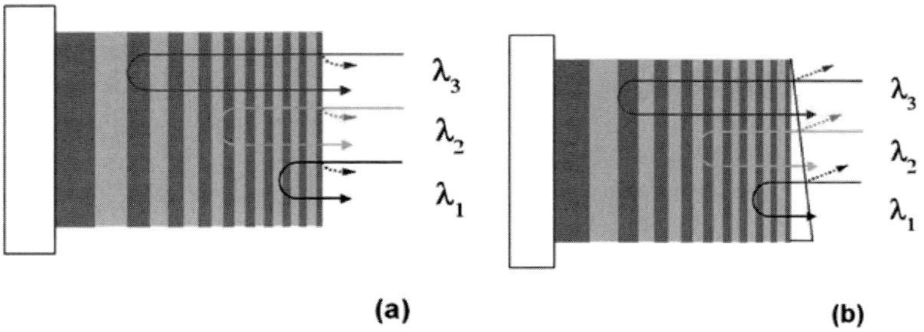

Fig. 4 Schematic representation of a standard CM **(a)** and of a TFI CM **(b)**. In order to achieve negative GDD upon reflection, the inequality $\lambda_1 < \lambda_2 < \lambda_3$ must be fulfilled. Rays reflected at the front interface (represented by dots) can interfere with rays reflected within the multilayer (full lines) giving rise to GTI-like GDD oscillations in the case of the standard CM. The wedge attached to the TFI mirror prevents this effect

To demonstrate the power of this concept, we present a tilted-front-interface chirped mirror (henceforth briefly TFI mirror) that provides dispersion control up to (and including) the fourth order over one octave, between 500 and 1000 nm (Fig. 5). Employing several reflections off such a mirror, the spectral chirp imposed on a broadband wave packet upon travelling several millimeters in e.g. fused silica and tens of centimeters in air can be compensated accurately enough for the generation of pulses with durations of around 3 fs in the near-single-cycle regime. A first attempt has already been made to manufacture this mirror. The squares in Fig. 5 depict GDD data measured by white light interferometry (Knox et al. 1988). Deviations of the measured GDD curve from the designed one are due to manufacturing errors, which need to be reduced in the future. The coating consists of 68 SiO_2 and TiO_2 layers.

Tilting the front interface with respect to the other interfaces of the multilayer results in perfect impedance matching of CMs to the environment over an arbitrarily broad spectral range. This benefits the design of broadband CMs in several respects: i) the bandwidth of CMs drawing on this concept is only limited by technological issues (number of layers that can be coated and accuracy of thickness control), ii) the sensitivity to manufacturing errors is reduced, iii) the transmittance outside the HR band becomes larger enabling the design of ultrabroadband dichroic mirrors (Tempea et al. 2001), iv) mirrors with very smooth GDD curves can be designed.

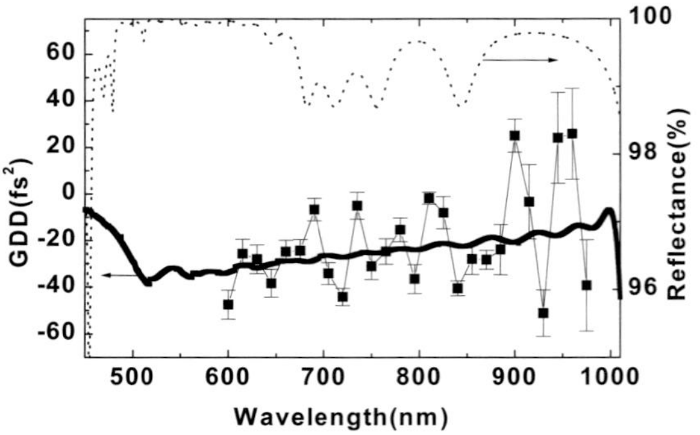

Fig. 5 Reflectance (dots) and GDD (full line) of a TFI-CM consisting of 68 SiO_2 and TiO_2 layers. The mirror precisely compensates the dispersion introduced by a physical path of 0.45 mm in fused silica plus 15 cm in air. The GDD could not be measured below 600 nm because of bandwidth limitations in the white light interferometer. The positive GDD and reflectance losses introduced by the glass wedge (which can be as thin as 50 μm if special technology is employed) were not taken into account

2 Design Methods

After their invention in 1994 (Szipöcs et al. 1994), CMs were rapidly improved with semi-analytical design methods (Szipöcs and Köházi-Kis 1997, Kärtner et al. 1997, Tempea et al. 1998) and became a key component in broadband delay lines used for sub-10-fs pulse generation (Morgner et al. 1999, Sartania et al. 1997, Baltuska et al. 1997, Xu et al. 1998).

As discussed in Sect. 1.3, CM designs can be significantly improved at the expense of additional complexity if air is replaced by glass as the medium of incidence. None of the design methods that will be described in the following takes into account the impedance mismatch between the coating and the environment, ab initio. At a first step this effect is neglected and subsequently counteracted either by means of an antireflection coating or an edge filter or by mere computer optimisation. For this reason we shall not separate the treatment of the design of standard CMs (having air as medium of incidence) from that of TFI-CMs and BASIC-CMs (employing glass as incidence medium). Rather, any of the design approaches presented in the following can be employed for calculating either of the two types of CMs.

The design methods proposed so far rely on Fourier synthesis, coupled-wave theory and on the properties of multilayers with periodically modulated thickness, respectively. We dubbed these approaches the Fourier-transform method (FTM), the double-chirping method (DCM) and the layer-thickness-modulation method (LTMM).

The three design methods are based on fundamentally different approaches. They all provide fairly simple and flexible "recipes" that, taking into account technological constraints and specifications, allow obtaining a pre-design that after limited computer optimisation closely fulfils the requirements of a wide range of applications. We shall restrict our discussion to providing insight into the phenomenological background of each method and to presenting the results needed for efficient CM design. The (often involved) derivation of the results is out of the scope of this chapter and the interested reader is referred to the relevant literature.

The "needles" technique (Tikhonravov et al. 1996), a powerful mathematical method for generating starting designs has also been successfully used for the design of chirped mirrors (Tikhonravov et al. 1999). This method is described in detail in the Chapter "Design of Optical Coatings".

2.1 Fourier Synthesis

In the seventies, a very versatile formula which allows the design of graded-index filters (also known as "rugate filters") having prescribed spectral amplitude *and phase* properties was derived (Sossi 1974):

$$\frac{1}{2}\int_{-\infty}^{\infty}\frac{d\ln[n(x)]}{dx}\exp(ikx)dx = Q(k)\exp[i\Phi(k)]\cdot \tag{3}$$

Here $n(x)$ is the refractive index, $k=2\pi/\lambda$ is the wave number in vacuum and x is twice the optical thickness from the centre of the graded-index medium to the physical position z. Partial integration and Fourier transform of (3) yield:

$$\ln\left[\frac{n(x)}{n_0}\right] = \frac{i}{\pi}\int_{-\infty}^{\infty}\frac{Q(k)}{k}\exp\{i[\Phi(k)-kx]\}dk\cdot \tag{4}$$

The Q function $Q(x)$ and the phase-factor $\Phi(k)$ are obviously related to the spectral amplitude and phase of the filter respectively, but they are not uniquely defined. As prior to the development of multilayer coatings for ultrafast optics the phase change on reflection was typically not specified, the phase factor $\Phi(k)$ was used as a degree of freedom that allowed improving the technical realisability of the designs (by e.g. reducing the index ratio or changing the overall thickness of the coating).

In 1994 Szipöcs and Köházi-Kis showed that Eq. (3) can be efficiently used for the design of dispersive rugate reflectors using a quite intuitive choice for Q and Φ:

$$Q(k) = |r(k)| \tag{5}$$

$$\Phi(k) = \varphi(k)\,. \tag{6}$$

Here $|r(k)|$ and $\varphi(k)$ are the amplitude and the phase respectively of the complex amplitude reflectance of the filter.

If we assume a second order polynomial expression for the phase factor

$$\Phi(k) = d_0 + d_1(k - k_0) + d_2(k - k_0)^2 \tag{7}$$

and a Gaussian shape for the amplitude reflectance

$$Q(k) = \exp\left[-\frac{(k - k_0)^2}{2\sigma^2}\right]. \tag{8}$$

Equation (4) can be integrated to yield the logarithmic refractive index profile of a reflector that introduces a phase shift on reflection almost identical to $\varphi(k)$. Deviations of the actual phase shift on reflection (that can be exactly calculated by dividing the graded index structure in very thin layers of constant index and using the transfer matrix formalism) from the designed one (specified by means of $\varphi(k)$) originate from the approximate nature of the expression (3) which neglects the finite dimensions of the medium and multiple reflections within the filter (which might arise due to a rapid variation of the medium's impedance). For this last reason the FTM cannot be used for the direct synthesis of filters consisting of discrete layers. In order to overcome these drawbacks of the FTM, it is assumed that both the substrate and the medium of incidence have a refractive index $n(-\infty) = n(\infty) = \sqrt{n_{\min} n_{\max}}$ ($n_{\min}$ and $n_{\max}$ being the extreme values of the index profile $n(x)$). Furthermore, the equivalent impedance of the structure has to be adiabatically matched to that of the surrounding medium, i.e. smooth apodization functions should be used for the envelope of the index profile $n(x)$ (Southwell 1989). This consideration makes a Gaussian reflectance (and consequently index envelope) function preferable to a super Gaussian one.

The rugate chirped mirror designs relying on the above-described design technique were converted into two material multilayers by a two step process (Szipöcs and Köházi-Kis 1997):

i) The rugate mirror design consists of a sinusoidal modulation of the index (with variable period) on which a slower amplitude modulation ("envelope") is superimposed. Each half-period of the sinusoidal modulation is replaced by a homogeneous layer having an index given by the envelope and thickness determined by the half-period of the sinusoidal function. In mathematical terms, sin(kx) is replaced by sign[sin(kx)].

ii) The multilayer consisting of layers of index arbitrarily varying between two constrained values (these of the low and high index material effectively available) is transformed into a two material stack by means of an equivalent stack method. These methods permit replacing a layer of arbitrary index by a stack consisting of two materials with given optical constants.

Two drawbacks of this conversion procedure should be mentioned. The "equivalent" stack is precisely equivalent to the replaced layer only at a certain wavelength and equivalent stacks often contain very thin layers that cannot be accurately manufactured. The first drawback can be overcome by means of numerical refinement. In order to avoid the occurrence of excessively thin layers in the equivalent stack, a suitable conversion method (Southwell 1985, Fabricius 1992) can be used for transforming multilayers consisting of N different index layers into a two component design having $2N$ layers, as discussed in (Szipöcs and Köházi-Kis 1997).

The multilayer obtained following steps (i) and (ii) exhibits broadband reflectance and introduces a group delay upon reflection oscillating with large amplitude around the target curve defined by Eq. (7). Computer optimisation is required to suppress these oscillations. A design example illustrating the calculation of the rugate mirror and the conversion to a homogeneous index multilayer has been presented previously (Szipöcs et al. 1997).

Apart from providing an analytical approach for the generation of a good starting structure, the Fourier synthesis concept was also used for deriving a frequency domain optimisation method (Szipöcs et al. 2000). The main advantage of the method resides in the reduced number of parameters that need to be optimised (much smaller than the number of layers), which significantly speeds up the refinement process.

2.2 Double Chirping Method

In 1997 Matuschek et al. showed that the coupled mode theory can be used for accurately describing the amplitude and phase behaviour of multilayer interference coatings with *arbitrary* strong index modulation if the coefficients determining the coupling between the forward and backward propagating fields are properly chosen. Although mathematically more involved as compared to the traditional transfer matrix formalism, this approach offers the advantage of providing deeper insight into the physical origin of the multilayers' phase behaviour. Furthermore, under certain circumstances the coupled mode calculation can be inverted and employed for the design of multilayer reflectors that introduce a given phase shift upon reflection.

It had been previously proposed to chirp the Bragg period of reflectors in order to control the penetration depth of different wavepackets and thus the spectral variation of the group delay. This method has not proven its worth in the design of CMs because interference between beams partially reflected off the front and intermediate layers give rise to pronounced modulations of the frequency dependent phase shift imposed upon reflection. These undesirable effects could only be removed within a limited bandwidth after extensive computer optimisation. The coupled mode approach allowed identifying the cause of this phase modulation in the excessively rapid variation of the characteristic impedance of a transmission line modelling the multilayer. By analogy to microwave engineering it was recognized that these spurious reflections could be eliminated by *slowly* changing the characteristic impedance, i.e. by adiabatically switching on the coupling between the forward and the backward propagating waves. This is equivalent to chirping not only the Bragg period of the multilayer, but also the ratio between the thickness of the high index layer and that of the low index layer within each period (Kärtner et al. 1997).

Coupled mode analysis (Matuschek et al. 1999) yields the following expression for the distance between the front interface and a given index step at which the Bragg wavenumber has the value $k_B=2\pi/\lambda_B$ for a mirror introducing a given (not necessarily constant) GDD:

$$m(k_B)=\frac{c^2}{2\pi}\left(1-\frac{|k_0|}{\pi}\right)\int_{k_B}^{k_B^{\max}}\bar{k}_B GDD(\bar{k}_B)d\bar{k}_B\cdot \tag{9}$$

Here the continuous variable m is related to the z-coordinate (see Fig. 1a) by $dm=dz/\Lambda(z_m)$ where $\Lambda(z_m)$ is the physical thickness of the index step at the position z_m.

The above formula implicitly describes how the Bragg wavenumber of the stack should change with the distance from the front interface in order to obtain a certain GDD, thus defining a chirp law. Usually only low order dispersion is desired, i.e. GDD(k_B) is most often a low order polynomial function. In such cases an explicit chirp law can be obtained analytically by inverting Eq. (9). Of particular interest is the simple case of constant GDD=D_0, in which the chirp law can be written as a square root:

$$k_B(m)=k_B^{\max}\sqrt{1-\frac{4\pi^2}{c^2D_0(\pi-|k_0|)(k_B^{\max})^2}|m|}\cdot \tag{10}$$

Equation (9) can be used for generating a chirped Bragg reflector having a GDD curve that oscillates with a large amplitude around the prescribed value. In order to suppress these oscillations, the characteristic impedance of a transmission line modelling the reflector has to be adiabatically varied through the front section of the mirror, which was shown to be equivalent to chirping the thickness of the high index layer over the first Bragg pairs (Kärtner et al. 1997). Thus, the thickness of the high index layers is increased over the first N index steps according to

$$d_{h,j}=\left(\frac{j}{N}\right)^{\alpha}\frac{\lambda_{B,N}}{4n_h} \tag{11}$$

where $\lambda_{B,N}$ is the Bragg-wavelength of the Nth index step, j is the number of the index step (counted from the front of the mirror), n_h is the refractive index of the high-refracting material and α is a small integer number.

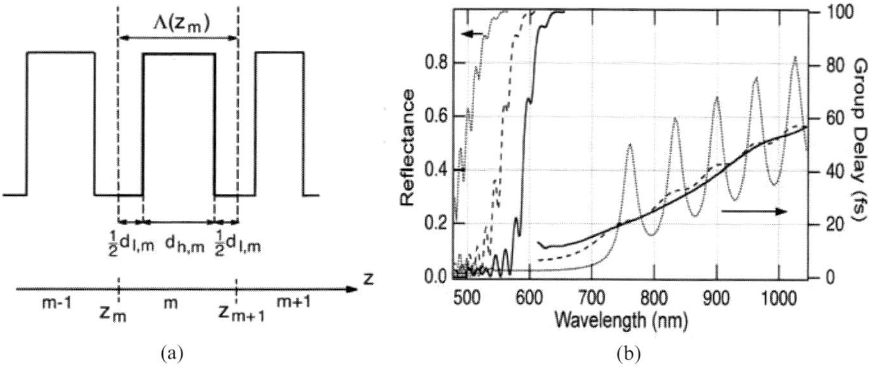

Fig. 6 (a) Schematic representation of the index profile, illustrating the symbols used in the text; $\Lambda(z_m) = d_{h,m} + d_{l,m}$ represents the physical thickness of the symmetrically defined index step, where $d_{h,m}$ and $d_{l,m}$ are the physical thickness of the high and low index layer respectively. **(b)** Reflectance and group delay corresponding to a linearly chirped design (dots) and double chirped multilayers in which the thickness of the high index layer was chirped linearly (dashed line) and quadratically (full line) over the first 12 layer pairs (courtesy of U. Keller and co-workers)

Figure 6b illustrates the DCM design method (Kärtner et al. 1997). A mirror consisting of 25 index steps (with refractive indices 2.5 and 1.5) was designed by chirping the Bragg wave number linearly between 6.8 μm^{-1} and 11 μm^{-1} over the first 20 index steps and keeping it constant over the remaining 5. As expected, the GD curve of the linearly chirped design exhibits large oscillations due to spurious resonances. These oscillations can be efficiently suppressed by chirping the thickness of the high index layers over the first 12 pairs according to the relationship (11) with N=12 and α=1 (dashed line) and α=2 (full line). A larger α results in a stronger chirp and a smoother GD and GDD curve. However, the thickness of the first high index layer will pose manufacturing problems. Choosing a smaller α (which results in a thicker top layer) and using computer optimisation for improving the mirror's performance can provide a more robust (i.e. less sensitive to manufacturing errors) design.

2.3 Layer Thickness Modulation Method

Multilayer mirrors with sinusoidally modulated layer thickness (Fig. 7a) exhibit multiple high reflectance (rejection) bands (Fig. 7b). Precise mathematical equations that relate the position and strength of the stop bands to the amplitude and period of the function modulating the layer thickness have been determined (Tempea et al. 1998, Perilloux 1998, Perilloux 1999). Drawing on this theory, an algorithm for the efficient synthesis of two component chirped mirrors with prescribed reflectance and phase characteristics has been derived (Tempea et al. 1998).

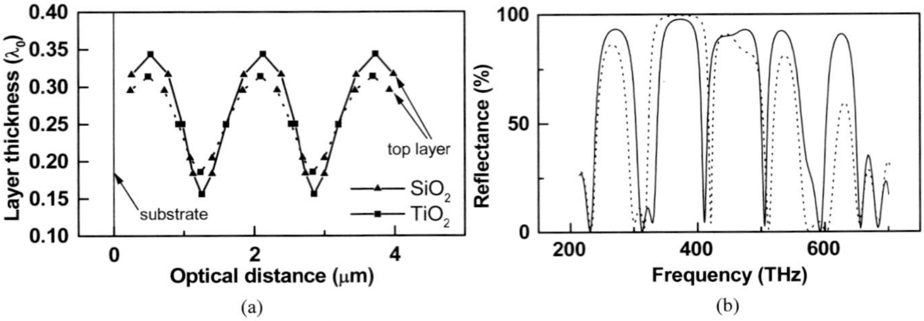

Fig. 7 Multilayer structures with a sinusoidal modulation imposed on the layer thickness for two different values of the modulation amplitude **(a)** and the respective reflectance versus frequency curves **(b)**. The dotted reflectance curve in panel **(b)** corresponds to a design with layer thickness modulated by a function depicted by the dotted line in panel **(a)**

The performance of chirped mirrors critically depends on the accuracy achievable in controlling the penetration depth of the different spectral components. In a multilayer structure subjected to a simple linear chirp this accuracy is severely impaired by the insufficient reflectance of the interfaces, resulting in many partially reflected/transmitted beams of comparable intensity. Interference between these partial waves messes up the dispersion curve of the mirror. This should be avoidable by composing the mirror of narrow band stop filters with gradually shifted centre wavelengths, each consisting of more than two layers to provide high reflectance.

Such a multiple band stop filter structure can be constructed by chirping the period of the sinusoidal layer thickness modulation from the substrate towards the top layer of the two component multilayer. This leads to merging of the rejection peaks in Fig. 7b, resulting in a continuous high reflectance band. The stacks that reflect different wavelength components of the HR range are placed at different distances from the substrate, thus allowing control of the wavelength dependence of the penetration depth. The generic formula describing the variation of the layer thickness in such a system is given by

$$t(x) = t_0(x) + A(x)\sin\left(2\pi \frac{x}{\Lambda(x)}\right) \tag{12}$$

where x measures the optical distance of the respective layer from the substrate and the number of layers within a modulation period $\Lambda(x)$ must be kept constant, i.e. $\Lambda(x) = \alpha t_0(x)$. The parameter α determines the number of layers within a modulation period and was set at $\alpha \approx 5$, i.e. equal to its minimum permitted value.

The parameters in Eq. (12) are determined by the frequency range $f_{min}=c/\lambda_{max}$, $f_{max}=c/\lambda_{min}$ over which high reflectance is required and by the prescribed group delay versus frequency function within this range. For a nearly constant negative GDD = D over the high reflectance range we require

$$t_0 = \frac{1}{4}\left(\frac{\lambda_{\min} - \lambda_{\max}}{d} x + \lambda_{\max}\right) \tag{13}$$

which implies, according to (12), a modulation period $\Lambda(x)$ that varies linearly with x. The total optical thickness of the coating d should fulfil

$$d \geq d_\tau = \frac{c}{2}\left[\tau(f_{\max}) - \tau(f_{\min})\right] \tag{14}$$

where $\tau(f_{\max}) - \tau(f_{\min}) = 2\pi D(f_{\max} - f_{\min})$.

It is usually difficult to achieve negative GDD at the shortest wavelength of the high reflectance range. This difficulty relates to the fact that the shortest wavelength components have a small penetration depth that tends to result in a near constant group delay. This problem can be alleviated by increasing the amplitude $A(x)$ of the modulation as given by $A=x(A_2-A_1)/d+A_1$, where A_1 and $A_2>A_1$ are the limits between which the amplitude is allowed to vary.

Eqs. (12-14) constitute a complete recipe for constructing a multilayer structure that, after some limited computer refinement, is expected to introduce constant negative GDD over an ultrabroad spectral range.

The efficiency of the method can be illustrated for a chirped multilayer mirror required to provide dispersion control over a bandwidth of more than 300 nm around λ_0 = 780 nm. The mirror structure has been generated by using a direct computer implementation of Eqs. (12-14) with the input parameters λ_{min} = 600 nm, λ_{max} = 950 nm, D= - 60 fs^2, A_1 = 12.5, A_2 = 62.5, d = 10.5 μm and employing TiO_2 ($n \approx 2.35$) and SiO_2 ($n \approx 1.45$) as layer materials.

The unoptimised design exhibits a high (R > 99%) reflectance over a bandwidth of about 170 THz in the wavelength range of 640 - 970 nm, as shown in Fig. 8, but the group delay is subject to large oscillations (dotted line in Fig. 8). These undesirable oscillations can be readily eliminated by computer optimisation. Using only 15 iterations of the Variable Metric algorithm of TFCalc (Software Spectra, Inc) with the design targets given by R> 99% and D = - 40 fs^2 between 655 nm (458 THz) and 950 nm (315 THz) results in a dramatic improvement of the linearity of the group delay curve (dashed line in Fig. 8). After optimisation (with the medium of incidence being air), the first section of the stack acts as an edge filter that reflects the shortest wavelength wavepacket while providing impedance-matching at longer wavelengths. The optimisation is completed after as few as 80 iterations, which takes merely a few minutes employing a standard personal computer.

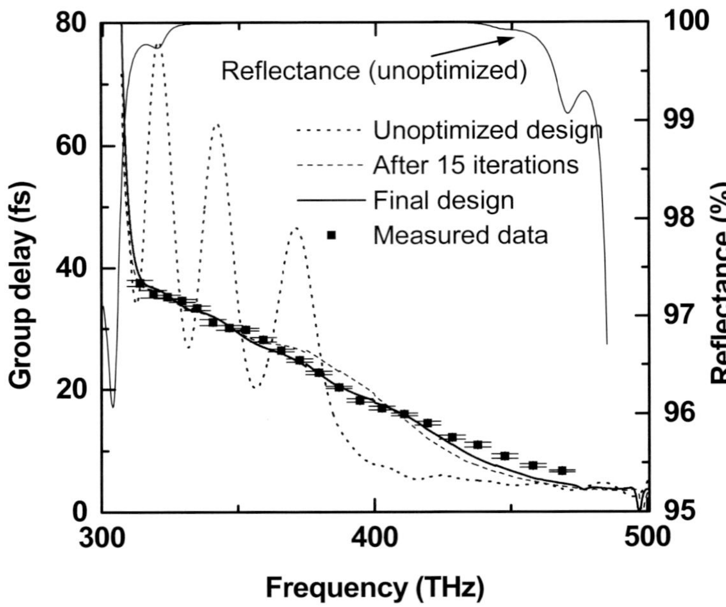

Fig. 8 Reflectance and group delay versus frequency of a chirped mirror before and after optimisation as described in the text. The full squares with error bars depict measured data obtained by white light interferometry (Knox et al. 1988) where the absolute value of the group delay was arbitrarily chosen. The spectral resolution of the measurement was 10 nm

3 Computer Optimisation of Dispersive Multilayers

Several quasi-analytic design methods were presented in Sect. 2. For most applications these methods can provide good starting designs that allow efficient numerical optimisation for approaching the dispersion and reflectance targets. However, there are complex chirped mirror design targets for which quasi-analytical pre-design methods fail to provide starting structures that are sufficiently close to a satisfactory solution. In this case global optimisation (a tailored implementation of which is presented in Sect. 3.2) outperforms the local refinement of the pre-design. Such complex design problems include the calculation of

i) chirped input couplers (mirrors that provide high reflectance and negative GDD but high transmittance over the gain and absorption band of the laser medium, respectively;

ii) output couplers that provide pre-defined GDD and limited, controlled transmittance between 5 and 30% over a wide spectral rage (used to couple out energy over a broad band from ultrafast laser oscillators);

iii) chirped mirrors with extremely large bandwidth, e.g. equal or greater than one optical octave.

Transmittance in a given wavelength range (as required for i), controlled leakage in the HR range (necessary for ii) or a trade-off between losses and bandwidth (necessary for achieving the bandwidth required by iii) cannot be explicitly taken into account by analytic pre-design methods. Numerical refinement with specially tailored merit functions is required for tackling these problems. A trade-off has to be made between the extent to which different requirements can be fulfilled. This trade-off is usually realised by means of the weighting factors of different targets in the merit function. In ultrafast optics not only the losses but also the change in duration experienced by a pulse on reflectance/transmittance on a multilayer is relevant. This makes the correct choice of weighting factors between GDD and reflectance/transmittance targets non-trivial. Time-consuming (and often not very successful) trial of different intuitively chosen weighting factors can be avoided by using a merit function (described in Sect. 3.1) that quantifies the quality of a probe pulse reflected by the mirror rather then directly evaluating the properties of the mirror.

The merit function of coatings consisting of many layers usually has many local extrema (Boudet et al. 1996). Global optimisation methods, in contrast to standard (gradient-like) optimisation techniques prevent the refinement process from being captured in a local optimum. Global optimisation algorithms have been successfully employed for solving several multilayer filter design problems (Boudet et al. 1996, Dobrowolski and Kemp 1990, Michielssen et al. 1992). Employing global optimisation for the design of CMs is not straightforward because these algorithms require a large number of evaluations of the merit function. This drawback becomes extremely manifest in the case of dispersive mirrors that are characterised by means of mathematically involved merit functions. An implementation of the memetic global optimisation algorithm that, in conjunction with a fast method for evaluating the merit function, enables fast and efficient numeric design of CMs (Yakovlev and Tempea 2002) is described in Sect. 3.2.

The GDD of CM designs (particularly that of broadband ones) is highly sensitive to small deviations of the layer thickness from the nominal values. Any manufacturing process introduces systematic and random errors in the optical thickness of layers. The systematic errors (like a uniform increase in the thickness of all layers) are usually less critical for broadband coatings because they only slightly change the reflectance and dispersion characteristics and/or shift them in the spectral domain, but the random ones pose a serious problem causing large oscillations of the GDD curve. In Sect. 3.3 we describe a stochastic algorithm which searches for a design with reduced sensitivity to small random perturbations of layer thickness (Yakovlev and Tempea 2002).

3.1. A Merit Function Based on the Simulation of Pulse Reflection

From a mathematical point of view the optimisation of a chirped mirror may be regarded as the minimisation of a merit function $f(x_1, x_2, \ldots, x_n)$ whose arguments are usually (but not necessarily) layer thickness. An ideal chirped mirror would reflect 100% of the incident radiation and introduce a GDD exactly equal to the prescribed one. Thus, the merit function commonly used for characterising the amplitude properties of multilayers has to be adapted for taking into account the GDD targets too (Szipöcs et al. 1994):

$$f = \frac{1}{m}\sum_{i=1}^{m}\left(v_i(1-R_i)^2 + w_i(\widetilde{D}_i - D_i)^2\right). \tag{15}$$

Here R_i and D_i are the reflectance and the GDD of the mirror at wavelength λ_i calculated from the parameters of the merit function $x_1,\dots,x_n$, $\widetilde{D}_i$ is the target GDD at this wavelength and v_i and w_i are the weights given to the reflectance and the GDD values, respectively.

Standard CMs can be efficiently designed by employing this merit function. Usually both a reflectance close to 100% and accurate GDD-control over ≈ 300 nm can be achieved with a moderate (<50) number of layers. Extreme design specifications (e.g. bandwidth approaching an optical octave and/or high transmittance at the pump wavelength) call for a tradeoff between reflectance/transmittance and GDD deviations making the choice of the weights in Eq. (15) non-trivial. For example, at some iteration an optimisation algorithm may have to choose between two designs: one with an average reflectance of 98 % and an amplitude of GDD oscillations of 30 fs^2 and the other with an average reflectance of 97 % and GDD oscillations having an amplitude equal to 10 fs^2. Which of these designs will be preferred depends on the weights given to the reflectance and the GDD in (3.1), but which of them is really better for a particular application?

The merit function (15) may fail to evaluate the performance of the mirror when used for pulse manipulation (e.g. in a delay line for pulse compression). If the GDD curves of two different designs exhibit the same second order momentum deviations from the target GDD, the merit functions of the two designs will be equal. Still, reflecting the pulse off the mirrors with identical merit functions can affect the temporal shape of the pulse in very different ways. Mirrors introducing a residual GDD described by the full oscillating curve or by the dashed step-like curve in Fig. 9 (left plot) have the same figure of merit according to Eq. (15) (we assume that the reflectance is 100% in both cases). The effect of the residual chirp corresponding to the two GDD-curves on a 5-fs bandwidth-limited Gaussian pulse (having a centre wavelength of 800 nm) is also shown in Fig. 9 (right plot). The oscillating dispersion distorts the pulse shape to a much smaller extent, although both residual GDDs deliver exactly the same contribution to the merit function (15). This ambiguity and the difficulty in choosing the weight coefficients can be avoided by using a merit function that directly evaluates the effect of the amplitude and phase change introduced by the mirror on a reflected broadband pulse.

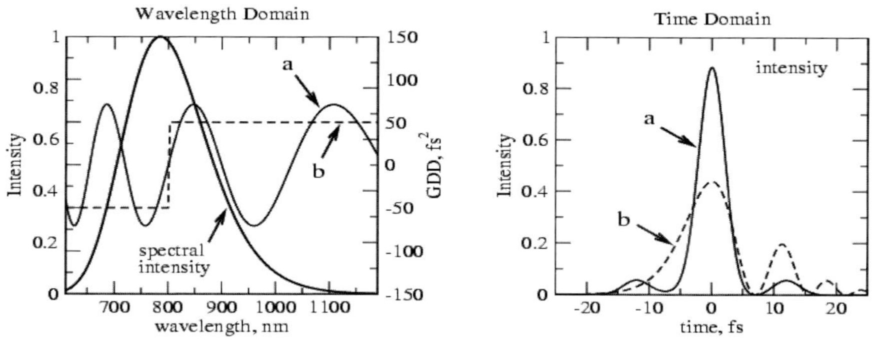

Fig. 9 Comparison of influences of oscillating **(a)** and step-like **(b)** deviations from the GDD-target on the shape of a 5-fs bandwidth-limited Gaussian pulse: The left panel depicts the spectrum of the pulse and GDD deviations from the target that produce equal contributions to the merit function given by Eq. (15). The step-like GDD distorts the pulse shape much more strongly compared to the oscillating GDD as revealed by the respective reflected temporal intensity profiles (right panel)

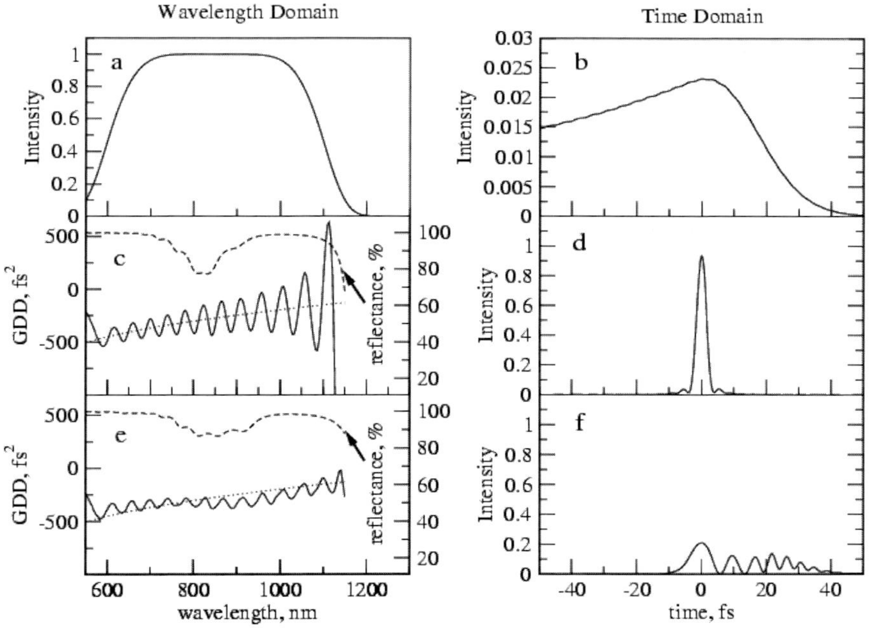

Fig. 10 Comparison of two TFI chirped mirrors. The incoming pulse (for the compression of which the mirrors were designed) is shown in panel **b**. It has the spectrum depicted in **(a)** and carries a chirp corresponding to a 10 mm propagation path through fused silica. Panels **(d)** and **(f)** show the calculated intensity obtained by simulating 10 bounces of the chirped

pulse off a mirror optimised for the peak intensity and off a mirror optimised with the conventional merit function, respectively. Comparison between **(d)** and **(f)** clearly demonstrates that compression with the mirror optimised with the conventional merit function **(e)**, which has smaller GDD oscillations and better reflectance, results in a much poorer pulse shape

The effect of the reflection on the temporal pulse shape can be calculated by multiplying the complex spectrum of the incident pulse by the transfer function of the mirror $H(\omega) = r(\omega)\exp[i\varphi_r(\omega)]$, where $r(\omega)$ and $\varphi(\omega)$ are the amplitude reflectance and the phase shift on reflection, respectively. Fourier transforming the product yields the temporal shape of the reflected pulse. Depending on the parameter most critical for the application, either the peak intensity, the (root mean square or full width at half maximum) pulse duration or a weighted combination of these parameters can be used to construct the merit function. Obviously, traditional targets can be additionally used, e.g. in order to impose high transmittance within the absorption band of the crystal.

The evaluation of the pulse reflection simulation merit function (PRSMF) consumes more time than the evaluation of (15) because it involves calculating a Fourier transform. Thus, the PRSMF should only be used for optimisation of sophisticated dispersive coatings (e.g. chirped input/output couplers, ultrabroadband CMs) that cannot be unambiguously characterised by the standard merit function (15). Figure 10 compares two chirped mirrors, one of which was optimised to yield maximal peak intensity of the reflected pulse and the other was optimised to minimise (15). Although GDD oscillations are much smaller and the reflectance is better in the second case, the reflected pulse is much longer.

3.2 Global Optimisation

Irrespective of its form, the merit function $f(x_1, x_2, \ldots, x_n)$ of a multilayer coating has many local extrema if the number of layers is large. Function f is usually constructed such that an improvement of the design corresponds to a decrease in the function's value. The parameters of the merit function $x_1, \ldots, x_n$ define a multidimensional search space. A local optimisation algorithm starting from an initial design, which corresponds to a point in the search space, follows a trajectory towards one of the nearest local minima. If the starting design is chosen close enough to the ideal design solution, the "global" minimum, then local optimisation will most likely converge to this minimum. If a good starting design cannot be found, global optimisation techniques usually outperform local refinement. Evaluation of the merit function of dispersive mirrors is more time consuming than in the case of amplitude-only filters since the GDD has to be calculated, and in turn, the number of wavelengths a coating is analysed at has to be increased. Global optimisation techniques require many evaluations of the merit function, which slows down extremely the optimisation of CMs. In what follows we describe a global optimisation algorithm that employs a faster method for evaluating the merit function in conjunction with partial local optimisation to ensure fast convergence of CM refinement (Yakovlev and Tempea 2002).

This algorithm belongs to the category of evolutionary algorithms (Michaelewicz 1996). It maintains a "population" of potential solutions. At each iteration a new population is built based on the previous one according to a strategy including both deterministic and stochastic elements. Like in all stochastic algorithms, there is no guarantee that within limited time the global optimum will be found, but evolutionary algorithms (the most widely-known example of which is probably the genetic algorithm) have proved to be effective in dealing with multidimensional search spaces with complicated merit function topology.

Memetic algorithms, a class of evolutionary algorithms in which local search plays a significant role (Radcliffe et al. 1994), appear to suit the optimisation of multilayer optical coatings very well. We adopt the terminology of evolutionary computation and call a set of trial designs a *population* and a particular design from this set a *member* of the population. Like in any genetic algorithm the memetic optimisation process evolves from population to population by means of crossover and mutation operators. The difference between the genetic and memetic algorithms is that in the latter each newly constructed member is locally optimised before it is included in the new population, i.e., constructing a new set of trial designs from the old one, we locally optimise each design that is going to be added to the new set. This local optimisation does not necessarily have to lead to the very (local) extremum. Yet, local optima can be approached more efficiently without severely compromising iteration speed. The effectiveness of the memetic algorithm depends to a large extent on the effectiveness of this kind of local optimisation, which we call *partial refinement*.

There is a simple trick that allows performing the partial refinement of multilayer coatings quickly and efficiently. The standard way to find the properties of a coating is the transfer matrix method (Macleod 1985). To calculate the reflectivity of n layers at a particular wavelength, one has to calculate the product of n transfer matrices M_k. This is usually a very time consuming part of computer optimisation. The calculations can be made faster in the case when the merit function has to be evaluated for several designs that differ from some particular previously characterised design by only one layer. Let us define matrices

$$L_k = \begin{cases} 1, k = 1, \\ \prod_{i=1}^{k-1} M_i, 1 < k \le n, \end{cases} \qquad R_k = \begin{cases} 1, k = n, \\ \prod_{i=k+1}^{n} M_i, 1 \le k < n. \end{cases}$$

When all the matrices are calculated, one can quickly find the reflectivity in the case when only the k-th layer is changed, because only one single matrix M_k has to be calculated and only two matrix multiplications instead of $n-1$ have to be performed:

$$\prod_{i=1}^{n} M_i = M_1 M_2 ... M_{k-1} M_k M_{k+1} ... M_n = L_k M_k R_k .$$

This opens a way for fast evaluation of the gradient of the merit function, swift optimisation of one particular layer and so on. There are a few ways of implementing the partial refinement utilising this method. For example, one can calculate the gradient and optimise the mirror along this direction (hill climbing) or optimise each layer one by one. The method becomes especially effective if the number of layers is large.

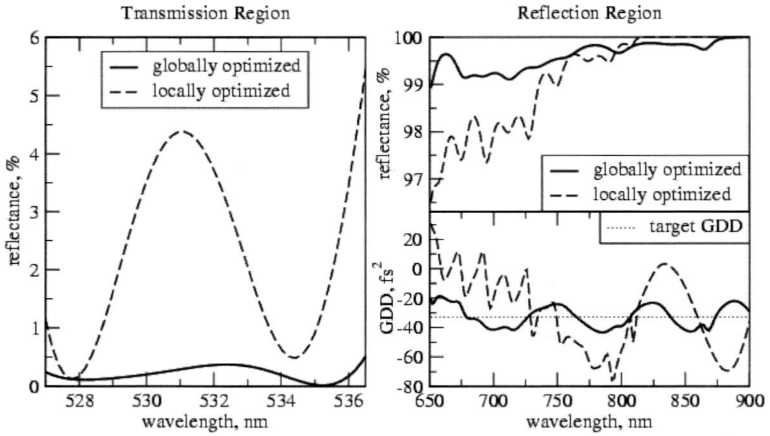

Fig. 11 Comparison of an input coupler design found by the memetic algorithm with that obtained by local optimisation starting from a double chirped structure. Large transmittance in the wavelength range 527-537 nm as well as high reflectance in the range 650-900 nm together with a target dispersion of –33 fs^2 were required

This technique implemented in the framework of the memetic algorithm results in fast and efficient global optimisation. Fig. 11 demonstrates its power for the example of a chirped input coupler that was optimised first by the memetic algorithm and then by conjugate gradient local optimisation starting from a double chirped design. Input couplers are dichroic mirrors employed in femtosecond oscillators. They must provide high reflectance and prescribed GDD within the laser gain band and high transmittance at the pump wavelength where the laser material efficiently absorbs radiation. Although double chirped mirror designs tend to provide high transmittance at wavelengths below the HR range, none of the predesign methods allows the exact spectral range of the pass-band and the required transmittance value to be directly taken into account. As the initial design tends to be far off the optimum one, global optimisation finds a better solution as compared to local refinement.

3.3 Sensitivity Optimisation

The problem of design sensitivity to small perturbations of layer thicknesses is particularly severe in the case of coatings for ultrafast optics because the GDD curve is affected by such errors to a much larger extent than the mirror's reflectance. Furthermore, CM designs consist of a large number of layers and often contain very thin layers (<15 nm). Coating errors of the order of few Angstroms can result in GDD oscillations with an amplitude of tens of fs^2. The sensitivities of the GDD curves to manufacturing errors of designs obtained by different analytic pre-design methods (Szipöcs and Köházi-Kis 1997, Matuschek et al. 2000) appear to be comparable. In this context, a numerical refinement method that allows reducing the sensitivity to manufacturing errors provides important help.

Let us assume that it was possible to optimise a coating by manufacturing each trial design and measuring its properties to calculate the merit function. Each manufactured design would have randomly perturbed layer thicknesses. Even by manufacturing the same design twice, two slightly different values of the merit function would be obtained. It would look as if the merit function were corrupted with "noise". To find a minimum of such a function would mean to find a point where its average value (determined from many manufacturing simulation runs) is minimal. This solution would obviously correspond to the most robust design. This consideration indicates that special optimisation algorithms suitable for noisy functions could be used for the design of robust coatings. The "noise" corrupting the merit function can be simulated by calculating the merit function each time not exactly for the nominal layer thicknesses, but for randomly perturbed ones. The random perturbation simulates the deviations from the nominal thicknesses introduced by the manufacturing process. The method presented here obviates the need for explicitly calculating the sensitivity of the complicated merit functions of CMs at each optimisation step, which would require prohibitive computational effort.

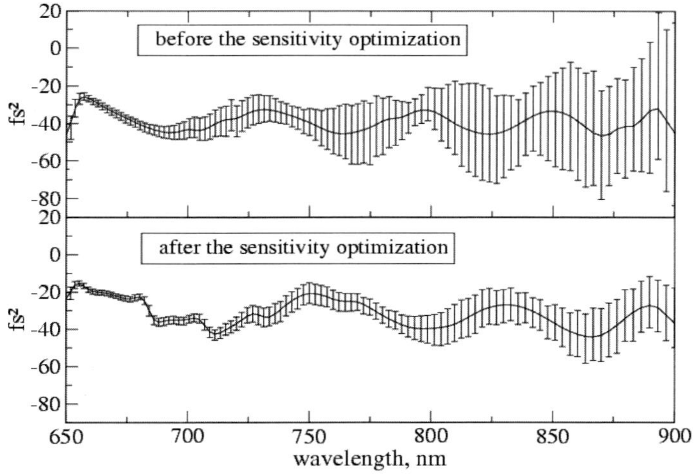

Fig. 12 Sensitivity optimisation of a chirped mirror (49 layers, SiO_2 and TiO_2). The error bars represent the rms deviations obtained by averaging the GDD over 1000 curves calculated for multilayers with perturbed layer thicknesses. The thickness perturbations were simulated by an additive noise affecting all the layers in the same way; the distribution was taken to be Gaussian with a standard deviation of 1 nm

A similar problem of optimisation of noisy functions was solved in the field of adaptive optics by means of the "parallel stochastic perturbative gradient descent" algorithm (Vorontsov et al. 2000), which draws on the stochastic quasi-gradient algorithm (Ermoliev et al. 1998). The underlying idea is simple: at each iteration we estimate the gradient of the merit function and make a step in the direction opposite to the gradient in an attempt to minimise the merit function. It resembles the steepest descent method, but the stochastic nature of the merit function requires specialised techniques to estimate the gradient and places certain restrictions on the sequence of step sizes. If all the necessary conditions are met, the algorithm is guaranteed to converge to a stationary point with probability one, though the convergence may be very slow (Ermoliev et al. 1998). One of the advantages of this method is that the sensitivity does not have to be explicitly calculated in order to be optimised. The same algorithm works also for chirped mirrors. To estimate the gradient of the merit function, the simultaneous perturbation approximation (Spall 1992) was found to be particularly efficient.

An example illustrating the sensitivity optimisation is shown in Fig. 12. A chirped mirror was optimised first for a constant negative GDD without any requirements regarding its sensitivity, using the memetic algorithm. The design obtained was further optimised with the aid of the stochastic quasi-gradient algorithm. As a result, the sensitivity to manufacturing errors (quantified by the error bars in Fig. 12) was substantially reduced. Although sensitivity optimisation tends to decrease the quality of the ideally manufactured design (with the layer thicknesses exactly equal to the prescribed ones), non-negligible random manufacturing errors make the robust design preferable.

4 Conclusions and Outlook

Conventional delay lines (gratings and prisms pairs) are increasingly replaced with dispersive multilayer mirrors in ultrafast optics applications that require accurate compensation of small amounts of dispersion over large bandwidths. Owing to the advent of quasi-analytical design methods (a few of which were summarized in Sect. 2), CMs providing high reflectance and GDD control over about 160 THz at 800 nm were designed and manufactured. These mirrors enabled routine generation of few-cycle laser pulses directly from Ti:sapphire oscillators (Xu et al. 1998, Sutter et al. 1999) as well as from extra-cavity compressors (Cheng et al. 1998, Baltuska et al. 1997, Sartania et al. 1997).

Fundamental limitations in the design of CMs can be overcome by either employing glass as the medium of incidence (Tempea and Krausz 2000, Gallman et al. 2000) or by optimising mirrors such that the *overall* GDD of a mirror pair satisfies the specifications (Laude and Tournois 1999, Kärtner et al. 2001). Both these techniques allow the design of all-reflective delay lines having a bandwidth of one optical octave (e.g. between 500 nm and 1 μm). Such mirrors were already employed in Ti:sapphire resonators, resulting in generation of pulses having octave-spanning spectra and durations in the two cycle range (Ell et al. 2001, Matuschek et al. 2000). The capability of such mirrors to compress the continuum generated by means of guided nonlinear propagation has remained unexploited so far. The bandwidth limited duration of experimentally generated continua closely approaches the theoretical limit duration: the oscillation period of the optical field at the carrier wavelength (i.e. 2.7 fs at 800 nm). TFI and BASIC mirrors as well as matched mirror pairs are very close to providing accurate GDD control over the corresponding bandwidth. Thus, improvements of readily available techniques might lead to single-cycle pulse generation in the very near future.

References

Baltuska A, Wei Z, Pshenichnikov MS, Wiersma DA, Szipöcs R (1997) All-solid-state cavity-dumped sub-5-fs laser. Appl. Phys. B 65: 175-188

Boudet T, Chaton P, Herault L, Gonon G, Jouanet L, Keller P (1996) Thin-film designs by simulated annealing. Applied Optics **35:**6219-6226

Cheng Z, Tempea G, Brabec T, Ferencz K, Spielmann Ch, Krausz F (1998) Generation of intense diffraction-limited white light and 4-fs pulses. In: Elsässer T, Fujimoto JG, Wiersma DA, Zinth W (eds) Ultrafast Phenomena XI .Springer, Berlin, pp. 8-11

Dobrowolski JA, Lowe D (1978) Optical thin film synthesis program based on the use of Fourier transforms. Appl. Opt. **29**: 3039-3050

Dobrowolski JA, Kemp RA (1990) Refinement of optical multilayer systems with different optimisation procedures. Applied Optics **29**: 2876-2893

Dobrowolski JA, Tikhonravov AV, Trubetskov MK, Sullivan BT, Verly PG (1996) Optimal single-band normal-incidence antireflection coatings. Appl. Opt. **35**: 644-658

Druessel J, Grantham J, Haaland P (1993) Optimal phase modulation for gradient-index optical filters. Opt. Lett. **18**: 1583-1585

Ell R, Morgner U, Kärtner FX, Fujimoto JG, Ippen EP, Scheuer V, Angelow G, Tschudi T, Lederer MJ, Boiko A, Luther-Davies B (2001) Generation of 5-fs pulses and octave-spanning spectra directly from a Ti:sapphire laser. Opt. Lett. **26**: 373-375

Ermoliev Yu, Wets RJB, eds. (1998) Numerical techniques for stochastic optimisation. Springer

Fabricius H (1992) Gradient-index filters: conversion into a two-index solution by taking into account dispersion. Appl. Opt. **31**: 5216-5220

Fork RL, Cruz CH, Becker PC, Shank CV (1987) Compression of optical pulses to 6 femtoseconds by using cubic phase compensation. Opt. Lett. **12**: 483-485

Gallman L, Matuschek N, Sutter DH, Scheuer V, Angelow G, Tschudi T, Steinmeyer G, Keller U (2000) Smooth dispersion compensation: novel chirped mirrors with supressed dispersion oscillations. In: Elsässer T, Mukamel S, Murnane MM, Scherer NF (eds) Ultrafast Phenomena XII. Springer, Berlin, pp. 62-64

Gires F, Tournois P (1964) Interférometre utilisable pour la compensation d'impulsions lumineuses modulées en fréquence. C. R. Hebd. Acad. Sci. **258**: 6112-6115

Golubovic B, Austin RR, Steiner-Shepard MK, Reed MK, Diddams SA, Jones DJ, Van Engen AG (1997) Double Gires Tournois interferometer negative-dispersion mirrors for use in tunable mode-locked lasers. Opt. Lett. **25**: 275-277

Kärtner FX, Matuschek N, Tschibli T, Keller U (1997) Design and fabrication of double chirped mirrors. Opt. Lett. **22**: 831-833

Kärtner FX, Morgner U, Ell R, Schibli T, Fujimoto JG, Ippen EP, Scheuer V, Angelow G, Tschudi T (2001) Ultrabroadband double-chirped mirror pairs for generation of octave spectra. JOSA B **18**: 882-885

Knox WH, Pearson NM, Li DK, Hirlimann CA (1988) Interferometric measurements of femtosecond group delay in optical components. Opt. Lett. **13**: 574-576

Laporta P, Magni V (1985) Dispersive effects in the reflection of femtosecond optical pulses from dielectric mirrors. Appl. Opt. **24**: 2014-2020

Laude V and Tournois P (1999) Chirped-mirror-pairs for ultrabroadband dispersion control. In: Conference on Lasers and Electro-optics (CLEO/US) , OSA Technical Digest Series, Optical Society of America, Washington, D.C., paper CTuR4

Macleod HA (1985) Thin-Film Optical Filters. Hilger A, ed., Bristol

Martinez OE, Gordon JP, Fork RL (1984) Negative group velocity dispersion using refraction. JOSA A **1**: 1003-1006

Matuschek N, Kärtner FX, Keller U (1997) Exact coupled-mode theories for multilayer interference coatings with arbitrary strong index modulations. IEEE J. Quantum Electron. **33**: 295-302

Matuschek N, Kärtner FX, Keller U (1999) Analytical design of double-chirped mirrors with custom-tailored dispersion characteristics. IEEE J. Quantum Electron. **35**: 129-137

Matuschek N, Gallmann L, Sutter DH, Steinmeyer G, Keller U (2000) Back-side-coated chirped mirrors with ultra-smooth broadband dispersion characteristics. Appl. Phys. B **71**: 509-522

Mayer EJ, Möbius J, Euteneuer A, Rühle WW, Szipöcs R (1997) Ultrabroadband chirped-mirrors for femtosecond lasers. Opt. Lett. **22**: 528-530

Michalewicz Z (1996) Genetic Algorithms + Data Structures = Evolution Programs. 3rd rev. and extended ed. Springer

Michielssen E, Ranjithan S, Mittra R (1992) Optimal multilayer filter design using real coded genetic algorithms. IEEE Proceedings-J **139:** 413-420

Morgner U, Kärtner FX, Cho SH, Chen Y, Haus HA, Fujimoto JG, Ippen EP, Scheuer V, Angelow G, Tschudi T (1999) Sub-two-cycle pulses from a Kerr-lens mode-locked Ti:sapphire laser. Opt. Lett. **24**: 411-413

Moulton PF (1986) Spectroscopic and laser characteristics of Ti:Al_2O_3. JOSA B **3**: 125-133

Nisoli M, De Silvestri S, Svelto O (1996) Generation of high energy 10 fs pulses by a new pulse compression technique. Appl. Phys. Lett. **68**: 2793-2795

Penselin S, Steudel A (1955) Fabri-Perot Interferometerverspiegelungen aus dielektrischen Vielfachschichten. Z. Phys. **142**: 21-41

Perilloux BE (1998) Discrete thin-film layer thickness modulation. Appl. Opt. **37**: 3527-3532

Perilloux BE (1999) Discrete thin-film thickness-modulated designs: spacing of all possible stopbands. Appl. Opt. **38**: 2911-2915

Radcliffe NJ, Surry PD (1994) Formal memetic algorithms. Evolutionary computing, AISB Workshop. Fogarty T, editor, Springer

Sartania S, Cheng Z, Lenzner M, Tempea G, Spielmann Ch, Krausz F, Ferencz K (1997) Generation of 0.1-TW5-fs optical pulses at a 1-kHz repetition rate. Opt. Lett. **22**: 1562-1564

Sossi L (1974) On the theory of synthesis of multilayer dielectric light filters. Eesti NSV Tead Akad Toim Fuus Mat **23**:229-237

Southwell WH (1985) Coating design using very thin high- and low-index layers. Appl. Opt. **24**: 457-460
Southwell WH (1989) Using apodization functions to reduce sidelobes in rugate filters. Appl. Opt. **28**: 5091-5094
Spall JC (1992) Multivariate stochastic approximation using a simultaneous perturbation gradient approximation. IEEE Trans. on Automatic Control **37:** 332-341
Spence DE, Kean PN, Sibett W (1991) 60-fs pulse generation from a self-mode-locked Ti:sapphire laser. Opt. Lett. **16**: 42-44
Sutter DH, Steinmeyer G, Gallmann L, Matuschek N, Morier-Genoud F, Keller U, Scheuer V, Angelow G, Tschudi T (1999) Semiconductor saturable-absorber mirror assisted Kerr-lensmode-locked Tisapphire laser producing pulses in the two-cycle regime. Opt. Lett. **24**: 631-633
Szipöcs R, Ferencz K, Spielmann Ch., Krausz F (1994) Chirped multilayer coatings for broadband dispersion control in femtosecond lasers. Opt. Lett. **19**: 201-203
Szipöcs R, Köházi-Kis A (1994) Proc SPIE **2253**:140
Szipöcs R, Köházi-Kis A (1997) Theory and design of chirped dielectric laser mirrors. Appl. Phys. B **65**: 115-135
Szipocs R, Kohazi-Kis A, Lako S, Apai P, Kovacz AP, DeBell G, Mott L, Louderback AW, Tikhonravov AV, Trubetskov MK (2000) Negative dispersion mirrors for dispersion control in femtosecond lasers: chirped dielectric mirrors and multi-cavity Gires-Tournois interferometers. Appl. Phys. B **70**: 51-57
Tempea G, Krausz F, Spielmann Ch, Ferencz K (1998) Dispersion control over 150 THz with chirped dielectric mirrors. IEEE JSTQE **4**: 193-196
Tempea G and Krausz F (2000) Dispersion management over one octave with tilted-front-interface chirped mirrors. In: Elsässer T, Mukamel S, Murnane MM, Scherer NF (eds) Ultrafast Phenomena XII. Springer, Berlin, pp. 65-67
Tempea G, Yakovlev V, Bakovic B, Krausz F, Ferencz K (2001) Tilted-front-interface chirped mirrors. JOSA B **18**: 1747-1750
Tikhonravov AV, Trubetskov MK, DeBell G (1996) Application of the needle optimisation technique to design optical coatings. Applied Optics **35**: 5493-5508 (1996)
Tikhonravov AV, Trubetskov MK, Keller U, Matuschek N (1999) Designing of coatings for femtosecond lasers with phase derivatives targets. Proceedings SPIE **3738**: 221-229
Tomlinson WJ, Stolen RH, Schank CV (1984) Compression of optical pulses chirped by self-phase modulation in fibres. JOSA B **1**: 139-149
Turner AF, Baumeister PW (1966) Multilayer mirrors with high reflectance over an extended spectral region. Appl. Opt. **5**:69-76
Valdmanis JA, Fork RL, Gordon JP (1985) Generation of optical pulses as short as 27 femtoseconds directly from a laser balancing self-phase modulation, group velocity dispersion, saturable absorption and saturable gain. Opt. Lett. **10**: 131-133
Verly PG, Dobrowolski JA (1990) Iterative correction process for optical thin film synthesis with the Fourier transform method. Appl. Opt. **29**: 3672-3684
Vorontsov MA, Carhart GW, Cohen M, Cauwenberghs G (2000) Adaptive optics based on analog parallel stochastic optimisation: analysis and experimental demonstration. JOSA A **17:** 1440-1453
Xu L, Tempea G, Spielmann Ch, Krausz F, Stingl A, Ferencz K, Takano S (1998) Continuous-wave mode-locked Ti:sapphire laser focusable to $5X10^{13}W/cm^2$. Opt. Lett. **23**:789-791

Yakovlev V, Tempea G (2002) Novel techniques for chirped mirror optimization, submitted for publication

Abbreviations

AR = anti-reflection
CM = chirped mirror
DCM = double-chirping method
fs = femtosecond = 10^{-15} seconds
FTM = Fourier-transform method
GD = group delay
GDD = group delay dispersion
HR = high reflectance
LTMM = layer thickness modulation method

Symbols

τ = group delay
R = (intensity) reflectance
ω = angular frequency
n = refractive index
λ = wavelength in vacuum
c = speed of light in vacuum
k = wavenumber= $2\pi c/\lambda$

Optical Coatings for Displays

Li Li

National Research Council of Canada, Ottawa, Ontario, Canada

1 Introduction

Electronic displays are one of the fastest growing industries in the world. Displays provide a critical interface between human and electronic information. They are widely used in computer monitors, computer aided design, engineering and manufacturing, multimedia, presentations, electronic games, mapping, medical and military applications, television, digital cinemas, virtual reality, etc [Machover 1997]. Many different display technologies co-exist today [Tannas 1985; Karim 1999; Goede 1999; Stupp and Brennesholtz 1999; MacDonald and Lowe 1997]. Each has its advantages and disadvantages and hence targets different applications.

The light sensors in most display applications are human eyes. This is why the requirements for displays, such as colour, luminance, luminance contrast and speed, are all directly linked to the optical properties of the human eye. Optical coatings are eminently suitable for controlling light through the use of the phenomenon of light interference in thin films [Macleod 2001; Thelen 1989; Dobrowolski 1995]. They have played a critical role in many types of displays [Stupp and Brennesholtz 1999; Schmidt 1999; Keller 1997]. This chapter discusses in details various kinds of optical coatings for display applications.

In Sect. 2 of this chapter, the principles of major display technologies as well as some background information are briefly described. Various optical coatings used to enhance the luminance contrast of displays are given in Sect. 3. In Sect. 4 different colour filters are described that are used to separate white light into three primary colours, or to combine colour images into a single picture. Polarizers and polarizing beam-splitters that are used in liquid crystal displays are discussed in Sect. 5. In Sect. 6 other optical coatings for display applications are given. Finally, conclusions are presented in Sect. 7.

2 Electronic Displays

2.1 Relationship Between the Human Eye and Electronic Displays

The human eye is the light sensor in most electronic display applications. Like any other optical sensing system, it has certain specific optical properties. It responds only to electromagnetic radiation in the visible part of the spectrum 400 – 700nm (the photonic luminance response curve $V(\lambda)$ has non-zero values for the wavelengths λ=380-780nm); it has red, green and blue photo receptors and thus it can perceive colours; it has a limited aperture and thus a spatial resolution of about 1 arc minute; it has a response time of about 0.1s; it can detect light with a certain minimum illuminance; and it can only distinguish objects with a certain luminance or colour contrast [Wyszecki and Stiles 1982]. The most important display properties are directly linked to and determined by the properties of the eye. For example, the following properties of displays should meet certain standard values in order for the eye to see proper images: the luminance (in cd/m^2) or brightness in a perceptive term, the colour performance (RGB colours), the resolution (size of the pixels or total number of pixels horizontally and vertically), the display speed of images, and the luminance contrast (the luminance ratio of the "on" and "off" pixels). In addition, the efficiency (lumens/watt) of a display is also important because it indicates how much electrical power (watts) can be converted into optical power (lumens). Other significant display properties are the size and weight, durability and cost.

2.2 How to Obtain Colours in Displays

In order to see colour images, the eye must be able to receive light of three primary colours, such as red (R), green (G) and blue (B). Spatial, time-sequential or simultaneous colour schemes are commonly employed to obtain full colour displays (Fig.1). In the spatial colour scheme, three RGB sub-pixels form a full colour pixel. Since these sub-pixels are relatively small at the normal viewing distance and cannot be resolved by the eye due to its limited resolution, they appear as a single pixel with an area-integrated colour. The spatial colour scheme is simple and is often used in direct-view or compact projection displays. However, it is less efficient because only 1/3 of the area is used and furthermore requires three times as many pixels.

In the time-sequential colour scheme the RGB images are directed at the eye in sequence at a rate that is much faster than its response time. As a result, the eye cannot tell apart individual RGB images but sees a time-integrated colour image. The time-sequential scheme is also simple and often used in compact projectors with one or two display panels. However, it is also less efficient because only 1/3 of the total available time is used for each colour and it also requires very fast displays. A modified time-sequential colour scheme, called the scrolling colour scheme, may be able to overcome the efficiency problem by recycling the 2/3 of the rejected colour light [Dewald et al. 2000; Shimizu 2001].

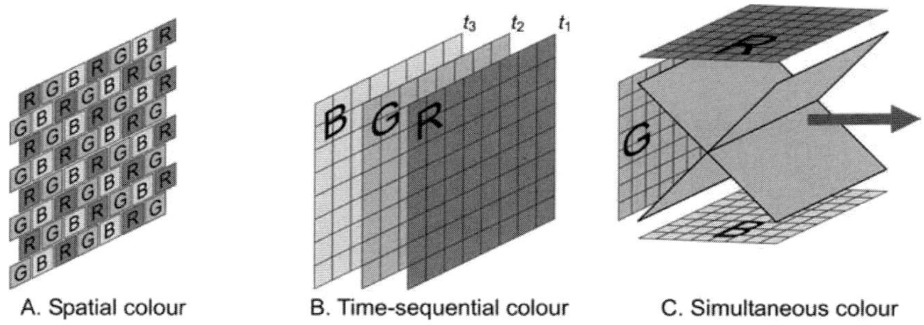

Fig.1. Three different schemes of forming colour images.

In the simultaneous colour scheme, RGB colour images are combined by a colour combiner and are simultaneously directed to the eye. Therefore, no area or time is lost - this is most efficient scheme of the three. But the simultaneous colour scheme is more complicated and expensive. It is mainly used in medium- to high-end projectors and requires three or more display panels.

2.3 Major Display Technologies

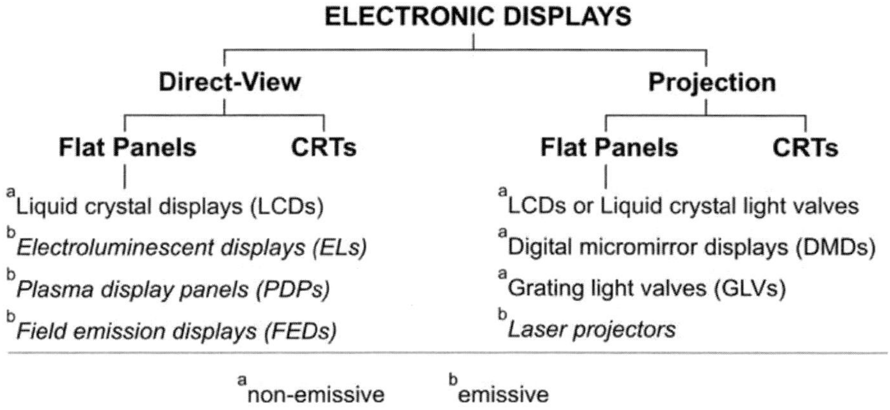

Fig. 2. Major display technologies.

Electronic displays are often classified into two categories: direct-view and projection, as shown in Fig.2. In a direct-view display, images are formed directly on the screen that is being viewed. Such a screen has a relatively small size and is suitable only for a few viewers, an exception is that light-emitting diode (LED) displays in which each individual LED is a pixel and they are often used in large indoors or outdoor stadiums. In projection displays images are first formed on small panels and from there they are projected onto a screen with an optical lens. In presentation or electronic cinema projectors the projection screen can be large and viewed by a large audience; in computer monitors the projection screen can be small. Both cathode ray tubes (CRTs) and flat panel displays can be used in direct-view and projection displays. Flat panel displays have the advantage of being thin, lightweight and having low power consumption while CRTs have a good performance and a low cost.

Displays can also be categorized as being emissive or non-emissive. Emissive displays emit light directly and are usually more robust and have a better angular viewing performance than non-emissive displays. Non-emissive displays act as light modulators or switches and they require an additional light source.

2.3.1 Cathode Ray Tubes (CRTs)

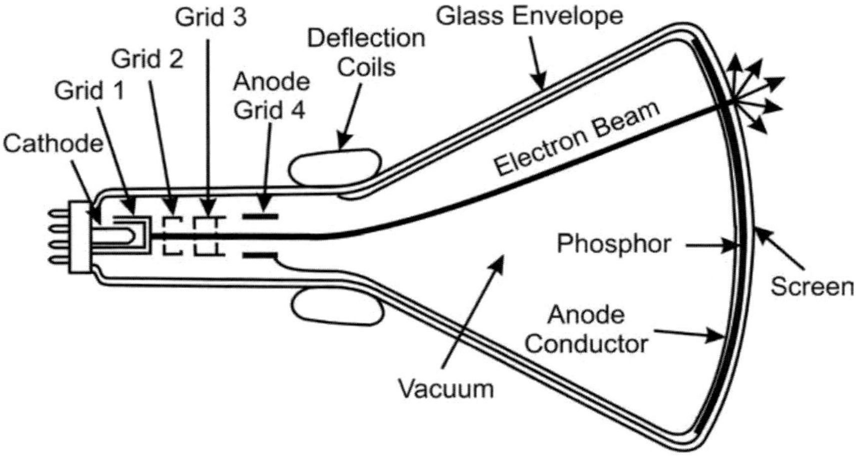

Fig. 3. A monochrome cathode ray tube [Goede 1999; Iki 1999].

CRTs have the longest history in electronic displays. The basic structure of a monochrome cathode ray tube is shown in Fig.3. An electron beam from the cathode is focused by grids 1-4 and then accelerated towards the anode that is coated with a phosphor. The phosphor converts the high energy of electrons into visible light. To form an image, the deflection coils (or yoke) are used to scan the whole screen line by line at a fast speed. Colour CRTs have three electron guns, each focus the electron beam through a mask on the appropriate pixels on the screen coated with RGB phosphors.

CRTs produce very good images with high brightness and vivid colours at low cost. They are commonly used in TV sets and computer monitors. However, because of the long distance between the cathode and the anode required for the acceleration of the electrons, CRTs are much bulkier and heavier than their flat panel counterparts, especially for displays with large screen sizes. CRTs with screen sizes greater than 35 inches measured along the diagonal are usually based on the projection approach. In a projection CRT three small and bright red, green and blue CRTs are used to produce the original colour images and a projection optical system is used to combine and enlarge the colour images on the viewing screen. The size and brightness of projection CRTs are limited by the amount of light emitted by the original monochrome CRTs.

2.3.2 Liquid Crystal Displays (LCDs)

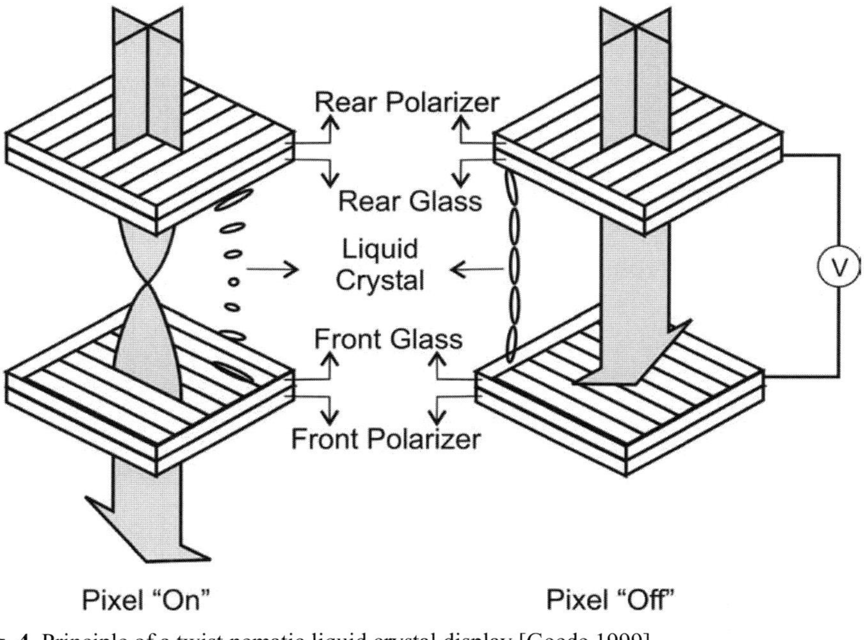

Fig. 4. Principle of a twist nematic liquid crystal display [Goede 1999].

LCDs are based on the electro-optical properties of liquid crystals [Yeh and Gu 1999; Chigrinov 1999]. A typical twisted nematic LCD pixel consists of a liquid crystal layer and two transparent electrodes placed between two glass substrates (Fig.4). The pixel is sandwiched between the first (rear) polarizer and the second (front) polarizer. The transmitting axes of the two polarizers are either perpendicular (normally white mode) or parallel (normally black mode) to each other. Unpolarized incident light passes through the first polarizer and becomes linearly polarized. When no voltage is applied to the liquid crystal layer, it acts as a half-wave plate and rotates the polarization of the incident polarized light by 90°. In the normally white mode the second polarizer therefore transmits the light and the pixel is in the "on" state. When a voltage is applied to the liquid crystal pixel, the liquid crystal molecules align themselves with the electric field and the light does not undergo a rotation of its plane of polarization. In this instance, the second polarizer blocks the light and the pixel is in the "off" state.

LCDs are normally addressed using a matrix scheme [Lueder 2001]. An exception is the liquid crystal light valves for projection displays that use small CRTs to address the panels, instead of patterning the panels with many pixels. LCDs operate either in a transmissive mode, as shown in Fig.4, or in a reflective mode in which the LC layer acts as a quarterwave plate and where the light passes through it twice. Reflective LCDs have thinner LC layers and therefore they can respond to electric field changes more rapidly, resulting in a faster display [Wu and Yang 2001]. In addition, reflective LCDs can have a larger aperture because the addressing matrix, which consists of metal electrodes, is behind the pixels and does not block the light. Furthermore, if silicon substrates are used, the driving electronics for the liquid crystal can be integrated on the silicon [Lu and Yang 2000]. These displays are called liquid crystals on silicon (LCOSs). LCOSs are micro-displays with small panel sizes and are intended for use with projection lens. Another type of LCDs, the polymer dispersed LCD (or PDLCD, for short), is based on light scattering rather than on polarization switching. LCDs dominate the direct-view flat panel markets. They also share a big portion of the projection display market.

2.3.3 Digital Micromirror Devices (DMDs)

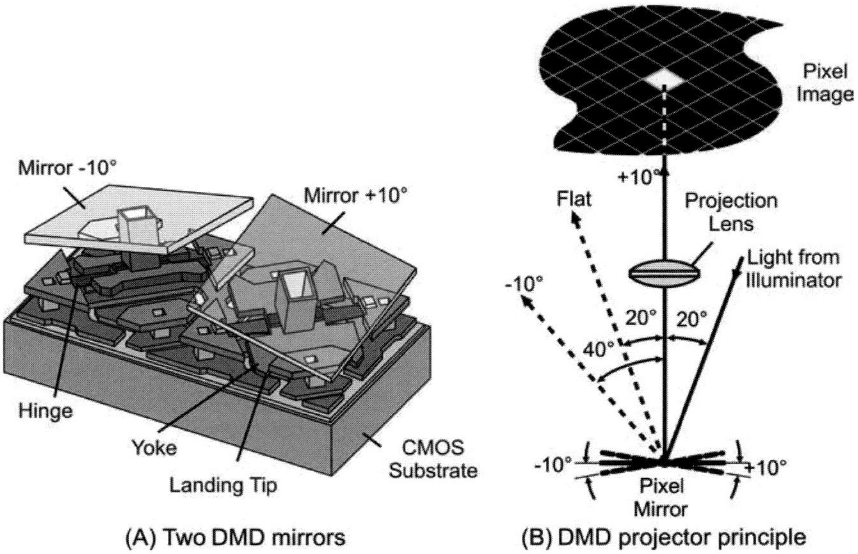

Fig. 5. Principle of a DMD display [Sampsell 1993; Hornbeck 1997].

A DMD consists of rows and columns of electronically controllable aluminum mirrors or pixels on a semiconductor substrate that has built-in addressing electronics (Fig.5) [Sampsell 1993; Hornbeck 1997]. Each mirror has two fixed positions of -10° or +10° which reflect the incident light in one of the two directions. An electronic static force is used to tilt the mirrors. Because the pixels are very small (about 17μm in size), DMDs are rarely used as direct-view displays but are commonly used in projectors, as shown in Fig.5. The reflected light by a +10° mirror will pass through a projection lens and is projected onto a screen; thus, the pixel is called "on". Light reflected by an –10° mirror takes a different path and is absorbed by a light sink, thus the pixel is called "off". The mirrors can be switched "on" or "off" in microseconds; hence, it is capable of displaying rapidly changing images. DMDs are also called by digital light processing (DLPs). DLPs also share a big portion of the projection display market.

2.3.4 Plasma Display Panels (PDPs)

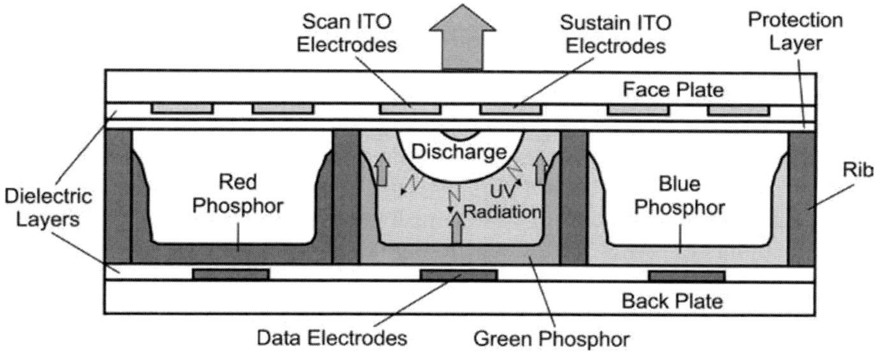

Fig 6. Basic structure of a plasma display panel [Goede 1999].

In a PDP, each pixel behaves like a small fluorescent light tube and consists of a cathode, an anode and a phosphor layer sealed with inert gas (Fig.6) [Weber 1996; Mikoshiba 1999]. When a high voltage is applied to the electrodes, a discharge or plasma is formed which radiates UV light. The phosphors convert the UV radiation into visible RGB light. PDPs are made by a screen-printing process and thus their pixels cannot be very small. PDPs target the 40-60" diagonal direct-view display market, such as wall-mounted TVs. They are much thinner than CRTs. However, the light efficiency is not high due to the small distances between the cathode and the anode, unlike in fluorescent tubes.

Electroluminescent Displays (ELs)

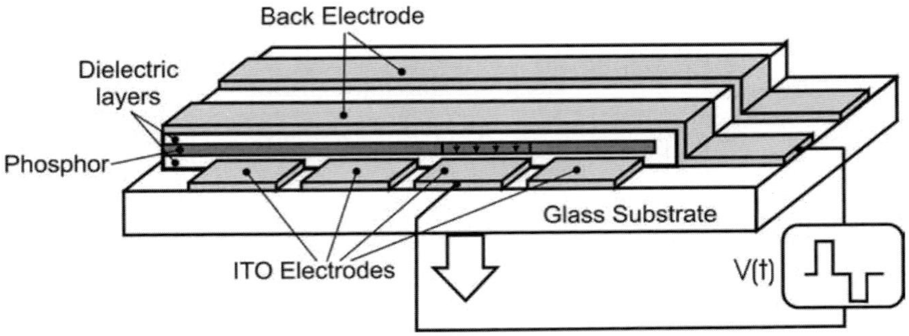

Fig. 7. Basic structure of a thin film electroluminescent display.

EL displays include inorganic thin film EL (TFEL) displays and organic EL displays [Mueller 1995; Miller 1997; Friend 1999]. Currently TFEL displays are more commercially successful than organic EL displays. A conventional TFEL display consists of an active phosphor layer surrounded by two dielectric layers (insulators), a transparent front electrode (ITO) and a back electrode (aluminum), as shown in Fig.7. Light emission is the result of the impact excitation of luminescent centres by charge carriers that are accelerated by high electric fields within the thin film structure. ZnS doped with Mn is often used as the phosphor and it emits yellow-orange colour. TFEL displays have good viewing angles and are very robust and can tolerate a wide temperature range. However, it is difficult to obtain colours in TFEL displays due to the lack of efficient blue phosphors. Thus, they are often used in niche markets such as military applications.

2.3.5 Field emission Displays (FEDs)

A FED is similar to a CRT in that electrons from the cathodes or field emitters are accelerated through a vacuum and are directed towards the anodes that are coated with phosphors [Gnade 1995; Dworsky and Chalamala 1999]. The phosphors convert the high energy of electrons into visible light. Unlike CRTs, in a FED the field emitters are cold cathodes and are made of silicon or metal micro-tips; each pixel consists of many micro-tips and is addressed by the row and column electrodes; the distance between the cathodes and anodes are much shorter, thus, thin FEDs can be made. Similar to CRTs, FEDs have good viewing angles, relatively low power consumption and good colours.

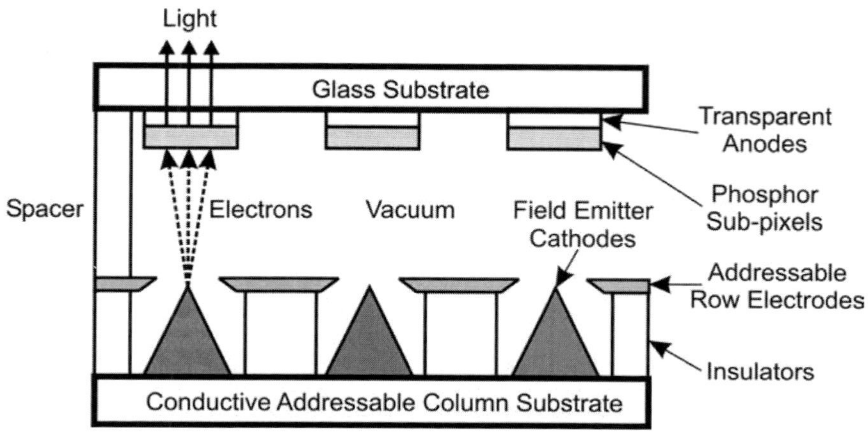

Fig. 8. Basic structure of a field emission display [Keller 1997].

3 Contrast Enhancing Coatings

A common problem for all displays is that their images may be severely degraded or completely disappear when they are viewed under strong ambient illumination. The degradation is caused by a reduction of the image contrast, which, in turn, depends on the contrast of the display panel itself, on reflections from the surfaces of any added components, as well as on the intensity of the ambient light. Many approaches have been used in the past to enhance the contrast of the display. In the following section, a general theoretical analysis of the problem is presented and examples are given to illustrate how thin film optical coatings can be used to enhance the contrast of displays.

Any display system, regardless of how complicated it may be, can be imagined to consist of three parts: the front, middle and back parts (Fig.9), in turn comprising all the components which are located before, within and behind the display panel, respectively. Each part has associated with it a transmittance, reflectance and absorptance: (T_1, R_1, A_1), (T_2, R_2, A_2), (T_3, R_3, A_3). Here, T, R and A take into account all possible factors that could affect them, such as the size of pixel apertures, the specular or diffused reflection, and scattering. In emissive displays such as CRTs, TFELs, FEDs and PDPs, the image light is emitted from the panel directly. In non-emissive and transmissive displays, the image light originates from an illuminator that is placed at the back, as in the case of transmissive LCDs. In a non-emissive reflective display the image light comes either from the ambient or from an illuminator in front of the display, as in the case with DLPs or LCOSs.

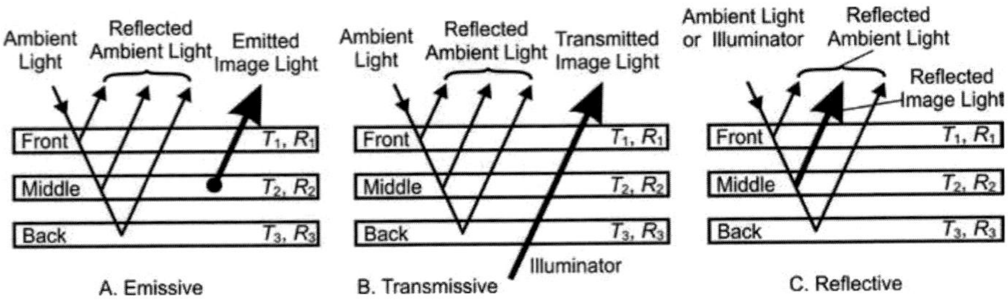

Fig. 9. Simplified display structures.

The luminance contrast of a display is defined to be the ratio of the total luminance of the light from the "on" pixel to the total luminance of the light from the "off" pixels. Under ambient illumination, the contrast ratio expressions for each display type depicted in Fig.9 can be written as:

$$CR_A = \frac{\int_{\lambda_1}^{\lambda_2} VI_{on}T_1 d\lambda + \int_{\lambda_1}^{\lambda_2} VI_0\left(R_1 + T_1^2R_2 + T_1^2T_2^2R_3\right)d\lambda}{\int_{\lambda_1}^{\lambda_2} VI_{off}T_1 d\lambda + \int_{\lambda_1}^{\lambda_2} VI_0\left(R_1 + T_1^2R_2 + T_1^2T_2^2R_3\right)d\lambda} \tag{1A}$$

$$CR_B = \frac{\int_{\lambda_1}^{\lambda_2} VI_1T_1T_{2on}T_3 d\lambda + \int_{\lambda_1}^{\lambda_2} VI_0\left(R_1 + T_1^2R_{2on} + T_1^2T_{2on}^2R_3\right)d\lambda}{\int_{\lambda_1}^{\lambda_2} VI_1T_1T_{2off}T_3 d\lambda + \int_{\lambda_1}^{\lambda_2} VI_0\left(R_1 + T_1^2R_{2off} + T_1^2T_{2off}^2R_3\right)d\lambda} \tag{1B}$$

$$CR_C = \frac{\int_{\lambda_1}^{\lambda_2} VI_0T_1^2R_{2on} d\lambda + \int_{\lambda_1}^{\lambda_2} VI_0\left(R_1 + T_1^2T_{2on}^2R_3\right)d\lambda}{\int_{\lambda_1}^{\lambda_2} VI_0T_1^2R_{2off} d\lambda + \int_{\lambda_1}^{\lambda_2} VI_0\left(R_1 + T_1^2T_{2off}^2R_3\right)d\lambda} \tag{1C}$$

where I_0 and I_1 are the spectral intensities of the ambient light and of the illuminator respectively; I_{on} and I_{off} are the intensities of the emitted light from the "on" and "off" pixels in Fig.9A; T_{2on} and T_{2off}, and R_{2on}, R_{2off},, are the transmittance and reflectance of the "on" and "off" pixels in Fig.9B and C, respectively; λ_1 and λ_2 are the lower and upper wavelengths of the visible spectrum; and $V(\lambda)$ is the photopic response curve.

In the nominators and denominators of Eqs. (1A, B, C), the first integral terms represent the contribution of the "on" and "off" pixels, and the second integral terms represent the influence of the ambient light. It is clear, that to enhance the contrast, it is necessary 1) to increase the first integral terms; 2) to reduce the second integral terms; or, 3) to reduce both the first and second integrals, but with a higher attenuation for the second term. Table 1 lists some of the optical coatings used for this purpose. Similar analyses can be also applied to other methods for contrast enhancement, such as those based on circulator polarizers or angular diffraction holograms [Trout et al. 2000].

Table 1. Optical coatings for contrast enhancement

	Emissive	Non-emissive Transmissive	Non-emissive Reflective
Reduce R_1	AR coating	AR coating	AR coating
Reduce R_2	AR or black coating	AR coating	AR coating
Reduce R_3	AR or black coating	AR coating	AR or black coating
Reduce T_1	NDF	NDF	-
Reduce R_1 & T_1	AAR, LRNDF, LRCF, LRBF	AAR, LRNDF, LRCF, LRBF	-

AAR: absorbing AR coatings; NDF: neutral density filters; LRNDF: low-reflectance neutral density filters; LRCF: low-reflectance cut-off filters; LRBF: low-reflectance bandpass filters.

AR coatings are most commonly used for enhancing the display contrast. They remove specular reflection from the surfaces or interfaces in a display system [Tong and Hu 1998]. For example, the front surfaces of CRTs, PDPs and FEDs, the surfaces of the sheet polarizers in LCDs, the surfaces of the lenses used in projectors are all normally treated with AR coatings. By doing so, the total amount of image light and hence the efficiency of the display are also increased. Because $T+R+A=1$, a reduction in R will result in an increase in T providing that A does not change. The design of AR coatings is rather straightforward. However, as always, a balance between the performance and cost must be made. If a wet deposition process is used to apply the AR coatings, one- to two-layer designs are preferred. If a vacuum deposition process is used, designs consisting of more than two-layer can be selected to achieve a better result. Anti-static layers (made of conducting ITO or metal films) are sometimes incorporated into the AR coatings in order to remove built-up static charges in the display screens [Ono et al. 1992; Onodera et al. 1994; Oyama et al. 1998; Tohda et al. 1992]. In addition, absorbing AR coatings have also been proposed whose additional function is similar to that of a low-reflectance neutral density filter (see below) [Hanaoka et al. 2000].

The use of black layer coatings is a very effective way of removing unwanted reflected ambient light [Dobrowolski et al. 1992]. For example, black layer coatings based on thin film interference are used to enhance the contrast of conventional TFEL displays. In a conventional TFEL display the back electrodes are made of aluminum films and are thus highly reflective. Under strong ambient light, a common operational environment for TFEL displays, the reflected ambient light can greatly exceed the light emitted by the image. As a result, the image can be completely overwhelmed by the ambient. By introducing a two-layer black layer coating consisting of layers of Inconel™ and SiO_2 on aluminum, the luminance reflectance of a conventional TFEL display was dramatically reduced from about 75% down to less than 5% Fig.10. The use of more layers can reduce the reflectance still further. Black layer coatings do not interfere with the normal operation of the TFEL panels.

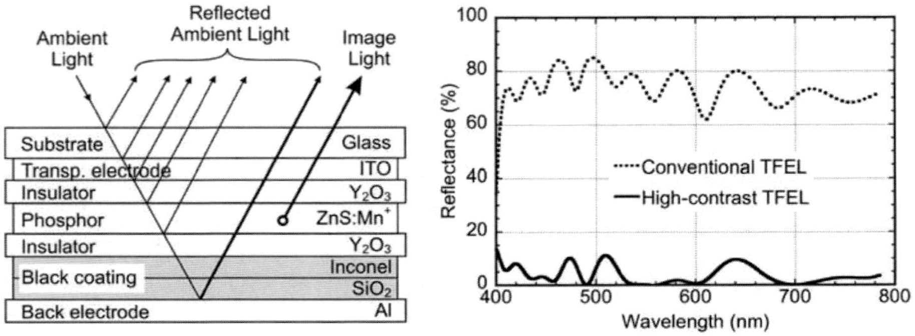

Fig. 10. High contrast TFEL display with a black layer coating [Li et al. 1994].

Neutral density filters (NDF) that have a uniform attenuation across the whole visible spectrum are often used as add-on components in front of displays in order to enhance the contrast because they attenuate the ambient light (T_1^2) more than the image light (T_1). The drawback is that they also reduce the luminance or brightness of the display. Therefore, they are more commonly used in brighter displays such as CRTs. In addition to neutral density glasses, single thin films of nickel or similar metals can also be used as good neutral density filters. The attenuation can be easily controlled by changing the thickness of the nickel films. More complicated NDF designs consisting of several nickel and dielectric layers can simultaneously attenuate the light and reduce the reflectance (Fig.11A). They provide much better results than AR coatings or neutral density filters alone.

Similarly, low-reflectance cut-off and bandpass filters can enhance the contrast of the display if the image light stems from only a narrow region of the visible spectrum (Fig.11B and C). This is the case with lasers, LEDs or $ZnS:Mn^+$ monochrome TFEL displays that only emit yellow light. Conventional all-dielectric cut-off or bandpass filters cannot be used for this purpose because they reflect any light that is not transmitted. A systematic method for designing low-reflectance coatings was described in detail by Dobrowolski et al [Dobrowolski et al. 1995].

The above are by no means the only solutions to the contrast problem. Many other examples exist. It is important to consider such solutions during the design of display panels and systems.

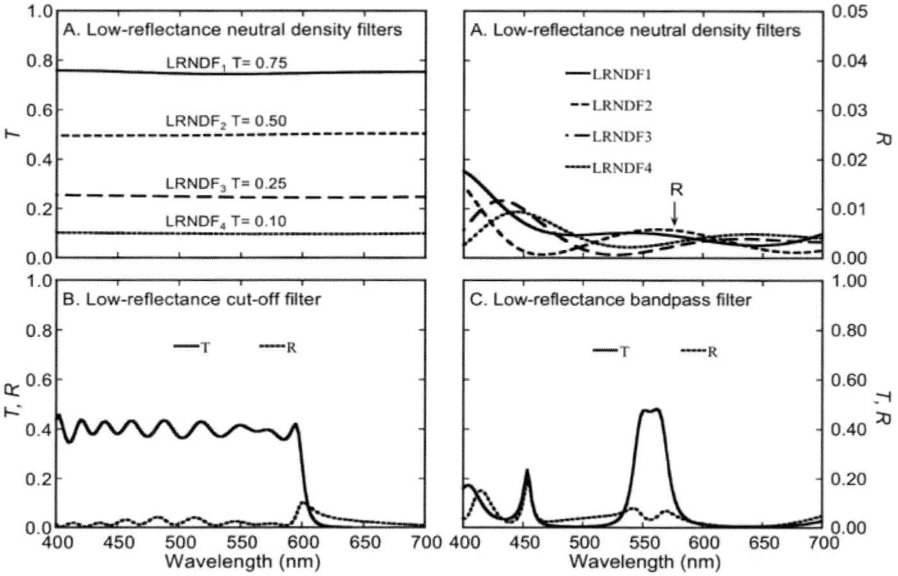

Fig. 11. Low-reflectance neutral density (**A**), short-wave pass (**B**) and bandpass filters (**C**).

4 Colour Coatings

As already mentioned in Sect. 2, it is essential to generate three primary colour images from which the full colour images can be formed. Colour filters are essential components in non-emissive displays that use additional light sources, such as LCDs and DLPs. Colour filters are less important in emissive displays since these displays usually emit RGB colour light directly. Pigment-dispersed or dye colour filters are often used in direct-view displays or in compact projection LCDs with the spatial colour scheme. They selectively let one part of the spectral light pass and absorb the rest and they can be applied to the display panels by photolithographic patterning or printing [Tani 1995]. Colour filters based on selective spectral absorption are very simple, but they absorb more than 70% of the incident light and are not stable when exposed to excessive heat or humidity.

High power arc lamps are often used in projection displays. To minimize any light loss, time-sequential or simultaneous colour schemes are often used. For such applications non-absorbing optical interference colour coatings, or dichroic filters, as they are usually called, are more suitable. Indeed, they have played a major role in projection displays. Dichroic filters consist of multilayers of low- and high-refractive index coating materials such as TiO_2, SiO_2, Al_2O_3, ZrO_2, Nb_2O_5, and Ta_2O_5 [see Chapter "Coating Materials" by Friz and Waibel]. Energetic deposition processes, such as ion-assisted deposition, RF or AC magnetron sputtering, are often used to deposit them [see Chapter "Thin Film Deposition Techniques" by Pulker]. Dichroic filters produced in this way are dense, very durable and stable.

Figure 12 shows the transmittances of two sets of typical transmissive and reflective dichroic filters. They transmit or reflect R, G and B light respectively. Reflective dichroic filters are also called dichroic mirrors and in transmission they have a cyan, magenta or yellow colour. The performance of dichroic filters is characterized by the position and sharpness of the cut-off edges and by the transmittance or reflectance in the pass- or stop-bands at certain angles of incidence for *s*- or *p*-polarized light. Dichroic filters are used to perform at least one of the following three major functions in projection displays:

1. they separate white light into RGB colour light (colour separator);
2. they combine RGB colour images (colour combiner); or,
3. they perform both the above functions (colour separator/combiner).

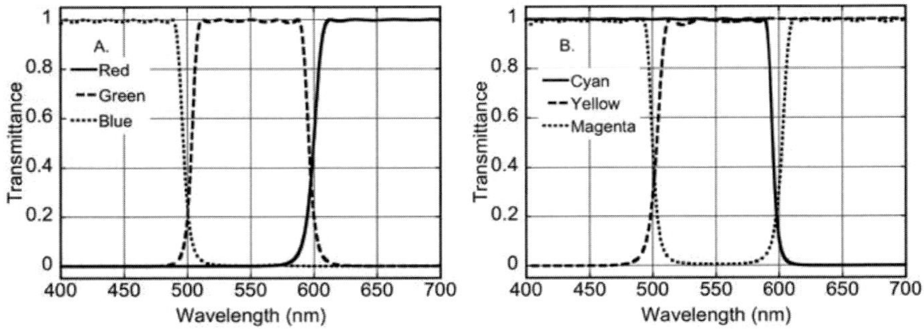

Fig. 12. Two sets of dichroic filters: (**A**) red, green and blue; (**B**) cyan, yellow and magenta.

Figure 13 illustrates various ways in which dichroic filters can be used to achieve the above goals. It can be seen that most dichroic or colour filters (except A) are operated at oblique angles of incidence in order to physically separate or combine the different colour beams. In addition, they have to work over a range of angles of incidence because projection systems use optics with relative aperture f[2]/2.8 to f/5.6, which corresponds to an angular field of ±11° to ±5.1° in air. Furthermore, for some applications such as DLPs and LCOSs with colour separator/combiners, these filters are required to work with both *s*- and *p*-polarized light.

[2] f: focal length.

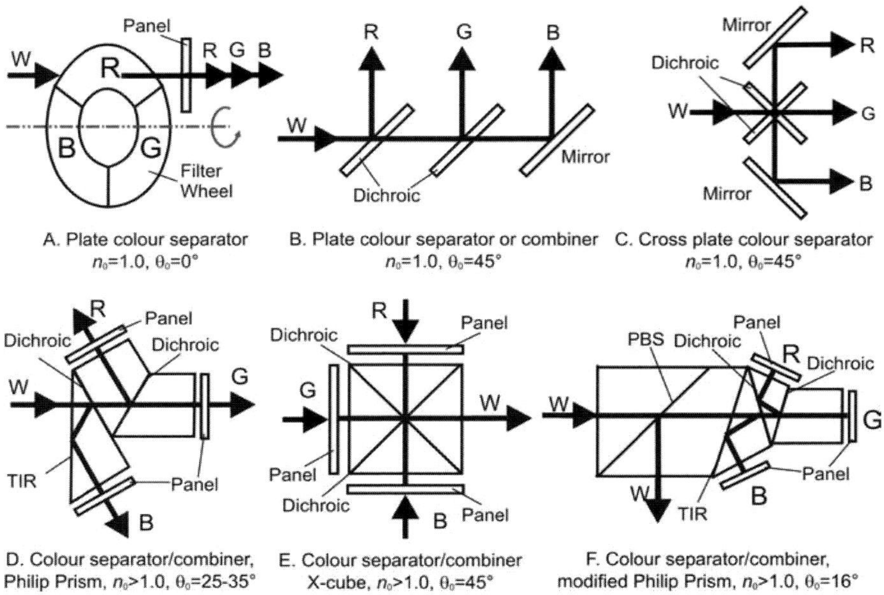

Fig. 13. Different ways of using dichroic filters in projection displays, W: white light.

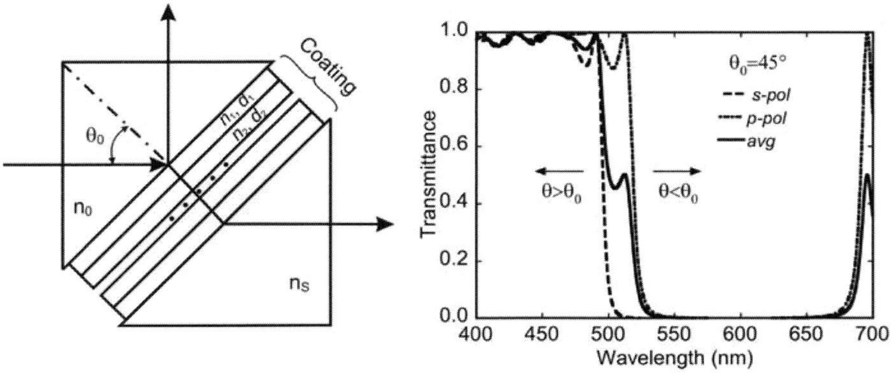

Fig. 14. A dichroic filter: (**A**) schematic layer structure; (**B**) calculated performance.

The design of dichroic filters is rather straightforward if they are intended to operate at normal incidence, or if they operate at oblique angles of incidence with either a small angular field or with light of one polarization only. They are essentially cut-off edge filters mostly composed of quarterwave layers with some non-quarterwave matching layers at both ends [Macleod 2001; Thelen 1989; Dobrowolski 1995]. However, the design and manufacture of dichroic filters is rather challenging if these filters are to operate over a wide angular field for both polarizations. It can be seen from Fig.14, that the cut-off edges of a typical colour filter do not coincide for *s*- and *p*-polarized light. In addition, these cut-off edges also shift with a change in the angle of incidence.

The above problems can be explained with the aid of the following equations:

$$\begin{cases} \eta_S = \eta_i \cos\theta_i \\ \eta_P = \eta_i / \cos\theta_i \end{cases} \tag{2}$$

$$\delta_i = \frac{2\pi}{\lambda} n_i d_i \cos\theta_i \tag{3}$$

$$C = n_0 \sin\theta_0 = n_1 \sin\theta_1 = n_2 \sin\theta_2 = n_S \sin\theta_S \tag{4}$$

where, η_S and η_P are the admittances for *s*- and *p*-polarized light; n_i, d_i, δ_i and θ_i are the refractive index, physical thickness, phase thickness and angle of incidence of layer *i*; λ is the wavelength; n_0 and θ_0, n_1 and θ_1, n_2 and θ_2, n_S and θ_S are the refractive indices and angles of incidence in the incident medium, the low- and high-refractive index layers and the exit medium, respectively.

Note that the admittance η is different for *s*- and *p*-polarized light: this is why there is a separation of the cut-off edges for *s*- and *p*-polarized light. The shift of the cut-off edges is due to the fact that the phase thickness δ changes with changes in the angles of incidence. To keep δ constant, λ has to shift to shorter wavelengths when θ increases, and vice versa.

To minimize both the above problems, the angles of incidence θ_i inside each layer should be minimized. The obvious solutions are to reduce C=$n_0\sin\theta_0$ or to increase n_1 or n_2. Since n_2 usually corresponds to the highest refractive index coating material available in the visible part of the spectrum, it is more practical to increase n_1, especially since $\theta_1>\theta_2$ and that therefore the low refractive index layers contribute more to the problem than the high refractive index layers. However, this approach reduces the refractive index ratio n_2/n_1 and, as a result, more layers will be required in the design to obtain a solution. The diagrams A-E in Fig.13 are arranged in an increasing order of the values of $n_0\sin\theta_0$. We can predict that, for the same number of layers, the quality of the performance of the devices will be ranked in the reverse order. The plate design A (Fig.13.A) has the best performance; unfortunately it can only be applied to displays with colour filter wheels (time-sequential colour scheme), for example, in a single DLP or a ferro-electric LCD projector. The next best performance is that of plate designs for an angle of incidence θ_0=45° (Fig.13B, C). They are often used to separate white light, or to combine colour images in transmissive LCD projectors, but not both. Dichroic cross-plates can be only used as a colour separator, and not as a colour image combiner because they cause image distortion due to their finite substrate thicknesses (Fig.13C). In principle, plate designs for angles of incidence θ_0<45° (not shown in Fig.13) should have a better performance. However, they take too much space and are difficult to assemble. Philip prisms are relatively compact and can be used to separate white light into its RGB components as well as to combine colour images (Fig.13D) [De Lang and Bouwhuis 1965]. However, total internal reflection (TIR) is needed to reflect one or more channels in Philip prisms. TIR causes light to be depolarized, which may be a problem in displays based on polarized light such as LCDs [Yamamoto and Atsuchi 1997; Rosenbluth et al. 2000; Greenberg and Bryars 2000]. To minimize the angles of incidence inside the colour coatings further, modified Philip prism in combination with a PBS for LCOS projectors was proposed (Fig.13F) [Li et al. 1999a]. The last is the X-cube design (Fig.13E). It is the most compact device; unfortunately, it also has the worst performance. It can be used as a colour separator as well as a combiner and is very suitable for use in reflective displays such as LCOSs.

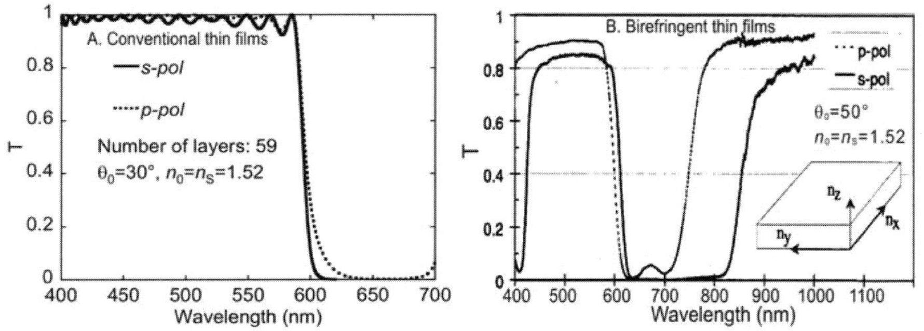

Fig. 15. Non-polarizing cyan filters: (**A**) conventional thin films; (**B**) birefringent thin films [Weber et al. 2001].

Thelen described a method for designing non-polarizing edge filters [Thelen 1989]. This method was used to select a starting point for the numerical design of a non-polarizing cyan colour coating in a Philip prism with n_0=1.52, n_S=1.52, n_1=1.45, n_2=2.35 and θ_0=30°. The layer thicknesses of the starting design were optimized using a computer program. Figure 15A shows the calculated final results. The cut-off edges of the filter for *s*- and *p*-polarized light are very close to each other and they stay close even there is a change in the angle of incidence. Another way to minimize the polarization separation and cut-off edge shift problems would be to use more than three coating materials in the colour filter design. However, this will result in an even more complicated layer system that is more expensive to manufacture because most coating systems are set up for the deposition of two materials only.

A different type of non-polarizing colour filter is based on alternative positive and negative birefringent multilayer polymer films [Weber et al. 2001]. Because each birefringent layer may have a different refractive index depending on the propagation direction and polarization state of the incident light, this gives an extra degree of freedom in the design of colour filters to match the edges of *s*- and *p*-polarized light. In practice, the selection of positive birefringent materials is rather limited, therefore, birefringent polymer colour filters usually consist of alternating negative birefringent and isotropic layers. Figure 15B shows the calculated performance of one such filter. The drawback of birefringent colour filters of this type is that the refractive index ratio of the two alternating films are rather small, thus, to achieve a certain performance, several hundred layers of films will be required. In addition, these filters usually do not stand up too well to heat.

Another development in colour filters is the patterned colour filter disk used for the scrolling colour scheme [Dewald et al. 2000]. This filter disk is patterned in such a way that, through the use of a combination of a special addressing scheme and a light integrator, the rejected two-thirds of the colour light can be recycled and thus the light loss can be minimized. It was proposed for use in a single-panel DLP projector whose efficiency is comparable to that of a three-panel DLP. The coating design for such a colour filter disk is similar to that shown in Fig.13A.

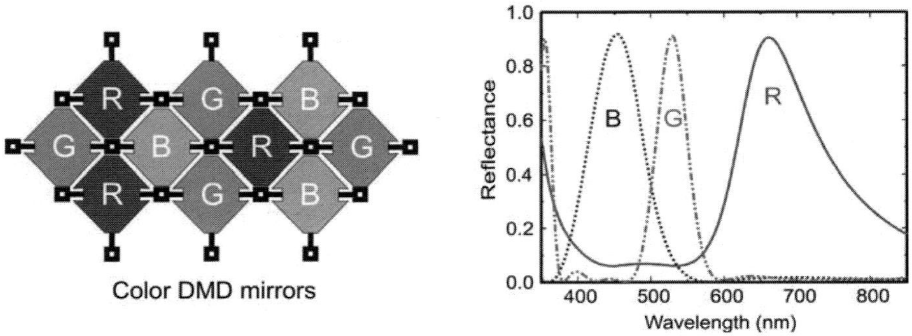

Fig. 16. Reflective metal-dielectric colour filters for colour DMDs.

Besides non-absorbing all-dielectric dichroic filters, the use of reflective metal-dielectric colour filters was also proposed for the construction of full colour DMDs (Fig.16) [Li et al. 1997]. The same principle can also be applied to reflective LCDs. Unlike dyes or pigment colour coatings, these metal-dielectric colour coatings specularily reflect light of one part of the spectrum and absorb light of all other wavelengths. They are more durable than dye or pigment colour coatings and they consist of far fewer layers (2-4) than all-dielectric colour filters. Figure 16 shows the calculated performance of three filters that match the National Television System Committee (NTSC) colour specifications.

5 Thin Film Polarizers and Polarizing Beam-Splitters (PBSs)

LCDs require the use of polarized light (Sect. 2). Since the light sources used in LCDs are mostly un-polarized, polarizers and polarizing beam-splitters are required to generate polarized light from these sources. In addition, polarizers and PBSs are also used to block polarized light from "off" pixels in LCDs. Polarizers are used either in transmission or reflection while PBSs use both the reflected and the transmitted polarized light beams. The important properties of polarizers for display applications are the working wavelength range, the transmittance or reflectance of the desired polarization, the angular field and the extinction ratio. In the case of PBSs, the above properties must be satisfied for both the reflected and transmitted light. The performance of a LCD is directly linked to these properties. For example, the contrast of a LCD depends on the contrast of the LCD panel as well as on the extinction ratio of the polarizers or PBSs used. The efficiency of the LCD depends on the amount of the incident un-polarized light that is used, as well as on the aperture of the optical system of the display that, in turn, is determined by the angular field of the polarizer or PBS. Other significant properties of polarizers and PBSs for display applications are their durability, sizes and costs.

Table 2. Properties of different types of polarizers and PBSs used for display applications

	Polaroid Polarizer	**Metal Wire-Grid Polarizer**	**Birefringent Multilayer Polarizer**	**MacNeille Polarizer**	**Li Li PBS**
Principle	Selective absorption	Selective reflection from metallic grids	Birefringence & thin film interference	Brewster angle & thin film interference	Above critical angle & thin film interference
Bandwidth	Broad	Broad	Broad	Broad	Broad
Angular Field	Large	Large	Large	Small	Large
Transmitted Beam	Yes	Yes	Yes	Yes	Yes
Reflected Beam	No	Yes	Yes	Yes	Yes
Extinction Ratio	T: high	T: high R: low	T: high R: low	T: high R: low	T: high R: high
Comments	>50% light absorbed	Some absorption	Less durable, light scattering	Small angular field	High index substrates

There exist many different types of polarizers and PBSs [Bennett 1995; Dobrowolski 1995; Li and Dobrowolski 2000]. However, only a few of these can be used in displays because of the strict requirements of this application. A brief comparison of the properties of polarizers or PBSs most often used for display applications is given in Table 2. The performance of these polarizers and PBSs are shown in Fig.17. In this section polarizers and PBSs based on thin films are emphasized because they are more suitable for use in project displays. Some of the ways in which these devices are implemented in displays are illustrated in Fig. These polarizer or PBSs are used to replace the second sheet polarizer P2 (Fig.18D and E), or to partially replace the first sheet polarizer P1 in polarization converters (Fig.18F and C), or to replace both the first and the second sheet polarizers (Fig.18G).

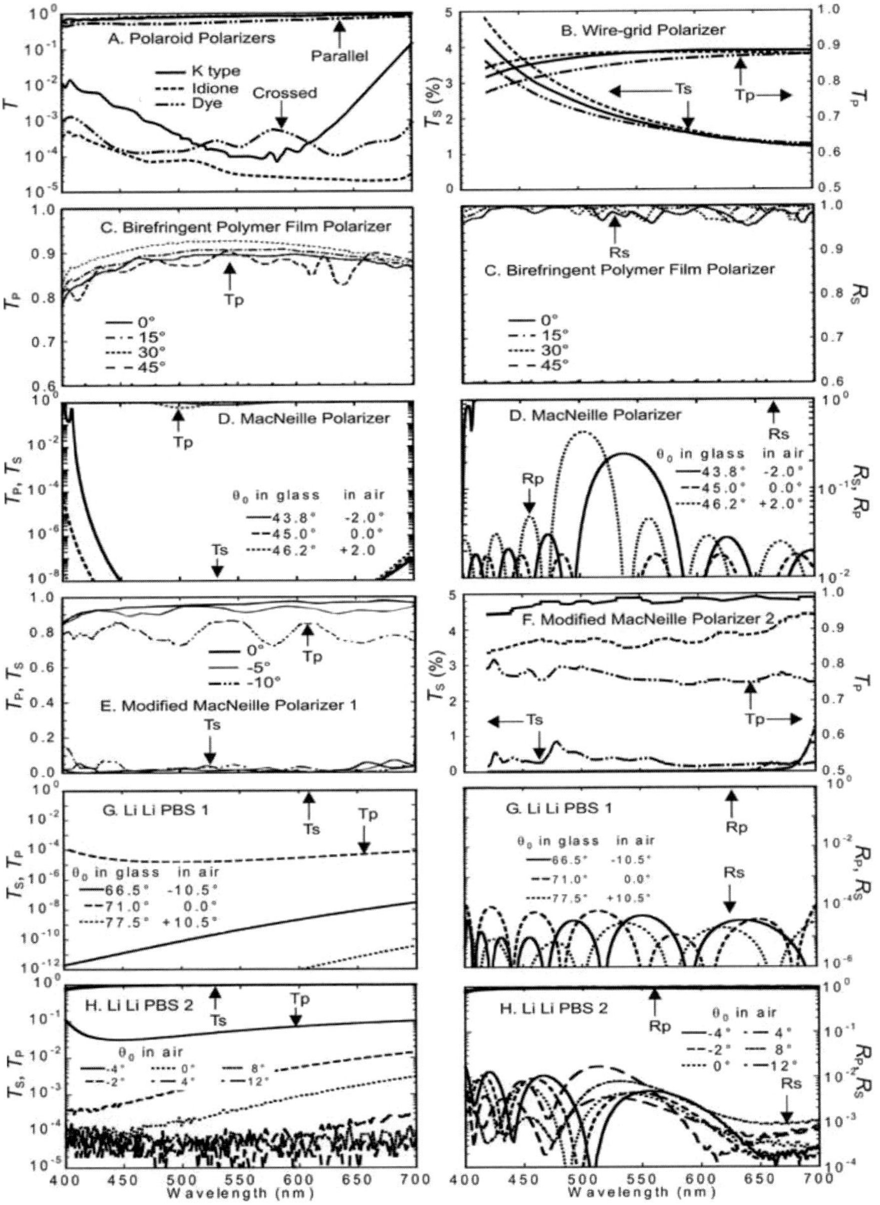

Fig. 17. Performance of different polarizers and PBSs.

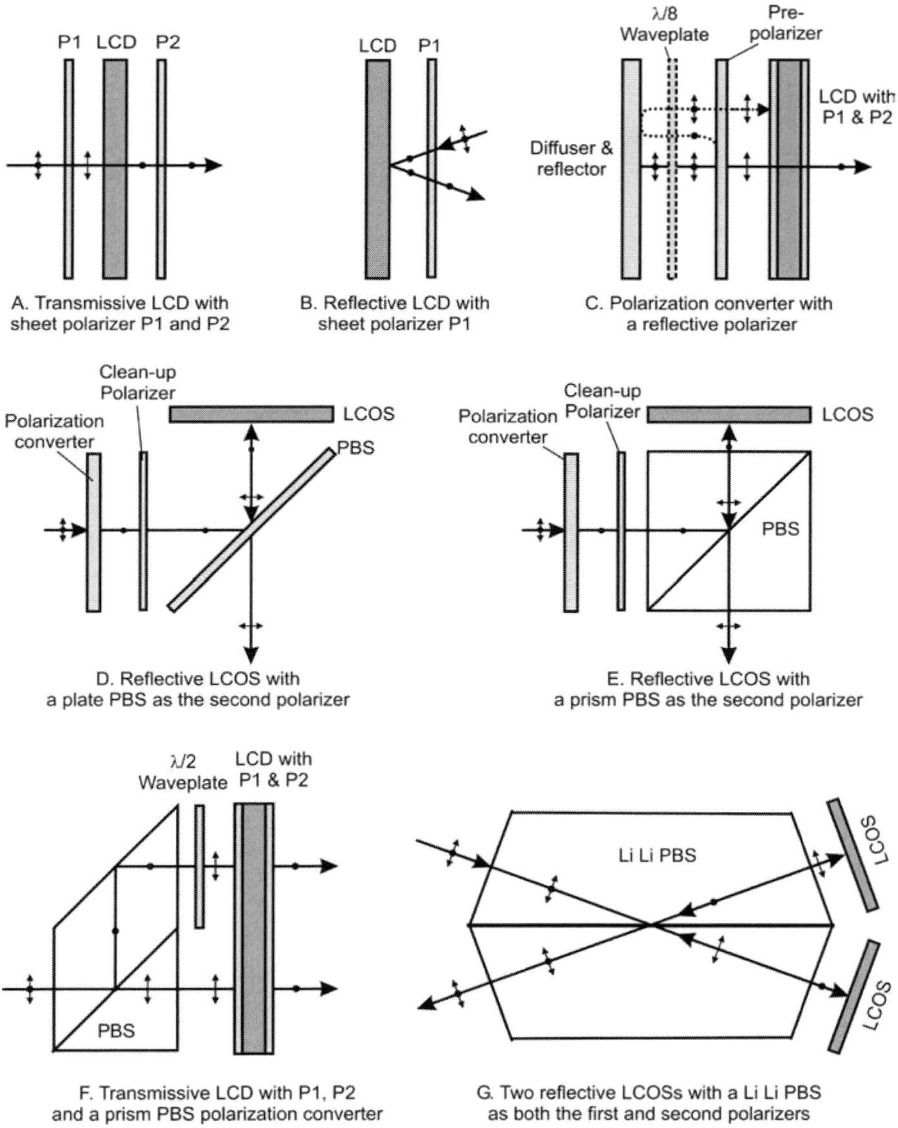

Fig. 18. Various ways of using polarizers and PBSs in displays.

5.1 Polaroid Polarizers

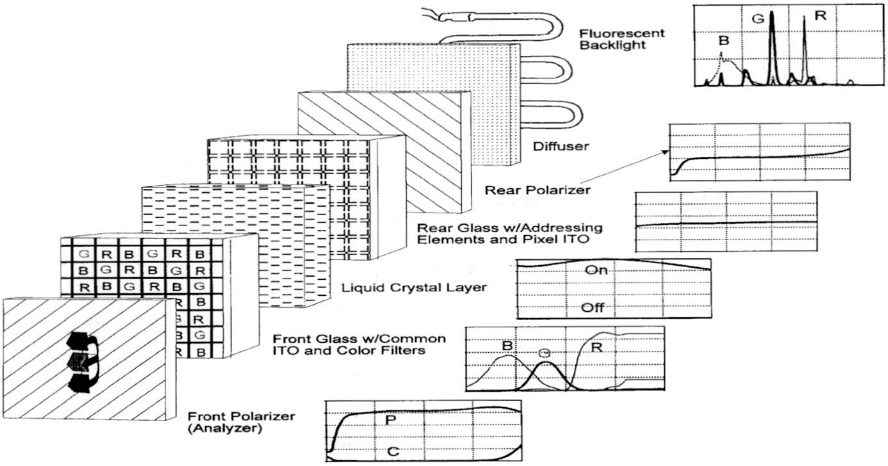

Fig. 19. Structure of a direct-view active matrix LCD [Silverstein 1995].

Polaroid polarizers are commonly used in LCDs to act as the first and second polarizers because they have a rather broad bandwidth, good extinction ratios and a wide angular field (Fig.17A). Figure 19 shows most of the components used in a typical direct-view active matrix LCD. It can be seen that each component has a limited optical efficiency. The total light output of a typical LCD is rather low and is only about 3-5%. Clearly, the first sheet polarizer causes the biggest problem because it allows only 40-46% of the incident light to pass and absorbs the rest. In addition, the second polarizer also absorbs light from the "off" pixels. All the light absorbed by the two polarizers will be eventually transformed into heat. This can be a serious problem in projectors that use arc lamps because these are much more powerful than the fluorescent lamps that are employed in direct-view LCDs. Under such conditions the performance of the absorbing sheet polarizers can deteriorate quickly due to excess heating (the new K-type sheet polarizers are relatively more stable [Cael et al. 1997]).

5.2 Metal Wire-Grid Polarizers and PBSs

Recent developments in metal wire-grid polarizers have made such devices available for use in the visible spectrum [Pentico et al. 2001; Arnold et al. 2001]. Until recently grid polarizers were used only in the IR and microwave regions because of the difficulty of making the small-sized metal grids required for operation at visible wavelengths. Figure 17B shows the performance of one such polarizer [Moxtec 2001]. It has a good performance in transmission, but it is not a good PBS. It works at normal as well as oblique angles of incidence. It has been proposed for use with direct-view as well as projection displays. However, it absorbs some light due to the use of aluminium wires. If light interacts with it twice, such as in a PBS configuration (Fig.18D), significant light will be lost.

5.3 Birefringent Thin Film Polarizers and PBSs

Thin film polarizers or PBSs that are based on a stack of alternating stretched birefringent plastic films of two different materials have been described recently [Wortman 1997; Weber et al. 2001; Jonza et al. 2001]. The stretching process causes the films to become birefringent and to have different refractive indices in different directions in the plane of the film. By carefully choosing the materials and aligning their relative orientations in such a way that the refractive indices of all the layers in the stack have the same value for one polarization, light of this polarization will be transmitted. For the other polarization, the refractive indices of the alternating layers are different; as a result, light of that polarization will be partially reflected. In order to increase the reflectance, one has to use hundreds of layers because the refractive index differences in the stretched films are rather small compared to those of conventional thin film coatings. Bandpass polarizing coatings or colour polarizing filters can be obtained in this way too. In order to construct a broadband polarizer for the whole visible spectrum, even more layers are needed with chirped thicknesses. Figure 17C shows the performance of such a polarizer [Wortman 1997]. It is not a good PBS but a reasonably good polarizer in transmission [Jonza et al. 2001]. The polarizer works at normal angles of incidence as well as at oblique angles.

One application of such a birefringent polarizer is as the reflective polarizer used in a polarization converter for direct-view LCDs (Fig.18C). The unwanted polarized light is recycled and eventually it is partially converted to light of the desired polarization and passed by the reflective polarizer. Since the reflective polarizer does not have a good extinction ratio, the first Polaroid polarizer is still needed to clean up the light. Another application of such a birefringent polarizer is to replace the second Polaroid polarizer in a LCOS projector (Fig.18D). Since it has a free-standing film structure with no supporting substrates, maintaining a good flatness could be a problem. To overcome this, the polarizer can be embedded between two glass substrates. Potential problems for such birefringent sheet polarizers are their durability and their stability with respect to heat and humidity.

5.4 MacNeille and Modified MacNeille Polarizers

The MacNeille polarizer is based on the Brewster angle effect as well as on interference of light in thin films [MacNeille 1946]. When light is incident at the Brewster angle $\theta_B = \tan^{-1}(n_1/n_2)$ upon an interface between two media of refractive indices n_1, n_2, all *p*-polarized light is transmitted and part of the *s*-polarized light is reflected. To increase the reflectance for the *s*-polarized light, a quarterwave multilayer stack consisting of layers with refractive indices n_1 and n_2 can be embedded between two prisms of refractive index n_0. The design angle θ_0, and the refractive indices n_1, n_2 and n_0 must satisfy the following condition:

$$n_0 \sin\theta_0 = \frac{n_1 n_2}{\sqrt{n_1^2 + n_2^2}} \tag{5}$$

Once n_1 and n_2 are fixed, only one parameter, either θ_0 or n_0, can be freely chosen. Figure 17D shows the calculated transmittance and reflectance of a MacNeille polarizer designed for θ_0=45°, with n_0 = 1.68, n_1=1.38, n_2=2.35. At the design angle it is a good polarizer only for the transmitted light. Because the Brewster angle condition is not satisfied at the substrate interfaces, some *p*-polarized light is present in the reflected beam even at the design angle. Once the angle of incidence θ_0 departs from the design value (even if only by as little as ±2°), the Brewster angle condition is no longer satisfied and the performances of both the transmitted and the reflected beams degrade rapidly. Significant amounts of *p*-polarized light are also reflected. The MacNeille polarizer is really a small angular field polarizer and not a polarizing beam-splitter. In the form described above, the MacNeille polarizer is not much used in displays.

However, the shortcomings of the MacNeille polarizer can be at least partially removed by modifications of the basic design. One can either optimize all layer thicknesses or use more than two different coating materials in the polarizer design. This later approach results in more than one Brewster angle, hence, the angular field can be broadened [Li and Dobrowolski 1996]. One such modified MacNeille polarizer is a broadband, relatively wide-angle and low extinction ratio PBS. It is best suitable for use in Fig.18F as a polarization converter [Imai et al. 1992]. The PBS separates the incident light into two beams of polarized light; a half-wave plate is then used to convert the *s*-polarized light into *p*-polarized light. Light of both polarizations can therefore be used for imaging and thus the emitted light is utilized more efficiently. Often polarization converters consisting of arrays of small PBSs and lenses are used: this reduces the PBS size and improves the uniformity of the illumination [Itoh et al. 1997]. Figure 17E shows the spectral performance of such a PBS. Because it does not possess a good extinction ratio, a clean-up Polaroid polarizer is still necessary to remove any unwanted polarized light. Such a modified MacNeille PBS can also be used in combination with polarizing colour filters [Robinson et al. 2000; Chen et al. 2001].

The goal of another modification of the classical MacNeille polarizer design is to obtain a true polarizer in transmitted light, without considering the performance for the reflected beam [Li et al. 2000; Unaxis 2002]. This goal can be achieved by using similar design approaches, but only at the expense of lowering the transmittance at higher or lower angles of incidence (Fig.17F) [Li et al. 2000]. Such a polarizer can then be used as the second polarizer in LCOSs to remove light from "off" pixels (Fig18E). To produce a more efficient display, it must be used in combination with a polarization converter and a clean-up Polaroid polarizer.

5.5 Li Li PBS

The Li Li PBS is a truly broadband, wide-angle, high-extinction ratio device [Li and Dobrowolski 1999; Li and Dobrowolski 2000]. A detailed comparison of the Li Li PBS and the MacNeille polarizer is given in reference [Li and Dobrowolski 2000]. It works at angles greater than the critical angle and is based on interference of light in thin films. A starting design for a Li Li PBS usually consists of a symmetrical structure $n_0 \mid [\ (n_1, d_1)\ (n_2, d_2)\ (n_1, d_1)\ ]^N \mid n_0$; here N is the number of periods and n_1, d_1 and n_2, d_2 are the refractive index and metric thickness of the low- and high-refractive index layers. The values of d_1, d_2 are small compared to the wavelength. The relationship between the construction parameters of the coatings and the design angle θ_0 is given by:

$$n_2 < n_0 \le n_1 \text{, or, } n_1 < n_0 \le n_2 \tag{6}$$

$$d_2 = \frac{(n_1^2 - n_0^2)}{(n_0^2 - n_2^2)} 2d_1 \tag{7}$$

$$n_0 \sin\theta_{LL} = \frac{n_1 n_2}{\sqrt{(n_1^2 + n_2^2 - n_0^2)}} \text{, } \theta_C < \theta_{LL} \le \theta_0 < 90° \tag{8}$$

where θ_{LL} is the lower-limit angle of the design angle θ_0, and $\theta_{\mathrm{C}}=\sin^{-1}(n_1/n_0)$ is the critical angle for the low refractive index layers.

As long as Eqs. (6 and 7) are satisfied, the symmetrical structure acts like a perfect AR coating for s-polarized light, and it is independent of wavelength and angle of incidence. Providing that Eq. (8) is satisfied, the symmetrical structure also acts like a perfect metal layer (negative imaginary equivalent admittance) for p-polarized light of all wavelengths and any angle of incidence between θ_{LL} and 90°. If the coating is thick enough or if N is large enough, all p-polarized light is reflected.

In practice, the starting design is usually simplified by optimizing the layer thicknesses and by reducing the number of layers in the system. Figure 17G shows the calculated performance of a 33-layer Li Li PBS based on n_1=1.45, n_2=2.35, and n_0=1.75. The operational wavelength range is between 400 to 700 nm and the angular field is ±10.5° measured in air (compared to ±2° in a MacNeille polarizer). The extinction ratios are better than 10^4:1 for both the transmitted and reflected beams. Figure 17H shows the measured performance of a prototype PBS based on a simpler design than that of Fig. 17G. Better results should be possible in practice, if desired.

The Li Li PBS can be used as the first, second, or both polarizers in LCDs (Fig.18G) [Li et al. 1998; Li et al. 1999c; Li et al. 1999b]. It is the only PBS capable of replacing both the first and second plastic sheet polarizer. Figure 20 shows the layout of a projection system based on the Li Li PBS that uses two reflective LCOS panels. The un-polarized incident beam of light from the lamp is separated into two beams: the p-polarized beam is reflected and the s-polarized beam is transmitted. The polarization of the light reflected from the "on" pixels of the LCOS 1 in the reflection arm is rotated by 90° and therefore it becomes s-polarized. The light then passes through the PBS and is projected onto the screen. Similarly, in the transmission arm, the reflected light from "on" pixels of LCOS 2 is p-polarized and is reflected by the PBS and is also projected onto the screen. As a result, light of both polarizations is used for the imaging and no light is lost. Therefore, high efficiency can be achieved. Full colour displays can be achieved by using a time-sequential colour scheme such as a colour filter wheel or through the use of two full colour LCOS panels. In an even more elaborate configuration, the simultaneous colour scheme could be used with six monochrome LCOS panels to obtain a very efficient full colour display.

The conventional LCD panels with the Li Li PBS can also be used to form a simple 3D display. In Fig.20 the light arriving on the screen from LCOS 1 is s-polarized, while that from LCOS 2 is p-polarized. If left- and right-eye images are fed to the two panels, 3D images can be viewed with the aid of polarizing glasses. This configuration is very compact and flexible and can display either 2D or 3D images, depending on the type of images that are fed to the LCOS panels. Of course, the Li Li PBS can be used to form 3D displays not only with LCOSs, but also with PDLCDs or DLPs and other types of display panels in which retardation films are needed to convert the polarization [Li et al. 1999b].

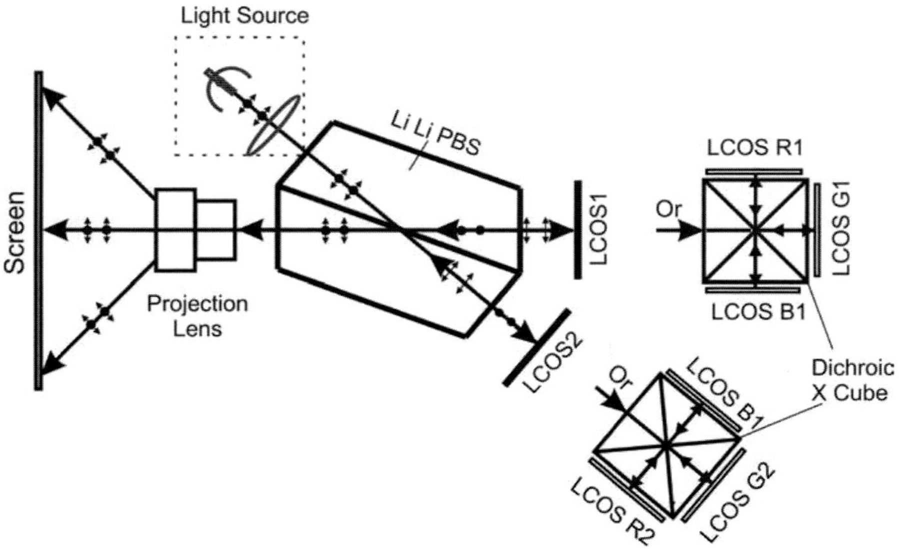

Fig. 20. High efficiency 2D/3D displays with two or six LCOSs.

A number of different prototype Li Li PBSs have been constructed and the experimental measurements were in very good agreement with the theoretically calculated performance. The device shows great potential for use in display devices but has, as yet, not been incorporated in a commercial product. Difficulties to be overcome include the finding of a reliable method for the bonding of the two high refractive index prisms in a mass production environment. Progress in this area is being made. Another issue need to be addressed is the thermal stress induced birefringence which could degrade the performance of prism based polarizers and PBSs.

Another type of thin film polarization converters for direct-view LCDs has been proposed (Fig.18C) [Pang and Li 1999]. The PBS used in this converter is similar to the Li Li PBS. However, instead of operating at angles greater than the critical angle in order to achieve negative equivalent admittance, alternating silver films and dielectric layers are used [Li and Pang 2001]. The advantage of such metal-dielectric PBSs is that they consist of only 3-7 layers that are sandwiched between glass prisms. Such a device can also be used in polarization converters (Fig.18C and E) as well as to replace the second sheet polarizers.

6 Other Optical Coatings for Displays

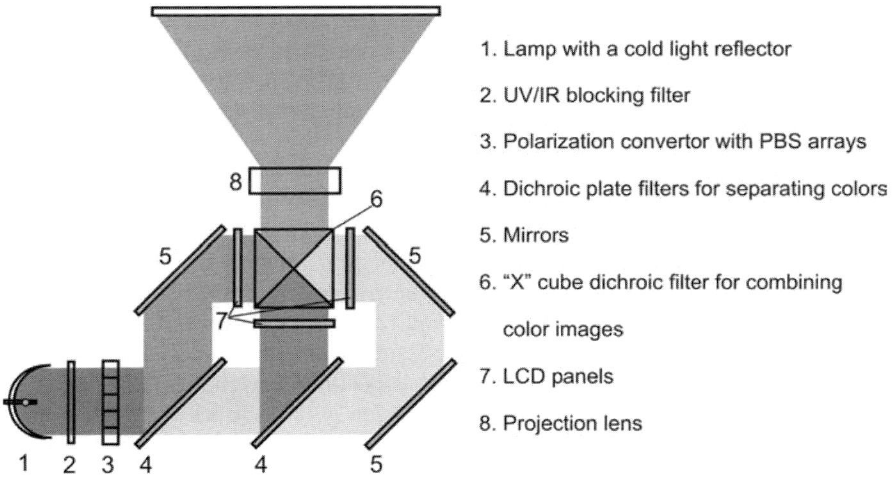

Fig. 21. A typical LCD projector with various thin film components.

Other types of thin film optical coatings have been used in both direct-view and projection displays. Figure 21 shows a typical LCD projector layout. Besides the previously described colour filters and polarization converters, cold and hot mirrors are commonly used in projection displays to remove UV and IR radiation from arc lamps. They are made of all-dielectric multilayer coatings. Aluminum or enhanced aluminum mirrors are also commonly used to redirect light beams. In addition, the light intensity of the RGB colours in a display system may not have the right balance for achieving good colour images. This can be caused by the emission spectrum of the lamps, or by the properties of the components used in the system. As a result, colour-balancing coatings may sometimes be needed to correct this problem.

7 Conclusions

This chapter has briefly reviewed the major display technologies currently in use. It has described in some detail various aspects of optical coatings for display applications. It demonstrated how AR coatings, low-reflectance neutral density filters, cut-off and bandpass filters and black layer coatings can be used to enhance display luminance contrasts. In addition, it described various ways of using colour optical coatings to separate white light or to combine colour images in order to obtain full colour displays. The coating design issues for plate dichroic filters, Philip prisms and X-cubes were discussed. Furthermore, the principles of various polarizers and polarizing beam-splitters for LCD applications were discussed and their performance compared. These included Polaroid absorbing sheet polarizers, wire-grid polarizers, birefringent polymer film polarizers, MacNeille and modified MacNeille polarizers as well as Li Li polarizers and PBSs. It has been shown that the Li Li PBS is the only broadband and wide-angle device that has high extinction ratios in both the transmitted and reflected beams. It can replace both the first and the second Polaroid polarizers to form high efficiency 2D and 3D displays. Finally, other optical coatings used in displays were demonstrated with the help of a typical LCD projector layout.

Wet and vacuum deposition processes are commonly used to fabricate optical coatings for display applications. Vacuum processes can produce complicated multilayer coatings. They include electron-beam evaporation, ion-assisted e-beam evaporation, advanced plasma system, ion plating, magnetron sputtering, ion-beam sputtering, etc. These processes have been used to fabricate AR coatings, dichroic filters and thin film polarizers or polarizing beam-splitters. For more information on thin film coating manufacturing issues, interested readers are referred to Chapters "Coating Materials", "Thin Film Deposition Techniques", "Large Area deposition" and "Characterization and Monitoring". As for future developments, it will be very desirable to use vacuum deposition systems that are capable of depositing more than two different materials. In addition, the systems must have very good layer thickness control. Such systems will make it possible and more affordable to use more complicated coating designs with better performance such as the ones shown in this chapter.

Displays, especially projectors, are quite complicated optical systems that use many different coating components. It is essential for display optical system designers to understand the principles as well as the advantages and disadvantages of these coating components. It is also important for optical coating designers to understand the principles of displays and optical system design. In the future, we will see more innovative projector layouts that take the advantages of both. It is expected that some of the better performance optical coatings described in this chapter will be implemented in future display systems. In addition, as the current display technologies evolve and new display technologies emerge, the need for better or new types of thin film coatings will arise.

Acknowledgement

The author would like to thank Dr. J. A. Dobrowolski for his help during the preparation of this book chapter.

References

Arnold S, Gardner E, Hansen D, Perkins R (2001) An improved polarizing beamsplitter LCOS projection display based on wired-grid polarizers. SID Digest, pp 1282-1285.

Bennett JM (1995) Polarizers. In: Bass M, Van Stryland E, Williams DR, Wolfe WL (eds) Handbook of Optics. McGraw-Hill, New York, pp. 3.1-3.70.

Cael JJ, Nkwantah G, Trapani G (1997) High-durability KE polarizer for LCD applications: design considerations. SID Digest, pp 809-812.

Chen J, Robinson MG, Sharp G (2001) Two-panel architecture for reflective LCD projector. SID Digest, pp 1084-1087.

Chigrinov V (1999) Liquid crystal devices : physics and applications. Artech House, Boston.

De Lang H, Bouwhuis G (1965) Optical system for a color television camera. US Patent 3,202,039.

Dewald DS, Penn SM, Davis M (2000) Sequential color recapture and dynamic filters: a method of scrolling. SID Digest, pp 40.2.

Dobrowolski JA (1995) Optical properties of films and coatings. In: Bass M, Van Stryland E, Williams DR, Wolfe WL (eds) Handbook of Optics. McGraw-Hill, New York, pp. 42.1-109.

Dobrowolski JA, Sullivan BT, Bajcar RC (1992) Optical interference, contrast-enhanced electroluminescent device. Appl. Opt., 31: 5988-5996.

Dobrowolski JA, Li L, Kemp R (1995) Metal/dielectric transmission interference filters with low reflectance: I. Design. App. Opt., 34(25): 5673-5683.

Dworsky LN, Chalamala BR, 1999. Field-emission displays, SID Seminar Lecture Notes, pp. F-1.

Friend R, 1999. Organic electroluminescent displays, SID Seminar Lecture Notes, pp. F-4.

Gnade B, 1995. Field-emission displays, SID Seminar Lecture Notes, pp. M-8.

Goede W, 1999. Electronic information display perspective, SID Seminar Lecture Notes, pp. M-1.

Greenberg MR, Bryars BJ (2000) Skew Ray compensated color separation prism for projection display applications. SID Digest, pp 88-91.

Hanaoka H, Ikui K, Watanabe K, Yamauchi H (2000) The transmittance control of film laminated CRTs by light absorbent anti-reflective coating and pigmented adhesive. SID Digest, pp 1054-1057.

Hornbeck LJ (1997) Digital light processing for high-brightness, high-resolution applications. SPIE, 3013, pp 27-40.
Iki T, 1999. CRTs, SID Seminar Lecture Notes, pp. M-2.
Imai M, Sakamoto M, Nishida N (1992) A polarization converter for high-brightness liquid crystal light valve projector. International Display Research Conference, pp 235-238.
Itoh Y, Nakanura JI, Yoneno K, Kamakura H, Okamoto N (1997) Ultra-high-efficiency LC projector using a polarized light illuminating system. SID Digest, pp 993-996.
Jonza JM, Weber MF, Ouderkirk AJ, Stover CA (2001) Optical film. US Patent 6,296,927 B1.
Karim MA (1999) Electro-optical displays. Optical Engineering. John Wiley & Sons, Chichester.
Keller PA (1997) Electronic display measurement concepts, techniques and instrumentation. John Wiley & Sons, New York.
Li HF, Huang HC, Kwok HS (2000) Broad, large acceptance angle polarizing beam splitter for reflective LCD projectors. SPIE, 3954, pp 186-196.
Li HF, Zheng ZR, Gu PF, Liu X, Cheng PW, Chen J, Huang HC, Kwok HS (1999a) Improved projection optics for reflective silicon CMOS light valves. SPIE, 3634, pp 123-131.
Li L, Dobrowolski JA (1996) Visible broad-Band, wide-angle thin film multilayer polarizing beam-splitter. Appl. Opt., 35(13): 2221-2225.
Li L, Dobrowolski JA (1999) Thin film polarizing device. U. S, Patent 5,912,762.
Li L, Dobrowolski JA (2000) High-performance thin-film polarizing beam splitter operating at angles greater than the critical angle. Appl. Opt., 39: 2754-2771.
Li L, Pang Z (2001) Thin film polarizing device having metal-dielectric films. US Patent 6,317,264.
Li L, Sullivan BT, Dobrowolski JA (1999b) High efficiency projection displays having thin film polarizing beam-splitters. US Patent 5, 982, 541.
Li L, Dobrowolski JA, Grant PD, Sullivan BT (1997) Color deformable mirror device having optical thin film interference color coatings. U. S, Patent 5,619,059.
Li L, Dobrowolski JA, Sullivan BT, Pang Z (1999c) Novel thin film polarizing beam-splitter and its application in high efficiency projection displays. Proceedings of Projection Displays V, SPIE, 3634, pp 52-62.
Li L, Dobrowolski JA, Sullivan BT, Bajcar B, Simpson R (1994) High-contrast TFEL displays with additional optical interference components. SID Digest, 25, pp 140.
Li L, Dobrowolski JA, Sullivan BT, Marshall G, Pang Z (1998) High efficiency projection displays with novel thin film polarizing beam-splitters. SID Digest, pp 687-689.
Lu M, Yang KH (2000) Reflective nematic LC devices for LCOS applications. SID Digest, pp 1-7.
Lueder E (2001) Liquid crystal displays: addressing schemes and electro-optic effects. SID Series in Display Technology. John Wiley & Sons, Chichester.
MacDonald LW, Lowe AC (Editors), 1997. Display systems design and applications. John Wiley & Sons, Chichester.
Machover C (1997) How applications have driven display requirements. In: MacDonald LW, Lowe AC (eds) Display systems design and applications. John Wiley & Sons, Chichester, pp. 3-16.
Macleod HA (2001) Thin-film optical filters. Institute of Physics Publishing, Bristol and Philadelphia.
MacNeille SM (1946) Beam splitter. US Patent 2,403, 731.

Mikoshiba S, 1999. Color plasma displays, SID Seminar Lecture Notes, pp. M-4.
Miller MR, 1997. Inorganic electroluminescent displays, SID Seminar Lecture Notes, pp. F-1.
Moxtec I (2001) Product literature: Polarizing beam-splitter.
Mueller GO, 1995. Electroluminescent devices, SID Seminar Lecture Notes, pp. M-6.
Ono Y, Ohtani Y, Hiratsuka K, Morimoto T (1992) A new antireflective and antistatic double-layered coating for CRTs. SID Digest, pp 511-514.
Onodera M, Matsuda H, Mori H, Ito T (1994) A color display tube with a high-contrast and anti-reflection coating. SID Digest, pp 823-826.
Oyama T, Ohsaki H, Hayashi Y, Tachibana Y (1998) A new layer system for wideband anti-reflection coatings designed for CRTs. SID Digest, pp 262-265.
Pang Z, Li L (1999) Novel high-efficiency polarizing backlighting system with a polarizing beam-splitter. SID Digest, pp 916-919.
Pentico C, Gardner E, Hansen D, Perkins R (2001) New, high performance, durable polarizers for projection displays. SID Digest, pp 1287-1290.
Robinson MG, Korah J, Sharp G, Birge J (2000) High contrast color splitting architecture using color polarization filters. SID Digest, 24, pp 92-95.
Rosenbluth AE, Lu M, Yang K-H, Ho K, Singh RN, Nakasogi T (2000) Correction of contrast in projection systems by means of phase-controlled prism coatings and band-shifted twist compensators. SPIE, 3954, pp 63-103.
Sampsell JB (1993) An overview of Texas Instruments digital micromirror device (DMD) and its applications to projection displays. SID Digest, 24, pp 1012-1015.
Schmidt TC, 1999. Projection displays, SID Seminar Lecture Notes, pp. F-6.
Shimizu JA (2001) Scrolling color LCOS for HDTV rear projector. SID Digest, pp 1072-1075.
Silverstein L, 1995. Color in electronic displays, SID Seminar Lecture Notes, pp. F-7.
Stupp EH, Brennesholtz MS (1999) Projection displays. SID Series in Display Technology. John Wiley & Sons, Chichester.
Tani M, 1995. LCD color filters: characteristics and future issues, SID Seminar Lecture Notes, pp. F-5.
Tannas LE (Editor), 1985. Flat-panel displays and CRTs. Van Nostrand Reinhold, New York.
Thelen A (1989) Design of optical interference coatings. McGraw-Hill Optical and electro-optical engineering series. McGraw-Hill, New York.
Tohda H, Arimoto N, Hayama H, Takahashi T (1992) Anti-glare, anti-reflection and anti-static (AGRAS) coating for CRTs. International Display Research Conference, pp 289-292.
Tong H-S, Hu C-M (1998) Recent developments in surface coatings for CRTs. SID Digest, pp 253-256.
Trout TH, Gambogi WJ, Steijn KW, Machara ST (2000) Volume holographic components for display applications. SID Digest, pp 202-205.
Unaxis (2002) Product literature: HELF™ - polarizing beamsplitter.
Weber L, 1996. Color plasma displays, SID Seminar Lecture Notes, pp. M-6.
Weber MF, Strharsky R, Gilbert LR, Stover CA, Nevitt TJ, Ouderkirk AJ (2001) Birefringent multiplayer polymer beamsplitters. Technical Digest of Optical Interference Coatings, pp FA5-1-3.
Wortman DL (1997) A recent advance in reflective polarizer technology. International Display Research Conference, pp M98-M106.

Wu S-T, Yang D-k (2001) Reflective Liquid Crystal Displays. SID Series in Display Technology. John Wiley & Sons, Chichester.

Wyszecki G, Stiles WS (1982) Color science: concepts and methods, quantitative data and formulae. John Wiley & Sons, Toronto.

Yamamoto H, Atsuchi N (1997) Prism system and liquid crystal projection device. US Patent 5,595,591.

Yeh P, Gu C (1999) Optics of liquid crystal displays. Wiley series in pure and applied optics. Wiley, New York.

Abbreviations and Symbols

CRT	-	cathode ray tube
DMD	-	digital micromirror display
DLP	-	digital light processing
EL	-	electroluminescent display
TFEL	-	thin film electroluminescent display
FED	-	field emission display
LCD	-	liquid crystal display
LCOS	-	liquid crystal on silicon
PDLCD	-	polymer dispersed liquid crystal display
PDP	-	plasma display panel
PBS	-	polarizing beam-splitter
λ	-	wavelength
$V(\lambda)$	-	photonic response curve of human eye
n	-	refractive index
d	-	layer thickness
δ	-	layer phase thickness
η	-	optical admittance
θ	-	angle of incidence
θ_C	-	critical angle
θ_{LL}	-	lower-limit angle
I	-	spectral intensity of light
CR	-	luminance contrast ratio
T	-	transmittance
R	-	reflectance
A	-	absorptance
s-	-	*s*-polarization
p-	-	*p*-polarization
N	-	number of layers or periods

Optical Thin Films for Spontaneous Emission Control

Hervé Rigneault

Institut Fresnel, Domaine Universitaire de St Jérôme, Marseille, France.

1 Introduction

Spontaneous emission control is aimed at controlling spectrally and spatially light emitted from very localized emitters such as atoms, molecules or any other kind of tiny optical source. Because this type of localized emitters usually radiate moderate light power, there is great interest in applications to concentrate the emitted radiation into a narrow angular direction or into a well defined spectral bandwidth.

The biggest application of such a concept is probably in the LED market (Craford 1996) where any lost photon directly affects the device efficiency. The holy grail is to concentrate 100% of the photons emitted by the semiconductor junction into the user angular and spectral windows (Hunt et al. 1995).

The other field of interest of spontaneous emission control is in ultra-sensitive fluorescence applications. This field is aimed at filtering and detecting very dim radiation and is usually associated with the photon counting regime in which the interaction of a light wave with a high sensitivity detector gives birth to photoelectrons which, after subsequent amplification, can be digitally counted. A large number of such applications are found in spectroscopy (Lakowicz 1999) with a special interest in biology and more recently in optical communications (Bouwmeester et al. 2000). Here again every collected photon is welcome to improve the signal to noise ratio.

At first, it seems than conventional optical materials and designs can be simply implemented in such applications to control, filter and detect optical radiation; the only straightforward requirement being transparency at the emitter wavelength. Among them, optical thin films play and important role since they can filter, reflect or transmit wavelengths in a desired way. Nevertheless, any thin film designer is well aware than spectral and angular requirements are difficult to fulfill if large angle and broad band spectrum optical properties are required (see also

Chap. 4 "Design of Optical Coatings" by Tikhonravov). This is indeed precisely the case when one investigates the optical properties of a coating located in the very vicinity of a tiny broadband emitter whose radiation pattern is intrinsically isotropic.

In relation to this later point, optical thin films are usually designed to work with travelling electromagnetic waves; i.e. waves that exhibit a non zero Poynting flux through the plane of the layers. This chapter shows that is it possible to control the light emitted by an optical source located in the very vicinity of a multilayer coating. In this specific case, the coating 'feels' the evanescent waves which take place around the source location. These waves can carry energy in a plane perpendicular to the layers thus reducing the energy classically detected through travelling waves. Some questions relative to the coating properties for these high spatial frequency evanescent waves remain important to clarify (see also Chap.2 "Photonic Structures as Interference Devices" from Ribbing).

In this framework, the goal of this chapter is to present the problems and the potential of optical thin films to control and filter the light emitted from dim sources located in the very vicinity of optical coatings. Starting from the basic physical ideas and illustrated with experimental implementations, we will see that optical thin films can give original solutions to the problem of controlling and detecting the scarce photons emitted by tiny emitters such as ions and molecules. These non conventional properties may find important applications in the quickly expanding fields of biotechnology and optical communications.

This chapter is organized as follows:

- Section 2 presents how thin film stacks can strongly modify the radiation pattern of emitters located in their very vicinity. This leads to the concept of 'spontaneous emission control' which extends the conventional applications of thin films. After having presented the basic concepts and the main lines of an electromagnetic theory which can predict this radiation pattern change, sect. 2.3 gives some experimental illustrations in the case of luminescent rare earth ions located in thin films stacks. The key result is the demonstration than very directed emission is possible from emitters located in Fabry-Perot like microcavities. Nevertheless, it turns out that a large part of the emitted light remains trapped in the multilayer structure.
- Section 3 investigates the potential of corrugated thin films to extract the trapped guided light power. After a brief theoretical introduction which emphasizes the physical picture, sect. 3.4 presents experimental results for molecular emitters located in what we call a 'waveguide grating structure'.

2 Planar Thin Films to Control the Luminescence from a Single Emitter

2.1 The Spontaneous Emission Problem

In its standard textbook description, spontaneous emission is the irreversible emission of a photon by an atom into the free space modes of the electromagnetic field, accompanied by a transition of the atom from an electronic state of energy E_2 to one of lower energy E_1. This phenomenon is an intrinsically quantum mechanical process and its proper description requires the quantization of both the atoms and the field. In free space, since the mode density of the electromagnetic field is isotropic an excited atom can emit a photon in any direction with the same probability leading to an isotropic radiation pattern in a plane perpendicular to the atomic dipole vector Fig. 1A.

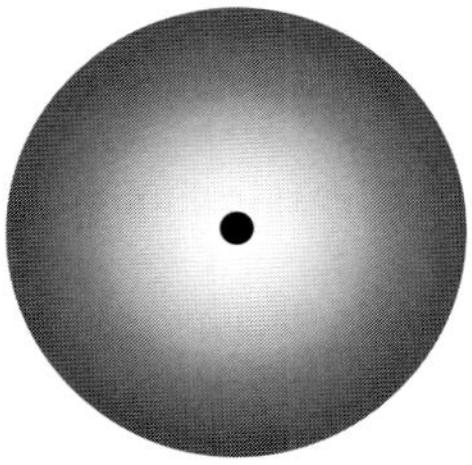
A

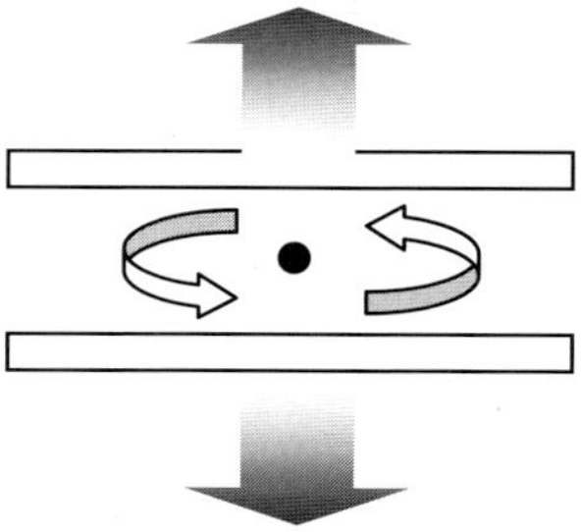
B

Fig. 1. Single atomic isotropic radiation pattern in free space (A) – Modification of the atomic spontaneous emission inside a cavity constituted of two planar mirrors (B).

In recent years it has become increasingly clear that this description of spontaneous emission is not general and that the spontaneous emission is not an intrinsic atomic property: rather, it can be modified by tailoring the electromagnetic environment that the atom can radiate into. For example this is the case if the atom is located between two mirrors at a distance which is of the order of the atomic transition wavelength (Fig. 1B). This system was first studied by Purcell (Purcell 1946) who showed that the spontaneous emission rate can be enhanced or inhibited depending on the atom location inside the cavity.

More recently, a lot of works have been devoted to planar microcavities to demonstrate fundamental changes in the spontaneous emission rates and spectra of emitting media. Cavity effects have been demonstrated at radio frequencies (Goy et al. 1983) and resonators of millimeter size have been shown to modify the spontaneous emission lifetime of ions. This concept has been extended to planar microcavities at optical wavelengths and experiments using flowing dye (De Martini et al. 1987), dye-doped films (Suzuki et al. 1991) and semiconductors (Björk et al. 1991) as active media have demonstrated significant lifetime changes. Erbium ions have been introduced into a Si/SiO_2 planar resonant microcavity and strong enhancement of the luminescence intensity has been reported (Hunt et al. 1995; Vredenberg et al. 1993).

While there is no doubt about the reality of these various possible aspects of spontaneous emission in cavities, they are by no means trivial. In particular, one might wonder how an atom can possibly know that it is inside a cavity and accordingly modify its radiative properties. This question was first addressed by Parker and Stroud (Parker and Stroud 1987), who carried out multimode calculations of the interaction between an atom and a cavity field and found that it takes the atom a time of the order of the transit time of light across the cavity to precisely determine its environment and adapt its further evolution. More recently a fully transient quantum analysis has been performed showing that an atom initially in its excited state starts to emit a wave packet which propagates away from the atom(Gieβen et al. 1996). When this wave packet encounters the cavity walls, it is reflected back toward the atom and adds either constructively or destructively to the light instantly radiated by the atom. As a result, the atomic decay is either enhanced or inhibited.

This simple picture is in qualitative agreement with the classical picture of spontaneous emission (Deppe et al. 1994)(Rigneault and Monneret 1996) which explains the changes of the radiative properties in terms of a self-driven dipole due to the radiation reaction of the reflected field at the location of the dipole. Fig. 1B gives a schematic view of this process.

In this chapter we will not describe the 'strong-coupling regime' where the atom and the resonant electromagnetic mode of the cavity interchange energy much faster than the leakage time of the cavity. From the simple description of Fig.1B, this would be the case if the emitted photon wave packet would bounce back and forth between the mirrors for a time long enough so the atom could reabsorb it (Rabi oscillations). This process has been observed in semiconductor planar microcavities due to the strong coupling of the exciton with the resonant electromagnetic mode of the cavity (Weisbuch et al. 1992).

The present description is restricted to the case where the cavity decay time is much shorter than the atom-cavity mode interaction time. This so called 'weak-coupling regime' takes place with dye molecules or rare earth atoms inside dielectric microcavities.

Although much attention has been paid to the theoretical formulation of spontaneous emission modifications in cavities, only a few studies have treated the rigorous electromagnetic problem of sources radiating in realistic multi-dielectric planar microcavities like those studied experimentally.

The next section summarizes a modal theory of atoms radiating inside dielectric multilayer microcavities.

In order to follow closely the experimental situation, the stacks under consideration can support guided modes, a point which turns out to be very important as guided modes can carry energy.

2.2 Modal Analysis of Spontaneous Emission in Dielectric Multilayer Structures

Modal Fields

The complete set of modes of a lossless multilayer dielectric structure includes an infinite number of radiation modes and evanescent modes, as well as a finite number of guided modes (Rigneault and Monneret 1996). The evanescent modes are neglected in this analysis since they do not carry power far away from the guide.

Restricting this brief presentation to the TE polarization state, the electric modal fields have particular amplitude distributions of the form

$$\mathbf{F}(\mathbf{r},\mathbf{k}_p) = E(z,k_p)\exp i(\mathbf{k}_p\mathbf{r})\mathbf{u}(\mathbf{k}_p) \qquad (1)$$

where $E(z,k_p)$ stands for the z distribution of the electric field in the depth of the stack (see Fig. 2), $\mathbf{k}_p = \alpha\,\mathbf{e}_x + \beta\,\mathbf{e}_y$ is the planar part of the **k** vector, which labels the field $\mathbf{F}(\mathbf{r},\mathbf{k}_p)$ and $\mathbf{r} = x\,\mathbf{e}_x + y\,\mathbf{e}_y + z\,\mathbf{e}_z$ is the spatial vector. $N_{eff}=\left|\mathbf{k}_p\right|/k_0$ is the spatial frequency and has the same value in all media ($k_0=2\pi/\lambda$ where λ stands for the vacuum wavelength); $\mathbf{u}(\mathbf{k}_p)$ is the TE unit electric field vector which lies in the xy plane.

Because of their modal properties, the fields that concern the radiation modes must correspond to standing waves in the direction normal to the layers. Thus, they only vary by a phase factor when they propagate along the x direction (see Fig. 2). These fields can be described by a superposition of two z-contrapropagative plane waves, that have the same cross power through the xy plane (Fig. 2A).

Two types of radiative modes are usually distinguished. The fully radiative modes $\mathbf{F}_{fr}$ ($0<N_{eff}<n_a$) radiate both in the air and in the substrate and are degenerated twice in $\mathbf{F}_{fr+}$ and $\mathbf{F}_{fr-}$, whereas the substrate radiation modes F_{sr} ($n_a<N_{eff}<n_s$) only radiate in the substrate (assuming $n_a<n_s$). Figs. 2A and 2B give schematic views of these modes. Fig. 2C shows a guided mode F_g, for which ($n_s<N_{eff}<n_H$), where n_H is the highest refractive index of the stack. These guided modes m, with specific modal frequencies N_{effm}, show evanescent waves in the outer media and can propagate without losses within the stack.

It can be shown (Rigneault and Monneret 1996) that these modal fields satisfy orthonormalization relations of the type

$$\int \mathbf{F}_i(\mathbf{r},\mathbf{k}_p)\mathbf{F}_i(\mathbf{r},\mathbf{k}_p')^* d\mathbf{r} = (2\pi)^3\delta(\mathbf{k}_p - \mathbf{k}_p') \qquad (\mathrm{i} = \mathrm{fr},\mathrm{sr},\mathrm{g}) \qquad (2)$$

where the Dirac function $\delta(\mathbf{k}_p\text{-}\mathbf{k}_p')$ must be replaced by the Kronecker symbol $\delta_{\mathbf{k}_p,\mathbf{k}_p'}$ in the case of guided modes (i = g). * stands for the complex conjugate.

At this level, the coupling of these modal fields with sources located within the stack can be either formulated in the framework of classical (Rigneault and Monneret 1996) or quantum electrodynamics (Rigneault and Monneret 1997a). For the sake of simplicity, we only describe here the classical approach which considers the radiation of a dipole of charge q oscillating along the **e** direction with an amplitude *a*.

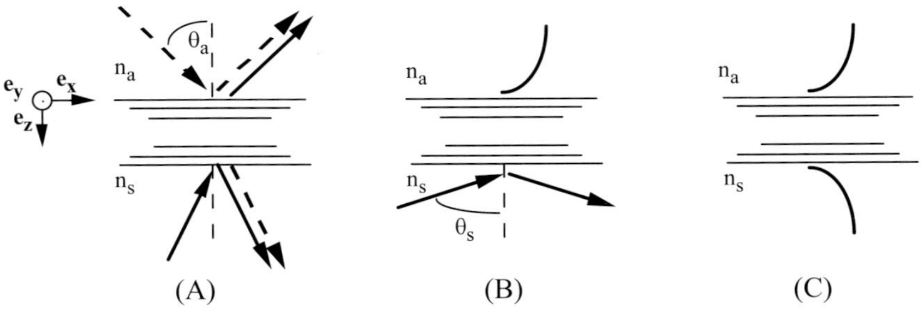

Fig. 2. A: Full radiative mode ($0 < N_{eff} < n_a$), B: substrate radiative mode ($n_a < N_{eff} < n_s$), C: guided mode ($n_s < N_{eff} < n_H$).

Classical Analysis

In the classical context, the total field $\mathbf{E}(\mathbf{r},t)$ emitted by a dipole $qa\mathbf{e}$ is expanded on the modal functions $\mathbf{F}_{fr}$, $\mathbf{F}_{sr}$ and $\mathbf{F}_g$

$$\begin{aligned}\mathbf{E}(\mathbf{r},t) &= \sum_{\gamma_g} \int_{O_g} c_g(t,\mathbf{k}_{pg})\mathbf{F}_g(\mathbf{r},\mathbf{k}_{pg})d\mathbf{k}_{pg} \\ &+ \int_{O_{fr}} \left[c_{fr-}(t,\mathbf{k}_p)\mathbf{F}_{fr-}(\mathbf{r},\mathbf{k}_p) + c_{fr+}(t,\mathbf{k}_p)\mathbf{F}_{fr+}(\mathbf{r},\mathbf{k}_p)\right]d\mathbf{k}_p \\ &+ \int_{O_{sr}} c_{sr}(t,\mathbf{k}_p)\mathbf{F}_{sr}(\mathbf{r},\mathbf{k}_p)d\mathbf{k}_p \\ &+ \text{TM term}\end{aligned} \tag{3}$$

where the integrals are evaluated in the **k** space regions O_g, O_{fr} and O_{sr} respectively associated with the full radiative, substrate radiative and guided modes. Knowledge of the *c* coefficients is sufficient to compute the total dipole electric field.

To achieve such a task, the problem is reduced to the resolution of the two coupled Eqs (4) and (5):

$$\Delta\mathbf{E}(\mathbf{r},t) - \varepsilon(z)\mu_0 \frac{\partial^2 \mathbf{E}(\mathbf{r},t)}{\partial t^2} = \mu_0 \frac{\partial^2 \mathbf{p}(\mathbf{r},t)}{\partial t^2} + \mu_0 \frac{\partial \mathbf{j}(\mathbf{r},t)}{\partial t} \tag{4}$$

$$\frac{\partial^2 a(t)}{\partial t^2} + \omega_0{}^2 a(t) = \frac{q}{m}\left[\mathbf{e}.\mathbf{E}(\mathbf{r}_0,t)\right]. \tag{5}$$

Here Eq.(4) is the classical total electric field evolution equation with a dipolar source term $\mathbf{p}(\mathbf{r},t) = q\ a(t)\ \delta(\mathbf{r}-\mathbf{r}_0)\ \mathbf{e}$ located at $\mathbf{r}=\mathbf{r}_0$ and a dissipative current $\mathbf{j}$. Eq.(5) stands for the dipole evolution equation located at $\mathbf{r}=\mathbf{r}_0$ and whose amplitude is a(t). From Eqs.(3), (4) and (5) and in the low coupling regime it is possible to compute the amplitude of the total field emitted by the dipole in the various modal fields.

In the case of a lossless stack, it possible as well as to compute the total power P emitted in air, in the substrate and into the guided modes. More precisely, this total power P is calculated using Eq.(5) where the right hand side term is a damping contribution to the dipole evolution. It is therefore possible to express the dipole damping rate which appears as a sum of the damping rates associated with each modal field. Finally, the electromagnetic power P is simply obtained by multiplying the dipole damping rate by the dipole mechanical energy (Rigneault and Monneret 1996).

Lifetime Alteration

It is interesting to focus on the dipole lifetime defined as the inverse of its damping rate. If τ_0 is the vacuum dipole lifetime, the altered dipole lifetime τ in the stack is simply given by the ratio of the total vacuum power divided by the total emitted power in the stack:

$$\frac{\tau}{\tau_0} = \frac{P_0(\omega_0)}{P}. \tag{6}$$

It is therefore possible to alter the emitter lifetime when located in a multilayer stack. Nevertheless it can be shown that this effect is weak in planar structures (Deppe et al. 1994)(Belarouci et al. 1999) and the main effect is rather to redirect the dipole emission into the various modal fields depending on the multilayer structure under consideration.

In order to illustrate this brief theoretical presentation, Sect.2.3 focuses now on the practical case of a luminescent source located in a realistic dielectric multilayer microcavity.

2.3 Luminescence of Erbium and Praseodymium Ions Located into Multilayer Dielectric Microcavities

2.3.1 Experimental Apparatus

Before reporting experimental results, let us describe the apparatus (Fig. 3). The spontaneous emission of the erbium (Er) green transition ($^4S_{3/2}$ -> $^4I_{15/2}$, $\lambda_0 = 540$ nm) or the praseodymium (Pr) blue ($^3P_0$ ->$^3H_4$, $\lambda_0 = 490$ nm) transition are investigated under the excitation provided by the λ_p (488 nm for Er, 457.9 nm for Pr) lines of a cw Ar^+ laser. These ions are incorporated into the dielectric matrix using the ion implantation technique. The pump beam is focused on the sample, exciting a 0.3 mm diameter active spot. A fiber optics bundle connected to a 77K cooled spectrometer optical multichannel analyser displays the spectrum and the intensity relaxed by the rare earth atoms in every direction θ. This first experimental configuration is designed to get the radiation pattern from the air side (upper semi-infinite outer medium).

In order to obtain the amount of light emitted into the guided modes, a coupling out prism ($SrTiO_3$) is placed against the sample 2 mm away from the active spot (see also Chap.11 "Optical Thin Film for Microcomponents" by Escoubas and Flory). The prism movements are controlled manually in the horizontal and vertical directions with an accuracy of 1 μm. It has a 100% coupling out efficiency (Rigneault et al. 1997b). The fiber bundles can rotate around the prism's edge, for each angle θ' associated with a guided direction giving the spectrum and amount of light relaxed into this particular guided mode.

2.3.2 Sample Preparation

All the layers considered here are made of tantalum pentoxyde (Ta_2O_5) and silicon dioxide (SiO_2) respectively for high (H) and low (L) refractive index materials (see also Chap.5 "Coating Materials" by Friz). These materials are deposited on fused silica substrates by the ion plating technique (see also Chap.6 "Film Deposition Methods" by Pulker) in a Balzers BAP 800 plant. Er and Pr implantations are performed in the Ta_2O_5 material. The implantation fluence is a few 10^{15} ions/cm^2 and the energy is 300 keV. This leads to an Er and Pr peak concentration of a few 10^{20} ions/cm^3. Post implantation thermal annealing is carried out on the samples at 400°C during 1 hour.

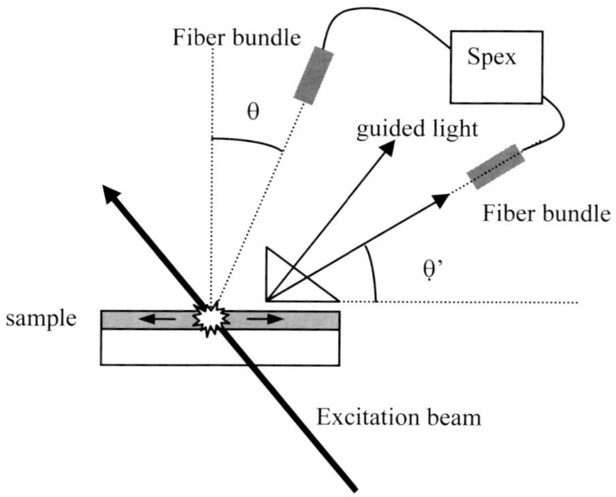

Fig. 3. Experimental apparatus to collect the full radiative and the guided light.

2.3.3 Erbium Implanted Sample

Figure 4 gives a schematic representation of the three samples considered in this section, and the position of the Er concentration peak (dashed line) is specified.

Sample (A) is a tantalum pentoxide (Ta_2O_5) single layer whose optical thickness is $6\lambda_1/4$ (λ_1 = 632.8 nm, n_H = 2.23). It is deposited on a fused silica substrate and the Er concentration peak lies 27 nm under the film/air interface.

Sample (B) is a half-microcavity structure coated on BK7 glass. Its formula is HLHLHL 2H, where H and L stand respectively for a high and low refractive index layer whose optical thickness is $\lambda_0/4$ (λ_0 = 532 nm, n_H = 2.27 ; n_L = 1.48). Here again, the Er concentration peak is located 27 nm under the film/air interface.

Sample (C) is a Fabry-Perot type microcavity structure coated on BK7 glass. Its formula is HLHLHL 2H LHLHLH and its quality factor is given by $Q = \lambda_0/\Delta\lambda$ = 40. This sample was manufactured in two steps; first a sample identical to sample (B) was coated and implanted. Second, this coating was overcoated with the LHLHLH mirror.

Full Radiative Emission Measured from the Air Side

Figure 5 shows the theoretical (plain) and experimental (doted) radiation patterns for samples (A), (B) and (C) measured from the air side and the associated pump powers (λ_p=488 nm) necessary to get the same luminescence signal at normal incidence.

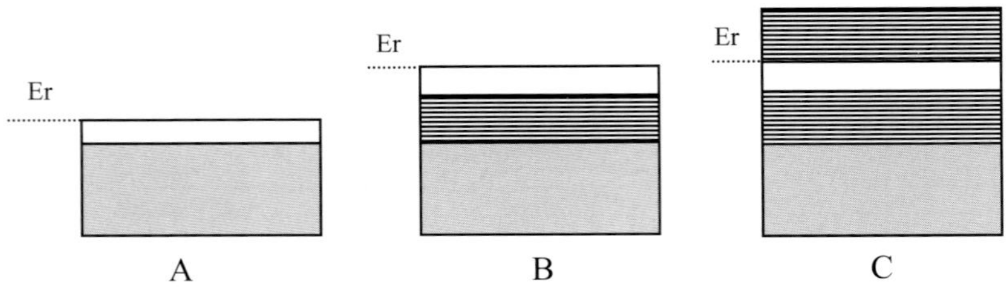

Fig. 4. Erbium implanted samples (A), (B) and (C). The peak Er concentration is indicated (dashed line).

Experimental shapes are in good agreement with theory for all samples. The radiative emission of atoms located in sample (C) is more directive than in samples (A) and (B). The pump beam intensity is 100 times smaller for sample (C) than for sample (A) and 3 times smaller for sample (C) than for sample (B). These results illustrate clearly the resonant effect of the microcavity. They confirm that photoluminescence is spatially affected by the cavity structure. In particular, the more pinched the radiation pattern is, the higher the emitted intensity in the normal direction .

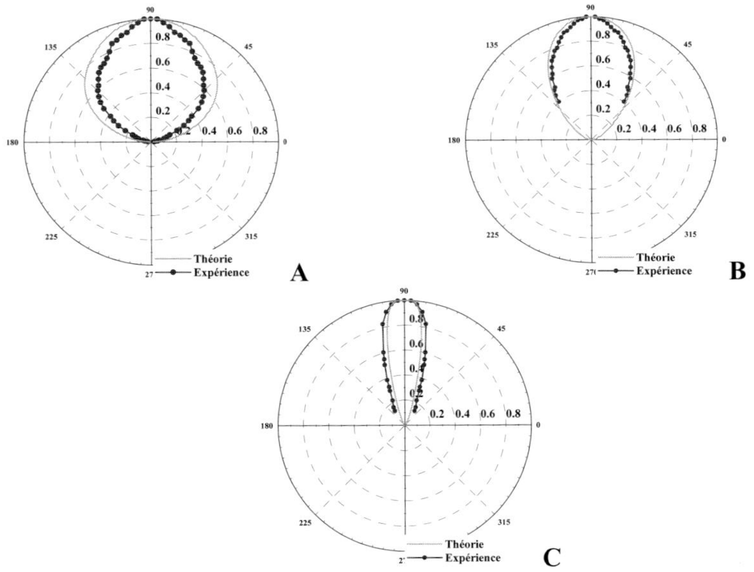

Fig. 5. Normalized theoretical (plain) and experimental (doted) radiation pattern for samples (A), (B) and (C). The excitation power was (A) 294mW, (B) 6mW, (C) 2.5mW.

Guided Wave Emission

By the way, only the light emitted into the full radiative modes in air has been considered. To illustrate the guided wave emission, Fig. 6 gives the measured (Min exp – Max exp) luminescence signal in the five guided modes supported by sample (A). , We have plotted the theoretical prediction on the same graph. Overall good agreement is found between theory and experiment. In order to sharpen our investigation, it is interesting to focus our attention on the ratio $R = \frac{P_{GM}}{P_{FRM}}$, where P_{FRM} and P_{GM} are the powers emitted in the full radiative modes and in the guided modes respectively, that are actually accessible with our experimental apparatus. Table 1 gives the R ratio for the various samples considered. In all cases, the agreement between computations and experimental measurements is good showing that most of the power is emitted into the guided modes of the structures. This important result can be explained because of the sharp electromagnetic resonances associated with the guided modes of the planar multilayers.

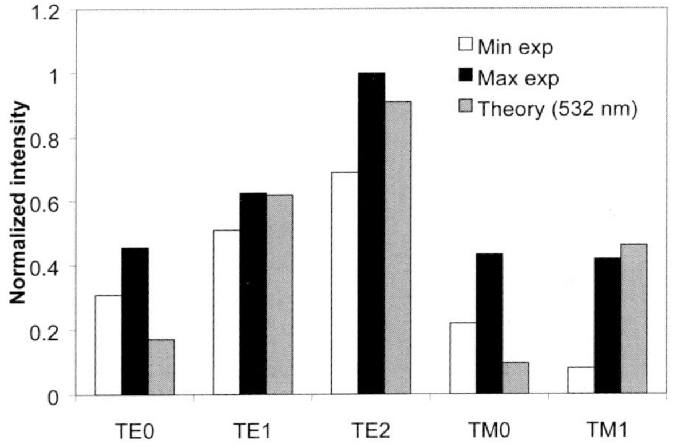

Fig. 6. Experimental and theoretical guided power – Sample A.

Table 1. Experimental and theoretical ratio $R=P_{GM}/P_{FRM}$ where P_{GM} and P_{FRM} are respectively the guided and the radiative luminescence powers.

Sample	(A)	(B)	(C)
Theoretical ratio R	33	11	15
Experimental ratio R	38	7	16

2.3.4 Praseodymium Implanted Samples

In the previous section some results concerning the luminescence of Er ions implanted into various type of microcavities were presented. Similar results can be obtained with Pr ions with the possibility of making blue emitting devices which are important for applications. We focus here on the $^3P_0 \rightarrow {}^3H_4$ at $\lambda_0 = 490$ nm. The ions are excited using the $\lambda_p = 457.9$ nm line of an argon laser.

Although the microcavity (C) presented achieves good spontaneous emission control in a direction normal to the layers, this symmetric structure (see Fig. 4C) allows the emitted light to escape equally from the air side (see Fig. 5C) and from the substrate side. For the applications one expects to design a structure which emits its light in only one half space. As an example of such an emitter, Fig. 7 presents a non symmetric structure implanted with Pr ions and the associated measured radiation pattern.

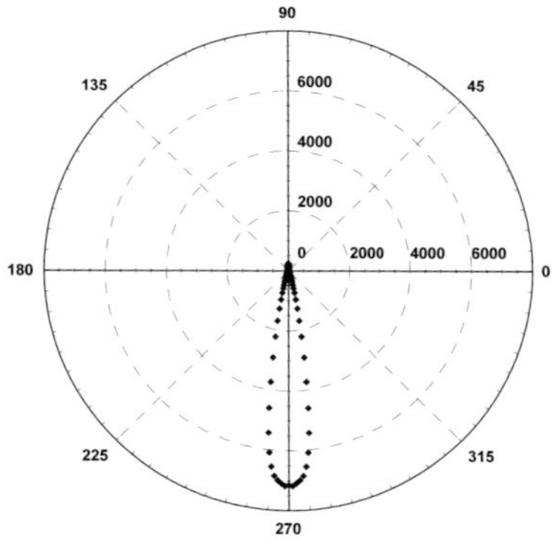

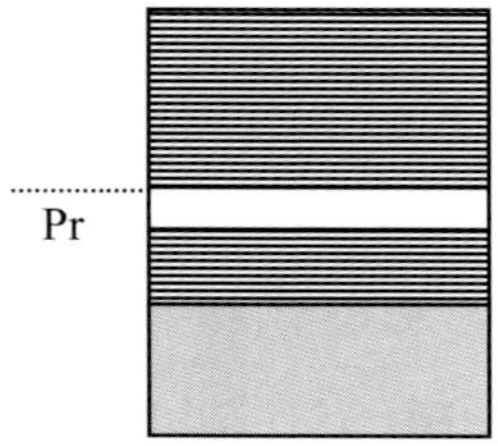

Fig. 7. Non-symmetric Pr implanted microcavity – 30% of the total emitted luminescence is emitted on the substrate side.

The considered structure formula is HLHLHLHL 2H LHLHLHLH LHLHLHLH at $\lambda_0 = 490$ nm which appears as a simple Fabry-Perot symmetric microcavity coated with a top LHLHLHLH mirror. Unlike the Er microcavity sample (C), the
emission at $\lambda = 490$ nm at normal incidence is much stronger on the substrate side than on the air side. Indeed, from the calculations, we have shown that more than 30% of the light is emitted around a direction normal to the layers in the substrate half space. This confinement comes from a strong overall resonance of the structure and cannot be simply explained by the reflection of the coated top mirror. It is interesting to mention that the very directive confinement in a direction normal to the sample appears only for the cavity resonance at $\lambda_0 = 490$ nm. For the other wavelengths in the Pr inhomogeneous spectrum the resonance is shifted to larger angle for $\lambda < 490$ nm whereas it remains centered at normal incidence but decreases significantly for $\lambda > 490$ nm (Robert et al. 1998).

2.4 Optimal Spontaneous Emission Control With Planar Structures?

Sects. 2.2 and 2.3 have presented a theoretical description and some experimental investigations of the spontaneous emission of rare earth ions located in various type of planar dielectric structures. Although spontaneous emission is a purely quantum process, it has been shown that classical antenna calculations can give an accurate description of the process in very good agreement with experiments.

Both for Er and Pr ions, strong enhancement of the spontaneous emission in a direction normal to the layer can be achieved if the ions are located in the spacer of a Fabry-Perot type structure. Furthermore, we have shown that a non symmetric structure can confine more than 30% of the emitted light into one half space.

Nevertheless, these planar structures still suffer the emission into the guided modes which propagate in the plane of the layers. Attempts to decrease this guided wave power by locating the emitters in a low refractive index medium turns out to fail (Rigneault and Monneret 1996). Alternatively, the use of metallic mirrors to increase the angular reflection range of dielectric mirrors is useless because the expected extracted power improvement into the full radiative modes is ultimately absorbed in the metal (Rigneault et al. 1999a).

Despite these unsuccessful attempts, there is some hope of controlling this amount of light trapped in the structure by processing the thin film surface. The idea is to first maximize the power emitted into a guided mode and then to couple out this trapped power using surface corrugations. Once processed in such a way, the film becomes a waveguide grating structure (WGS)

3 Corrugated Thin Films for Light Extraction: the Waveguide Grating Structure Case

3.1 Introduction

In order to increase the amount of extracted light from planar microcavities it seems interesting to use corrugation to couple the guided light to travelling waves that propagate outside the structure. This idea has been already presented and experimentally confirmed by Kitson *et al.* who have shown that it is possible to scatter light from guided surface plasmon modes into travelling waves for dye molecules on metallic grating (Kitson et al. 1996).

Although corrugation can be applied in principle to any type of microcavity supporting trapped or guided modes, the analysis is restricted to planar structures where coating techniques and etching processes are now well controlled on nanometer scales.

The use of gratings to extract guided light has been extensively treated in the domain of integrated optics (Nevière et al. 1973 Tamir 1975). Although most of the reported works deal with gratings with a periodicity along one direction, it is necessary to consider crossed gratings (or doubly periodic gratings), which exhibit periodicity along two orthogonal directions, in order to couple out the total guided mode power whatever the azimuth direction ϕ of the guided mode. This point will be discussed in Sect. 3.2 which presents the wave-vector diagram analysis and gives a simple graphic understanding for the light propagation into singly or doubly periodic shallow gratings. Beside these simple considerations, it is possible to give an accurate electromagnetic theory of sources radiating into crossed gratings. This point is beyond the scope of this book section and can be found in (Rigneault et al. 2000). Finally, Sect. 3.3 focuses on the experimental implementation of such waveguide grating structures (WGS) and presents radiation pattern results for molecular emitters located in crossed gratings with shallow and deep grooves.

3.2 Simple Description of the Extraction Problem into Crossed Gratings: the Wave-Vector Diagram

This section presents a physical insight of the radiation mechanisms for a source located in a periodic structure that supports guided modes. As presented in (Zengerle 1987), light propagation into planar structures with shallow corrugation can be physically understood from the wave-vector diagram (WVD) which plots in **k** space the areas associated with the various kinds of modes existing in the structure of interest. If the wave-vector coordinates along the three axes of unit vectors ($\mathbf{e}_x$,$\mathbf{e}_y$,$\mathbf{e}_z$) of a right hand Cartesian coordinate frame are denoted by α, β and γ so that $\mathbf{k}=\alpha\,\mathbf{e}_x+\beta\,\mathbf{e}_y+\gamma\,\mathbf{e}_z$, the WVD in the ($\alpha$,$\beta$) plane of a simple single mode slab waveguide (refractive index n_2) coated on a substrate (refractive index n_3) (Fig. 8A) is divided into three areas corresponding to the three types of electromagnetic modes that exist inside and outside the guide (see Fig. 8B). As previously, $\mathbf{k_p}$ **is** the planar part of the **k** vector, $\mathbf{k_p}=\alpha\,\mathbf{e}_x+\beta\,\mathbf{e}_y$.

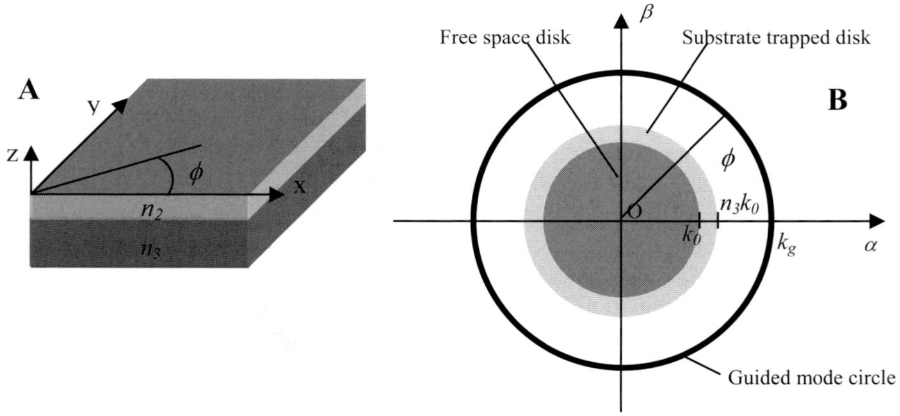

Fig. 8. A: Mono-mode slab waveguide; B: Associated WVD.

The *free space disk* corresponding to travelling waves which propagate inside and outside the guide is defined for $0<|\mathbf{k}_p|<k_0$ where $k_0=2\pi/\lambda=\omega/c$ and λ is the wavelength of radiation. For $k_0<|\mathbf{k}_p|< n_3\, k_0$ where n_3 is the refractive index of the substrate, the light remains trapped in the substrate and for $|\mathbf{k}_p|=|\mathbf{k}_g|$ the light is guided into the fundamental guided mode with the specific wave-vector $\mathbf{k}_g$. As can be seen from Fig. 8B, there is no way to couple out the guided mode in this simple slab since the $|\mathbf{k}_g|$ circle never intercepts the free space disk.

To extract the guided mode, whatever its direction of propagation (i.e. for any azimuthal angle ϕ), we introduce periodic modulations of the slab along both the x and the y axis with periods d_x and d_y (see Fig. 9A).

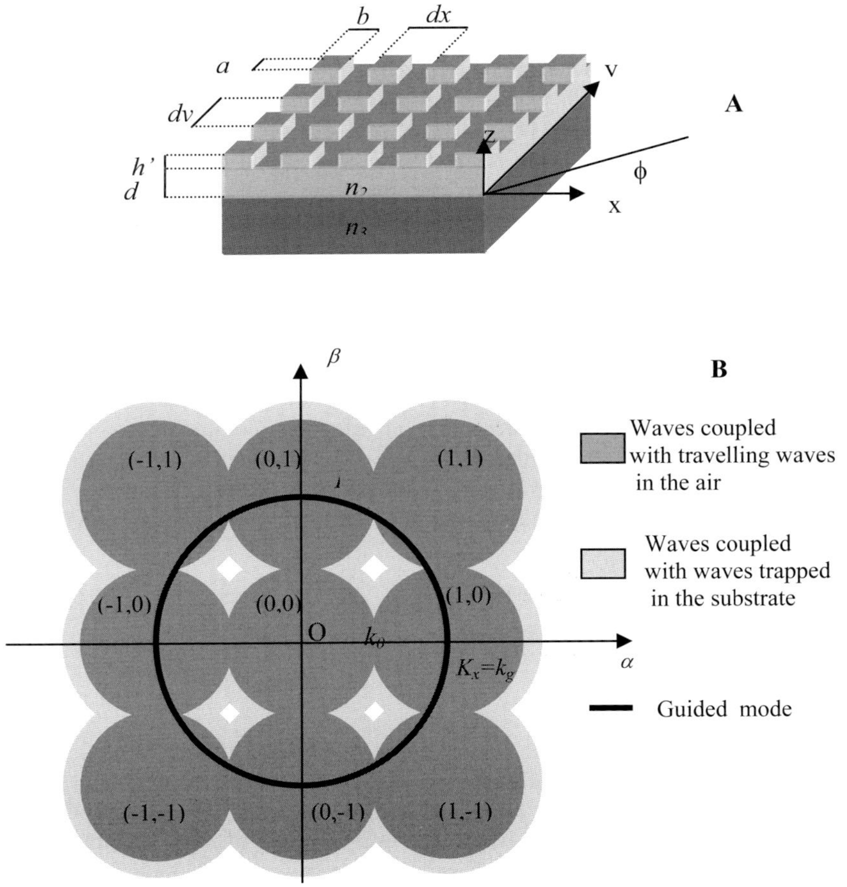

Fig. 9. A: Crossed grating structures etched in the monomode waveguide of Fig. 8, B: associated WVD.

Thus, any wave-vector $\mathbf{k_p}$ is coupled, through the grating, to all the waves of planar wave vectors $\mathbf{k_p}^{m,n}$ with $\mathbf{k_p}^{m,n} = \mathbf{k_p} + mK_x\,\mathbf{e}_x + nK_y\,\mathbf{e}_y$ (m and n positive or negative integers) where $K_x=2\pi/d_x$ and $K_y=2\pi/d_y$. Guided mode extraction is possible if there is a couple (m',n') and a free space wave vector $\mathbf{k_p}^{inc}$ ($0<|\mathbf{k_p}^{inc}|<k_0$) that satisfy $\mathbf{k}_g = \mathbf{k_p}^{inc} + m'K_x\,\mathbf{x} + n'K_y\,\mathbf{y}$. This coupling phenomenon is illustrated in Fig. 9B where the free space disk along the α and β axis has been duplicated periodically (with periods K_x and K_y). In our specific configuration, the guided mode circle now always intercepts one of the duplicated free space disks. Hence, one may expect total extraction in free space of the guided mode power (whatever its direction of propagation). Note that in this plot we have chosen $K_x = K_y = |\mathbf{k}_g|$,which implies that $d_x<\lambda$ and $d_y<\lambda$ (subwavelength scale grating).

From this simple graphic representation, it is possible to obtain qualitatively the radiation pattern from a source located inside the grating. Assuming that the source emits most of its power into the fundamental guided mode, we know from Fig. 9B that this guided light is extracted in free space via the coupling with the grating waves of order (m',n'). The extracted free space wave vector will therefore satisfy kpout = kg + m'Kx x + n'Ky y. Fig. 12 (solid and dash lines) presents a graphic representation of the radiation pattern of the system by plotting into the free space disk all the arc shape intersections of the guided mode circle with the duplicated free space disks (see Fig. 9). Note that this simple analysis holds only for shallow grating where the corrugation acts solely as a coupler without disturbing the guided mode wave vector. This is not the case for deep gratings that strongly modify the dispersion relation of the guided mode (in this case, we observe a grating-induced anisotropy). Moreover, although helpful for understanding the radiation pattern, WVD does not provide any quantitative data concerning the amount of light emitted into the arcs or the angular width of such energy concentrations. To accomplish such a task it is necessary to model rigorously a dipole emitting into a doubly periodic structure (Rigneault et al. 2000). The theory is also a valuable tool to explore the deep grating case for which the previous WVD simple graphic analysis is inadequate.

3.3 Light Emission from Europium Chelates Located in WGS

Sample Preparation

Emitter

For the source, an organic lanthanide ion complex is used on account of the narrow emission bands and the large Stokes'shift between absorption and emission bands (more than 250 nm) offered by this category of light emitting molecules (Gudgin Dickson et al. 1995). These two spectroscopic characteristics allow easy selective narrow-band wavelength filtering for the specific chelate signal. Moreover, compared to the conventional organic dyes, most of lanthanide chelates have higher photochemical stability and lower sensitivity to variations in environment and temperature. The binuclear europium chelate (C-Eu) (Galaup et al. 1996) used in this study shows an absorption band around 330 nm in methanol

(MeOH) solution and a principal emission band around 618 nm (width 3 nm) (corresponding to the 5Do-7F2 transition of the europium ion). This narrow band transition will be used for the experiments.

WGS

Two samples denoted S (Shallow grating) and D (Deep grating) have been manufactured to study light emission from C-Eu embedded in the structures. For sample S, we start from a coating of tantalum pentoxyde (Ta_2O_5) deposited by electron beam evaporation with ion assistance on fused silica substrates. The coating of sample D is titanium oxide (TiO_2) with the ion plating technique, allowing a higher refractive index of the slab (n_g=2.39).

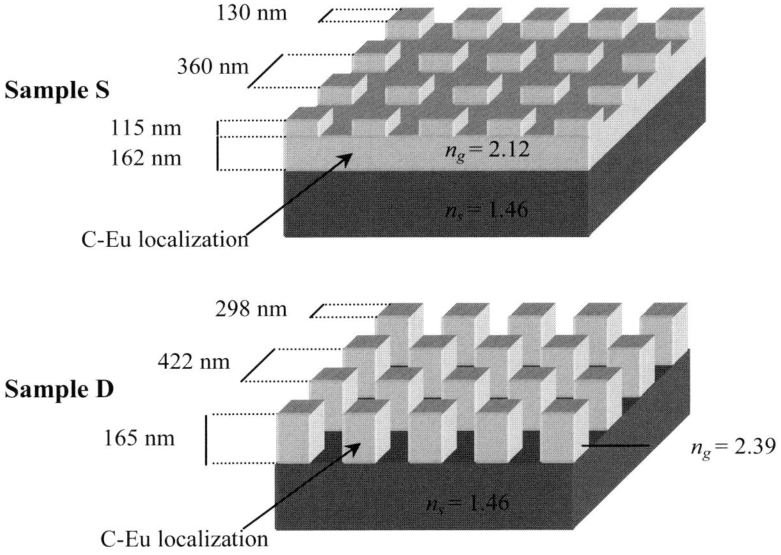

Fig. 10. Geometry and opto-geometric parameters of sample S (shallow WGS) and D (deep WGS).

In order to incorporate the C-Eu into the dielectric slab, the coating process is stopped halfway and a drop of 10 µl of C-Eu in methanol at 10^{-4} M is placed on the sample. After solvent evaporation the sample is put back into the evaporation plant and overcoated. The result is a coating of highly transparent dielectric material containing a thin layer of emitters at its center. The thickness of the film is adjusted to support two guided modes (one for each polarization) for the peak C-Eu emission wavelength at 618 nm. At this stage, the refractive indexes of the layers are accurately measured using the m-line technique (see also Chap.11 "Optical thin film for microcomponents" by Escoubas and Flory). This information is important to design the grating geometry precisely in order to match the coupling between the guided modes and the radiative modes. To complete the process, the slab is patterned and etched using electron beam photolithography and reactive ion etching techniques. The periods of the gratings obtained are respectively 360 nm and 422 nm for samples S and D which are zero order gratings for the emission wavelength (λ=618 nm). The periods are chosen in such a way that the first order, for an incident wave normal to the sample (λ=618 nm), corresponds to the waveguide fundamental guided mode (K_x=k_g). The detailed geometries of these two samples are depicted on Fig. 10.

Experimental Set-up

To measure the radiation patterns of the two samples S and D a fiber bundle connected to a spectrometer (resolution 1 nm) can rotate around the WGS active spot (Fig. 11 – angle θ). Furthermore, to describe the full radiation pattern the sample can rotate around its normal axis (Fig. 11 – angle φ). The pumping beam is a CW HeCd laser at 325 nm. The pump angle is set to 30° in order to keep the transmitted pump beam from entering the fiber bundle. This UV protection is reinforced with a color glass filter located in front of the spectrometer entrance slit. The illuminated area on the sample is 1 mm^2 and the fiber bundle is set 5 cm from the grating active spot. In order to reduce bleaching of the C-Eu molecules, the incident UV power is reduced to 100 μW. To obtain a radiation pattern in air for each direction (θ,φ), we measure the emitted power at the peak emission wavelength of 618 nm. The data are then processed to map the radiation pattern in the reduced **k** space (α' =cosφ cosθ, β' =sin φ cosθ).

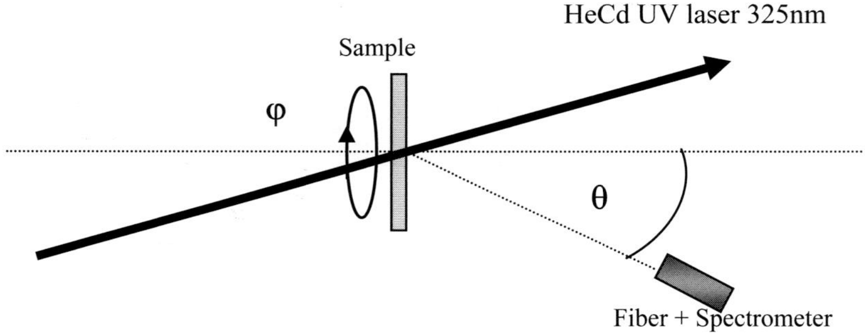

Fig. 11. Experimental setup to measure C-Eu emission into WGS.

Results

Shallow Grating S

Fig. 12 shows the superposed calculated and measured C-Eu radiation pattern in the (α', β') space for the shallow grating sample S. As expected from the calculations, we clearly measure a concentration of power into arcs that are interpreted as guided modes coupled to propagating waves (Rigneault et al. 1999b). More precisely, the full line corresponds to the TE guided mode whereas precisely, the full line corresponds to the TE guided mode whereas the dotted line corresponds to the TM mode which can escape the structure, being coupled to radiative waves that propagate away from the grating structure. From Fig. 12 both TE and TM arcs are visible. Although accurately predicted by the full electromagnetic calculation, the background signal is similar to the free space radiation pattern of a dipole located in a simple dielectric layer (Rigneault et al. 1997b). As discussed in Sect. 3.3, the shallow grating radiation pattern can be simply interpreted in terms of the wave vector diagram (WVD). The case of the deep grating is less intuitive and requires numerical calculations as guided and propagative modes are strongly coupled by the deep corrugation.

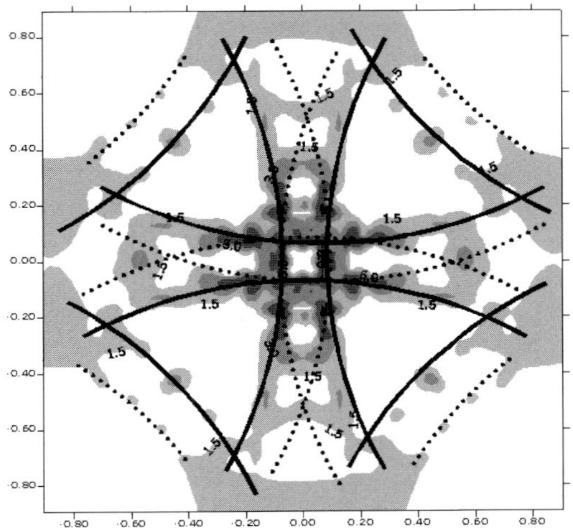

Fig. 12. Predicted (full and dotted lines) and measured radiation pattern in the (α, β) space, for C-Eu located in the shallow grating (sample S). The full line corresponds to the TE guided modes whereas the dotted line corresponds to the TM mode.

Deep Grating D

Figure 13A gives the calculated radiation pattern in the (α, β) space for a C-Eu molecule located in sample D whereas Fig. 13B gives the result of measurements. Although the agreement is qualitative, in both cases we observe, broad arcs associated with a strong emission around the direction normal to the sample (Fig. centre $\alpha'=\beta'=0$). This point is interesting for applications in the case of solid state emitters located in high index slabs. Sample D corresponds to the case where the film is totally etched. In this case of maximal perturbation of the waveguide structure, the radiation pattern is far from being intuitive since the WVD perturbative approach is no longer valid.

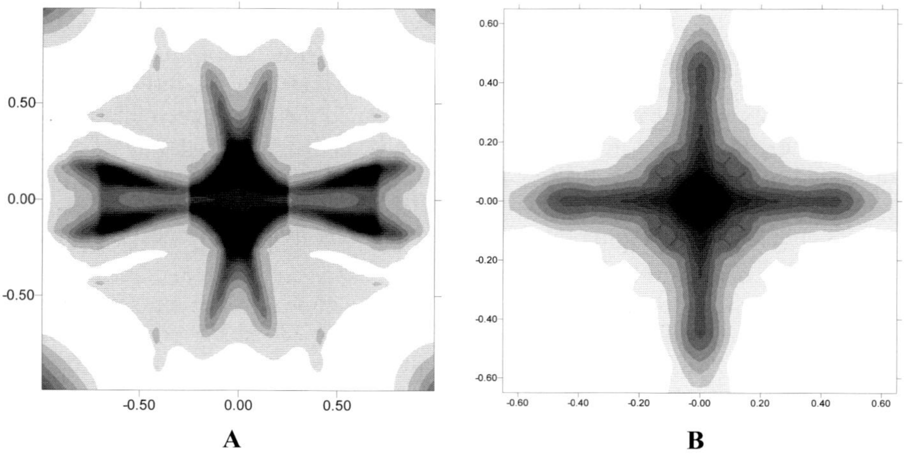

Fig. 13. Predicted (A) and measured (B) radiation pattern for C-Eu molecules located in the fully etched waveguide sample D (arbitrary units). We note a maximum of radiated intensity in the normal direction (α=β=0).

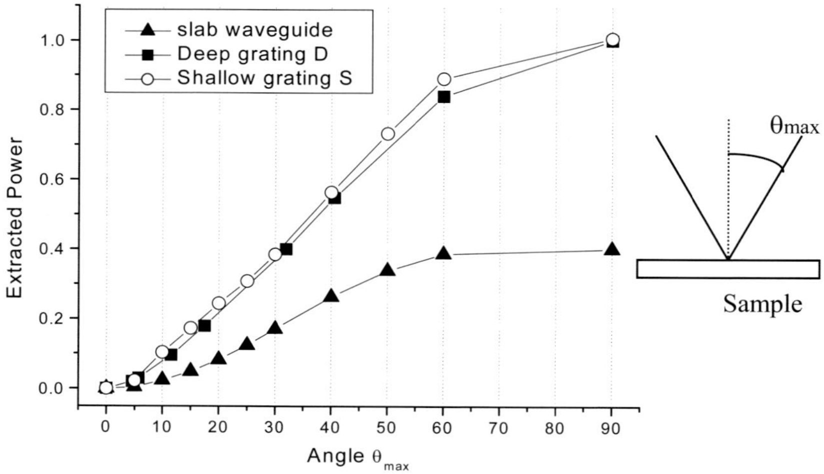

Fig. 14. Extracted power versus collection angle for a simple slab waveguide, samples D and S.

Comparison with a Slab Waveguide

As addressed in Sect. 3.2, Figs. 12 and 13 show that corrugations are interesting for extracting the guided mode power. In order to get practical quantitative results it is interesting to compare the integrated collected power for various angles. More precisely, the experimental situation is often a collection optics with a numerical aperture NA=sinθ_{max} located above the luminescent sample (Fig. 14 inset). From the measurements presented in Figs. 12 and 13, we have computed the integrated collected power in the range (0, θ_{max}), θ_{max} being varied between 0° and 90°. Results are shown in Fig. 14 for samples S, D and a simple slab waveguide containing C-Eu. The curves are normalised so that the total extracted power in the upper half space is unity for a dipole located on a bare silica substrate surface (no guided mode). From the calculation we find that for θ_{max}=90°, samples S and D reach the level of the bare substrate indicating that the corrugated structures can extract 100% of the guided mode power. In contrast, the slab waveguide extracted power reaches only 0.4 leaving 60% of the emitted power in the guided mode. Assuming that the two gratings do not significantly alter the spontaneous emission rate as compared to the bare silica substrate case, we find that the corrugated structures are much more efficient in extracting the power from an emitter located in a high index layer. Still under the constant rate assumption, and for a given θ_{max}, the deep grating D does not allow extracting a larger power than the shallow grating S. Nevertheless, the transverse mode structure of sample D (see Fig. 13) seems to be better adapted to further optical processing such as coupling into fiber optics.

4 Conclusion

This chapter has first presented a theoretical study and an experimental investigation of spontaneous emission for rare earth ions located in various types of planar dielectric structures. Although spontaneous emission is a purely quantum process, it has be shown that classical antenna calculations can give an accurate description of the process in very good agreement with experiments.

For both Er and Pr ions, strong enhancement of the spontaneous emission in a direction normal to the layer can be achieved if the ions are located in the spacer of a Fabry-Perot type structure. Although the emitter lifetime is not significantly affected in these planar stacks, we have shown that a non-symmetric structure can confine more than 30% of the emitted light in one half space. Nevertheless, these planar structures still suffer from emission into the guided modes which propagate in the plane of the layers.

To circumvent this limitation, we have presented a corrugated film a so-called 'Waveguide Grating Structure', which is able to extract the total guided mode power whatever the polar direction ϕ. A simple graphic analysis of the arc shape radiation pattern at infinity in terms of a 'Wave Vector Diagram' has been presented and is valid for shallow corrugation. Although rigorous modeling is possible, this simple analysis has been compared to experimental investigations of C-Eu emission into shallow and deep gratings and the radiation patterns obtained are in good agreement with the numerical predictions. When compared to a slab waveguide, the corrugated structures appear to extract the guided mode power much more efficient even for low numerical aperture collection optics.

Undoubtedly, many new basic science 'surprises' and novel applications will emerge from the spontaneous emission investigation of emitters located in this type of microstructure with potential applications in ultrasensitive fluorescence applications and light emitting devices such as LEDs.

References

Belarouci A, Jacquier B, Moretti P, Robert S, Rigneault H (2001) Praseodymium-doped planar multidielectric microcavities: induced lifetime changes over the emission spectrum. J. Opt. Soc. Am. B 18:832-838

Belarouci A, Menchini F, Jacquier B, Moretti P, Rigneault H, Robert S (1999) Luminescence properties of Pr3+-doped optical microcavities. J. of Luminescence 83-84:275-282

Björk G, Michida S, Yamamoto Y, Igeta K (1991) Modification of spontaneous emission rate in planar dielectric microcavity structures. Phys. Rev. A 44:669-681

Bouwmeester D, Ekert A, Zeilinger A (2000) The Physics of Quantum Information. Springer-Verlag, Berlin

Craford MG (1996) Commercial light emitting diode technology: status, trends, and possible futur performance. In: Rarity J, Weisbuch C (eds) Microcavities and Photonic Band Gaps : Physics and Applications. Vol. 324: NATO ASI Series. Kluwer Academic Publishers, Dordrecht, pp 323-332

Deppe DG, Lei C, Lin CC, Huffaker DL (1994) Spontaneous emission from planar microstructures. J. of Modern Opt. 41:325-344

De Martini F, Innocenti G, Jacobovitz GR, Mataloni P (1987) Anomaleous spontaneous emission time in a microscopic optical cavity. Phys. Rev. Lett. 59:2955 - 2958

Galaup C, Picard C, Cazaux L, Tisnes P, Aspe D, Autiero H (1996) Synthesis and luminescence of EU3+ complexes derived from novel receptors containing a tetralactam unit. New J. Chem. 20:997-999

Gieβen H, Berger JD, Mohs G, Meystre P (1996) Cavity-modified spontaneous emission: from Rabi oscillations to exponential decay. Phys. Rev. A 53:2816-2821

Goy P, Raimond JM, Gross M, Haroche S (1983) Observation of cavity-enhanced single atom spontaneous emission. Phys. Rev. Lett. 50:1903-1906

Gudgin Dickson EF, Pollak A, Diamandis EP (1995) Time-resolved detection of lanthanide luminescence for ultrasensitive bioanalytical assays. J. Photochem. Photobiol. B: Biol. 27:3-19

Hunt NJ, Schubert EF, Sivco DL, Cho AY, Kopf RF, Logan RA, Zydzik GJ (1995) High-efficiency narrow spectrum resonant cavity light emitting diodes. In: Weisbuch C, Burstein E (eds) Confined Electrons and Photons - New Physics and Applications. Vol. 340. Plenum, New York, pp 701

Kitson SC, Barnes WL, Sambles JR (1996) Photoluminescence from dye molecules on silver gratings. Opt. Commun. 122:147-154

Lakowicz JR (1999) Principles of Fluorescence Spectroscopy. Kluwer Academic/Plenum Publisher, New York

Nevière M, Petit R, Cadilhac M (1973) About theory of optical grating coupler-waveguide grating systems. Opt. Commun. 8:113-117

Parker J, Stroud CR (1987) Transient theory of cavity-modified spontaneous emission. Phys. Rev. A 35:4226-4237

Purcell EM (1946) Spontaneous emission probabilities at radio frequencies. The Physical Review 69:681

Rigneault H, Lemarchand F, Sentenac A (2000) Dipole radiation into grating structures. J. Opt. Soc. Am. A 17:1048-1058

Rigneault H, Amra C, Begon C, Cathelinaud M, Picard C (1999a) Light emission for sources located within metallodielectric planar microcavities. Appl. Opt. 38:3602-3609

Rigneault H, Lemarchand F, Sentenac A, Giovannini H (1999b) Extraction of light sources located inside waveguide grating structures. Optics Letters 24:148-150

Rigneault H, Monneret S (1997a) Field Quantization and spontaneous emission in lossless dielectric multilayer structures. JEOS, Quantum and Semiclassical Optics 9:1017-1040

Rigneault H, Robert S, Begon C, Jacquier B, Moretti P (1997b) Radiative and guided wave emission of Er3+ atoms located in planar multidielectric structures. Phys. Rev. A 55:1497-1502

Rigneault H, Monneret S (1996) Modal analysis of spontaneous emission in a planar microcavity. Phys. Rev. A 54:2356-2368

Robert S, Rigneault H, Lamarque F (1998) Spontaneous emission of Pr ions located in planar dielectric microcavities. J. Opt. Soc. Am. B 15:1773-1779

Suzuki M, Yokoyama H, Brorson SD, Ippen ED (1991) Observation of spontaneous emission lifetime change of dye-containing Langmuir-Blodgett films in optical microcavities. Appl. Phys. Lett. 58:998-1000

Tamir T (1975) Integrated Optics. Springer Verlag, New York

Vredenberg AM, Hunt NEJ, Schubert EF, Jacobson DC, Poate JM, Zydzik GJ (1993) Controlled atomic spontaneous emission from Er3+ in a transparent Si/SiO2 microcavity. Phys. Rev. Lett. 71:517

Weisbuch C, Nishioka M, Ishikawa A, Arakawa Y (1992) Observation of the coupled exciton-photon mode splitting in a semiconductor quantum microcavity. Phys. Rev. Lett. 69:3314

Zengerle R (1987) Light propagation in singly and doubly periodic planar waveguide. J. Mod. Opt. 34:1589-1617

Symbols

$\mathbf{k}_p = \alpha\, \mathbf{e}_x + \beta\, \mathbf{e}_y$: planar projection of the **k** wave vector
$\mathbf{k}_g$: guided mode wave vector
$\mathbf{r} = x\, \mathbf{e}_x + y\, \mathbf{e}_y + z\, \mathbf{e}_z$: spatial coordinate;
$\mathbf{F}(\mathbf{r},\mathbf{k}_p)$: modal electric field
$E(z,k_p)$: Electric field modulus along z axis
$N_{eff}=|\mathbf{k}_p|/k_0$: normalised spatial frequency
$\mathbf{u}(\mathbf{k}_p)$: Electric field vector
λ: wavelength
$k_0=2\pi/\lambda$: wave vector module in vacuum.
$\mathbf{F}_{fr}$: full radiative mode
$\mathbf{F}_{sr}$: substrate radiative mode
$\mathbf{F}_g$: guided mode
$\delta(\mathbf{k}_p-\mathbf{k}_p')$: Dirac function in **k** space
$\delta_{\mathbf{k}_p,\mathbf{k}_p'}$: Kronecker symbol
O_g : guided mode region in the **k** space
O_{sr} : substrate radiative mode region in the **k** space
O_{fr} : full radiative mode region in the **k** space
$\mathbf{p}(\mathbf{r},t)$: time dependent dipole
q : elementary charge
$a(t)$: time dependent dipole amplitude
e : dipole unit vector
j : dissipative current
τ_0 : emitter's lifetime in free space
τ : emitter's lifetime in a photonic microstructure
P_0 : dipole electromagnetic radiated power in free space
P : dipole electromagnetic radiated power in a photonic microstructure
P_{FRM}: full radiative mode power
P_{GM}: guided mode power
d_x : grating periodicy along x axis
d_y : grating periodicy along y axis
$K_x=2\pi/d_x$: grating wave vector along x axis
$K_y=2\pi/d_y$: grating wave vector along y axis
$\mathbf{k_p}^{inc}$: incoming wave vector coupled to the guided mode
$\alpha' = \cos\varphi \cos\theta = k_x/k_0$: reduced k_x
$\beta' = \sin\varphi \cos\theta = k_y/k_0$: reduced k_y

Epilogue

Thin film optical coatings are an enabling technology. Achievement of the optical mulitlayer design goals may ultimately determine the overall system performance. Coatings produce reflection, transmission and absorption responses in almost any fashion imaginable. However, understanding the limits of design, materials properties, and process control allow those responsible for the design of these optical films to successfully negotiate achievable system performance goals.

The information assembled here provides the tools for the thin film and system designers to mark the limits of our capabilities, set realistic performance goals and chart new investments required to extend the state of the art. We have presented a comprehensive summary of this vital and challenging technology.

We have reviewed the implementations of this technology in various micro and opto-electronic applications.

While our technological capability has improved significantly over the last decade, the demands on coating performance are continuing to expand. Thousands of passive optical coatings have been successfully developed to collect, distribute and modify optical radiation from the X-rays to microwaves. In addition to meeting prosaic reflection, transmission and absorption spectral demands, performance is increasingly required over multiple wavelength regions and for complementary performance targets. Some of these regions are tightly spaced in wavelength space and others extend over broad spectral regions. Typically it is easier to design these coatings than to produce them.

Coatings are increasingly multi-functional. In addition to achieving specific optical properties, barrier properties may be required, or specific electrical properties must be integrated into the multilayer stack. Other applications require tight spatial patterns of the optical films. Finally, while meeting stringent optical performance requirements, producers of optical coatings are under strong pressure to meet reduced fabrication costs.

Characterization techniques continue to provide critical feedback to the designers and process engineers. Armed with this knowledge, coatings can be optimized to closely match design performance.

Thin film researchers are well positioned for advancing and exploiting the emerging fields of nanotechnologies. These promising fields demand atomic level control in three dimensions. Increased characterization capabilities, and improved process control allow for the design and fabrication of these amazing three-dimensional structures on the nanoscale.

Patterning techniques will be used to provide the spatial control necessary to engineer new optical functionalities. Continued work in the field of chemistries of materials may allow utilization of controlled chemical reactions to produce self-assembling molecules to precisely pattern optical films.

Improved film properties, novel designs, and new materials will expand the areas of thin film utilization. New work with emerging processes such as Atomic Layer Deposition (ALD) may provide the control needed to provide layer-by-layer structural control.

Exciting new applications are on the horizon. Higher performance films, coatings meeting stringent multispectral specifications, tough films with enhanced durability, precisely patterned films, films on micro-optics and MEMS structures, biologically functional films and films integrated into opto-electronic subassemblies are emerging from laboratories around the world.

Optical thin films are a driving force in many new innovations. The future of thin film optical interference coatings is limited only by the imagination of the researchers in the various laboratories, institutes, and industrial concerns around the world. We look forward to reporting and updating these promising new accomplishments in our next edition.

Ric P. Shimshock
MLD Technologies
Mountain View, CA USA

Index

Biographies

Preface by Angus Macleod

Angus Macleod has more than 35 years of experience in optical coatings, both in manufacturing and in research. He was born and educated in Glasgow, Scotland, and worked both in industry and academia in Great Britain before joining the University of Arizona as Professor of Optical Sciences in 1979.
Since 1995, he has been full time with Thin Film Center, Inc., a software, training and consulting company in Tucson that he co-founded in 1986. He is the author of Thin Film Optical Filters, 3rd edition (Institute of Physics Publishing, 2001).

Natural Coatings
Peter Vukusic, School of Physics, University of Exeter, Exeter, UK

Peter Vukusic received an undergraduate degree in physics and postgraduate degrees in optics from Exeter University and Imperial College. He recently returned to Exeter in 1997 to begin intensive work on the understanding and characterisation of photonic effects in Nature. He has authored many articles detailing newly discovered structural photonic effects, principally in the order of Lepidoptera. Several of these have led to patents. In addition to understanding these natural structures and effects, his research addresses the application of such natural design protocols to a broad gamut of modern optical technologies. He was awarded a UK Biological Research Council advanced fellowship to further his work on photonics in invertebrates; the only physicist to receive such an award. As part of a genuine interest to further the public understanding of science, he regularly presents lectures, seminars and workshops on the subject of light and colour in Nature, to schools, societies and non-academic organisations.

Photonic Structures as Interference Devices

Carl G. Ribbing; Alt., Dept. of Materials Science, Uppsala, Sweden

The thesis of Carl G. Ribbing concerned the electronic structure of amorphous semiconductors. Parts of the research was made in Uppsala and parts at Stanford University, California. He received his PhD in 1973 and Docent degree in 1974 from Uppsala University,. Sweden where he was university lecturer from 1979-2000, when he was promoted professor. He is presently working part-time for Dept. of Functional Materials of theSwedish Defence Research Agency in Linköping. Since 2000 acting as Topical editor "Optical materials" for Applied optics ". He has studied bulk and thin film selective optical properties for various solar energy, window coating and space applications. His present interests are focussed on selective thermal emission control. He has published about one hundred scientific papers, conference contributions and patents in the fields mentioned.

Some Fundamentals of Optical Thin Film Growth

Norbert Kaiser, Fraunhofer Institute for Applied Optics and Precisions Engineering, Jena, Germany

Norbert Kaiser received his Diplom Physiker in 1974, his Dr. rer. nat. in 1983 and his Dr. habil in 1999 from the University of Jena. After 9 years of research on nucleation and growth of thin films, he joined the Optical Thin Film Group of the Physikalisch Technisches Institut, Jena, and was responsible for R&D on coatings for the UV. In 1991 he transferred to the vacuum laboratory of the Physikalisch Technische Bundesanstalt. Since 1992 he heads the Optical Thin Film Department and is deputy director of the Fraunhofer Institute for Applied Optics and Precision Engineering in Jena. He has authored a large number of papers and patents on nucleation, growth, and structure-related properties of thin optical films and laser induced damage phenomena.

Design of Optical Coatings

Alexander V. Tikhonravov, Research Computer Center Moscow State University, Moscow, Russia

Alexander Tikhonravov received his Master of Science degree from Physics Faculty of Moscow State University in 1970. In 1973 he received his Ph.D. degree and in 1986 his Doctor of Sciences degree in Physics and Mathematics from Moscow State University. He is now Professor of Theoretical Physics and director of Research Computing Center of Moscow State University. He has authored more than 200 publications, among them book "Basics of Optics of Multilayer Systems (together with Sh. Furman). Alexander Tikhonravov is the inventor of the needle optimization technique and one of the developers of OptiLayer thin film software family.

Coating Materials

Martin Friz, Merck KGaA, Gernsheim, Germany,
Friedrich Waibel, Umicore Materials AG, Balzers, Fürstentum Liechtenstein

Martin Friz received his Diplom Physiker in 1965 and his Dr. rer. nat in 1986 at the University of Tübingen. He worked for about 2 years on nuclear technology in Munich. Since 1971 he is with Merck KGaA in Darmstadt and Gernsheim in R&D laboratory working on evaluation and development of thin film evaporation materials.

Friedrich Waibel, born 1954 in Unterterzen, Switzerland, had different jobs in both industry and building trade before he was awarded the high school diploma through evening classes he attended from 1981 to 1984. Afterwards, he studied chemistry at the University of Zurich as a working student and graduated with the Diploma for Inorganic Chemistry in 1992. In the group of Prof. Dr. John R. Günter he performed his Ph.D. studies on „Thermally induced oxygen uptake and release in the Ba-Cr-O-system", and was awarded in 1999 by the University of Zurich with the title Dr. sc.nat. Since 2000 he has been working for Unaxis Materials AG, where he is responsible for the R&D on PVD materials for optical applications.

Film Deposition Methods

Hans K. Pulker, Institute of Ion Physics, University of Innsbruck, Austria

Hans K. Pulker was born in Austria and studied at the Universities of Vienna and Innsbruck where he received his Dr. phil. He worked many years in various thin film research and development laboratories of the Balzers AG in Liechtenstein.

1973 he received his Dr.habil. from the University of Innsbruck where he later became Professor at the Institute of Ion Physics and Head of the Thin Film Technology Group. 1998 he started to work full time at the University. He has more than 35 years experience in deposition and characterization of thin films. He wrote the monograph Coatings on Glass (2nd rev. edn. Elsevier, Amsterdam, 1999), authored many patents and published a large number of papers.

Large Area Deposition

Günter Bräuer, Fraunhofer Institute for Thin Films and Surface Engineering, Braunschweig, Germany

Günter Bräuer received his physics diploma in 1980 and his doctorate degree in 1984, both from the University of Giessen, where his research work was focussed in the field of ion beam sputtering. In 1984 he joined Leybold-Heraeus in Hanau as a member of the R&D group for thin film technology. He was mainly engaged in the development of reactive sputter processes for the production of dielectric layers on large areas. From 1988 through 1992, he was manager of R&D groups for coatings on data storage media and large area glass coating. From 1992 through 1999, he has headed the R&D department for thin film technology at Leybold. Since 1999 Günter is professor at the Technical University of Braunschweig and director of the Fraunhofer institute for thin films and surface engineering (IST) in Braunschweig as well as the Fraunhofer institute for electron beam and plasma technology (FEP) in Dresden.

Characterization and Monitoring

Detlev Ristau, Laser Zentrum Hannover, Hannover, Germany

Detlev Ristau was born in Hannover, Germany in 1957. He completed his diploma study at the University of Hannover in 1982 under the direction of H. Welling. From 1982 to 1983 he worked at the Department of Electrical Engineering of the Rice University, Houston, as a research scholar. Since 1983 he is active in the field of optical thin film technology, starting as a research assistant at the Institute for Quantum Optics, University of Hannover. After receiving his Ph.D. degree in 1988, he was employed as the leader of the Thin Film Group at the Institute for Quantum Optics. Since 1992 he is responsible for the Department of Thin Film Technology at the Laser Zentrum Hannover. He is involved in the development of characterization techniques and deposition processes for optical thin film components.

Mechanical Stress in Optical Coatings

Georg N. Strauss, Thin Film Technology Group, Institute of Ion Physics, University of Innsbruck, Austria

Georg N. Strauss was born in Styria, Austria and studied physics at the Institute of Experimental Physics at the University of Innsbruck. He received his Mag. rer. nat. in 1997 in the field of thin film technology and solid state physics with a work on stress in dielectric thin films. In 1999 he changed to the Institute of Ion Physics at the University of Innsbruck. During 1997 and 2002 Georg worked in several international projects concerning the investigation and analysis of the plasma of different ion and plasma enhanced pvd processes. At the moment he is in the last stage to finish his thesis. Georg authored a number of papers on mechanical and optical properties of thin films as well as on plasma properties of different pvd processes.

Optical thin film for microcomponents

Ludovic Escoubas and Francois Flory, Institut Fresnel, Marseille, France

Ludovic Escoubas graduated from the Ecole Nationale Supérieure de Physique de Marseille (ENSPM), France, in 1994 and received a PhD in Optics in 1997. Since 1998, he has been a senior lecturer teaching Optics at the ENSPM and doing his research work at the Fresnel Institute in the group "Microstructured Optical Components". His current research interests are integrated optics, characterizations using guided waves, diffractive optics, gas sensors, ion implantation and optical interference coatings, and more generally microstructured optical components. He has authored more than 40 papers and communications.

Francois Flory received his PhD in 1978 and his "thèse d'Etat" in 1985 in light scattering and in the relation between optical properties and the microstructure of thin films, respectively. He is now Professor in Optics at the Ecole Nationale Supèrieure de Physique de Marseille. He is editor of the book "Thin Films for Optical Systems" and of more than 100 papers and communications. His current research interest is now in the field of optical micro components.

Optical coatings for the DUV/VUV

Roland Thielsch, Southwall Europe, Dresden, Germany

Roland Thielsch studied physics at the Technical University Dresden. He received his Diplom Physiker in surface science and electron physics in 1978 and his Dr. rer. nat. in 1990 from the Technical University Dresden. From 1978 - 1993 Roland worked in a research group “Vacuum physics and vacuum technology” at the Physics department of the Technical University as staff research scientist in the various fields of thin film deposition technologies. In 1993 he joined the Institute of Solid state physics and material research in Dresden and in the spring of 1996 the Department of optical coatings of the Fraunhofer Institut für Angewandte Optik und Feinmechanik. From 1997 until he left the Fraunhofer Institut Roland worked as scientist in charge for the development of UV optical coatings. In January 2000 he joined Southwall Europe GmbH, where he held the position of a senior process engineer in the field of large area deposition of optical coatings onto plastic web.

Roland authored a number of papers and patents on optical and structural properties of thin films as well as on related hardware aspects.

Multilayer Coatings for EUV/X-ray Mirrors

Sergey Yulin, Kharkov State Polytechnic University, Kharkov, Ukraine and Fraunhofer Institute for Applied Optics and Precision Engineering, Jena, Germany

Sergey Yulin received his Diplom Physiker in 1989 and his Dr. rer. nat. in 1998 from the State Kharkov University. From 1989 to 1999 his work was focused on research and development of EUV/Soft X-ray optics on the base of artificial multilayer coatings. He has been a staff scientist at Fraunhofer Institute for Applied Optics and Precision Engineering in Jena since 1999, where his work has armed on Si and Sc-based multilayer-coated optics for different applications. Since 2000 he heads the Soft-X-Ray and EUV-Coatings group. He has published over 20 scientific papers and patents.

Laser Resistant Coatings

Christopher Stolz and François Y. Génin, Lawrence Livermore National Laboratory, National Ignition Facility, Livermore, CA, USA

Christopher Stolz received an undergraduate degree in Applied Mathematics, Engineering, and Physics from the University of Wisconsin – Madison and is currently pursuing a graduate degree in Material Science. He joined Lawrence Livermore National Laboratory in 1989 and has worked on laser coatings for both the Atomic Vapor Laser Isotope Separation (AVLIS) and Inertial Confinement Fusion (ICF) laser programs. He is currently responsible for the large-aperture high-fluence mirrors and polarizers to be installed on the National Ignition Facility (NIF). His research interests include laser resistant coatings and surfaces and laser damage characterization tools such as photothermal microscopy. He is a cochair for the annual Laser-Induced Damage in Optical Materials Symposium and has authored and coauthored over 40 papers.

François Y. Génin received his B.S. in Electrochemical and Electrometallurgical Engineering from the Institut Polytechnique de Grenoble in France in 1986. He was granted an IBM Fellowship to pursue graduate studies at Carnegie Mellon University that led to a Masters of Engineering and a Ph.D. in Metallurgical Engineering and Materials Science in 1988 and 1991, respectively. He joined Lawrence Livermore National Laboratory in 1991 as a Post Doctoral Scientist in the Chemistry and Materials Science Department. During his career at LLNL, he worked for the National Ignition Facility (NIF) on laser optical damage and advanced optics polishing. He also worked on laser-materials interactions and femtosecond laser processing. He currently holds a joint position in NIF and in the Medical Division of the Physics and Applied Technology Directorate. He has authored/co-authored about 50 papers in the area of thin films, laser processing, and biomaterials and hold several patents in medical technology. He also has co-organized symposia on femtosecond phenomena, and guest edited a special issue on Femtosecond Laser Processing in Journal of Applied Physics A. Finally, in 1996, he co-founded a medical device company, Berkeley Advanced Biomaterials Inc. that manufactures synthetic bone graft substitutes used in orthopedic and dental surgery.

Coatings for UV-Free Electron Lasers

Alexandre Gatto, Fraunhofer Institute for Applied Optics and Precision Engineering, Jena, Germany

Alexandre Gatto graduated from the Ecole Nationale Supérieure de Physique de Marseille in 1995. In 1996, he worked as a military scientist in Institut de Recherche sur les Phénomènes Hors Equilibre on UV photoablation process of nuclear materials. He obtained his PhD in 1999 after three years of research in Laboratoire d'Optique des Surfaces et des Couches de Minces de Marseille (Institut Fresnel). He joined in 1999 the Fraunhofer Institute for Applied Optics and Precision Engineering in Jena as project manager in the field of DUV-VUV resistant optics. He is involved in the development of radiation resistant VUV coatings for Excimers and Free Electron Lasers.

Optical Coatings on Plastics

Ludvik Martinu and Jolanta E. Klemberg-Sapieha, Ecole Polytechnique de Montreal, Montreal, Canada

Ludvik Martinu obtained his PhD degree in Experimental Physics from the Charles University in Prague, Czech Republic, in 1985. In 1988 he joined the Ecole Polytechnique in Montreal, Canada, where he is now Full Professor and Associate Director of the Thin Film Research Center (GCM). His main research interests are in the area of physics and technology of thin films, in which he particularly focuses on new plasma-based fabrication processes and new materials for optical and functional coatings, optical filters, and photonic devices. He is author or co-author of more than 200 publications. He is active in numerous scientific societies and organizing committees, in particular, the Society of Vacuum Coaters (SVC), International Conference on Metallurgical Coatings and Thin Films (ICMCTF), and Sigma Xi – The Scientific Research Society. He is also associate editor of "Plasmas and Polymers" published by Kluwer/Plenum.

Jolanta-Ewa Klemberg-Sapieha obtained her Master degree in Chemical Engineering and PhD degree in Materials Science from the Technical University in Lodz, Poland. Since 1978, she has been with the Department of Engineering Physics, Ecole Polytechnique, Montreal, Canada, where she pursues her research as a Senior Research Scientist. She has many years of experience in the area of plasma processing and surface and interface analysis of materials, in particular polymers. More recently, she focuses on micro- and nanomechanical properties of thin films and on hard and ultra-hard protective and metallurgical coatings. She is an author or co-author of more than 150 publications and member of numerous research societies.

Interference Coatings for Ultrafast Optics

Gabriel Tempea, Vladislav Yakovlev and Ferenc Krausz , Vienna University of Technology, Wien, Austria

Gabriel Tempea received his Dipl.-Ing. degree in technical physics from the Technical University of Bucharest in 1996 and his PhD-degree from the Vienna University of Technology in 1999. Since January 2000 he has been working as research assistant at the Photonics Institute in Vienna. He is currently at the CELIA Labs in Bordeaux, France for a half-year post-doctoral research stay. His research interests focus on strong-field physics and ultrafast phenomena, including the design of multilayer interference coatings for femtosecond lasers. He is author or co-author of more than 25 articles published in reviewed scientific journals.

Ferenc Krausz was born in Mor, Hungary on May 17, 1962. He received his M. S. in Electrical Engineering from the Budapest University of Technology in 1985, his Ph. D. in Quantum Electronics from the Vienna University of Technology in 1991, and his "Habilitation" degree in the same field from the same university in 1993. In 1998 he joined the Department of Electrical Engineering as an Associate Professor and became a Full Professor in the same department in 1999.

His research interests have included nonlinear light-matter interactions, femtosecond light pulse generation from the infrared to the x-ray spectral range, and studies of ultrafast microscopic processes. His contributions to these fields were recognized by the 1994 Physics Award (Fritz-Kohlrausch Award) of the Austrian Physical Society, by the Carl-Zeiss-Award of the Ernst Abbe Foundation in Germany in 1998 and by the Wittgenstein Prize of the Austrian Science Fund in 2002. He is co-founder of Femtolasers GmbH, a Vienna-based company specializing in cutting-edge femtosecond laser sources, and speaker of the Center of Excellence *"Advanced Light Sources"* of the Austrian Science Fund.

Ferenc Krausz is married, he and his wife, Angela, have two children: Anita, born in 1989 and Martina, born in 1993.

Vladislav Yakovlev received his BSc in physics from the Novosibirsk state university (Russia) in 1997 and his master's degree in 1999 from the same university. Since 2000 he has been working in the Institute of Photonics, Vienna, where he was involved in theoretical investigations in the field of ultrafast phenomena, including the design of dispersive mirrors.

Optical Coatings for Displays

Li Li, National Research Council, Ottawa, Ontario, Canada

Li Li received her Bachelor degree in optical engineering from Zhejiang University in 1984, and her Master and Doctor degrees in opto-electronics from Shanghai Institute of Technical Physics (SITP), the Chinese Academy of Sciences in 1986 and 1989, respectively. After graduation, she worked briefly as a research associate at SITP. Between Jan. 1990-May 1991, she was a visiting scientist at the National Research Council of Canada (NRCC). Since May 1991 she has been with the Thin Films Group of the Institute for Microstructural Sciences of NRCC. She is currently a senior research officer. Her major research interests are optical coatings and their applications in various fields such as electronic displays, optical storage and telecommunications. She is the inventor or co-inventor of eight issued US patents and has authored more than 30 papers in refereed journals and conference proceedings.

Optical thin films for spontaneous emission control

Hervé Rigneault CNRS Institut Fresnel, Marseille, France

Hervé Rigneault received his Engineer Diploma from the Ecole Nationale Superieure de Physique de Marseille in 1991, his PhD diploma in Optics in 1993. He joined the Centre National de la Recherche Scientifique in 1994 as a full time researcher in Optical Physics and worked for 5 years on spontaneous emission control for ions and molecules embedded into dielectric thin films. He received his Dr. habil in 2000 from the University of Marseille. Since 2000 he heads the 'Microstructured Optical Components' group of the Fresnel Institut. He field of research is now moving at the optics/biology interface where he develops ultra-sensitive apparatus to study molecular diffusion in living cells.

Epilogue

Ric P. Shimshock, MLD Technologies, Mountain View, CA USA

Ric Shimshock received his MSc. in Optical Engineering from the University of Arizona Optical Sciences Center (OSC) in 1987. His work at the OSC focused on the development and characterizations of transition metal nitrides for use as high temperature solar selective absorber coatings in photothermal power systems. After joining Optical Coating Laboratory Inc (OCLI) as a Project Engineer, he worked in the Government Aerospace group from 1979-1986 supporting various projects associated with space based remote sensing, infrared imaging systems and laser coatings. He was involved in the transfer of new materials and processes to the production environment. He worked at Deposition Sciences Inc. (DSI) as Program Manager from 1986-1995 where he developed high performance products using various processes such as LPCVD, PECVD, Long Throw Sputtering, Ion Assisted Deposition and Microwave Assisted Sputtering. He was hired by Southwall Technologies Inc in 1995 as the Director of Advanced Products and worked on supporting various web based coating technologies including sputtering AC and DC magnetron, PECVD and wet chemistries for display, solar control and architectural applications. He is currently employed at MLD Technologies working on scaling an ultra low loss IBS processes. He is active in the OSA, is currently on the Board of Directors of the SVC, and involved with the ICCG and AIMCAL organizations.

Springer Series in
OPTICAL SCIENCES

New editions of volumes prior to volume 60

1 **Solid-State Laser Engineering**
By W. Koechner, 5th revised and updated ed. 1999, 472 figs., 55 tabs., XII, 746 pages

14 **Laser Crystals**
Their Physics and Properties
By A. A. Kaminskii, 2nd ed. 1990, 89 figs., 56 tabs., XVI, 456 pages

15 **X-Ray Spectroscopy**
An Introduction
By B. K. Agarwal, 2nd ed. 1991, 239 figs., XV, 419 pages

36 **Transmission Electron Microscopy**
Physics of Image Formation and Microanalysis
By L. Reimer, 4th ed. 1997, 273 figs. XVI, 584 pages

45 **Scanning Electron Microscopy**
Physics of Image Formation and Microanalysis
By L. Reimer, 2nd completely revised and updated ed. 1998,
260 figs., XIV, 527 pages

Published titles since volume 60

60 **Holographic Interferometry in Experimental Mechanics**
By Yu. I. Ostrovsky, V. P. Shchepinov, V. V. Yakovlev, 1991, 167 figs., IX, 248 pages

61 **Millimetre and Submillimetre Wavelength Lasers**
A Handbook of cw Measurements
By N. G. Douglas, 1989, 15 figs., IX, 278 pages

62 **Photoacoustic and Photothermal Phenomena II**
Proceedings of the 6th International Topical Meeting, Baltimore, Maryland,
July 31 - August 3, 1989
By J. C. Murphy, J. W. Maclachlan Spicer, L. C. Aamodt, B. S. H. Royce (Eds.),
1990, 389 figs., 23 tabs., XXI, 545 pages

63 **Electron Energy Loss Spectrometers**
The Technology of High Performance
By H. Ibach, 1991, 103 figs., VIII, 178 pages

64 **Handbook of Nonlinear Optical Crystals**
By V. G. Dmitriev, G. G. Gurzadyan, D. N. Nikogosyan,
3rd revised ed. 1999, 39 figs., XVIII, 413 pages

65 **High-Power Dye Lasers**
By F. J. Duarte (Ed.), 1991, 93 figs., XIII, 252 pages

66 **Silver-Halide Recording Materials**
for Holography and Their Processing
By H. I. Bjelkhagen, 2nd ed. 1995, 64 figs., XX, 440 pages

67 **X-Ray Microscopy III**
Proceedings of the Third International Conference, London, September 3-7, 1990
By A. G. Michette, G. R. Morrison, C. J. Buckley (Eds.), 1992, 359 figs., XVI, 491 pages

68 **Holographic Interferometry**
Principles and Methods
By P. K. Rastogi (Ed.), 1994, 178 figs., 3 in color, XIII, 328 pages

69 **Photoacoustic and Photothermal Phenomena III**
Proceedings of the 7th International Topical Meeting, Doorwerth, The Netherlands,
August 26-30, 1991
By D. Bicanic (Ed.), 1992, 501 figs., XXVIII, 731 pages

Springer Series in
OPTICAL SCIENCES

70 **Electron Holography**
By A. Tonomura, 2nd, enlarged ed. 1999, 127 figs., XII, 162 pages

71 **Energy-Filtering Transmission Electron Microscopy**
By L. Reimer (Ed.), 1995, 199 figs., XIV, 424 pages

72 **Nonlinear Optical Effects and Materials**
By P. Günter (Ed.), 2000, 174 figs., 43 tabs., XIV, 540 pages

73 **Evanescent Waves**
From Newtonian Optics to Atomic Optics
By F. de Fornel, 2001, 277 figs., XVIII, 268 pages

74 **International Trends in Optics and Photonics**
ICO IV
By T. Asakura (Ed.), 1999, 190 figs., 14 tabs., XX, 426 pages

75 **Advanced Optical Imaging Theory**
By M. Gu, 2000, 93 figs., XII, 214 pages

76 **Holographic Data Storage**
By H.J. Coufal, D. Psaltis, G.T. Sincerbox (Eds.), 2000
228 figs., 64 in color, 12 tabs., XXVI, 486 pages

77 **Solid-State Lasers for Materials Processing**
Fundamental Relations and Technical Realizations
By R. Iffländer, 2001, 230 figs., 73 tabs., XVIII, 350 pages

78 **Holography**
The First 50 Years
By J.-M. Fournier (Ed.), 2001, 266 figs., XII, 460 pages

79 **Mathematical Methods of Quantum Optics**
By R.R. Puri, 2001, 13 figs., XIV, 285 pages

80 **Optical Properties of Photonic Crystals**
By K. Sakoda, 2001, 95 figs., 28 tabs., XII, 223 pages

81 **Photonic Analog-to-Digital Conversion**
By B.L. Shoop, 2001, 259 figs., 11 tabs., XIV, 330 pages

82 **Spatial Solitons**
By S. Trillo, W.E. Torruellas (Eds), 2001, 194 figs., 7 tabs., XX, 454 pages

83 **Nonimaging Fresnel Lenses**
Design and Performance of Solar Concentrators
By R. Leutz, A. Suzuki, 2001, 139 figs., 44 tabs., XII, 272 pages

84 **Nano-Optics**
By S. Kawata, M. Ohtsu, M. Irie (Eds.), 2002, 258 figs., 2 tabs., XVI, 321 pages

85 **Sensing with Terahertz Radiation**
By D. Mittleman (Ed.), 2003, 207 figs., 14 tabs., XVI, 337 pages

86 **Progress in Nano-Electro-Optics I**
Basics and Theory of Near-Field Optics
By M. Ohtsu (Ed.), 2003, 118 figs., XIV, 161 pages

87 **Optical Imaging and Microscopy**
Techniques and Advanced Systems
By F.-J. Kao, P. Török (Eds.), 2003, 133 figs., XIV, 475 pages

Springer Series in
OPTICAL SCIENCES

88 **Optical Interference Coatings**
By N. Kaiser, H.K. Pulker (Eds.), 2003, 203 figs., 50 tabs., XVI, 504 pages

Printing: Mercedes-Druck, Berlin
Binding: Stein+Lehmann, Berlin